Springer Series in Optical Sciences Volume 7

Editor David L. MacAdam

Springer Series in Optical Sciences

Laser Spectroscopy III

Proceedings of the Third International Conference
Jackson Lake Lodge, Wyoming, USA, July 4–8, 1977

Editors
J. L. Hall and J. L. Carlsten

With 296 Figures

Springer-Verlag Berlin Heidelberg New York 1977

Dr. John L. Hall
Dr. John L. Carlsten

Joint Institute for Laboratory Astrophysics
University of Colorado
Boulder, CO 80309, USA

Dr. David L. MacAdam

68 Hammond Street, Rochester, NY 14615, USA

Conference Chairmen: Dr. John L. Hall and Dr. John L. Carlsten

TICOLS Steering Committee:
A. Mooradian H. Walther V. S. Letokhov C. J. Bordé H. Takuma
R. G. Brewer T. W. Hänsch S. E. Harris G. Series S. Haroche
R. N. Zare M. O. Scully C. Rhodes

TICOLS Sponsors:
National Bureau of Standards
Joint Institute for Laboratory Astrophysics
National Science Foundation
U. S. Energy Research and Development Agency
U. S. Army Research Office
Office of Naval Research
Chromatix
Coherent Radiation
Laser Analytics
Spectra Physics
William Waldorf Astor Foundation

ISBN 3-540-08543-2 Springer-Verlag Berlin Heidelberg New York
ISBN 0-387-08543-2 Springer-Verlag New York Heidelberg Berlin

Offset printing: Beltz Offsetdruck, Hemsbach
Bookbinding: Konrad Triltsch, Graphischer Betrieb, Würzburg
2153/3130-543210

Preface

The Third International Conference on Laser Spectroscopy, affectionately known as TICOLS, was held at Jackson Lake Lodge, Wyoming, on July 4-8, 1977. As its predecessors held at Vail, Colorado, and Mêgêve, France, the conference aimed at providing an informal atmosphere in which leading international scientists could discuss the applications of laser techniques to spectroscopic problems of outstanding interest. Jackson Lake Lodge was chosen as the location for TICOLS because of its excellent conference facilities and remote location which would allow the participants ample opportunity to meet and discuss physics in an informal, relaxed manner.

The conference was truly international: 165 scientists represented 19 countries, including Algeria, Brazil, Canada, Denmark, England, France, Germany (DDR), Germany (FRG), Holland, Israel, Italy, Japan, New Zealand, Pakistan, Poland, Scotland, Sweden, the U.S.A., and the U.S.S.R.

Numerous people have contributed their time, effort, and enthusiasm, without which the success of this conference would not have been possible. In particular, we would like to thank our sponsors for their financial support and the members of the steering committee for their advice in planning the program. In addition, we express our appreciation to those members of the staff at the Joint Institute for Laboratory Astrophysics and the staff at the Bureau of Conferences and Institutes at the University of Colorado for their help with the numerous tasks involved in organizing this conference. Finally, we are grateful to Alice Levine for her rapid and efficient technical editing of the manuscripts, which allowed publication of this book less than four months after TICOLS was held.

J. L. Hall
August 1977 J. L. Carlsten

Contents

IX. Laser Wavelength Measurements

X. Postdeadline Papers

I. Fundamental Physical Applications of Laser Spectroscopy

PRESENT STATUS OF THE LASER REVOLUTION IN SPECTROSCOPY

B.P. Stoicheff

Department of Physics, University of Toronto
Toronto, Canada M5S 1A7

Introduction

I feel deeply honoured to have been invited to give the opening address to this illustrious group of spectroscopists at the third International Conference on Laser Spectroscopy. And this being July 4th, Mr. Chairman, on behalf of those of us visiting the United States, I should like to say "Happy 201st Birthday" to our American colleagues.

When I received the invitation for this talk, I was impressed with the spirit of the organizers who planned that "TICOLS should be like a marriage feast, celebrating the modern happy union between laser technology and the science of spectroscopy". On that occasion, I was reminded of the Conference on "Fundamental and Applied Laser Physics" [1] in Iran in 1971 when during a panel discussion on the Future Course of Quantum Electronics, I had said "...there has been a revolution in the field of light scattering in the past 5 or 6 years, ...and I can foresee that in a few years, a similar revolution will occur in infrared spectroscopy and probably in the visible region", and urged that efforts be made to open up the difficult vacuum ultraviolet region. Of course, in 1971, after a decade of rapid laser development, it did not take much imagination to predict such happenings, and especially not for those of us who have been encouraging the use of lasers in spectroscopy since the early 1960s. Progress in this direction has been extremely rapid since about 1970, and perhaps justifiably, there has been much talk about a "Laser Revolution in Spectroscopy". Hence the title of this talk on its present status, a review of some highlights of recent progress in spectroscopy with lasers.

Advances in Coherent Sources and Spectroscopic Techniques

In the past 5 or 6 years, there have been several very important conferences where new developments in lasers and their application to spectroscopy have been reported and discussed. The forerunner of these was perhaps the 1971 meeting in Esfahan, Iran, already mentioned. Then in 1973, the first in the present series on Laser Spectroscopy took place in Vail, Colorado [2], followed in 1975 by the second, in Megève, France [3]. A conference on "Spectroscopic Methods Without Doppler Broadening" was held in Aussois, France in 1974 [4], and one on "Tunable Lasers and Applications" in Loen, Norway, in 1976 [5]. The proceedings of these conferences have been published, and more detailed reviews of laser spectroscopy have recently appeared in texts edited by WALTHER [6] and SHIMODA [7].

The developments in new lasers and in the use of nonlinear processes for generating coherent radiation have been impressive. The frequency gap between the microwave and far-infrared regions has essentially been bridged, and coherent radiation is available through the infrared, visible and ultraviolet regions, to 380Å in the vacuum ultraviolet. In several wavelength regions, limited tunability is possible, some with linewidths of a few kilohertz. Stabilized lasers have been produced with one-second frequency instabilities as small as 5×10^{-13}, and laser frequencies to a few hundred terrahertz can be measured. Unprecedented optical resolving power, up to 5×10^{10}, has been achieved. Equally superb resolution in the time domain has been obtained with sub-picosecond pulses. And the increase in output power of cw, pulsed, and tunable lasers has kept pace with these remarkable advances.

With such sources, many of the techniques of spectroscopy have been improved and complemented by new methods. Moreover, the sensitivity of absorption and emission measurements has been increased manifold (to the point where only a few atoms or molecules suffice for detection of fluorescence). Thus, it is no wonder that new life and progress have come to many areas of atomic and molecular spectroscopy. Here might be mentioned experiments in: sub-Doppler spectroscopy using saturation absorption or two-photon absorption from opposed beams; infrared-microwave double resonance; laser magnetic resonance; negative-ion spectroscopy; highly-excited Rydberg atoms and molecules; quantum beat spectroscopy; time-resolved relaxation measurements; coherent transient effects; coherent Raman scattering; atoms in strong resonant fields; isotope separation, particularly the new process of collisionless dissociation of molecules by intense laser radiation; and, the search for parity violation in atoms caused by weak neutral currents.

The achievements to date in each of these areas of current interest are such that I could not do justice to even one of these topics in the time allotted. Rather, I should like to briefly sketch, for several experiments, the leap forward in capabilities from the pre-laser period (or from 1970 for some experiments) to those of the present day. For this purpose, I have selected the topics[1] of saturated absorption spectroscopy, coherent tunable VUV radiation, coherent Raman spectroscopy, and two-photon absorption.

Saturated Absorption Spectroscopy

Two excellent examples come to mind, namely, the infrared ν_3 band of CH_4 [8] [9], and the Balmer H_α and D_α lines of atomic hydrogen and deuterium, respectively [10]. The spectra of CH_4 are shown in Figs.1 and 2. At the top if Fig.1 is the first infrared spectrum of the ν_3 band of CH_4 obtained about 40 years ago by A.H. NIELSEN and H.H. NIELSEN, at a resolving power of 10^3. In (b) is a portion of this spectrum taken with a grating spectrometer by E.K. PLYLER, about 1960, at a resolving power of 10^5. In (c) are shown the saturated absorption peaks (or inverted Lamb dips) for the P(7) line; and in (d), the 13 Stark components obtained by BREWER with a resolving power of $\sim 5 \times 10^7$. HALL and BORDE used the method of frequency offset-locking to obtain the magnetic hyperfine components (Fig.2) with a resolution of 10KHz. Then by paying careful attention to pressure broadening, to transit-time broadening, to the quality of beam wavefront and to the laser linewidth, HALL and his colleagues obtained resolution better than 1KHz, or a resolving power of $\sim 5 \times 10^{10}$, and were successful in observing the splitting of each of the hyperfine structure components into doublets due to recoil (Fig.2). In summary, a factor of 10^7

---

[1] The references cited will generally be review articles, and not necessarily the original paper or papers reporting on specific experiments or discoveries.

in resolving power has been achieved at 3.39 μm, since the earliest infrared spectrum of CH_4, and a factor of about 10^3 since the early saturated absorption spectrum [8].

A He-Ne laser stabilized to the methane P(7) line was used by Evenson and his colleagues [11] to obtain the present value for the speed of light. They measured the frequency of CH_4 against the Cs frequency standard by

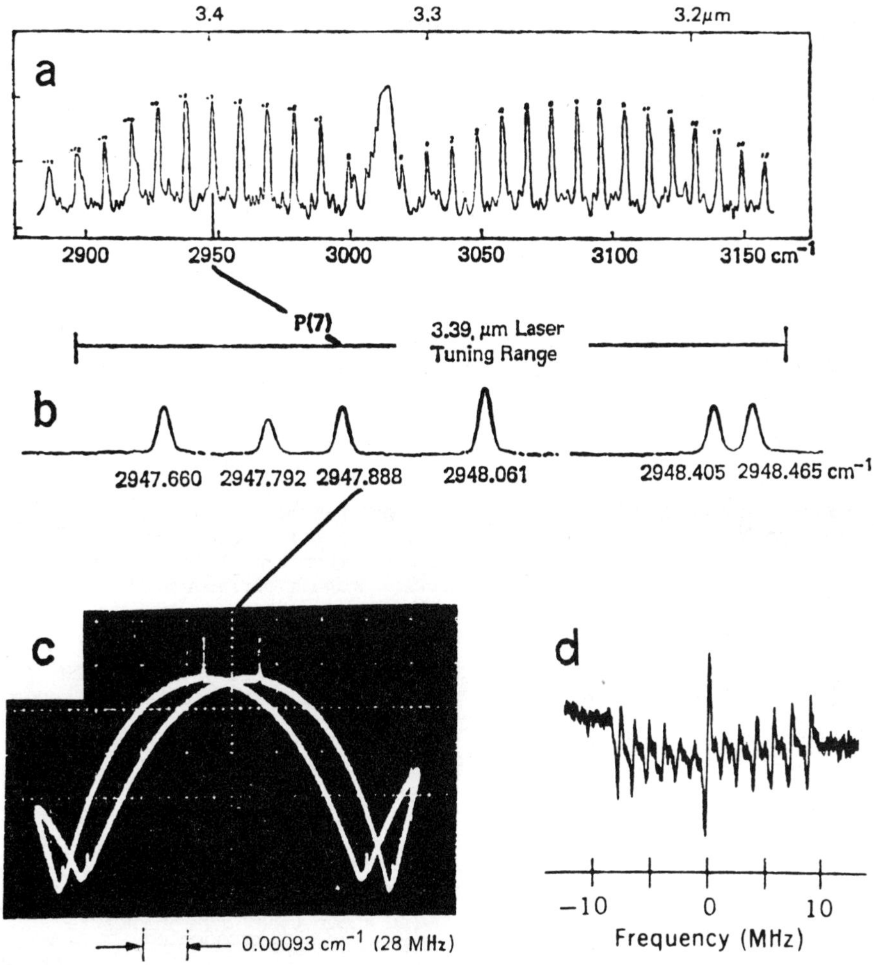

Fig.1 The ν_3 vibration-rotation band of CH_4 at 3.3 μm. (a) Low-resolution absorption spectrum. Each line is 24,000 MHz wide. (b) A higher resolution absorption spectrum showing the P(7) rotational fine structure. Each line is 260 MHz wide because of Doppler broadening. (c) A saturated absorption spectrum of the $F_2(2)$ component at 2947.888 cm^{-1}. Each "spike" is 400 KHz wide. (d) Derivative Lamb dip spectrum of the E component at 2947.792 cm^{-1} in the presence of a Stark field [8].

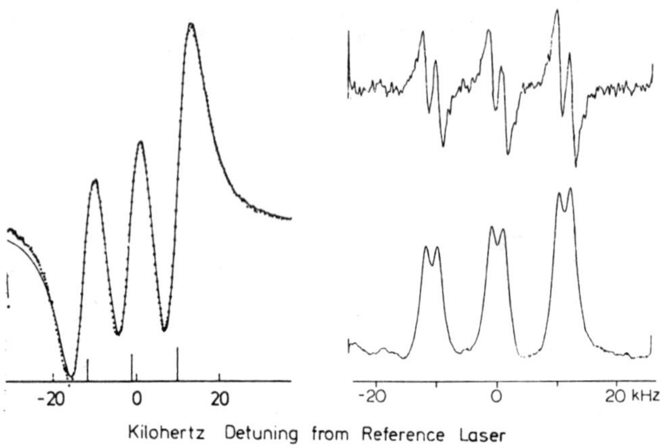

Fig.2 (a) Derivative spectrum of hyperfine structure ($\Delta F=-1$) of the $F_2(2)$ line. (b) Derivative spectrum (top) and doublets (bottom) due to recoil, observed with resolving power of 5×10^{10} [9].

using CO_2, H_2O, HCN laser oscillators, and klystrons, together with the ingenious diode frequency-mixer developed in JAVAN'S laboratory; and measured the wavelength with respect to the ^{86}Kr 6058 Å primary length standard. Their determination gives c=299,792,458 m/sec, to an accuracy of $4:10^{9}$. This represents an improvement of about 100 in the accuracy of the previously accepted value.

In Fig.3, a comparison is made of the Doppler-broadened H_α line at 300°K, with the resolved structure of the saturation spectrum clearly showing the Lamb shift for the n=2 level. By measuring the absolute wavelengths of the $2P_{3/2}-3D_{5/2}$ components for H_α and D_α, HÄNSCH and his colleagues obtained a new value for the Rydberg constant, $R_\infty=109737.3143(10)$ cm^{-1}, with an order of magnitude improvement in precision.

Coherent Tunable VUV Radiation

HARRIS and his co-workers [12] have shown the way to the generation of coherent UV and VUV radiation. In a series of papers beginning about 1971,

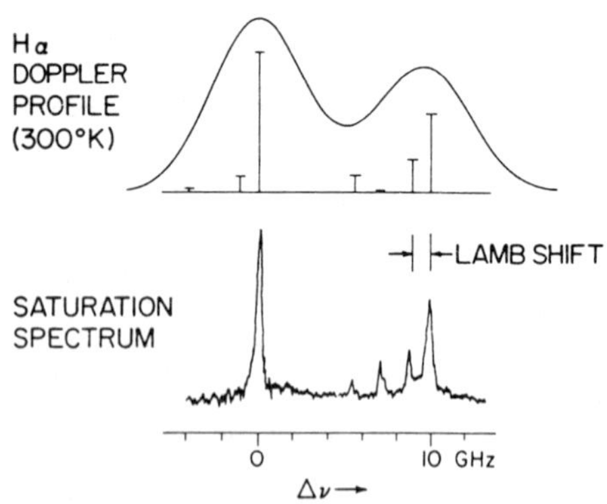

Fig.3 Doppler-broadened absorption profile of the red Balmer alpha line with theoretical fine structure. Below is shown a saturation spectrum of the line, with the Lamb shift indicated [10]. The most intense component is due to the $2P_{3/2}\rightarrow3D_{5/2}$ transition.

they discussed sum, difference, and harmonic generation using phase-matching and resonant enhancement in metal vapors and inert gases. They succeeded in producing tunable radiation over limited regions from 1950 to ~1200 Å, and the third harmonic of 2660 Å radiation at 887 Å. Recently, REINTJES and his colleagues [13] extended the techniques to produce the fifth and seventh harmonics of 2660 Å, at 532 and 380 Å.

In 1974, HODGSON, SOROKIN, and WYNNE[14] used selective resonance enhancement in 4-wave frequency mixing ($2\omega_1+\omega_2=\omega_3$) in Sr vapor to generate continuously tunable radiation from 1960 Å to 1770 Å. Similar methods with Mg vapor have extended the region from 1600 to 1230 Å [15]. The radiation from these coherent VUV sources has a linewidth of ~0.1 cm^{-1}, with up to 10^{12} photons per pulse (10^{-8}sec) being produced. Such intensity is sufficient for use in spectroscopy; the radiant intensity/cm^{-1} is larger than obtainable with synchrotrons; and the effective resolution is better than that available with the best grating spectrographs.

These sources have been used in two different studies. The IBM group has used the emitted radiation to probe the Sr autoionized levels and their band contours [14]. At Toronto, we have selectively excited individual rotational levels of CO in a region of strong perturbation (at λ 1550 Å) and measured their lifetimes by means of fluorescence decay [16].

Coherent Anti-Stokes Raman Spectroscopy

In 1971, I had the opportunity of reviewing the situation in Brillouin and Raman scattering with lasers [17]. There was no doubt that many important advances in Raman spectroscopy of liquids and solids had been made with laser excitation. However, in the area of high-resolution Raman spectroscopy of gases, only a factor of about three in resolution had been achieved over pre-laser spectra. While it had been shown that, in principle, scattering in the forward direction reduces broadening due to Doppler motion (Fig. 4), it was only after the development of coherent anti-Stokes Raman spectroscopy by TARAN [18] AKHMANOV [18], HARVEY [18] and their co-workers, amongst others, that major improvements in resolution were achieved.

Although the application of this method to high-resolution spectroscopy of gases is still very difficult, it is now possible to obtain good quality spectra with CW lasers, at resolving powers approaching 10^7. An example is

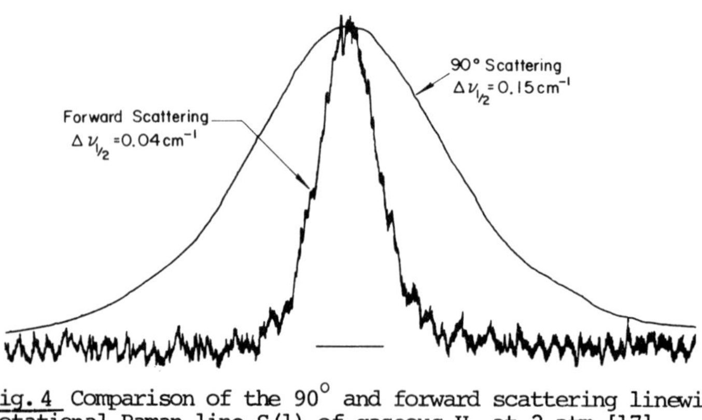

Fig. 4 Comparison of the 90° and forward scattering linewidths in the pure rotational Raman line S(1) of gaseous H_2 at 2 atm [17]

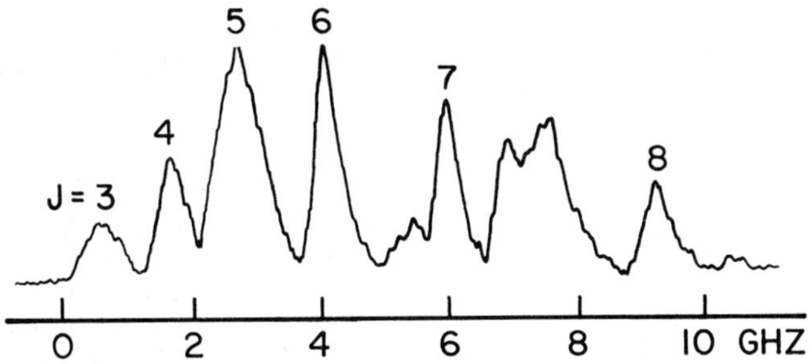

Fig. 5 Coherent anti-Stokes Raman spectrum of the ν_1 Q branch of CH_4 with the rotational structure resolved [19]. The gas pressure was 20 Torr, and the effective resolution ~ 60 MHz.

shown in Fig. 5. It is well-known that the Q branch of the totally symmetric ν_1 band of CH_4 is an extremely sharp, line-like feature (~0.3 cm^{-1} wide). Very recently, SMIRNOV and his colleagues [19] were able to resolve the rotational structure of this Q branch (Fig. 5) with the gas at 20 Torr, and with an instrumental resolution of 60 MHz. The accurate measurement of Raman frequency shifts, however, still remains a problem.

Doppler-Free Two-Photon Absorption Spectroscopy

Two-photon absorption with counterpropagating laser beams is a powerful new addition to other methods of sub-Doppler spectroscopy such as saturated absorption, quantum beats, level crossing, and Hanle effect. Since the first experiments in 1974, this method has become a flourishing field of study. Two-photon spectra of molecules as well as of atoms have been investigated, and spectra have been obtained in the infrared, visible, and ultraviolet. The present limitations to its broader use in spectroscopy are the small range of tunability and frequency instability of available dye lasers.

The majority of the investigations to date have been concerned with Rydberg states of alkali atoms and the measurement of their term values and fine structure splittings. Spectra of Na, K, and Cs have been observed with

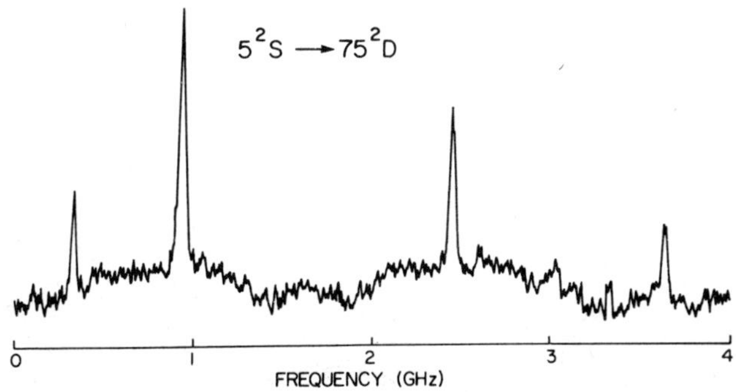

Fig. 6 Two-photon spectrum of ^{85}Rb and ^{87}Rb, showing the sharp components of the $5^2S \rightarrow 75^2D$ transition [20].

excited S and D levels up to principal quantum numbers n~20, using the ensuing cascade fluorescence for detection. In a recent study with Rb, excited D levels up to n=85 have been detected using a sensitive thermionic diode (Fig.6). An effective resolution of ~15 MHz was used, and the fine structure splittings were found to vary as $A/n_{eff}^3-B/n_{eff}^5$ [20]. With improvement in frequency stability to the kilohertz region, it should also be possible to determine hyperfine structure splittings of these highly-excited states.

One of the more interesting two-photon experiments is the study of the 1S-2S transition in atomic hydrogen, since in theory, the subsequent two-photon emission should produce a line of 1 Hz width (or a resolving power of 10^{15}). In a preliminary experiment (at a resolution of 50 MHz) HÄNSCH and his co-workers [21] have excited the 2S metastable state by two-photon absorption using pulsed radiation at 2430Å. They monitored the absorption

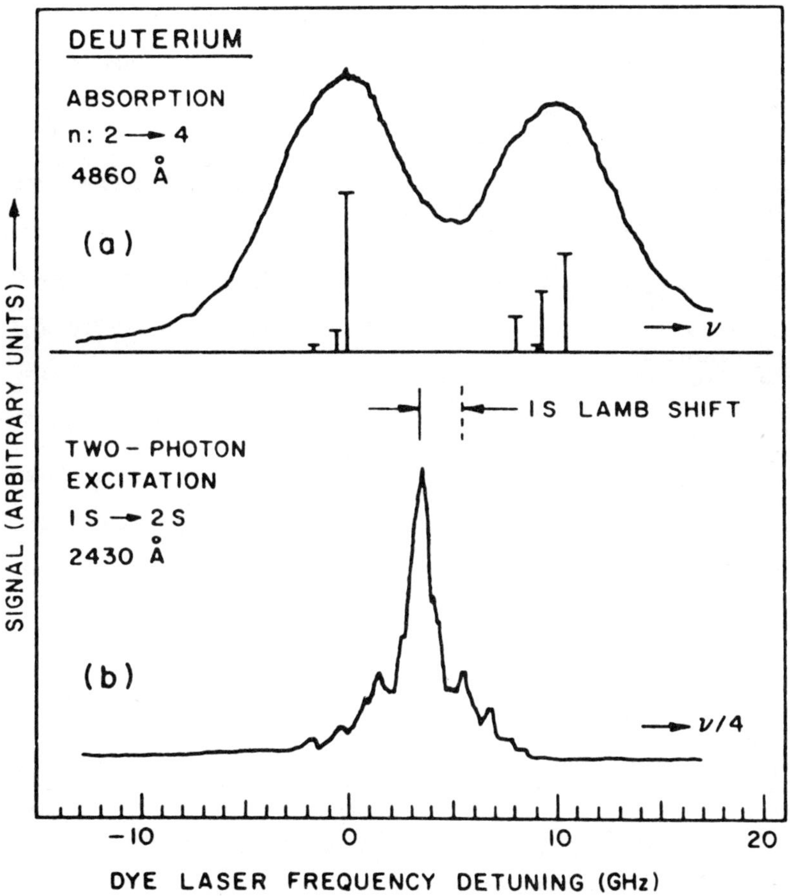

Fig.7 (a) Absorption profile of the deuterium Balmer-β line with theoretical fine structure; (b) simultaneously recorded two-photon resonance of the 1S→2S transition in deuterium.

by observing Lyman alpha (1216Å) fluorescence emission. They simultaneously observed the absorption spectrum of the Balmer beta line at 4860Å, and were able to measure the difference between $\nu(H_\beta)$ and $\frac{1}{4}\nu(1S\rightarrow 2S)$, this being essentially the Lamb shift of the 1S ground state (Fig.7). The values 8.3±0.3 GHz (Deuterium) and 8.6±0.8 GHz (Hydrogen) improved the accuracy of an earlier value by HERZBERG. I understand that we are to hear of considerable improvements in these values in a paper at this Conference.

With these few examples, I have tried to give an indication of the progress in laser spectroscopy in the last few years, and since the last conference in 1975 in Megève. And I look forward to learning of the most recent developments and applications in spectroscopy during this Conference.

References

1 "Fundamental and Applied Laser Physics", M.S. Feld, A. Javan, and N.A. Kurnit, eds., Wiley-Interscience, N.Y., 1973

2 "Laser Spectroscopy", R.G. Brewer and A. Mooradian, eds., Plenum Press, N.Y., London, 1974

3 "Laser Spectroscopy", S. Haroche, J.C. Pebay-Peyroula, T.W. Hänsch, and S.E. Harris, eds., Springer-Verlag, N.Y., Heidelberg, 1975

4 "Méthodes de Spectroscopie Sans Largeur Doppler de Niveaux Excités de Systèmes Moléculaires Simples", J.C. Pebay-Peyroula and J.C. Lehman, eds., Colloques Internationaux du C.N.R.S. No.217, Paris, 1974

5 "Tunable Lasers and Applications", A. Mooradian, T. Jaeger, and P. Stokseth, Springer-Verlag, N.Y., Heidelberg, 1976

6 "Laser Spectroscopy of Atoms and Molecules", H. Walther, ed., Springer-Verlag, N.Y., Heidelberg, 1976

7 "High Resolution Laser Spectroscopy", K. Shimoda, ed., Springer-Verlag, N.Y., Heidelberg, 1976

8 R.G. Brewer, Science 178, 247 (1972)

9 J.L. Hall, in [4], p.105; C.J. Bordé and J.L. Hall, in [2] p.125; J.L. Hall, C.J. Bordé, and K. Uehara, Phys. Rev. Lett. 37, 1339 (1976)

10 T.W. Hänsch, in [5] p.326

11 K.M. Evenson, F.R. Petersen, and J.S. Wells, in [2] p.143; K.M. Evenson and F.R. Petersen, in [6] p.349

12 S.E. Harris, J.F. Young, A.H. Kung, D.M. Bloom, and G.C. Bjorklund, in [2] p.59; See also [15]

13 J. Reintjes, C.Y. She, R.C. Eckardt, N.E. Karangelen, R.A. Andrews, and R.C. Elton, Appl. Phys. Lett. 30, 480 (1977)

14 R.T. Hodgson, P.P. Sorokin, and J.J. Wynne, Phys. Rev. Lett. 32, 343 (1974); P.P. Sorokin, J.A. Armstrong, R.W. Dreyfus, R.T. Hodgson, J.R. Lankard, L.H. Mangenaro, and J.J. Wynne, in [3] p.46

15 B.P. Stoicheff and S.C. Wallace, in [5] p.1

16 A. Provorov, B.P. Stoicheff, and S.C. Wallace, unpublished

17 B.P. Stoicheff, in [1] p.573

18 F. Moya, S.A.J. Druet, and J.-P.E. Taran, in [3] p.66; J.-P.E. Taran, in [5] p.378: S.A. Akhmanov, A.F. Bunkin, S.G. Ivanov, N.I. Koroteev, A.I. Kovrigin, and I.L. Shumay, in [5] p.389: A.B. Harvey, J.R. McDonald, and W.M. Tolles, in "Progress in Analytical Chemistry", Plenum, N.Y. 1977

19 V.G. Smirnov, private communication, unpublished

20 K.C. Harvey and B.P. Stoicheff, Phys. Rev. Lett. 38, 537 (1977)

21 T.W. Hänsch, S.A. Lee, R. Wallenstein and C. Wieman, Phys. Rev. Lett. 34, 307 (1975)

EXPERIMENTAL STUDY OF A HIGHLY FORBIDDEN MAGNETIC TRANSITION. SEARCH FOR PARITY VIOLATION

M.A. Bouchiat and L. Pottier

Laboratoire de Spectroscopie Hertzienne de l'Ecole Normale Supérieure
24, rue Lhomond, 75231 Paris Cedex 05, France

This paper is devoted to the experimental study of a forbidden magnetic dipole S-S transition, having an oscillator strength as small as 4×10^{-15}. We shall report here on the influence of a dc-electric field and outline the new method which allowed us to measure both the magnitude and the sign of the magnetic dipole of transition, in a given phase convention. This method makes use of the spin polarization created in the excited state by an interference between the magnetic dipole and the dc-field-induced electric dipole amplitudes [1] . If a small abnormal electric dipole amplitude $E_1^{p.v.}$ also exists, as a result of the parity violating electron-nucleus interaction induced by neutral currents, a similar method can be used to detect it. It is based on the search for an interference effect involving now the parity violating and the dc-field-induced electric dipole amplitudes, and exhibiting typical symmetry properties [2] . This is actually the ultimate purpose of the present experiment. So we shall first review the reasons motivating the achievement of this test at a sufficient level of accuracy.

It is now well known that several renormalizable models of the spontaneously broken symmetry gauge theory unifying weak and electromagnetic interactions (among which the most popular is the WEINBERG-SALAM model [3]) predict the existence of a component of weak neutral current interactions which violates parity reflection symmetry, and preserves time reversal invariance. There is presently no direct evidence for parity violation in neutral current interactions. The neutrino experiments performed near the accelerators are unable to provide an answer unless some supplementary hypothesis is made. The reason is that neutrino and antineutrino experiments are performed with particles belonging always to a single helicity state. In such conditions, it can be shown that a parity violating and a parity conserving hamiltonian both yield the same predictions, unless a definite assumption is made concerning the existence of one single heavy vector boson. So it now looks reasonable to expect that Atomic Physics may play a decisive role in solving this fundamental question.

The problem in Atomic Physics is to observe the effects of an electron-nucleus short-range weak interaction, induced by an electronic-nucleonic component of neutral currents, owing to its intrinsic symmetry properties different from that of the electromagnetic interaction. For instance, if it is possible in an experiment to define a handedness, one may expect a different answer for the right and left handed experiments. So, for instance, an unorientated vapour, in a zero magnetic field, may exhibit circular dichroïsm or rotatory power. The right-left asymmetry is given by the ratio between the weak and electromagnetic interaction amplitudes.

At first sight, Atomic Physics is not likely to be the ideal place for searching for neutral current effects.The naïve order of magnitude given below seems to suggest that it is a rather hopeless enterprise. For a given momentum transfer q typical of atomic processes $q \sim m_e \alpha c$, the amplitudes A_{weak} and $A_{e.m.}$ associated respectively with the exchange of the heavy vector boson Z_0 (short range interaction) and the massless photon (long range interaction) are given, up to constant factors, by the following expressions :

$$A_{weak} \propto \frac{g^2}{q^2 + M_{Z_0}^2 c^2} \sim \frac{g^2}{M_{Z_0}^2 c^2} \qquad A_{e.m.} \propto \frac{e^2}{q^2} \quad .$$

The assumption that the strength of the weak interaction induced by neutral currents is of the same order of magnitude as the ordinary weak interactions (like for instance β-decay) implies

$$\frac{g^2}{M_{Z_0}^2 c^2} \sim G_F = \text{Fermi constant (in units } \hbar = c = 1).$$

So we get an estimate of the amplitude ratio :

$$A_{weak} / A_{e.m.} \sim \frac{G_F q^2}{e^2} \sim G_F m_e^2 \alpha \sim 10^{-14} \quad .$$

Since searching for a right-left asymmetry of 10^{-14} is out of the question, it is crucial to find different ways for enhancing the weak amplitude and suppressing the electromagnetic one. A decisive point for starting several experiments in Atomic Physics has been the fact that there is an enhancement of the parity violating effects in heavy atoms [4] following a Z^3 law; this gives enhancements of the order of 10^6 to 10^7 for $Z > 50$. The physical interpretation of the Z^3 variation can be summarized as follows : one factor Z comes from the coherent effect of the nucleus; the second factor Z reflects the fact that the density of the valence electron increases like Z because of the Coulomb attraction of the nucleus; the third factor Z results from the proportionality of the parity violating potential to the velocity of the electron which grows like Z near the nucleus. Furthermore the electromagnetic amplitude can be partially hindered by working with a magnetic dipole transition instead of an electric dipole one. By choosing a *forbidden* magnetic transition like the Cesium 6S-7S transition, the hindrance factor affecting M_1 can be still increased by a factor 10^4 and the asymmetry to be searched reaches the level of 10^{-4}. So one can already see two important reasons which have contributed to our choice of a Cesium S-S transition (☆). Also very important for this choice were the simple atomic structure of the Cesium atom, with a single electron around a closed shell, and the existence of many, well interpreted, spectroscopic data. These conditions are favourable to compute the parity violating effects in a reliable way.

(☆) Let us mention that another experiment, very similar in principle to the present one, and also suggested in Ref. [4] , is presently being performed in Berkeley on the $6P_{1/2}$-$7P_{1/2}$ transition of Thallium [7] .

The Cesium experiment has proceeded in four stages, closely connected, which we now describe.

The first stage is the observation of the hyperfine structure of the single photon 6S-7S transition induced by a dc-electric field [5] . Cesium atoms in a vapour at 10^{-2} torr were excited by a single mode c.w. dye laser tuned at the resonant frequency. Resonance was detected by monitoring the fluorescence on the allowed $7S-6P_{1/2}$ transition in the direction perpendicular both to the field and to the beam. Figure 1 shows the fluorescence yield versus the laser frequency. Fig. 1-a corresponds to a linear polarization parallel to $\vec{E}_0$, fig. 1-b to a polarization at right angle; the fluorescence yield, plotted at a difference scale, is about 300 times smaller in the latter case. Interpretation of these facts can be found in the symmetry properties of the second-rank polarizability tensor T which relates the induced electric dipole $\vec{D}$ to the external field

$$\vec{D} = \vec{\vec{T}}\,\vec{E}_0 \tag{1}$$

and can be treated according to first order perturbation theory :

$$< 7S_{1/2}|\vec{D}|6S_{1/2} > = - < 7S_{1/2}|\vec{d}.\vec{E}_0 \frac{1}{E_{7S}-\mathcal{H}} \vec{d} + \vec{d} \frac{1}{E_{6S}-\mathcal{H}} \vec{d}.\vec{E}_0|6S_{1/2} > \tag{2}$$

where $\mathcal{H}$ is the unperturbed atomic hamiltonian.

Simple symmetry considerations imply the following decomposition of the effective dipole operator into two components :

$$\vec{D} = -\alpha\,\vec{E}_0 - 2i\beta\,\vec{S} \wedge \vec{E}_0 \quad , \tag{3}$$

α and β being real coefficients. It has to be noted that the polarization tensor operating between two S states cannot have any component of order 2; its scalar component α is expected to predominate since the vector component proportional to β is identically zero unless the spin-orbit coupling in the intermediate P states is included. So we expect that the ratio β/α is of the order of magnitude of the ratio between the P-state fine structure energy splitting and the energy difference between P state and S state. This ratio is about 1/10 for Cesium. In the following, we shall use again this property of the d.c.-field-induced dipole, namely to possess two distinct components : a spin-independent one along $\vec{E}_0$, and a spin-dependent one perpendicular to $\vec{E}_0$, 10 times smaller and $\pi/2$ out-of-phase.

For instance, in the interaction with a resonant linearly polarized light beam, if the polarization $\vec{\varepsilon}$ is parallel to $\vec{E}_0$, only the component of $\vec{D}$ along $\vec{E}_0$ acts on the atomic wave function and it does as a scalar operator : the total angular momentum $\vec{F} = \vec{I} + \vec{S}$ is conserved, only $\Delta F = 0$ transitions are allowed. Conversely, when $\vec{\varepsilon}$ and $\vec{E}_0$ are perpendicular, only the spin-dependent component of $\vec{D}$ operates, and it does as a spin component, all hyperfine transitions $\Delta F = 0, \pm 1$ become possible. This is what is expressed by the matrix element of the d.c.-field-induced dipole :

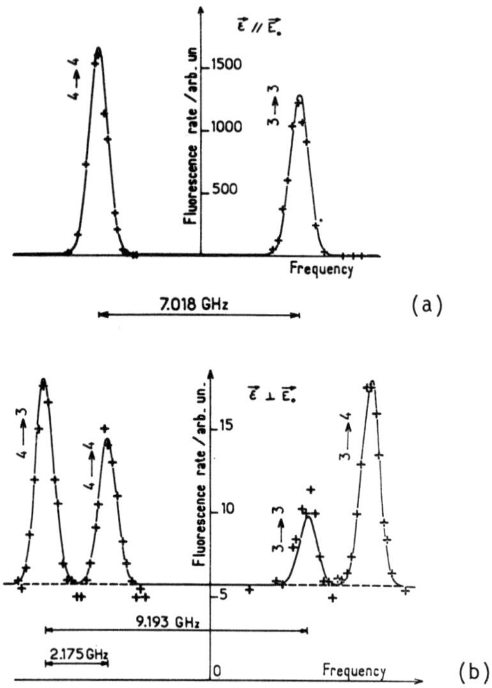

Fig. 1 Fluorescence rate (arbitrary units) versus incident laser beam frequency. Electric field 1000 V/cm. Solid line : theoretical spectrum; crosses : experimental points. (a) $\vec{\varepsilon}//\vec{E}_0$, (b) $\vec{\varepsilon} \perp \vec{E}_0$.

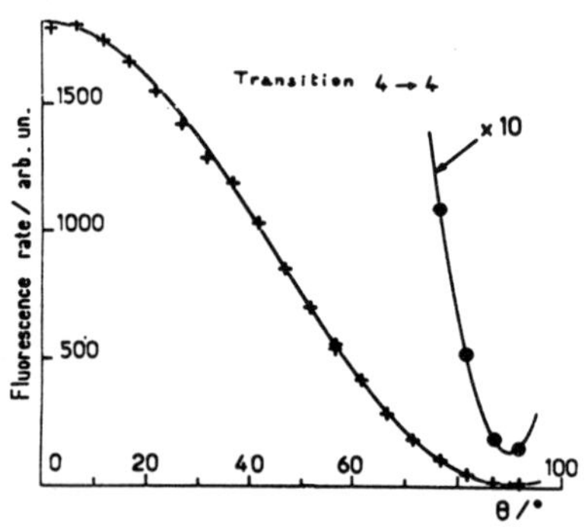

Fig. 2 Peak intensity of the fluorescence rate (arbitrary units) versus the angle θ, for the $6S_{1/2}$ (F=4) → $7S_{1/2}$ (F=4) transition. Solid line : theoretical curve; crosses : experimental points. The background is 4 units.

$$< 7S, m'_S |\vec{D}.\vec{\varepsilon}| 6S, m_S > = -\alpha \vec{E}_0.\vec{\varepsilon}\, \delta_{m'_S m_S} - 2i\beta\, (\vec{S} \wedge \vec{E}_0)_{m'_S m_S}.\vec{\varepsilon} \quad , \qquad (4)$$

and the typical selection rules for Cesium of nuclear spin 7/2 follow. Using the known h.f.s. splittings of the 6S and 7S states one can compute a theoretical spectrum and compare it with the experimental one : on Fig. 1 the curves are theoretical : they agree with the experimental points. We have also measured the fluorescence yield at resonance (peak height) as a function of the angle θ of $\vec{\varepsilon}$ with $\vec{E}_0$. The results for the 4-4 hyperfine component are shown on Fig. 2. The comparison between experiment and theory yields the magnitude of the ratio α/β [5] . We obtained $|\alpha/\beta| = 8.8 \pm 0.4$.

The second stage of the experiment is the observation of a spin polarization in the excited 7S state, resulting from the interference between the scalar and tensor components of the d.c.-field-induced electric dipole (see above). The direction of the incident light beam (perpendicular to $\vec{E}_0$) being taken for z axis, its circular polarization can be written

$$\vec{\varepsilon} = \frac{1}{\sqrt{2}} (\hat{x} + i\xi_i \hat{y}) \quad ,$$

with $\xi_i = \pm 1$. The $\pi/2$ phase difference between the two components of the light wave electric field can compensate the phase difference between the two components of the d.c.-field-induced electric dipole (Eq. 3), thus two real terms appear in the transition amplitude and they can interfere :

$$\vec{D}.\vec{\varepsilon} = - \frac{\alpha}{\sqrt{2}} E_0 + 2\, \xi_i\, \beta\, S_z\, E_0 \ . \qquad (5)$$

Assuming for simplicity a zero nuclear spin, we can write the density matrix of the upper state, up to a constant factor and to first order in β/α :

$$\rho_e \propto \frac{1}{2} \alpha^2 E_0^2 - 2\, \xi_i\, \alpha\beta\, S_z\, E_0^{\,2} .$$

The first term represents the population, the second expresses the fact that a spin polarization, which we shall denote $\vec{P}_e^{(2)}$, is created along the beam :

$$\vec{P}_e^{(2)} = 2 < S_z > \simeq - 2\, \xi_i \frac{\beta}{\alpha} \hat{k}_i \quad , \qquad (6)$$

where $\hat{k}_i = \hat{z}$ is the unit vector along the incident beam.

Let us note here that the magnitude of $\vec{P}_e^{(2)}$ does not depend on $\vec{E}_0$; it only involves the ratio β/α measured in the first stage of the experiment.

To detect a polarization $\vec{P}_e$ (e.g. $\vec{P}_e^{(2)}$) of the 7S state, we monitor the circular polarization of the fluorescence light. Let $\hat{k}_f$ denote the unit vector of the direction of detection. The intensity L_f transmitted through the circular analyzer is given by :

$$L_f = \frac{1}{2} (1 + \xi_f\, \vec{P}_e . \hat{k}_f)\, I_f \ , \qquad (7)$$

where I_f is the total intensity of the fluorescence, and $\xi_f = \pm 1$.

As can be seen from Eq. (7), this signal is sensitive only to a component of $\vec{P}_e$ along the direction of observation $\hat{k}_f$, which in our experiment is perpendicular to both $\vec{E}_0$ and $\hat{k}_i$. In the case of $\vec{P}_e^{(2)}$, if a magnetic field is applied along $\vec{E}_0$, then the spin polarization acquires, by Hanle effect, a component along $\hat{k}_f = \hat{k}_i \wedge \hat{E}_0$ ($\hat{E}_0$ = unit vector of $\vec{E}_0$). So we expect the variations of the signal versus the magnetic field to exhibit a dispersion line shape, of characteristic width ΔH :

$$S^{(2)} = \frac{H/\Delta H}{1 + (H/\Delta H)^2} \; P_e^{(2)} \; (\hat{H} \wedge \hat{k}_i) \cdot \hat{k}_f \qquad (8)$$

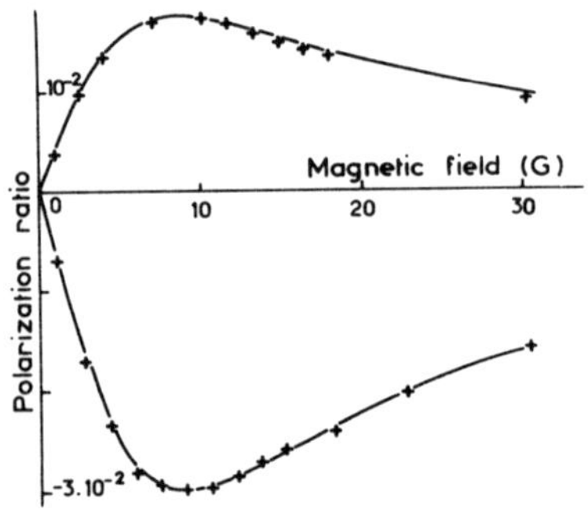

Fig. 3 Polarization ratio of the fluorescence light versus the magnetic field. Solid line : theoretical curve; crosses : experimental points. Upper curve $6S_{1/2}$ (F = 3) → $7S_{1/2}$ (F = 3) transition; lower curve $6S_{1/2}$ (F = 4) → $7S_{1/2}$ (F = 4) transition.

As shown on Fig. 3, this is what we have actually observed. The two curves are relative to the two $\Delta F = 0$ transitions. They are of opposite signs like the g-factors of the two hyperfine sublevels of the 7S state. From their signs, we can deduce the sign of β/α.

Let us insist on the fact that the signal associated with $\vec{P}_e^{(2)}$, whose magnitude is known since β/α has been measured, is very useful to calibrate other optical measurements of spin polarization of the upper state : in this way, the results become unaffected by a small depolarization due to collisions or by possible imperfections of the optical system. We currently use this method to calibrate our measurements on parity violation.

The third stage of the experiment is centered on the measurement, in magnitude and sign, of the ratio of the magnetic dipole to the d.c.-field-induced electric dipole. The sensitive technique illustrated by the present experiment is quite general and can be extended to many other cases [6] . Again, we measure the spin polarization in the upper state resulting from an interference between two amplitudes of transition, namely the magnetic dipole and the d.c.-field-induced electric dipole. Let us begin with a physical interpretation using a classical picture of the excitation process. To each emitting atom, we can associate two dipoles, an electric one along the applied electric field and a magnetic one along the atomic spin, both oscillating in phase at the optical frequency. From classical electrodynamics, we know the polarization of the plane wave radiated by each dipole in a direction of unit vector $\hat{k}$; the result is indicated on Fig. 4. If $\vec{\mu}$ and $\vec{D}$ were parallel (Fig. 4-a), the two radiated waves would be polarized at right angle and

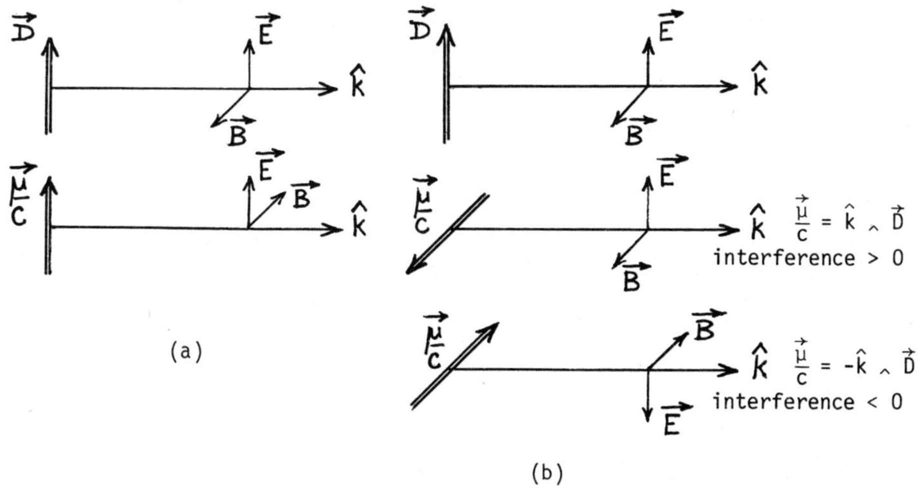

Fig. 4 Scheme illustrating the classical description of the emission process, for a mixed electric magnetic dipole transition. (a) no interference; (b) maximum interference.

could not interfere. But if $\vec{\mu}$ is at right angle with $\vec{D}$ (Fig. 4-b), then an interference can take place and it is maximum, positive or negative, for $\vec{\mu}/c = \pm \hat{k} \wedge \vec{D}$. As a result, the power radiated by an atom in the direction $\hat{k}$ is different depending whether its spin lies along the direction $\hat{k} \wedge \vec{D}$ or the opposite direction :

$$\frac{d\mathcal{P}}{d\Omega} = \frac{c}{8\pi} k^4 \left[|\vec{D} \wedge \hat{k}|^2 + |\frac{\vec{\mu}}{c} \wedge \hat{k}|^2 + 2\frac{\vec{\mu}}{c} \cdot (\hat{k} \wedge \vec{D}) \right] .$$

Since the probabilities for emission and absorption are the same in a given spin state, we also come to the conclusion that an atom, irradiated by an incident beam of direction $\hat{k}_i$, absorbs differently according to whether its spin is turned along $\hat{k}_i \wedge \vec{E}_0$ or in the opposite direction : more excited atoms are created in one spin state than in the other. The corresponding spin polarization $\vec{P}_e^{(1)}$ can be computed quantum mechanically. Within the condition $M_1 \ll E_1^{ind.}$, we obtain :

$$\vec{P}_e^{(1)} = \frac{8}{3} \frac{F(F+1)}{(2I+1)^2} \frac{M_1}{E_1^{ind.}} \hat{k}_i \wedge \hat{E}_o , \tag{9}$$

for a plane wave linearly polarized along $\vec{E}_o$ and a $\Delta F = 0$ transition.

The measurement of $\vec{P}_e^{(1)}$ yields directly the determination of $M_1/E_1^{ind.}$. The experiment is performed by modulating $\vec{E}_o$ at a low frequency ω_E; the fluorescence light intensity is monitored through a circular analyzer having its efficiency modulated at the frequency ω_P. The signal proportional to $\vec{P}_e^{(1)}$ appears at the sum and difference frequencies and is calibrated by aid of the known spin polarization $\vec{P}_e^{(2)}$ (Eq. 6). Note that $\vec{P}_e^{(1)}$ and $\vec{P}_e^{(2)}$ are easily distinguished, not only because they are created at right angle from each other, but also because $\vec{P}_e^{(1)}$ is an odd function of $\vec{E}_o$, while $\vec{P}_e^{(2)}$ is even.

From our measurements, we obtain the ratio between the magnitudes of the spin polarizations :

$$P_e^{(1)}/P_e^{(2)} = (2.32 \pm 0.13) \times 10^{-2} \text{ for } E_o = 1000 \text{ volt/cm}. \tag{10}$$

We also observe that $\vec{P}_e^{(1)}$ in a d.c.-field $\vec{E}_o$ is pointing along the vector $-\hat{k}_i \wedge \hat{E}_o$. From Eq. 9, it follows that the ratio $M_1/E_1^{ind.}$ is negative.

Since $E_1^{ind.}$ has been computed theoretically [4] to be :

$$E_1^{ind.} = 1.62 \times 10^{-5} \frac{|\mu_B|}{c} E_{oz} \text{ (volt/cm)} ,$$

we deduce M_1, in the same phase convention :

$$M_1 = (-4.24 \pm 0.34) \times 10^{-5} \frac{|\mu_B|}{c} . \tag{11}$$

This result is in agreement with a theoretical estimation [4] based on the remark that the combined effects of configuration mixing and spin orbit coupling, which give rise to a deviation Δg of the gyromagnetic ratio of the Cs ground state from that of the free electron, also contribute to M_1; so it was possible to express M_1 in terms of $\Delta g/g$, this last quantity being known from experiment.

The above result (Eq. 11) can also be expressed as an oscillator strength in zero electric field : $f_{6S-7S} = 4.05 \times 10^{-15}$, or as a single-photon M_1 decay rate : $\Gamma_{7S-6S} = 0.93 \times 10^{-6}\ s^{-1}$.

The last step of the experiment is the search for a parity violation induced by a time reversal invariant interaction. Time reversal invariance forbids the existence of a static electric dipole moment in a stationary atomic state but it allows the existence of an electric dipole of transition between two S states. The abnormal amplitude $E_1^{p.v.}$ has been computed in the framework of the WEINBERG-SALAM model [4] :

$$E_1^{p.v.} = -|e| < 6S,\ 1/2|z|7S,\ 1/2 > = -i\ 1.7 \times 10^{-11}\ |e|a_o . \tag{12}$$

Using the experimental result concerning M_1 (Eq. 11), a circular dichroïsm is then predicted for the resonant frequency of the 6S → 7S transition of Cesium :

$$P_c = 2\ \mathrm{Im}\ \{\ E_1^{p.v.}\ /\ M_1\ \} = 2.2 \times 10^{-4} \quad . \tag{13}$$

We now outline the method used to test the existence of $E_1^{p.v.}$. The basic principle is to search for a parity-violating d.c.-field-induced polarization, i.e. to search for a spin polarization created in the excited state as the result of an interference between two electric dipole amplitudes of transition, the parity violating amplitude $E_1^{p.v.}$ and the d.c.-field-induced amplitude $E_1^{ind.}$.

Let us retain, in the expression of the electric dipole moment of transition only the two components relevant to the present discussion :

$$\vec{D} = E_1^{ind.}\ \hat{E}_o - 2i\ \mathrm{Im}\ E_1^{p.v.}\ \vec{S}$$

Note that they are $\pi/2$ out-of-phase as a consequence of the symmetry properties under space and time reflection of the hamiltonian which is responsible for $E_1^{p.v.}$: i.e. $E_1^{p.v.}$ is pure imaginary in a phase convention where $E_1^{ind.}$ and M_1 are real [4] . If $\vec{D}$ interacts with an incident light beam of circular polarization

$$\vec{\varepsilon} = \frac{1}{\sqrt{2}}\ (\hat{x} + i\xi_i\hat{y}),$$

among the resulting amplitudes, two are real :

$$\frac{1}{\sqrt{2}}\ E_1^{ind.} + \frac{1}{\sqrt{2}}\ \xi_i\ \mathrm{Im}\ E_1^{p.v.}\ \sigma_y \quad ,$$

and their interference gives rise to a polarization in the upper state directed along the direction $\hat{k}_i \wedge \hat{E}_o$ and equal to :

$$\vec{P}_e^{\,p.v.} = \xi_i\ \frac{E_1^{p.v.}}{E_1^{ind.}}\ \hat{k}_i \wedge \hat{E}_o\ . \tag{14}$$

It is important to note the presence of the pseudoscalar quantity ξ_i in front. Since the vector product $\hat{k}_i \wedge \hat{E}_o$ has the transformation properties under space and time reflection of an angular momentum, the presence of a term of the form $\xi_i\ \hat{k}_i \wedge \hat{E}_o$ is a clear indication of the fact that the hamiltonian contains a P-odd T-even part; it receives contribution from the interaction involving vector/axial-vector neutral currents. Thus the abnormal spin polarization $\vec{P}_e^{\,p.v.}$ adds to the normal one $\vec{P}_e^{(1)}$ (Eq. 9) created by

M_1 $E_1^{ind.}$ interference, but it can be easily distinguished owing to its dependence on the spin state ξ_i $\hat{k}_i$ of the incident photon.

The method based on the search for a parity-violating d.c.-field-induced polarization has several important advantages:

1) The method leads to a complete elimination of the background coming from the radiative transitions induced by collisions :

$$\text{Photon} + \text{Cs}(6S) + \text{Cs}(6S) \rightarrow \text{Cs}(6S) + \text{Cs}^{\star}(7S) \ .$$

This is because the collision-induced dipole moment varies randomly in orientation from one collision to another and cannot contribute, on the average, to a macroscopic spin polarization.

2) The circularly polarized fluorescence intensity associated with $\vec{P}_e^{\,p.v.}$ is proportional to E_0. Thus it is possible to optimize the signal/noise ratio by an appropriate choice of the electric field magnitude. This means that the sensitivity of the method is increased, like in an heterodyne experiment.

3) Calibration of the sensitivity can be done during the course of the experiment, both conveniently and accurately, by using the known polarization $\vec{P}_e^{(2)}$ (Eq. 6) resulting from the interference between the two d.c.-field-induced dipole amplitudes. $\vec{P}_e^{(2)}$ is detected in exactly the same conditions than $\vec{P}_e^{\,p.v.}$, and easily distinguished by its different behaviour under the application of a magnetic field or reversal of the electric field.

4) Another advantage is that the right-left asymmetry under search is the same for an incident beam and the reflected beam passing back through the vapor. Indeed the parity violating signal involves the photon spin which is conserved by light reflection. This gives us the very interesting possibility of using efficient multipasses of the beam with the advantage of a large enhancement of the parity violating signal. Moreover it can be shown that multipasses also suppress several possible causes of false asymmetry, e.g. that coming from a spurious d.c. magnetic field along the beam. This may be interesting in the future; even though we have not observed such spurious effects yet.

The experiment is done in the following way : the circular polarization ξ_i of the incident light beam is modulated like sin $\omega_i t$ by making the laser beam go through a quarterwave plate rotating at the angular frequency $\omega_i/2$. The efficiency ξ_f of the circular analyzer is modulated like ξ_i but at a different frequency ω_f. So the $E_1^{p.v.}$ $E_1^{ind.}$ interference manifests itself through the apparition of a component at the frequencies $\omega_f \pm \omega_i$ in the photocurrent. (Note that the M_1 $E_1^{ind.}$ interference, independent of ξ_i, does not contribute to such a signal.). The signal is digitally integrated while the laser frequency is kept resonant.

Furthermore, we use the fact that the parity violating signal reverses when $\vec{E}_0$ does : we sequentially reverse the d.c.-field $\vec{E}_0$ at constant amplitude and substract the results obtained for two consecutive orientations. The mean value and standard deviation on the mean are calculated in real time.

The present results of the experiment from a 10 hours'data analysis give a measured value of the asymmetry compatible with a true value equal to zero :

$$|E_1^{p.v.} / M_1| = (0.56 \pm 1.84) \times 10^{-3} . \qquad (15)$$

The error quoted above is computed from the r.m.s. uncertainty to give a result at 90% confidence level.

The same result can be expressed in terms of an upper limit for the parity violating electric dipole amplitude :

$$|E_1^{p.v.}| < 2.8 \times 10^{-10} \; |e| a_0 \quad ,$$

or an upper limit for the average weak vector charge of the nucleons

$$\overline{C}_V < 6.2 \qquad \text{at 90\% confidence level.}$$

$\overline{C}_V$ being defined as :

$$\overline{C}_V = \frac{Z}{A} C_{Vp} + \left(1 - \frac{Z}{A} \right) C_{Vn} \; ,$$

C_{Vp} and C_{Vn} are the weak vector charges of the proton and neutron. In the WEINBERG-SALAM model, we have $C_{Vp} = - 1/2 + 2 \sin^2\theta$, $C_{Vn} = 1/2$.

The sensitivity of the present results has still to be increased by roughly one order of magnitude to reach the level at which the gauge models of weak interactions can be tested. It is important to note that *no problem of spurious asymmetries or systematic errors has arisen so far*. The problem is to improve the signal to noise ratio. This, we feel, can be achieved by using multipasses of the beam in a much more efficient way than it has been done up to now. We have already built a multipass cell with internal mirrors, in which the light intensity has been measured to be 120 times that of a single pass. Whereas, in the set-up that yielded the results just reported, multipasses were used between external mirrors; the intensity was thus only 3.5 that of a single pass, because of the reflection losses on the windows of the cell. Since our signal to noise ratio is presently proportional to the light intensity, it seems reasonable to expect that our multipass device will solve the signal to noise problem.

In conclusion, we have proposed a new method to search for parity violation in atoms, our main objective being to make a clean experiment. By working with a highly forbidden magnetic transition, we have minimized the importance of systematic errors. The drawback of a low counting rate has been compensated by several means, namely, detection of a polarized fluorescence intensity, heterodyne signal amplification and now use of multipasses of the beam. We believe that in its final version, this experiment will prove useful to solve the problem of parity violation in neutral current induced interactions.

References

1 M.A. Bouchiat and L. Pottier, J. Phys. Lettres, 37, L-79 (1976).
2 M.A. Bouchiat and L. Pottier, Phys. Letters, 62B, 327 (1976).
3 S. Weinberg, Phys. Rev. Lett., 19, 1264 (1967).
A. Salam, "Elementary Particle Physics", ed. N. Svartholm, p. 367 (1968).
4 M.A. Bouchiat and C.C. Bouchiat, Phys. Letters, 48B, 111 (1974); J. de Phys., 35, 899 (1974); J. de Phys., 36, 493 (1975).
5 M.A. Bouchiat and L. Pottier, J. Phys. Lettres, 36, L-189 (1975).
6 D.S. Bethune, R.W. Smith and Y.R. Shen, Phys. Rev. Letters, 38, 647 (1977).
7 S. Chu, E.D. Commins and R. Conti, Phys. Letters, 60A, 96 (1977)

SEARCH FOR OPTICAL ROTATION INDUCED BY THE WEAK NEUTRAL CURRENT

P.G.H. Sandars

Clarendon Laboratory, Oxford University
Oxford, England

1. Introduction

It is now clear that laser experiments on atoms can provide important information on one of the most crucial questions of elementary particle physics - the validity or otherwise of the unification of the weak and electromagnetic interactions in a single unified theory. The majority of the schemes put forward to date, and in particular the most basic and economical of them, the Weinberg-Salam [1, 2] model, predict the existence of additional weak interactions between the particles, electrons, protons and neutrons which make up an atom. These interactions are of order G_F the normal weak interaction coupling constant and therefore very weak since in atomic units, ($e = \hbar = m_e = 1$), $G_F = 2 \times 10^{-14}$. At first sight this small value would seem to rule out any possibility of an atomic experiment since it is clearly impossible to make a direct comparison between experiment and theory at the 1 part in 10^{-14} level. But this neglects the fact that the weak interaction violates parity while the electromagnetic interaction conserves it, and it may be possible to devise an experiment to look for a very small explicitly parity violating effect in the presence of a very much larger parity conserving background.

Following the pioneering paper of the BOUCHIATS [3] a great deal of effort and ingenuity has gone into the search for suitable experiments with fractional parity violation enhanced above the 10^{-14} figure implied by dimensional arguments. A very important ingredient in this search was the observation by the BOUCHIATS [3] that in heavy atoms the PNC effect would be enhanced by a factor of order Z^3. Thus most attention has been focussed on heavy atoms though a great deal of work is now being carried out on metastable hydrogen where the near degeneracy of the $2s_{\frac{1}{2}}$-$2p_{\frac{1}{2}}$ levels can compensate for the small Z. The heavy atom experiments can conveniently be divided into two groups (i)the search for circular dichronism, or a similar PNC effect, in highly forbidden M1 transitions e.g. $6s_{\frac{1}{2}} \rightarrow 7s_{\frac{1}{2}}$ in cesium [4] or $6p_{\frac{1}{2}} \rightarrow 7p_{\frac{1}{2}}$ in thallium [5] (ii)the search for optical rotation in an allowed M1 transition e.g. $6p^3_J \rightarrow 6p^3_{J'}$ in bismuth or $6p_{\frac{1}{2}} \rightarrow 6p_{3/2}$ in thallium.

The two types of experiment differ in that group (i) involves a relatively high ($\sim 10^{-3}$) fractional effect to be observed on an extremely highly forbidden transition; effects due to molecules and to pressure induced transitions constitute a serious background problem. Group (ii) on the other hand involves a smaller PNC effect ($\sim 10^{-7}$) to be observed on a much less highly forbidden transition. Optical problems are likely to be limiting here. In both cases, signal to noise considerations show that with modern laser techniques one should have adequate fundamental sensitivity to see effects of the predicted magnitude. To date only the optical rotation experiments to be described in this talk have reached this desired sensitivity but no doubt this is a temporary situation and in the long run the two types of experiment will prove complementary.

2. Optical Rotation

From simple optics we know that when plane polarized light is passed through a region of length L in which the refractive indices n_R and n_L for right and left circularly polarized light are unequal the plane of polarization of the light is rotated through an angle

$$\phi = \frac{\pi L (n^R - n^L)}{\lambda} \tag{2.1}$$

Near to a resonance where the refractive index is dominated by a particular transition $J_i \rightarrow J_f$, n^R and n^L are proportional to

$$\sum_{M_i, M_f} \left| \langle J_f M_f | D_q + iq M_q + i Q^2_q | J_i M_i \rangle \right|^2 \tag{2.2}$$

where $\underline{D}$, $\underline{M}$ and $\underline{Q}^2$ are the appropriate electric dipole, magnetic dipole and electric quadrupole transition operators and q = -1 for n^R and +1 for n^L. The sum over magnetic quantum numbers removes any interference terms between the dipole and quadrupole terms and if parity is conserved either $\underline{D}$ or $\underline{M}$ can have non-zero matrix elements, but not both in which case (2.2) becomes independent of q and $n^R = n^L$ and there is no optical rotation. If however parity is not conserved then both $\underline{D}$ and $\underline{M}$ can have non-zero matrix elements and interference between them leads to optical rotation. For a predominantly M1 transition into which is admixed a small E1 matrix element $E1^{PNC}$ one finds the following expression for the PNC optical rotation angle

$$\phi^{PNC} = -\frac{4\pi (n-1) L R \theta}{\lambda}, \qquad R = \frac{\mathrm{Imag.}\, E1^{PNC}}{M1} \tag{2.3}$$

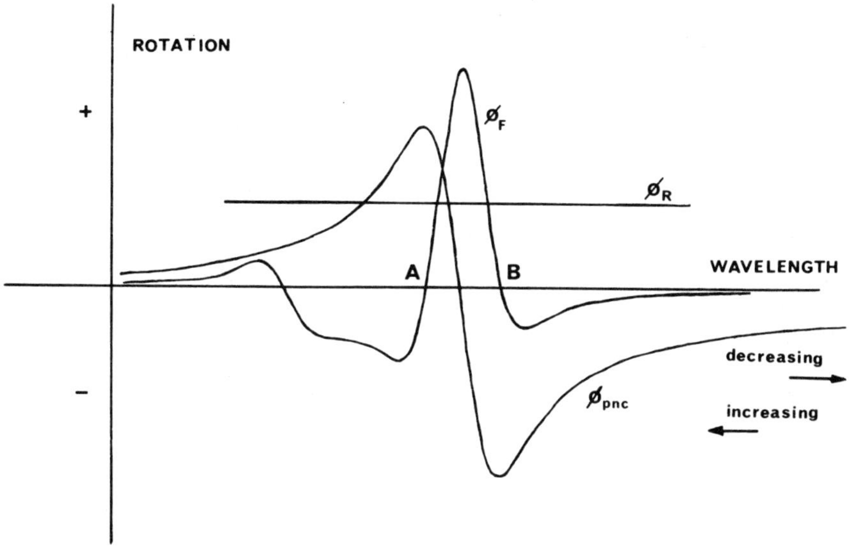

Figure 1. Wavelength dependence of the PNC optical rotation angle ϕ_{PNC}, the Faraday angle ϕ_F and residual polarizer angle ϕ_R for the 6 → 7 hyperfine component of the λ = 648 nm line in Bi as an example of wavelength behaviour near resonance.

Θ is a correction factor which takes into account the fact that a non-zero E2 matrix element can contribute to the overall refractive index but not to the optical rotation.

(2.3) implies that ϕ^{PNC} has the familiar dispersive dependence on wave-length of the refractive index (fig. 1). This rapid wave-length dependence near resonance is very important experimentally since it allows one to separate the optical rotation of interest from more slowly varying background effects. (2.3) also suggests that ϕ^{PNC} can be increased by lengthening the optical path length (n-1). But there is a limit to the extent to which this is possible because of the presence of absorption. A convenient measure is the optical rotation at one absorption length:

$$\phi^{PNC}_{1\,A.L.} = \frac{(\omega - \omega_0)\, R\Theta}{\Gamma/2} \tag{2.4}$$

where Γ is the resonance width and $(\omega - \omega_0)$ is the frequency off-set from resonance.

(2.4) suggests two alternative measurement philosophies: one can either move away from the resonance and look for a large angle, though with reduced wave-length dependence, or one can stay close to resonance and look for a smaller angle but with a sharper wave-length dependence. The choice will depend on detailed experimental considerations.

3. M1 Transitions in Bismuth

The choice of a suitable element and transition for an optical rotation experiment is determined by a number of requirements:

(i) The transition must involve either $s_{\frac{1}{2}}$ or $p_{\frac{1}{2}}$ electrons since the only non-zero matrix elements of H_{PNC} are between these states.

(ii) The atom should be heavy in order to take advantage of the Z^3 law.

(iii) In order to obtain one absorption length at reasonable atomic densities and cell lengths the transition should be allowed M1.

(iv) The transition should be in a spectral region where tunable lasers are available in order to make use of their excellent intensity and wave-length discrimination properties.

This combination of requirements is extremely restrictive and is only satisfied by a single element, bismuth, which has two suitable transitions $J = 3/2 \rightarrow 5/2$ $\lambda = 648$ nm, and $J = 3/2 \rightarrow 3/2'$ $\lambda = 876$ nm. The first which is being used in experiments at Oxford [6] and Novosibirsk [7] is in the range covered by continuous wave dye lasers, and the second which is the Washington [8] line has up to now required a pulsed parametric laser, though a dye for this region has recently been developed.

3.1 Calculations of $E1^{PNC}$

Detailed calculations of the PNC induced E1 matrix elements in bismuth have been made by a number of authors. The basic principle is to use perturbation theory in a relativistic central field model. In this scheme one can write

$$E1^{PNC} = C_{J'J} \sum_n \frac{\langle 6p_{3/2} | \underline{D} | ns_{1/2} \rangle \langle ns_{1/2} | H_{PNC} | 6p_{1/2} \rangle}{\varepsilon_{6p_{1/2}} - \varepsilon_{ns_{1/2}}} \qquad (3.1)$$

where $C_{J'J}$ is a known coefficient which represents the breakdown of jj coupling in the $6P^3$ shell. In (3.1) the states are solutions of a single particle Dirac equation

$$\left\{ \beta mc^2 + \underline{\alpha} . c \underline{p} - e\, \mathcal{U}(r) \right\} \phi = \varepsilon \phi \qquad (3.2)$$

and H_{PNC} is the usual PNC operator

$$H^{PNC} = \frac{G_F\, Q_W\, \rho_N(r)\, \gamma_5}{2\sqrt{2}} \qquad (3.3)$$

where Q_W is the weak charge introduced by the BOUCHIATS [3] and $P_N(r)$ is a suitable normalized nuclear density. BRIMICOMBE, LOVING, SANDARS [9] used a parametric potential fitted to observed energy levels and evaluated the infinite sum over excited states in (3.1) by turning the problem into the solution of a suitable inhomogenous Dirac equation. HENLEY and WILETS [10] used a rather similar method except that they solved a non-relativistic inhomogeneous equation which was fitted to the solution of the Dirac equation near the origin. NOVIKOV, SUSHKOV and KHRIPLOVICH [11] on the other hand used an extension of the FERMI-SEGRE method to relate the wave-functions at the origin required in the dominant 6S, 7S PNC matrix elements to observed energy eigenvalues and a semi-empirical method based on observations on transition rates in bismuth and thallium to obtain the desired matrix elements of r. The results of the calculations are given below.

	$R(876nm) \times 10^7$	$R(648nm) \times 10^7$
HENLEY and WILETS	-3.5	-4.4
BRIMICOMBE, LOVING and SANDARS	-2.3	-3.0
NOVIKOV, SUSHKOV and KHRIPLOVICH	-1.7	-2.1

The reasons for the disagreement between the author's results and HENLEY and WILETS are not entirely clear. The low value obtained by KHRIPLOVICH and co-workers follows from their belief based on empirical data that the straight-forward central field calculation overestimates the 6s-6p radial matrix element.

As will become apparent, the reliability of these calculations is a question of considerable importance. An approach adopted by the author is to use relativistic many-body perturbation theory. Applied in a straightforward way this method would rapidly become too complex to be useful since one is essentially concerned with third order perturbations involving the PNC operator, the residual two particle interaction and the dipole transition operator. But a very considerable simplification can be achieved by including H_{PNC} in the central field equation (3.2) which becomes

$$\left\{ \beta mc^2 + \underline{\alpha} \cdot c\underline{p} - eU(r) + G_F Q_W \rho_N(r) \gamma_5 / 2\sqrt{2} \right\} \phi = \varepsilon \phi \qquad (3.4)$$

so that the single particle functions while eigen-states of the angular momentum are no longer parity eigenstates. Use of (3.4) effectively reduces the order of perturbation by one with a consequent large reduction in the number of terms which need to be evaluated. Thus (3.1) is replaced by

$$E1^{PNC} = C_{J'J} \langle 6p_{3/2} | \underline{D} | 6p_{1/2} \rangle \qquad (3.5)$$

and a typical next order term is

$$\sum_{n'} \frac{\langle 6s_{1/2} | \underline{D} | n'd_{3/2} \rangle \langle n'd_{3/2} 6p_{3/2} | \frac{e^2}{r_{12}} | 6s_{1/2} 6p_{1/2} \rangle}{\varepsilon_{6p_{1/2}} + \varepsilon_{6s_{1/2}} - \varepsilon_{6p_{3/2}} - \varepsilon_{n'd_{3/2}}} \qquad (3.6)$$

A systematic evaluation of all terms of this order is under way and will be reported elsewhere [12]. These configuration interaction results will give some indication of the reliability of the lowest order calculations.

3.2 Calculations of M1

Calculation of the required M1 matrix elements is straightforward since a detailed analysis of the jj coupling structure of the $6P^3$ configuration has been given by LURIO and LANDMAN [13]. The accuracy should be comparable to the degree of fit to the diagonal magnetic moments and is much better than is required.

3.3 Hyperfine Structure

One complication which we have neglected so far is the presence of hyperfine structure. Instead of a single atomic transition one has a number of closely spaced ones. However their relative strengths are readily calculable and since both M1 and E1 are dipole matrix elements, the ratio is independent of the presence of the nuclear spin.

4. The Bismuth Experiments

Three optical rotation experiments on Bi are now in progress at Oxford [6], at the University of Washington in Seattle [7] and in Novosibirsk [8]. Since they contain many common features I shall describe them largely in terms of their similarities or differences to the Oxford experiment. This is not to claim any element of priority; as far as I am aware all three were initiated independently and more or less simultaneously.

4.1 The Basic Method

All three bismuth optical rotation experiments use the same basic method in which laser light is passed through a pair of crossed polarizers between which is placed an oven containing the bismuth vapour. The transmitted light intensity has the quadratic form

$$I = I_o \{ \phi + \phi^{PNC} \}^2 + C I_o \qquad (4.1)$$

where I_o is the full intensity and C is the extinction factor. ϕ is an angle which includes both residual misalignment of the polarizers and any non PNC rotation of the plane of polarization in the region between the polarizers.

Isolation of ϕ^{PNC} takes place in three stages: First, we obtain a signal which is linear in ϕ^{PNC} by making a known change in ϕ, $\delta\phi$ say. The corresponding change in intensity is

$$\delta I = 2 I_0 \, \delta\phi \{ \phi + \phi^{PNC} \} \tag{4.2}$$

In order to separate ϕ^{PNC} from ϕ we make an appropriate small change, $\Delta\lambda$ say, in the wave-length and make use of the rapid dispersive wave-length dependence of ϕ^{PNC} near resonance in comparison to the very much smaller wave-length dependence for ϕ (fig. 1). Thus

$$S = \Delta\delta I = 2 I_0 \{ \Delta\phi + \Delta\phi^{PNC} \} \tag{4.3}$$

With good apparatus design we may hope that $\Delta\phi \ll \Delta\phi^{PNC}$. But as a third stage we can require that the value of $\Delta\phi^{PNC}$ so determined depend correctly on the bismuth density and have the correct wavelength dependence.

The advantage of this crossed polarizer method stems from the signal to background. If, as is usual, we set $\phi^2 \approx C$ and $\delta\phi \approx \phi$, this is given by

$$\frac{S}{I} \approx \frac{\Delta\phi^{PNC}}{\sqrt{C}} .$$

With lasers and good polarizers values of C better than 10^{-6} can be obtained which implies, for example, that a 10^{-7} angle will produce a fractional change in intensity greater than 10^{-4}. In effect crossed polarizers give an enhancement factor $1/\sqrt{C}$.

4.2 The Oxford Experiment

The apparatus is illustrated schematically in fig. 2. The Spectra-Physics dye laser gives a few milliwatts of single mode radiation at 648 nm with a frequency stability much better than the Doppler width of a single bismuth hyperfine component. The two polarizers are of the Glan-Thompson type with an extinction ratio C $\approx 10^{-7}$. The detectors are standard low noise diodes. The bismuth is contained in an alumina tube which can be heated to 1500K to give vapour pressure yielding one atomic absorption length for the $F = 6 \rightarrow F = 7$ hyperfine component on which all our measurements have been made. Windows at either end of the oven are fixed rigidly to the same optical bench as the polarizers. Coating by the bismuth is made negligible by use of the buffer gas at approximately 0.2 atmospheres. The oven is surrounded by a mu-metal shield to reduce extraneous magnetic fields.

The controlled angle change $\delta\phi$ is produced by the FARADAY cell which is modulated at 328 Hz. The corresponding signal δI is detected by standard phase sensitive techniques. The wave-length discrimination is produced by switching the laser by 3 cavity modes (=1.2 GHz) from peak to trough of the dispersion curve with a repetition rate of 2.3Hz. The resulting change in δI is isolated by electronic switching. Unfortunately, the wavelength dependent angle $\Delta\delta\phi$ is not completely negligible but is comparable to the expected value of ϕ^{PNC}. Separation of the two effects is achieved by a series of 'sandwich' inter-comparisons with Bi present and absent, using a double oven system (fig. 2) which allows us to bring the Bi into or out of the laser beam without movement of any optical component.

An important feature of the Oxford experiment is the use of the Bi FARADAY effect. This has been investigated in detail for all the hyperfine components and is thoroughly understood (fig. 1). We utilize it both as a wave-length 'marker' and as a measure of the optical depth of Bi in the laser path. In order to avoid any possibility that the FARADAY effect produced by any residual stray magnetic field on the Bi might produce a rotation which could be confused with ϕ^{PNC} we switch the wave-length between the two points of equal, and zero, FARADAY rotation, which happily also correspond to the points of maximum difference of ϕ^{PNC} (fig. 1).

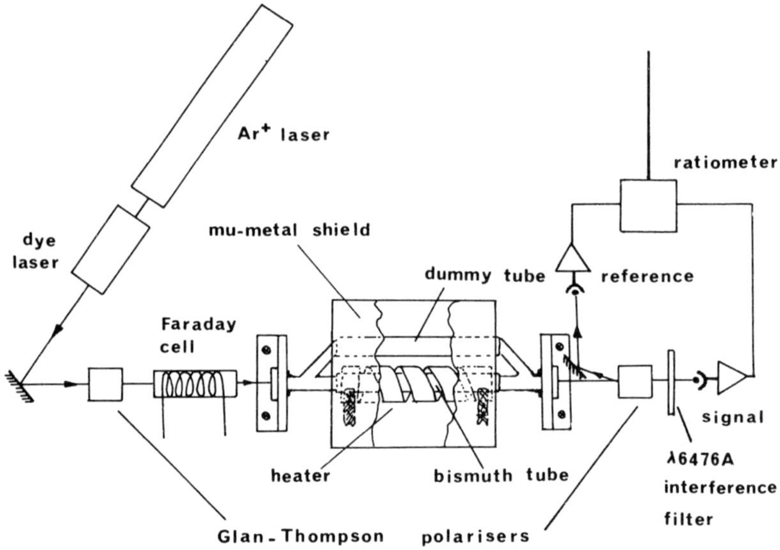

Figure 2. Schematic diagram of the Oxford optical rotation experiment.

4.3 The Washington Experiment

The two most fundamental differences between the Oxford and Washington experiments stem from the use of different transitions. The 876 nm Washington line has the advantage that it is free from molecular absorption and it is therefore possible to go to higher densities and larger angles than in the Oxford experiment. On the other hand dye lasers for the 876 nm line have only just been introduced and were not available when the Washington group constructed their apparatus. Instead they used a Chromatix pulsed parametric laser. This laser has the disadvantage that it has a line-width appreciably greater than the total width of the hyperfine structure.

In the Washington experiment the controlled angle $\delta\phi$ is also produced by a FARADAY modulator and is reversed in sign for alternate laser pulses. The wave-length dependence is extracted by repeatedly sweeping the laser profile over the atomic line with the data analysed by an on-line computer in a signal averaging mode. Calibration is in terms of the FARADAY effect with the total absorption as a check.

4.4 The Novosibirsk Experiment

This experiment on the 648 nm line is similar to the Oxford one in that a Spectra-Physics single mode dye laser is used, but it differs significantly in having no FARADAY modulator. Instead the wave-length dependence is the primary discriminant. The laser wave-length is modulated with amplitude equal to the width of a single hyperfine component at a repetition rate of 1kHz. The polarizer misalignment angle $\phi_R \sim 10^{-3}$ is reversed by hand with a period of order minutes. Use of ingenious phase sensitive detection and feedback circuits allows separation of the signal with the properties appropriate to a PNC rotation angle.

5. Results

Experience has shown that the tunable laser crossed polarizer method for measuring small wave-length dependent optical rotation angles has excellent sensitivity and signal to noise properties. All three Bi experiments are clearly capable of detecting angles of the size predicted by the WEINBERG-SALAM model. At present two of the experiments, at Oxford and at Washington have reached this level of sensitivity [14] and their present results are given below [15,16]

Oxford	:	$R(648\ \mathrm{nm}) = (+2.7 \pm 4.7) \times 10^{-8}$
Washington	:	$R(876\ \mathrm{nm}) = (-0.7 \pm 3.2) \times 10^{-8}$

(errors quoted at 90% confidence limit)

Comparison with the theoretical predictions based on atomic theory and the WEINBERG-SALAM model in section 3.1 shows that there is a serious discrepancy. The experiments find no rotation at the expected level.

This disagreement between atomic experiment and elementary particle theory would, if real, be of very considerable significance, and would throw into question widely held views on the nature of the unified theories of the weak and electromagnetic interactions. It is of the utmost importance that the atomic theory be improved and that more atomic experiments both optical rotation and other types be carried out.

References

1. S. Weinberg, Phys. Rev. Letters 19, 1264 (1967)
2. A. Salam, Proc. 8th Nobel Symposium, ed. N. Svartholm, Almkvist and Wiksell, Stockholm (1968)
3. M.A. Bouchiat and C.C. Bouchiat, Phys. Letters 48B, 111 (1974)
4. M.A. Bouchiat and L. Pottier, Phys. Letters 62B, 327 (1976)
5. E.D. Commins Private communication (1976)
6. P.E.G. Baird et al, Atomic Physics 5, ed. R. Marrus, M. Prior, H. Shugart, (Plenum Press, 1976) p 27.
7. L.M. Barkov, M.S. Zolotoryov Private communication (1977)
8. D.C. Soreide et al. Phys. Rev. Letters, 36, 352 (1976)
9. M.W.S.M. Brimicombe, C.E. Loving and P.G.H. Sandars, J. Phys. B9, L1 (1976)
10. E.M. Henley and L. Wilets, Phys. Rev. A14, 1411 (1976)
11. V.N. Novikov, O.P. Sushkov, I.B. Khriplovich ZhETF, 71, 1665 (1976)
12. C.E. Loving and P.G.H. Sandars to be published
13. D.A. Landman and A. Lurio, Phys. Rev. A1, 1330 (1970)
14. P.E.G. Baird et al, E.N. Fortson et al Nature 264, 528 (1976)
15. P.E.G. Baird et al submitted for publication
16. E.N. Fortson et al submitted for publication

INFRARED HETERODYNE SPECTROSCOPY IN ASTRONOMY

A.L.Betz and E.C. Sutton and R.A. McLaren

Department of Physics
University of California
Berkeley, CA 94720, USA

Department of Astronomy
University of Toronto
Toronto, Ontario, Canada

1. Introduction

The introduction of laser heterodyne techniques to astronomical spectroscopy now makes possible resolving powers heretofore difficult, if not impossible, to achieve with conventional instrumentation. Especially at infrared frequencies, where the heterodyne spectrometer maintains a strong sensitivity advantage for resolving powers greater than 10^5, the potential for such high resolution is especially appreciated in the study of molecular spectra. Generally, the narrowest spectral features will be found in cool low-pressure sources, such as the upper parts of planetary atmospheres. Stellar atmospheres, while tenuous, are extremely hot and have large convective motions which broaden the spectral lines, and therefore they usually do not require resolving powers much above 10^5. However some cooler infrared sources do show unresolved spectra which may become future targets for relatively low-resolution heterodyne spectroscopy. For the present, however, the study of planetary atmospheres appears to be the most promising application for the technique. This is true not only for practical considerations such as detectable signal strengths and relatively low Doppler shifts, but also for more specifically scientific reasons.

Accurately determined line shapes of molecular absorption lines can provide a great deal of information on the vertical pressure-temperature structure of a planetary atmosphere. In the case of Mars, an unsaturated CO_2 absorption line near 1000 cm^{-1} has a linewidth of only $\sim$ 0.001 cm^{-1}. A detailed analysis of the profile therefore requires a resolving power in excess of 10^6, which makes heterodyne spectroscopy the most practical approach. From a purely technical point of view, the resolution of a heterodyne spectrometer is limited only by the long-term frequency stability of the laser, which for a stabilized CO_2 laser can easily be 10 kHz. However, only enough resolution is really needed to resolve the details of the line structure. Since both the thermal and pressure-broadening effects on a CO_2 line of Mars are on the order of 30 MHz each, a resolution element size of 5 MHz seems quite adequate for the purpose. With respect to the 30 THz line frequency, this resolution bandwidth implies a resolving power of 6×10^6.

2. Techniques

It has only been recently with the rapid development of CO_2 lasers and infrared photodetector technology that heterodyne techniques have been successfully extended up from the microwave and into the infrared frequency range. Infrared heterodyne spectroscopy requires the source radiation, collected by a telescope, to be mixed with the coherent output of a CW laser serving as a local oscillator. The mixing element is a high speed infrared photodiode which produces a difference frequency spectrum at radio frequencies. The spectral analysis is then performed electronically in a multichannel filter bank system to achieve the desired resolution simultaneously in a number of contiguous and independent channels. In a sense, a heterodyne instrument can thus have a "multiplex" advantage over a single channel spectrometer, at least within the spectral range of the difference frequency bandwidth. However, because of the limited range of this difference band out of high-sensitivity photomixer diodes, the laser frequency must closely match that of the spectral line in the planetary atmosphere. In this regard, it is somewhat fortunate that the most technically suitable local oscillator, the CO_2 laser, oscillates on transitions of the molecule of principal scientific interest in the predominantly CO_2 atmospheres of Mars and Venus. Additional practical incentives for observing the CO_2 bands near 1000 cm^{-1} are that the Earth's atmosphere is relatively transparent at these frequencies and that the thermal continuum signals from both Mars and Venus are sufficiently strong to enable good absorption line spectroscopy to be accomplished. At higher frequencies not only will the blackbody continuum radiation from the planets rapidly decrease, but the fundamental quantum noise associated with heterodyne detection will naturally increase, thereby drastically limiting the achievable signal-to-noise ratio on these sources.

When a heterodyne receiver is properly matched to an astronomical telescope, the angular field-of-view is simply the diffraction pattern of the telescope aperture. For an 81 cm aperture, the size which was used in this work, the field-of-view is about 3.3 arcsec at a wavelength of 10 μm. This beamwidth is much smaller than the angular sizes of Mars and Venus during favorable observing times, and thus allows good spatial resolution over the surfaces of the planets. However, as far as detectable signal power is concerned, once the telescope beamwidth is smaller than the angular size of the planet, which is approximated as a blackbody source, then the signal level is independent of further increases in telescope aperture [1]. Of course, the increased angular resolution available from a larger telescope may be desired, but eventually a limit near 1 arcsec is reached at good observing sites because of turbulence in the Earth's own atmosphere. This means that telescope diameters up to about 2 m may be expected to be useful for heterodyne observations at a wavelength of 10 μm.

For a blackbody source at temperature T which fills the field-of-view of the heterodyne receiver, the signal-to-noise ratio for heterodyne detection is given by:

$$S/N = \eta \, (Bt)^{\frac{1}{2}} \, (e^{h\nu/kT}-1)^{-1} \quad , \qquad (1)$$

where B is the bandwidth of the filter element and t is the integration time [2]. The factor η signifies the overall efficiency of the detection process, and may for convenience of expression include optical transmission losses, detector quantum efficiency, and the relative noise contribution of the intermediate frequency amplifier with respect to the laser-induced shot noise (quantum noise). Practical receivers are now able to achieve $\eta = 0.20$, which implies that present system sensitivities are only a factor of 5 away from the theoretical quantum noise limit [3].

For the planet Mars, with a peak blackbody temperature, T, close to 270 K, (1) gives a signal-to-noise ratio of 17 for t = 60 seconds, B = 5 MHz, and ν = 30 THz. However, several practical limitations tend to further degrade this performance. In order to discriminate against unwanted background radiation, a common observational technique in infrared astronomy is to alternately and rapidly view the source and a nearby "empty" region of the sky and then to synchronously demodulate the signal at this "chopping" frequency. In this way, offset signals produced by thermal radiation from both the sky and the telescope structure are cancelled, since they remain constant in the two beams. However, this chopping reduces the average true signal level by a factor of 2. Also, losses in atmospheric and telescope transmission add an additional small reduction in received signal strengths. The final observed signal-to-noise ratio on the continuum from Mars is closer to 6, for the conditions outlined above, and Venus at 235 K is approximately one-half as strong as Mars. For cooler planets, such as Jupiter at an atmospheric temperature of 135 K, the signal levels fall off rapidly. Currently, continuum temperatures stronger than 200 K are required to permit good quality absorption line spectroscopy at 10 μm. As heterodyne systems operating at even longer infrared wavelengths become obtainable with the detection efficiency of present 10 μm systems, then spectroscopy on the cooler planetary atmospheres will be possible.

3. Instrumentation

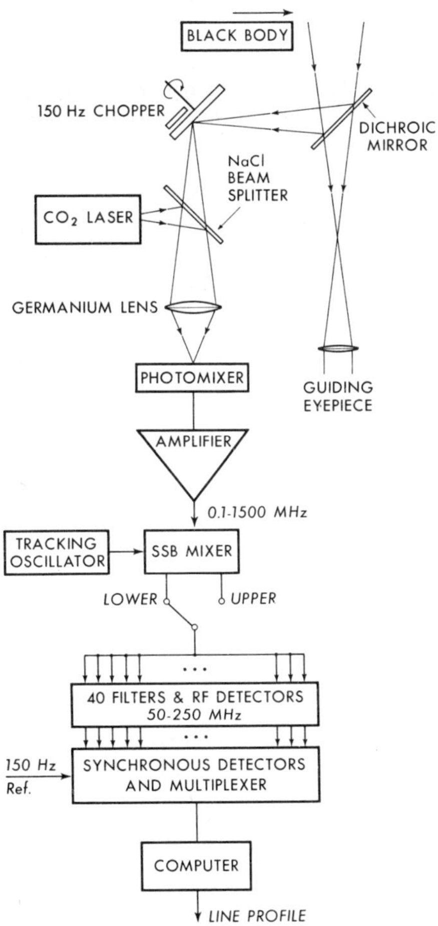

The spectrometer is currently installed at the McMath Solar Telescope of Kitt Peak National Observatory and utilizes one of the two 81 cm auxiliary solar telescopes. As illustrated in Fig.1, the source radiation collected by the telescope falls on a dichroic mirror which passes the visible image to an eyepiece and reflects the infrared image to a dual-beam focal-plane sky chopper. The infrared beam is then transmitted through a NaCl beamsplitter, where it is combined with the local oscillator signal from a stabilized CO_2 laser. The laser oscillates on selectable vibration-rotation transitions of CO_2 in the 890 to 980 cm^{-1} region. Once combined, the collinear beams are focused onto a cooled (77 K) mercury-cadmium telluride (HgCdTe) photodiode mixer, which generates an intermediate frequency (IF) signal extending from 0 to 1500 MHz. (Fortunately, the Doppler shifts of CO_2 lines from Mars and Venus are almost always within this range.) After amplification, a selected 200 MHz segment of this IF spectrum is first converted into the band of 50 to 250 MHz by an image rejection (single-sideband) mixer and then directed into a multichannel radio frequency filter bank. The local oscillator frequency for this second mixer is adjusted to center the desired

Fig.1 Experimental apparatus

line profile in the range of the filter bank. Also, in order to compensate for variations in the Doppler shift of the line caused by both the rotation of the Earth and changes in interplanetary velocity during the course of an observation, the frequency of this tracking oscillator is updated at regular intervals. The filter bank analyzes the spectrum of 50 to 250 MHz into 40 independent channels of 5 MHz each. The 40 detected power outputs are then synchronously demodulated at the sky-chopping frequency (150 Hz) and multiplexed into an observatory computer for on-line integration and analysis. After each integration period, a blackbody calibration source is inserted into the infrared signal beam. In this way the detection sensitivity is measured to be 2.2×10^{-16} W per 5 MHz channel for a signal-to-noise ratio of 1 after a 1-second integration. However, the detection of only one source polarization and the loss of half the signal due to sky-chopping can be viewed as effectively degrading this sensitivity by an additional factor of 4.

The simple, low power CO_2 laser used in this spectrometer will oscillate on only about 10 individual transitions, almost all P-branch, in the 00^01-$(10^00, 02^00)_I$ band. However, either $^{12}C^{16}O_2$ or $^{13}C^{16}O_2$ may be employed in the lasing medium to alter the spectral coverage. It is also possible to observe other CO_2 bands with transitions in close coincidence with the available laser frequencies. Freed [4] has measured the frequencies of some transitions in the 01^11-$(11^10, 03^10)_I$ band of $^{12}C^{16}O_2$ relative to known frequencies in the 00^01-$(10^00, 02^00)_I$ band of $^{13}C^{16}O_2$ near 900 cm^{-1}. In particular, he finds that the P(31) line in the higher band of $^{12}C^{16}O_2$ lies only 910.4 MHz lower in frequency than the P(16) transition of $^{13}C^{16}O_2$ in the 00^01-$(10^00, 02^00)_I$ band. As can be seen in the vibrational energy diagram of Fig.2, this fortunate coincidence is important for the spectroscopy of the Martian atmosphere in that it permits the observation of an absorption feature with a lower level

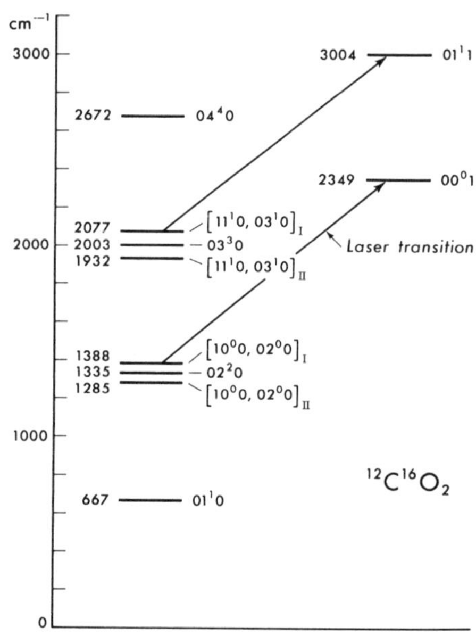

Fig.2 Vibrational energy-level diagram for $^{12}C^{16}O_2$ showing the two observed absorption bands

2464 cm^{-1} above the ground state. At 240 K, an appropriate temperature for the lower atmosphere, the Boltzmann factor for the population of the P(31) lower level is e^{-15}. A change in gas temperature of only 10 K is necessary to alter the lower level population by a factor of 2, thus making this line a sensitive indicator of atmospheric temperature.

4. Observations

As examples of the type of absorption line spectra which can be obtained with this spectrometer, Figs.3 thru 5 show the results from three different CO_2 bands on Mars. The exact lineshapes of these features have been fitted by a simple model atmosphere calculation to yield information on four particular atmospheric parameters. The large difference in line strengths for the three features makes each one particularly sensitive to at least one of these four parameters.

For relatively unsaturated line profiles such as the P(20) line of $^{13}C^{16}O_2$ in Fig.3, the continuum level beyond the pressure-broadened wings of the line is easily measured and thus determines the first parameter, the planetary surface temperature. The shape of the profile near line center is influenced mainly by low pressure gas at higher altitudes and helps fix the second variable, a quantity called "lapse rate", which is the change in atmospheric temperature as a function of altitude. The third parameter, the atmospheric

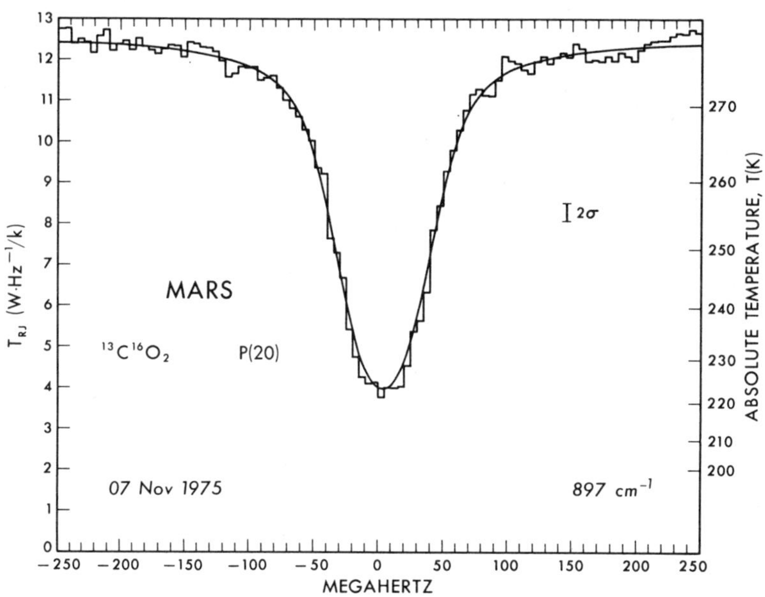

Fig.3 P(20) line of $^{13}C^{16}O_2$ in the 00^01-$(10^00, 02^00)_I$ band. The profile is reconstructed from three partially overlapping 40-channel integrations of 32 min. each. A subsolar surface temperature near 275 K and a lapse rate of 2 K/km is predicted by the model fit.

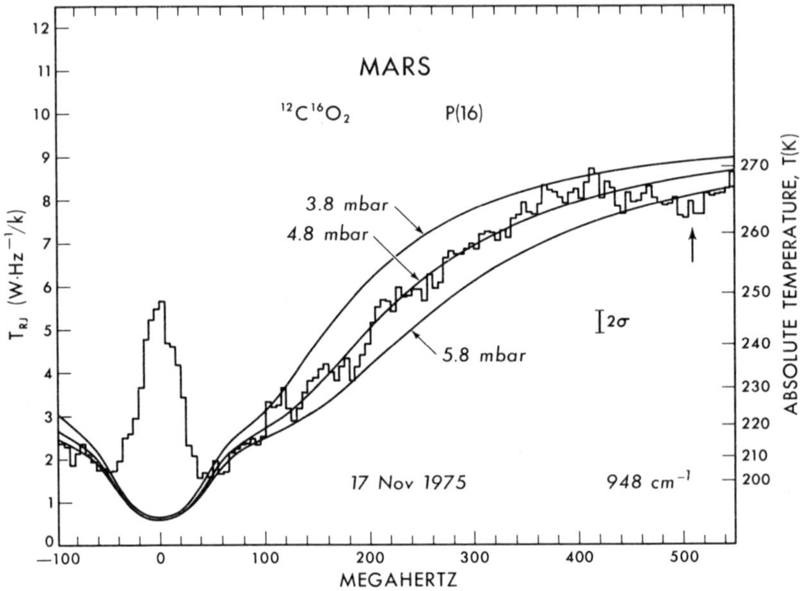

Fig.4 P(16) line of $^{12}C^{16}O_2$ in the 00^01-(10^00, 02^00)$_I$ band. The three illustrated model predictions differ only in the value for surface pressure. The absorption feature indicated by the arrow is the nearby P(23) transition of $^{12}C^{16}O^{18}O$ in the 00^01-(10^00, 02^00)$_I$ band.

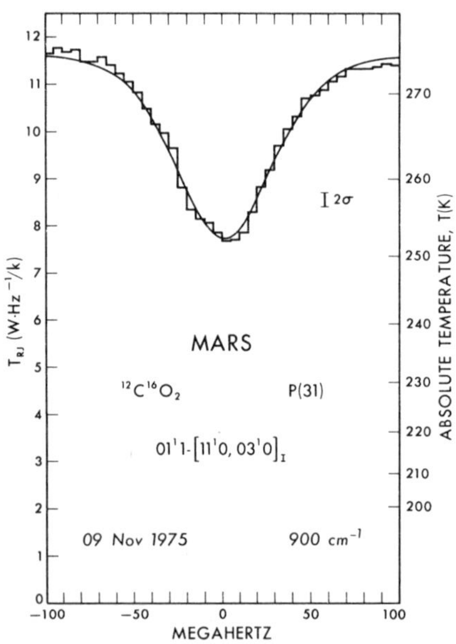

Fig.5 P(31) line of $^{12}C^{16}O_2$ in the 01^11-(11^10, 03^10)$_I$ band observed after 32 minutes of integration. The best model fit requires atmospheric temperatures of 235 to 240 K near the surface.

pressure near the surface, is clearly visualized by the pressure-broadened wing of the P(16) profile of $^{12}C^{16}O_2$ in Fig.4. Finally, the best estimate of the fourth quantity, the atmospheric temperature near the surface, comes of course from the observed strength of the $^{12}C^{16}O_2$ P(31) "hot band" line illustrated in Fig.5. With the large variation in line strengths available from these features, and similar lines which were also observed, it is then not difficult to fit a unique atmospheric model to all the data [5], as illustrated by the smooth curves drawn in the three figures.

Subsequent to these observations, direct measurements of these atmospheric conditions have now been accomplished by the two Viking spacecraft for two locations on the surface of Mars. However, the scientific effectiveness of high resolution Earth-based observations is still quite favorable. For example, in Fig.4, the significant departure of the observed spectrum from the model prediction at line center indicates an unusual level of non-thermal emission from the upper atmosphere. The 35 MHz half-width of this Gaussian emission profile indicates a CO_2 kinetic temperature of 170 K. Additional observations of the P(12) through P(28) transitions of this band also indicate a similar value for the rotational temperature. The strength of the emission on strong lines, such as the P(16), however, exceeds that from even a blackbody at 170 K by a factor of 10. An examination of possible excitation mechanisms for this non-LTE behavior seems to indicate pumping of the near infrared CO_2 bands by solar radiation, with subsequent reemission at the longer 10 µm wavelengths. On Mars, this process occurs in the upper atmosphere at an altitude near 75 km, where collisional de-excitation is not dominant [6]. This excitation hypothesis is strongly supported by subsequent heterodyne observations on the predominantly CO_2 atmosphere of Venus. The solar intensity on Venus is approximately 4 times as intense as on Mars, and so the non-thermal emission should be proportionally stronger. This is just as observed, as shown in Fig.6, which shows the emission from 3 positions receiving different solar intensities. On Venus, the linewidth and rotational temperatures are closer to 200 K, and the altitude of peak emission is near 115 km.

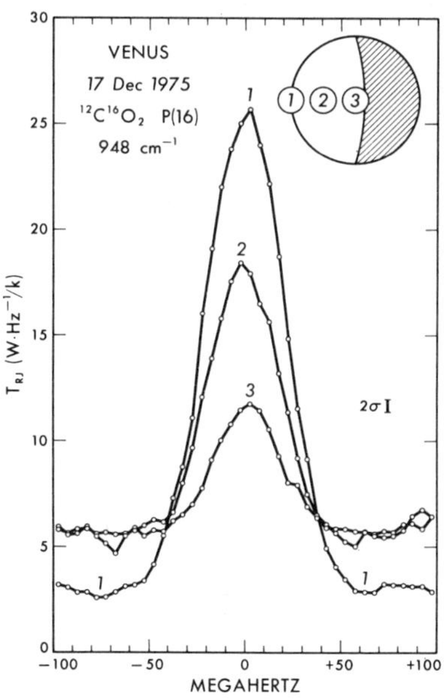

Fig.6 Emission lines of $^{12}C^{16}O_2$ in the $00^01-(10^00, 02^00)_I$ band observed from three illuminated regions on the 20 arcsec disk of Venus. Integration times are 16 minutes for positions 1 and 2, and 32 minutes for position 3.

Another natural application of heterodyne spectroscopy is in measuring the absolute Doppler shifts of spectral lines. These shifts can then be used to study the dynamical conditions of various regions in the planetary atmosphere. For example, on Venus studies of the $^{12}C^{16}O_2$ emission features at various positions on the planet have provided unique information on the atmospheric circulation at elevations near 115 km [7]. Statistical accuracies of a few m/sec can now be easily achieved for these motions after only about 15 minutes of integration. This statistical accuracy is not compromised by any long-term systematic drifts of the spectrometer, since the absolute stability of the instrument is governed by the laser frequency, which can easily be stabilized by saturation-resonance techniques to an accuracy of 10 kHz (0.1 m/sec). Lower regions of the atmosphere have also been probed by monitoring the shifts of weak $^{13}C^{16}O_2$ absorption lines, which are formed at altitudes near 60 km. Because of the much weaker intensity of these absorption features, however, the statistical accuracy of the Doppler velocity is proportionally worse, typically 20 m/sec after a 30 minute integration.

5. Conclusion

These results show the practical applicability of laser heterodyne techniques in astronomical spectroscopy. The full potential of the method, however, will only be realized with the development of tunable continuous-wave lasers able to serve as local oscillators in the 1000 cm^{-1} region. Since technical progress in this direction is encouraging, it seems fairly certain that within the next few years tunable systems will be operating, and that a much greater variety of astronomical sources will be observed.

Acknowledgements. We are grateful to C.H. Townes and M.A. Johnson for their contributions to this research, and to D.L. Spears for providing the infrared photodiodes. The observations were conducted with the cooperation of the staff of Kitt Peak National Observatory. This work was supported in part by NASA under Grants NGR 05-003-452 and NGL 05-003-272, and the NSF under Grant AST 75-20353.

References

1. A.E.Siegman: Proc.IEEE 54, 1350 (1966)
2. T.G.Blaney: Space Sci. Rev. 17, 691 (1975)
3. A.L.Betz: Ph.D. Dissertation, University of California (1977)
4. C.Freed, R.G.O'Donnell, A.H.M.Ross: IEEE Trans. Instrum. Meas. IM-25, 431 (1976)
5. A.L.Betz, R.A.McLaren, E.C.Sutton, M.A.Johnson: Icarus 30, 650 (1977)
6. M.A.Johnson, A.L.Betz, R.A.McLaren, E.C.Sutton, C.H.Townes: Astrophys.J. 208, L145 (1976)
7. A.L.Betz, M.A.Johnson, R.A.McLaren, E.C.Sutton: Astrophys.J. 208, L141 (1976)

PRECISION MEASUREMENT OF THE GROUND STATE LAMB SHIFT IN HYDROGEN AND DEUTERIUM [1]

C. Wieman [2] and T.W. Hänsch

Department of Physics, Stanford University
Stanford, CA 94305, USA

The measurement of the $2S_{1/2}$ state Lamb shift in hydrogen was one of the main stimuli for the development of quantum electrodynamics. Since the first experiment by W. E. LAMB and R. C. RETHERFORD [1] the Lamb shift has been measured for a great many hydrogenic states, and such measurements are among the most precise tests of QED [2]. One measurement which was conspicuously missing from this list, however, was the Lamb shift of the hydrogen ground state. This has only been measured in the past few years [3,4] and the experiment described here is a continuation of those measurements to higher precision by the use of several technical innovations. We have also made an improved measurement of the Lyman-α hydrogen-deuterium isotope shift sufficiently precise to give the first experimental confirmation of the predicted relativistic nuclear recoil correction. The technical improvements mentioned are the development of a highly sensitive new form of Doppler-free spectroscopy, polarization spectroscopy, and the use of a single frequency cw dye laser at 4860Å with a high power pulsed amplifier system.

The shift of the 1S state energy from the value predicted by simple Dirac theory, which we call the Lamb shift, is calculated to be 8149 MHz [5]. QED corrections are the dominant contribution, but the shift also includes some small nuclear effects. Although this is eight times larger than the shift of the 2S state it cannot be measured using the traditional RF approach because there is no nearby P state reference level. Also there are several difficulties in measuring the Lamb shift by determining the Lyman-α wavelength to sufficient precision using conventional absorption or emission spectroscopy. The most serious of these are the 40 GHz Doppler width at room temperature, and the difficulty in making precision wavelength measurements in the vacuum ultraviolet region (1215Å). Moreover, the present uncertainty in the Rydberg constant precludes any such Lamb shift measurement to better than one part in 10^3.

The recent development of several nonlinear laser spectroscopic techniques, and improved tunable dye lasers has made it possible to determine the 1S Lamb shift in quite a different manner, however. This approach, as first described in [3], uses laser spectroscopy to compare the Balmer-β (n=2 to n=4) transition energy with 1/4 the Lyman-α energy. The comparison is made

---

[1] Work supported by the National Science Foundation under Grant NSF-14786 and by the U.S. Office of Naval Research under Contract No. N00014-15-C-0841.

[2] Hertz Foundation Fellow; Present address: Department of Physics, University of Michigan, Ann Arbor, Michigan.

using a high power tunable narrowband laser at 4860Å, whose frequency doubled output at 2430Å excites the two-photon transition from the 1S ground state to the metastable 2S state. Simultaneously the Balmer-β line is observed using the fundamental laser output. If the Bohr formula were correct this interval would be exactly 1/4 the Lyman-α interval and we would find the two resonances at exactly the same fundamental frequency. The actual displacement is due to the ground state Lamb shift plus small well measured relativistic and QED corrections to the excited state energies. So an accurate comparison of the two optical energy intervals allows one to measure the unknown 1S Lamb shift. This approach avoids the technical problems of VUV spectroscopy since one only measures visible wavelengths. In addition, the use of counterpropagating beams to excite the 1S-2S two-photon transtion eliminates the Doppler broadening, as discussed in [6]. Because this measurement involves only the comparison of two transitions in hydrogen the uncertainty in the Rydberg constant does not matter.

The first experimental value [3] for the 1S Lamb shift was obtained by exciting the 1S-2S transition just as described, and recording a simple Doppler-broadened laser absorption spectrum of the Balmer-β line as the reference. Substantial improvement was obtained by using the technique of saturated absorption spectroscopy with a pulsed laser to observe the Balmer-β line [4]. This gave sub-Doppler resolution, but the limitation of the quoted Lamb shift accuracy was still the poor resolution of the Balmer-β line.

The present experiment, which is shown schematically in Fig. 1, measured the Balmer-β line with enormously improved resolution by using a single frequency cw dye laser and the technique of polarization spectroscopy. The Balmer-β portion of the experiment has been previously described in [7], but will be briefly reviewed.

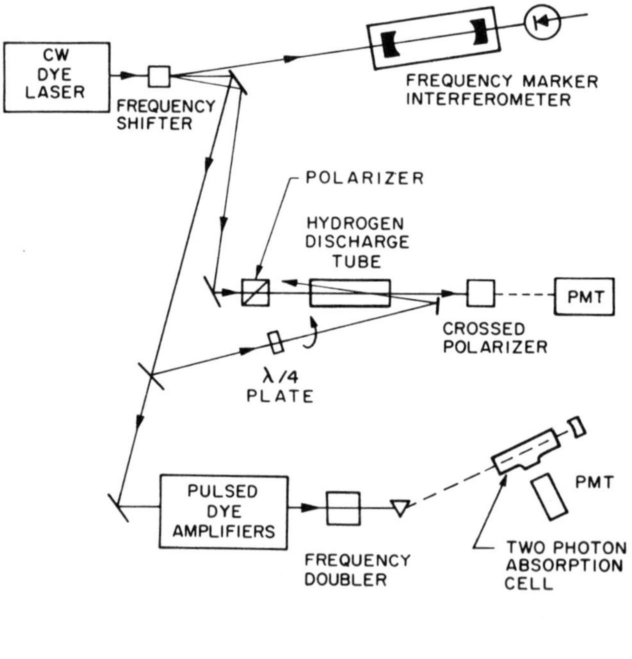

Fig. 1. Schematic of experimental apparatus

The laser was a cw jet stream dye laser (modified Spectra-Physics Model 375) using coumarin-one in ethylene glycol pumped by a UV argon laser (Spectra-Physics #171). Single frequency operation was achieved by having three intracavity etalons. At 1.5 W pump power the laser provides a single mode output of about 10 mW near 4860Å with a linewidth of ~ 3 MHz and continuously tunable over 4 GHz.

The technique of polarization spectroscopy uses induced optical anisotropy as a sensitive means to obtain Doppler-free line profiles. The basic schematic is shown in the middle portion of Fig. 1. A weak linearly polarized probe beam from the cw laser is sent through a Wood's type hydrogen discharge tube. Only a small fraction of this beam reaches a photodector after passing through a nearly crossed linear polarizer. Any optical anisotropy which changes the probe polarization will alter the light flux through the polarizer and can be detected with high sensitivity. Such an anisotropy can be induced by sending a second, stronger, circularly polarized beam from the laser in nearly the opposite direction through the discharge tube. Because the beams are counterpropagating their interactions are Doppler free. The improved resolution obtained using this technique enabled us to measure and correct for the effects of Stark and pressure shifts in the discharge.

Although the cw laser was highly advantageous in observing the Balmer-β line its power was much too low to excite the 1S-2S transition. Therefore, it was necessary to send the cw beam through a series of three pulsed dye amplifiers pumped by a 600 KW nitrogen laser (Molectron UV 1000). With a 1 mW input these gave output pulses 7 nanoseconds long with ~ 100 kW peak power and about 5% pulse to pulse amplitude fluctuation. The fundamental linewidth was about 120 MHz which gives two-photon linewidths less than one quarter as wide as those obtained with the previously used pulsed oscillator-amplifier system [4].

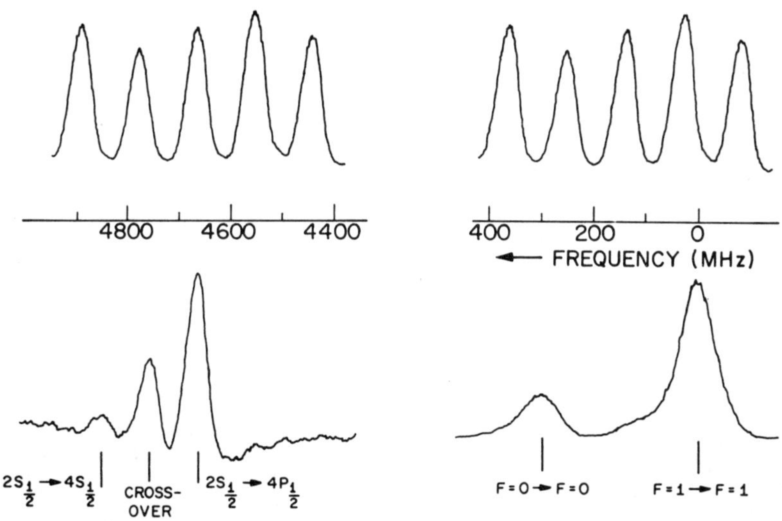

Fig. 2. Portion of the hydrogen Balmer-β polarization spectrum (left) and 1S-2S two-photon spectrum (right)

The two-photon spectrometer, as shown at the bottom of Fig. 1, is identical to that described in [3,4]. The output of the pulsed amplifier system goes into a lithium formate frequency doubling crystal which is angle tuned to phase match near 4860Å. The output of this crystal goes through a quartz prism which separates the doubled light from the fundamental. The doubled light passes through a hydrogen absorption cell, a small arm attached to the side of a Wood dicharge tube, and is reflected back on itself to obtain the Doppler-free 1S-2S excitation. This excitation is monitored by observing the collision induced 2P-1S fluorescence at 1215Å, through a MgF_2 window, with a photomultiplier.

In Fig. 2 we show typical spectra which were obtained. On the lower left is the portion of the Balmer-β transition used as a reference, while on the lower right is the 1S-2S two-photon spectrum. On the top of the figure are shown the frequency calibration lines obtained by sending a portion of the cw laser beam through the semiconfocal calibration interferometer shown at the top of Fig. 1.

The experimental results for hydrogen are shown in Table I.

TABLE I. Hydrogen 1S Lamb Shift

Uncorrected measurement:	8133.2 ± 4.8 MHz
Corrections:	
1S Two-Photon Spectrum:	
AC Stark Effect	3.2 ± .8 MHz
Amplifier shift/line shape	- 40.0 ± 28.0
Balmer-β line:	
Pressure, Voltage	57.2 ± 6.8
AC Stark effect	5.6 ± 2.0
Total	26.0 ± 29.0
Experimental 1S Lambs shift	8159.2 ± 29.0
Theory	8149.43 ± .08

As shown by this table, corrections must be made for a number of systematic shifts of the data, and much of the effort in this experiment was spent in determining these shifts. For the Balmer-β transition the significant corrections are the lineshifts caused by pressure and electric field perturbations in the discharge, and the AC Stark effect, or light shift, due to the polarizing laser beam. For the 1S-2S transition there is a small correction for the AC Stark effect, and a larger correction, the uncertainty of which is the main limitation in this experiment, to compensate for the frequency shifts introduced by the pulsed amplifiers. This shift is measured by using an interferometer to obtain simultaneous spectra of the cw beam and the pulsed output, but asymmetry and "chirping" in the pulsed laser spectrum causes substantial uncertainty in this correction.

For the ground state Lamb shifts we have obtained 8149. ± 29 MHz in hydrogen and 8182.6 ± 29 MHz for deuterium, with similar systematic corrections. These numbers are in excellent agreement with the theoretical results of 8149.43 ± .08 MHz and 8172.23 ± .12 MHz [5].

Because the two-photon lines are narrower and less noisy than the previous measurement [4] it was possible to also make a substantially improved measurement of the Lyman-α isotope shift between hydrogen and deuterium. The result was 670992.3 ± 6.3 MHz which agrees well with the theoretical value of 670994.96 ± .81 [5] and gives the first experimental confirmation of the 11.9 MHz correction due to the recoil of the nucleus [8].

The possibilities for further dramatic improvements in high resolution laser spectroscopy of hydrogen are by no means exhausted. The 1S-2S two-photon transition has a natural line width of only 1 Hz, and attempts to improve the spectral resolution by cw laser spectroscopy of an atomic beam, or by optical Ramsey-fringe spectroscopy with separate oscillatory fields are presently underway [9]. Only a twentyfold improvement in the accuracy of the H-D isotope shift is needed to obtain a new precise value of the electron/proton mass ratio. Simultaneous observation of the Balmer-β line in a beam of metastable hydrogen atoms promises important future improvements in the accuracy of the ground state Lamb shift, and may well turn such a measurement into one of the most stringent present tests of QED calculations.

References

1. W. E. Lamb and R. C. Retherford, Phys. Rev. 72, 241 (1947); Phys. Rev. 79, 549 (1950).
2. J. Brodsky and S. Drell, Annual Rev. of Nuclear Sci., 20, 147 (1970).
3. T. W. Hänsch, S. A. Lee, R. Wallenstein, and C. Wieman, Phys. Rev. Letters 34, 307 (1975).
4. S. A. Lee. R. Wallenstein, and T. W. Hänsch, Phys. Rev. Letters 35, 1262 (1975).
5. G. Erickson, private communication.
6. N. Bloembergen and M. Levenson, *High Resolution Laser Spectroscopy* (Springer-Verlag, 1976), pg. 315.
7. C. Wieman and T. W. Hänsch, Phys. Rev. Letters 34, 1170 (1976).
8. H. A. Bethe and E. E. Salpeter, *Quantum Mechanics of One and Two Electron Atoms* (Academic Press, New York, 1957), Sec. 42.
9. T. W. Hänsch, Physics Today, 30, 34 (1977).

SELECTIVE SINGLE ATOM DETECTION IN A 10^{19} ATOM BACKGROUND [1]

G.S. Hurst, M.H. Nayfeh, J.P. Young, M.G. Payne, and L.W. Grossman

Chemical Physics Section, Health and Safety Research Division
Oak Ridge National Laboratory, Oak Ridge, TN 37830, USA

1. Historical Introduction

Early work at the Cavendish Laboratory established the important features of the process in which high energy charged particles lose their energy in a gas. The work there by RUTHERFORD and GEIGER [1] in which collisions were used to "magnify" the number of electrons produced in a gas established the basis of the first single electron counters. One cannot avoid noting that if the laser had been available to RUTHERFORD, so that individual atoms could be selectively ionized, he most likely would have started the counting of individual atoms.

Subsequent to the discovery of the JESSE effect [2], i.e., the extreme sensitivity of total ionization by a charged particle in helium gas to traces of impurities, detailed studies were made of total ionization [3] and of time-resolved emission of vacuum ultraviolet radiations [4]. The objective of such studies was to learn the details of the intricate steps followed as the gaseous system returns to equilibrium following charged particle excitation. These steps consist of elementary radiation and collisional energy transfer processes initiated when a charged particle produces excited states in a gas. The understanding of these energy pathways is not complete, even in the simple case of He gas. Here, however, we proposed [5] an energy pathways model which appears to account for most of the known experimental and theoretical information. In the model it was suggested that a key step

$$He(2^1P) + He(1^1S) \rightarrow He(2^1S) + He(1^1S) \qquad (1)$$

had an unusually large cross section due to an accidental curve crossing in order to explain the time-resolved emission data. This led to an obvious need to make an independent measurement of the population of $He(2^1S)$ as a function of time following charged particle excitation of $He(2^1P)$.

2. Resonance Ionization Spectroscopy

In a search for more direct information, we conceived [6] of a method in which each atom in a selected quantum state would be converted to an ion pair by the absorption of two photons, one of which is resonant with an intermediate state. Two-photon ionization processes are well known and have been used more recently

[1] Research sponsored by the Energy Research and Development Administration under contract with Union Carbide Corporation.

by a number of groups for isotope separation; see the recent article by LETOKHOV [7].

Imagine the passage of charged particles (e.g., 2.5-MeV photons) through a gas in a short pulse at t=0 (see Fig.1). We wish to find the number of excited species in quantum defined states as a function of t. To do this, we tune a pulsed dye laser so that resonance photons excite the selected quantum state to an intermediate state lying more than half the distance to the ionization continuum. With modest energy per pulse and with a laser linewidth of several Angstroms, the intermediate state comes into quasiequilibrium with the excited state. Other photons from the same laser pulse photoionize the intermediate state.

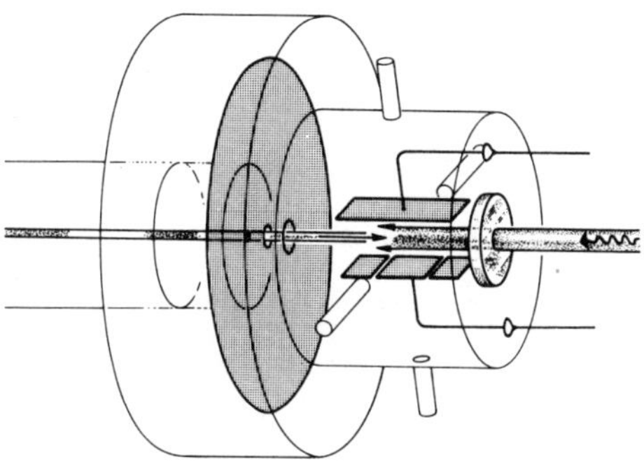

Fig.1 Schematic of resonance ionization spectroscopy of excited states. A pulsed beam of protons excites He gas in a parallel plate ionization chamber. After an arbitrary time, a laser pulse sweeps out quantum selected excited states

To measure $He(2^1S)$ we can use $He(3^1P)$ as the resonance intermediate state; thus, the laser [8] is centered at 5015 Å. If the laser delivers a pulse of 0.7 J in a 1-cm^2 beam, a photon fluence of 1.8×10^{18} photons cm^{-2} is delivered at 5015 Å. The product of photon fluence and photoionization cross section [approximately 4.4×10^{-18} cm^2 for $He(3^1P)$ and at 5015 Å] is considerably greater than unity; therefore, one may hope to saturate the ionization. If so, each excited state is converted to an ion pair. Thus, resonance ionization spectroscopy (RIS) is an absolute and a very sensitive method for the measurement of excited-state populations.

Figure 2 shows a typical recording for 0.6 Torr He. Positive-ion signals were collected in about 8 µsec, in agreement with ion mobility and diffusion data. The electron pulses were collected very quickly and are not resolved in the figure. A simple ratio of pulse heights gives the number of quantum-selected atoms per ion pair created in the initial event. Fig.3 shows this ratio to be saturated and in good agreement with theory (27%).

Studies in helium were continued [9] at higher pressures (15 to 100 Torr) to measure 1) the absolute number of excited states and 2) the lifetime of the $He(2^1S)$ as a function of the pressure. The lifetimes agreed with measurements of PHELPS [10] and with the conclusions by PAYNE et al. [5]. Another

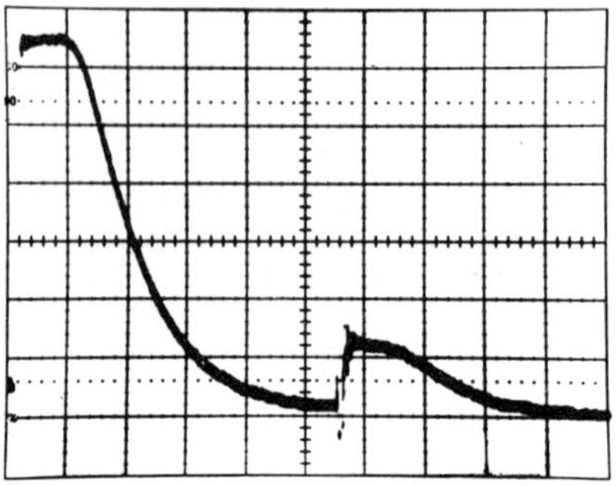

Fig.2 Ionization signal versus time. Positive ion signals due to direct ionization and resonance ionization of He(2^1S), 22 μsec later, in helium at 0.6 Torr.

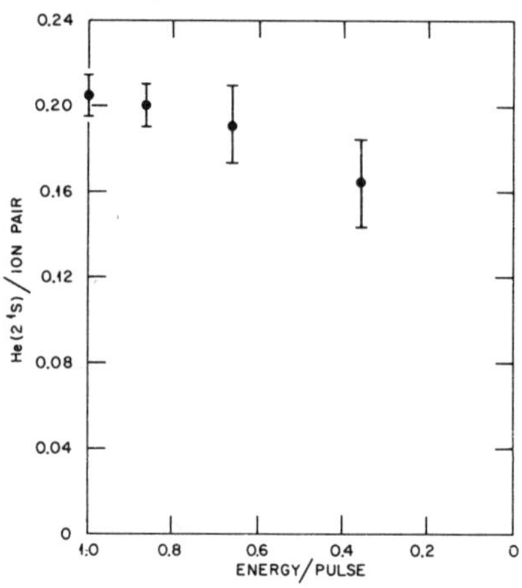

Fig.3 The ratio of He(2^1S) to direct ionization as a function of laser energy per pulse. Maximum energy was 700 mJ/pulse

important result followed from the helium studies at the higher pressures; namely, saturation of the resonance ionization required less laser energy per pulse per unit area, because He(3^1P) was converted to $(He)_2^+$ by a HORNBECK-MOLNAR reaction.

3. Resonance Ionization of Atoms in Their Ground States

In a few cases an element can be excited and then ionized by using one laser, i.e., the two-photon ionization process illustrated on the left of Fig.4; with two synchronized lasers, approximately one-half of the elements can be subjected to the RIS process, using lasers which are now commercially available.

In order to saturate (i.e., to convert each atom in ground state to an ion pair) the process illustrated in Fig.4, several conditions must hold. With only modest laser energy per pulse, the resonance steps are saturated, i.e., an equilibrium is established amongst all of the states connected with the

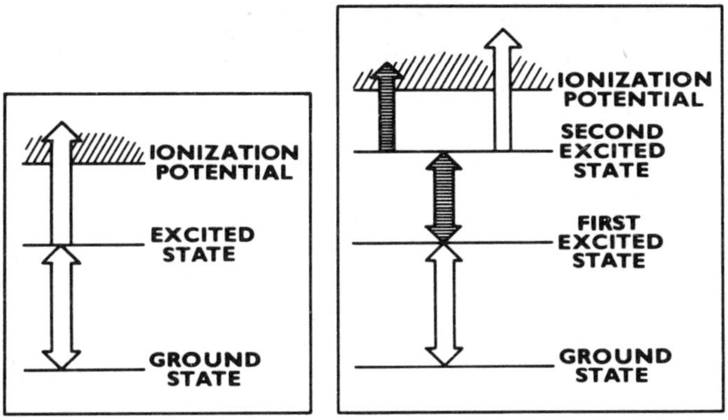

Fig.4 Resonance ionization of an atom using one laser (left sketch) and two lasers (right sketch). With lasers producing approximately 1 Joule of photons per cm^2, the resonance ionization process can be saturated by using a two-step process as shown in the left sketch or in a three-step process as shown in the right sketch

photon fields. With the levels in quasiequilibrium, several radiation and collisional processes are in competition for the states, and these in turn are all in competition with photoionization, the desired result. To ensure saturation, two conditions must be met: 1) the rate of photoionization of the coupled levels must be greater than the sum of the rates of all losses due to collisions and due to radiation, and 2) the product of effective photoionization cross section with photon fluence in a pulse must be substantially greater than unity.

Collisional line broadening does not change the conditions for saturation, but could cause a loss of selectivity. For instance, if the line center absorption coefficient is sufficiently large that a number of the resonance absorption-stimulated emission cycles are completed during a laser pulse, a few cycles can occur even when the laser is substantially detuned in wavelength, $\Delta\lambda$. As a consequence of the fact that nearly every absorption process will lead to photoionization by the detuned photon field, these rare absorption events can be observed far in the wings. This sensitivity was put to advantage to study the Cs-Ar interaction forces over a wide range [11].

Studies were made of two-photon ionization of Cs-Ar mixtures with the broadened Cs(7p) as an intermediate state [12,13]. Since Cs(7p) lies more than halfway to the ionization limit, two photons at 4555 Å can photoionize the Cs ground states. A Chromatix [14] (1.0-μsec FWHM, 3 cm^{-1} FWHM bandwidth, and up to 30 pps repetition rate) flashlamp pumped dye laser was used as a photon source. Persistence of ionization signals for detunings as large as 75 Å, due to collisional broadening in Ar at 1 atm, was observed. In some cases selectivity in single atom detection is reduced due to pressure broadening. However, low pressure proportional counters, or even channeltrons working in a vacuum, can be used where extreme selectivity is essential. Pressure broadening can be put to advantage in studies of rare events where there is a lack of accuracy on knowledge of the energy levels.

4. Photodissociation of Molecules: Time-Resolved Sources of Free Atoms

In some recent work [15] we showed that a spatially defined population of atoms in their ground states could be produced at a well-defined time. Such sources of atoms can be combined with time- and space-resolved one-atom detectors in a number of useful arrangements. We showed that all of the Cs molecules in the central portion of a laser beam could be dissociated to neutral Cs and I atoms in their ground states.

The experimental arrangement (Fig. 5) utilizes a parallel plate ionization chamber mounted inside a vacuum system. Even when the CsI sample is heated up to 700°K, the concentration of CsI molecules is quite low; nevertheless, it is more than adequate. A pulsed laser [14] is used to dissociate CsI in a narrow beam (i.e., about 0.5 mm in diameter). A second pulsed laser [8] having a much larger beam, about 7 mm in diameter, is coaxial with the first narrow beam and is used to detect those free atoms which were liberated on the axis of the detector beam at time $t = 0$ and which are still contained in the detector cylinder at the arbitrary time $t > 0$. It will be shown below that the photon fluence (ϕ_S) associated with the source beam is large enough to dissociate all of the CsI molecules contained in the central portion of a volume swept by the source beam, and the photon fluence (ϕ_D) associated with the detector beam is large enough to selectively remove one electron from each of the liberated atoms in the detector volume.

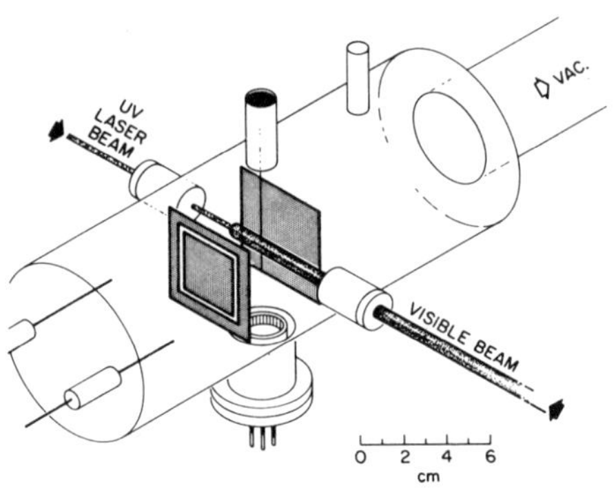

Fig.5 Experimental arrangement for the study of saturated photodissociation of alkali-halide molecules. The pulsed uv laser is used to dissociate CsI molecules at time $t = 0$, and the pulsed visible laser is used to detect Cs atoms at $t > 0$

In the initial experiments the wavelength of the source laser was set at 3175 Å near the CsI dissociation peak [16]. To detect the neutral Cs, the detector laser was set to produce photons at 4593 Å which promote $Cs(6^2S_{1/2})$ to $Cs(7^2P_{1/2})$; and, as appropriate for two-photon RIS, a second photon of the same wavelength photoionized $Cs(7^2P_{1/2})$ (see Fig.6).

Figure 7 shows the ionization signal as a function of the number of photons in a single pulse of the source laser, both for an unfocused beam and for a beam which was focused with a 50-cm focal length lens. The focused beam signals continue to rise gradually because of a nonuniform source beam. To obtain the Cs photoproduction cross section, σ, the following analysis was made.

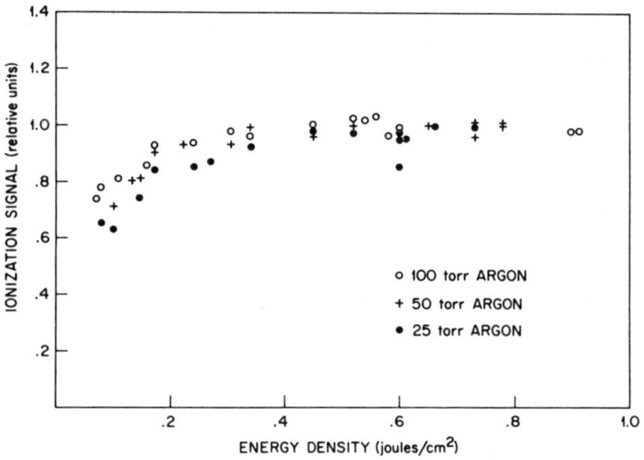

Fig.6 Measured signals due to Cs atoms as a function of the energy density of the detector laser and at a fixed energy density for the uv laser used to dissociate CsI. Saturation of the ionization signal occurs at energy density greater than 200 mJ/cm^2, indicating that every Cs atom produced in the source beam was photoionized

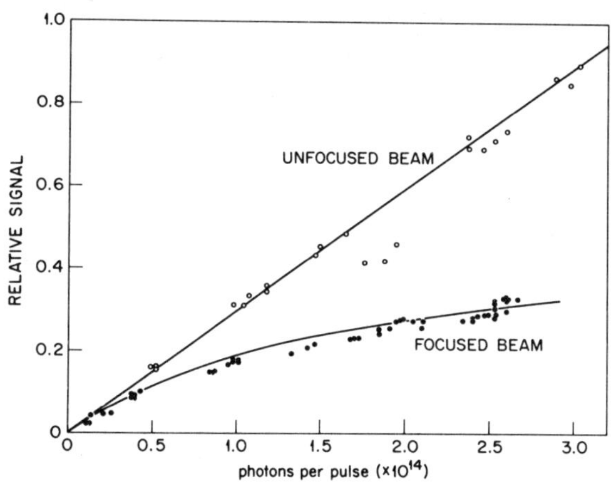

Fig.7 Signals due to Cs atoms as a function of the number of photons in a single pulse of the source laser (unfocused and focused). The function $F(\sigma\phi_0)$ in Eq. (4) is fitted to the experimental data and is the curve drawn through the focused data points. An analysis of the data (see text) leads to the absolute cross section for photoproduction of Cs without a prior knowledge of the concentration of CsI molecules

Assuming a Gaussian beam profile, we write

$$\phi(\rho) = \phi_0 e^{-\rho^2/R^2}, \tag{2}$$

where ϕ_0 is the fluence when the radius $\rho = 0$, and R is a constant. Since each atom is detected, the measured signal is proportional to

$$\lambda = N \int_0^\infty 2\pi\rho d\rho \left[1 - \exp\left(-\sigma\phi_0 \, e^{-\rho^2/R^2}\right)\right], \tag{3}$$

where λ is the number of atoms dissociated per unit of length and N is number density of the CsI molecules. It can be shown that

$$F(\sigma\phi_0) = \frac{\lambda}{N\pi R^2} = \gamma + \ell n \; \sigma\phi_0 + E_1(\sigma\phi_0), \tag{4}$$

where E_1 is the exponential integral and γ is Euler's constant (0.577....). The ratio of the focused beam signal to the unfocused beam signal (see Fig.7) is just $F(\sigma\phi_0)/\sigma\phi_0$, since in the limit $\sigma\phi_0 \to 0$, $F(\sigma\phi_0) = \sigma\phi_0$. To find σ, one sets $F(\sigma\phi_0)/\sigma\phi_0$ equal to an experimental ratio. The fluence ϕ_0 was determined experimentally by measuring the energy transmitted through a small aperture with a joule meter. In this way we found at 3175 Å a value of 2.9×10^{-17} cm^2 for the Cs photoproduction cross section.

Figure 8 shows the cross section for the production of Cs neutral from CsI as a function of wavelength. The present results have a functional form which is similar to that for photoabsorption [16], and the cross sections at the peak agree to within a few percent. The present measurements were made

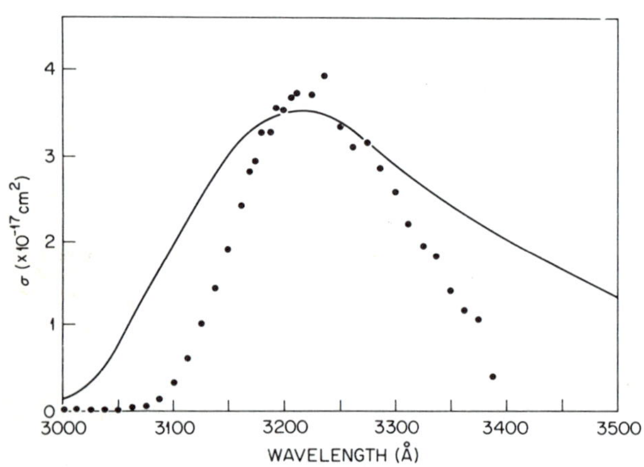

Fig.8 Cross section for the photoproduction of Cs from CsI. Solid line is photoabsorption data of [16]

at 325°K, while the photoabsorption data were taken at about 1000°K; the reduction in vibrational population at the lower temperature most likely accounts for the reduced width. A knowledge of the vapor pressure of CsI was not required to obtain our absolute photodissociation cross sections.

The method introduced (Fig.5) for the study of alkali halide photodissociation is well suited to rather direct measurements of the diffusion of a small number of alkali atoms in various gases and to a remarkably simple determination of the rate of reaction of free atoms with various other atoms or molecules [17]. By repeating the source and detector laser pulses at various time delays, i.e., $t > 0$, the number of atoms which have not escaped the cylinder by diffusion and which have not reacted chemically within the volume of the cylinder can be measured quite directly as a function of time. The measured fraction of surviving atoms, $\gamma(t)$, contains both the diffusion coefficient and the cross section for chemical reaction, as we now show.

Suppose a line source of atoms, λ, per unit length is created along an infinitely long line at time $t = 0$. When $t > 0$, the dissociated atoms spread by diffusion (coefficient D) through an inert gas while reacting chemically (at rate β) with a reactive species. If n is the number density of the alkali atom,

$$\frac{\partial n}{\partial t} = D\nabla^2 n - \beta n. \tag{5}$$

The solution for an unbounded medium (for a good approximation, see Fig.5) is

$$n(\rho,t) = \frac{\lambda}{4\pi Dt} \exp\left(- \frac{\rho^2}{4Dt}\right) e^{-\beta t}, \tag{6}$$

where ρ is the radial distance from the line source. If we define $\gamma(t)$ as the fraction of the number of ions contained in the cylinder of radius R, then

$$\gamma(t) = \frac{1}{\lambda} \int_0^R 2\pi\rho d\rho \; n(\rho,t) = \left[1 - \exp\left(- \frac{R^2}{4Dt}\right)\right] e^{-\beta t}. \tag{7}$$

Experiments were performed on the diffusion of Cs atoms in Ar. Referring to Fig.5, the CsI sample was heated to 629°K, producing about 3×10^8 CsI molecules per cm^3 in the region of the laser beams, where the measured temperature was 325°K. The source laser photodissociated nearly all of the molecules in a small volume of 10^{-2} cm^3 (area = 2.5×10^{-3} cm^2, length = 4 cm)--thus about 3×10^6 Cs atoms. The detector laser, operated at 4593 Å, ionized all of the Cs atoms by first exciting to the $Cs(7^2P_{1/2})$ level. With argon pressure at 100 Torr, we found that the fraction $\gamma(t)$ was diffusion controlled; i.e., reactive gas impurities were at sufficiently low concentration that $e^{-\beta t}$ was close to unity. At higher argon pressures we found it difficult to keep impurity levels low enough that the diffusion process would control the loss of atoms. We show in Fig.9 a fit of Eq. (7) to the data where $\beta = 10$ sec^{-1} and $D = 1.8$ cm^2 sec^{-1} at 50 Torr--thus, 0.12 cm^2 sec^{-1} at 1 atm.

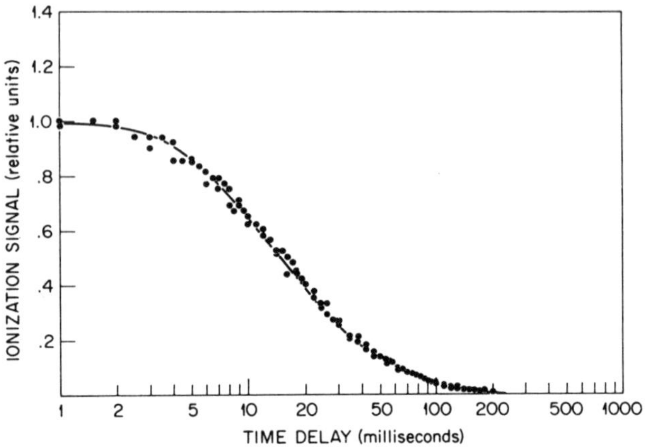

Fig.9 Relative Cs atom signal as a function of time in pure argon. A comparison of experimental data is made with diffusion theory

After adding small concentrations of O_2, the functions $\Upsilon(t)$ were controlled almost entirely by the exponential term $e^{-\beta t}$, and

$$\beta = 3.9 \times 10^3 \, P_{Ar}(P_{O_2})^{0.87 \pm 0.15} \tag{8}$$

for P_{Ar} between 25 and 150 Torr and P_{O_2} between 0.001 and 0.3 Torr.

Time resolution of both the atomic source and the atomic detector, as well as the ability to detect all of the surviving atoms, makes possible sensitive measurements of chemical kinetics and chemical transport processes in general in a very direct and simple way.

5. Demonstration of One-Atom Detection

We elected [18] to carry out the demonstration of one-atom detection with a proportional counter (Fig.10). Arbitrarily, low concentrations of Cs in the laser beam can be prepared in the following way. A Cs source is situated as shown in Fig.10 so that the distance from the source to beam can be adjusted. Whenever highly reactive atoms, such as Cs, are at very low concentrations in a counting gas, the steady-state concentration, δ, as a function of the diffusion distance, L, from a source emitting S_0 atoms/sec is [19]

$$\delta = \frac{S_0}{4\pi L^2 \bar{v}} e^{-\beta t} , \tag{9}$$

where $\bar{v}$ is about 4×10^4 cm/sec for Cs-Ar and S_0 is about 5×10^{14} for a 1-cm^2 source of Cs at 300°K. Using Eq. (8) for β and replacing t with $L^2/4D = L^2 P_{Ar}/3040\, D_1$, where D_1 is the diffusion coefficient at 760 Torr, one finds

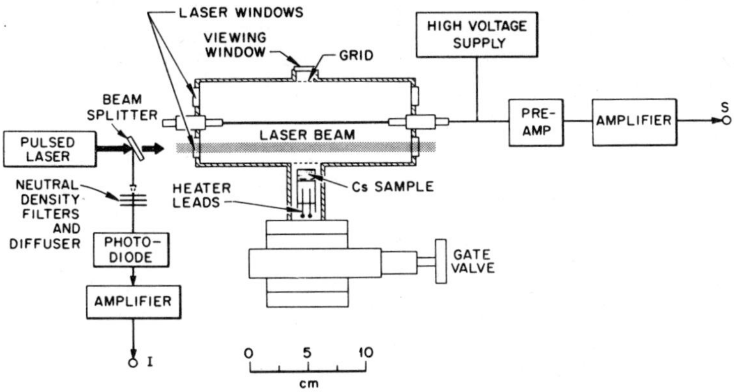

Fig.10 Experimental arrangement used to demonstrate one-atom detection

$$\delta = \frac{S_0}{4\pi L^2 \bar{v}} \exp\left(-1.3\, L^2\, P_{Ar}{}^2\, P_{O_2}/D_1\right). \tag{10}$$

Let $\delta = 1$ atom/cm^3, $L = 2$ cm, $P_{Ar} = 200$ Torr, and $D_1 = 0.12$ cm^2/sec; then we find that $P_{O_2} = 10^{-5}$ Torr, which is $P_{O_2}/P_{Ar} = 5 \times 10^{-8}$. In other words, to get even one atom of Cs per cm^3, the system must be free of O_2. We take advantage of these transport properties to control the concentration of Cs atoms in the beam simply by changing the counting gas pressure or the source-to-beam distance, L.

To obtain sufficient fluence (ϕ) to saturate, the 3-mm diameter laser beam was focused below the center of the counter wire by using a lens of 25-cm focal length. With 1-mrad divergence, the beam focused to 0.025-cm diameter and was approximately 0.10-cm diameter under both ends of the counter wire, creating a volume of about 5×10^{-2} cm^3. Under these conditions and with the field tubes decoupled from the wire to suppress photoelectrons from the quartz windows, there was no measurable photoelectron background. Backgrounds due to external radiations, such as cosmic rays, were eliminated by using electronic time-gating techniques.

In Fig.11 we plot a pulse height distribution for a large number of Cs atoms with the condition that the laser always saturated the ionization. From the known fact that each 6.4-keV X ray produces 250 electrons in P-10 gas, we find that the Cs peak in Fig.12 corresponds to 1.4×10^4 Cs atoms. Similar distributions were measured for populations ranging from 10^2 to 10^6 atoms. One electron in a proportional counter produces an exponential-like pulse height distribution. Fig.12 was obtained by using a Hg lamp to release single photoelectrons from the inner walls of the counter at a low rate. With the Cs sample in its lowest position and a gas pressure of 200 Torr, we obtain a nearly identical distribution due to the ionization of a Cs atom. In Fig.12 the fluctuations in the distributions are due to counting statistics, which do not reflect uncertainty in the total number of atoms counted. By integrating the total number of pulses produced above the normal electronic and laser transient noise, we find that it is not difficult to count directly 95% of the one-atom pulses. At all population levels, including one atom, we observed that the signals vanished when the laser was detuned from the Cs transition.

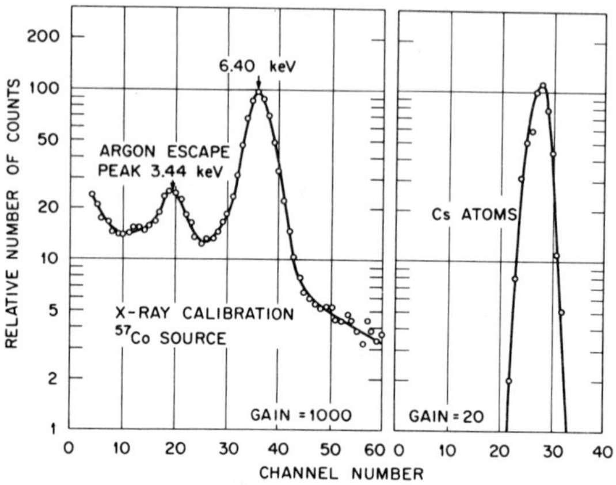

Fig.11 Pulse height spectrum due to resonance ionization of Cs atoms

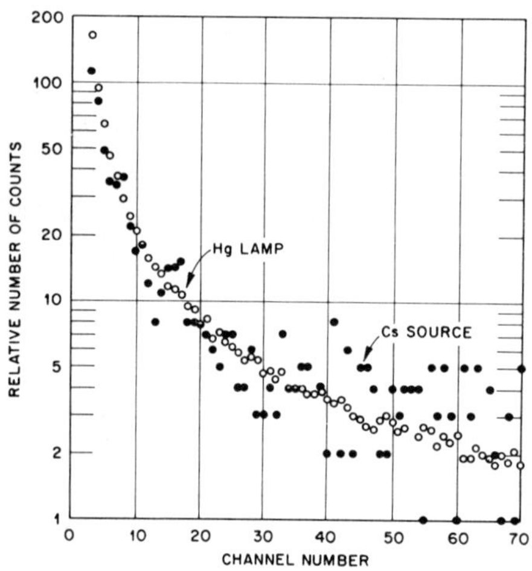

Fig.12 Pulse height distribution for the case where one Cs atom was counted per 20 laser pulses, compared with a one-electron distribution produced by an incoherent light source. Sensitive volume was 0.05 cm^3 per laser pulse. With available lasers, the volume per pulse could be greater than 100 cm^3. Background gas was about 10^{19} Ar atoms per cm^3 and 10^{18} CH_4 molecules per cm^3

Applications [19] of one-atom detection may include the detection of daughter atoms from radioactive decay, characterization of a few atoms of the higher actinides, slow evaporation of atoms from surfaces, slow transport processes, and the rate of reaction of the elements in various chemical environments. Analytical applications can be made even with molecular substances, e.g., the classification of a compound by dissociation and subsequent identification of resultant atoms. Several such applications are being pursued in our laboratory.

References

1 E. Rutherford and H. Geiger, Proc. R. Soc. A 81, 141 (1908).

2 W. P. Jesse and J. Sadauskis, Phys. Rev. 88, 417 (1952).

3 T. E. Bortner and G. S. Hurst, Phys. Rev. 93, 1236 (1954); J. E. Parks, G. S. Hurst, T. E. Stewart, and H. L. Weidner, J. Chem. Phys. 57, 5467 (1972).

4 N. Thonnard and G. S. Hurst, Phys. Rev. A 5, 1110 (1972); D. M. Bartell, G. S. Hurst, and E. B. Wagner, Phys. Rev. A 7, 1068 (1973).

5 M. G. Payne, C. E. Klots, and G. S. Hurst, J. Chem. Phys. 63, 1422 (1975).

6 G. S. Hurst, M. G. Payne, M. H. Nayfeh, J. P. Judish, and E. B. Wagner, Phys. Rev. Lett. 35, 82 (1975).

7 V. S. Letokhov, Phys. Today 30, 23 (1977).

8 Coaxial lamp dye laser, model 2100C, Phase-R Co., New Durham, NH 03855.

9 M. G. Payne, G. S. Hurst, M. H. Nayfeh, J. P. Judish, C. H. Chen, E. B. Wagner, and J. P. Young, Phys. Rev. Lett. 35, 1154 (1975).

10 A. V. Phelps, Phys. Rev. 99, 1307 (1955).

11 M. H. Nayfeh, G.S. Hurst, M. G. Payne, and J. P. Young, "Collisional Line Broadening Using Laser Excitation and Ionization," to be published.

12 A convenient energy diagram showing lifetimes and transition rates in Cs was compiled by Prof. Ray Hefferlin and reported in Ref. 19.

13 For a calculation of photoionization cross sections for several excited states of Cs, we are indebted to Prof. S. T. Manson. See also A. Msezane and S. T. Manson, Phys. Rev. Lett. 35, 374 (1975).

14 Model CMX-4, Chromatix Corp., Mountain View, CA 94043.

15 L. W. Grossman, G. S. Hurst, M. G. Payne, and S. L. Allman, Chem. Phys. Lett. (in press).

16 P. Davidovits and D. C. Brodhead, J. Chem. Phys. 46, 2968 (1967).

17 L. W. Grossman, G. S. Hurst, S. D. Kramer, M. G. Payne, and J. P. Young, Chem. Phys. Lett. (in press).

18 G. S. Hurst, M. H. Nayfeh, and J. P. Young, Appl. Phys. Lett. 30, 229 (1977).

19 G. S. Hurst, M. H. Nayfeh, and J. P. Young, Phys. Rev. A 15, 2283 (1977).

LASER FREQUENCY MEASUREMENTS: A REVIEW, LIMITATIONS, EXTENSION TO 197 THz (1.5μm)

K.M. Evenson, D.A. Jennings, F.R. Petersen, and J.S. Wells

Precision Laser Metrology Section, National Bureau of Standards
Boulder, Colorado, USA

Abstract

The CO_2 and He-Ne lasers stabilized respectively with CO_2 and CH_4 are now accepted as frequency and wavelength standards at 9 to 12, and 3.39 μm. This is due to their excellent frequency stability and accuracy, and the direct measurement of their frequencies. The measurement of both the frequency and wavelength of the CH_4 device yielded a hundred fold increase in the accuracy of the speed of light. This paper reports the extension of direct frequency measurements to the 197 THz (1.5 μm) cw line of a He-Ne laser, and reviews the current status of laser stabilization and speed of light measurements.

Introduction

With the advent of the laser, coherent sources of radiation became available from the microwave to the visible portion of the electromagnetic spectrum. Thus, one could think of measuring photon energy via either frequency or wavelength metrology techniques. The coherence of the cw laser's radiation is most accurately measured via direct frequency measurements, which do not suffer from the limitations of wavelength measurements: diffraction, mirror curvature, and phase shift at the mirror surface. These limit the accuracy in wavelength measurements to about $\pm 1 \times 10^{-10}$ [1]. At present, the absolute comparisons of wavelength are limited by uncertainties in the realization of the definition of the meter to $\pm 4 \times 10^{-9}$ [2]. In contrast, the absolute accuracy of laser frequency measurements is limited presently by the stabilities of lasers used in frequency synthesis. Eventually, the limitation will be in the accuracy of the sources which are now $\pm 0.8 \times 10^{-13}$ for the time standard [3] and $\pm 2 \times 10^{-13}$ for the 3.39 μm CH_4 stabilized He-Ne laser [4]. Most laser frequency measurements, so far, have been made through the use of a non-linear device: a tungsten nickel catwhisker point contact diode. This paper will discuss the stability of laser sources; the techniques of laser frequency measurement; the non-linear devices used in laser frequency measurements; laser speed of light measurements; the redefinition of the meter; the extension of laser frequency measurement to 197 THz; and some limitations of the techniques used so far in laser frequency measurements.

Stabilization of Lasers

Frequency measurements are mankind's most accurate measurements, with the accuracy limited only by the coherence of the sources themselves. For this reason it is appropriate to begin this discussion by considering some of the various types of very coherent sources, specifically those which can be considered as possible frequency standards themselves.

Although the spectral purity of a free running laser has been shown to be as good as a few Hertz [5], without some means of controlling the length of the cavity, the frequency can vary over the Doppler width of the laser transition from a few megahertz in the far infrared to over a thousand megahertz in the visible. (We are specifically discussing Doppler width instead of gain band width since most stable lasers use a gaseous gain medium.) This same Doppler width generally allows the oscillation of several longitudinal modes in the cavity, so that several frequencies spaced by c/2L, where c is the speed of light and L is the length of the cavity, will be oscillating at one time. To force single frequency oscillation, one generally chooses a laser cavity sufficiently short so c/2L is greater than the gain width. Then the laser frequency can be locked to some reference by controlling the length of the cavity. This reference can be either a frequency synthesized from other stable and known sources, or, some Doppler-free spectral feature. In the latter category, the lasers locked to these "Lamb dip" type of features become independent frequency or wavelength references. The best known of these are: a saturated fluorescence in CO_2 for locking each CO_2 line [6]; a saturated absorption in CH_4 to lock the 3.39 μm (88 THz) line of the He-Ne [7], saturated absorbtion in iodine to lock the 632.8 nm He-Ne [8], and a molecular beam technique to lock the argon laser [9]. The short term line width, the stability, and the accuracy capability of these radiation sources are shown in Table 1.

The two HeNe lasers locked to their respective sources have been compared with the present standard of length sufficiently accurately so that values of their wavelengths have been recommended as secondary standards by the Comite Consultatif pour la Definition du Metre [20]. The values are: methane, P(7), band ν_3; 3. 392 231 40 x 10^{-6} m and, iodine 127, R(127), band 115, component i; 0.632 991 399 x 10^{-6} m.

For absolute frequency measurements the present standard, Cs, as can be seen in Table 1, is limited to ± 8×10^{-14}. Other sources, such as the hydrogen maser may be more stable, but problems with the wall shift have excluded it from consideration as a primary standard.

The 3.39 μm He-Ne laser, although still in its early stages of development, is second only to Cs in accuracy capability and even surpasses it in stability at 100 sec. Thus, this He-Ne laser is an excellent secondary frequency and wavelength standard in the infrared. However, before a frequency standard at this high frequency can be considered, more efficient and simpler techniques of frequency synthesis must become more commonplace. A great deal of work is underway on new techniques, but what simplification will occur is not yet certain.

Frequency Measurements

Frequencies may be directly counted up to about 500 MHz by timed gating of electronic counters. At higher frequencies, a heterodyne technique is generally used whereby the difference between two nearly equal frequencies is generated and directly counted. One of the frequencies is synthesized from known frequencies by some sort of non-linear device which can either sum two or more different frequencies or generate harmonics, or both. The synthesis at laser frequencies presents special challenges since the frequencies are so high. An excellent review paper summarizes the application of non-linear devices to optical frequency measurement [21]. The silicon diode which generates many different orders of harmonics so well in the microwave ceases to respond at about one terahertz, approximately the frequency of the HCN laser, one of the lowest frequency lasers. Bulk mixers have been used in laser frequency synthesis for some time; however, due to low efficiency in these devices, phase matching must be used; thus, for cw radiation they are limited to 2nd harmonic generation or two waves mixing.

Table 1 Frequency Sources

Source	Ref	λ, μm	ν, THz	Stabilizing Source	Short-Term Line Width Unlocked 10^{-3}sec	Stability 100 sec	Accuracy Capability
H maser	10, 11, 12		0.0014	H. Active		10^{-15}	$1x10^{-12}$
H maser	13			H. passive		10^{-13}	$1x10^{-12}$
Cs beam	3		0.0092	Cs beam		$7x10^{-14}$ (estimated)	$8x10^{-14}$
N_2O laser	14	10-11	27-30	N_2O Sat. Fl.			$\simeq 1x10^{-9}$
$^{12}C^{16}O_2$	15	9.1 to 11	27-33	CO_2 Sat. Fl.	$2.5x10^{-13}$	$6x10^{-13}$	$\simeq 10^{-10}$
CO_2 isotope laser	15	9.0 to 12.4	24-33 all isotopes	CO_2 Sat. Fl.	$2.5x10^{-13}$	$6x10^{-13}$	$\simeq 10^{-10}$
He-Ne laser compact intracavity device	16	3.39	88	CH_4 Sat. Ab.			$2x10^{-11}$
He-Ne	17, 4, 18	3.39		CH_4	$2x10^{-13}$	$5x10^{-15}$	$2x10^{-13}$
He-Ne laser	19	0.63	475	I_2 Sat. Ab.		10^{-12}	$\simeq 10^{-10}$
Ar laser	7	0.5145	550	I_2 beam	$4x10^{-8}$	$2x10^{-13}$	$1x10^{-11}$

Another device, a MIM (metal-insulator-metal) diode, first used at laser frequencies in A. Javan's laboratory [22], similar in use to the silicon diodes used in the microwave region, is the main device used in laser frequency synthesis. It is similar to the catwhisker diode used in the early radio receivers; however, it uses only metals and the natural oxide that forms in the nickel base as the insulator. It has been used to generate as high a harmonic as the twelfth with an HCN laser [23] and has been used to generate harmonics to 148 THz. Later in this paper, we will describe its use as a 3rd and 4th order mixer at 197 THz (1.52 μm).

The catwhisker is fabricated from a 10 to 25 μm diameter tungsten wire electrochemically sharpened to a tip radius of about 0.05 μm. The straight portion of the whisker from the tip to a right angle bend serves as the antenna [24] to concentrate the focused laser field at the tip where the non-linear processes occur at the 1nm natural nickel oxide layer.

The unknown frequency, ν_x, is determined from the synthesis obtained when all of the lower frequency radiations impinge on the diode. That is: $\nu_x = l\nu_l \pm m\nu_m \pm n\nu_n \pm \nu_b$ where ν_l, ν_m, and ν_n are known frequencies, l, m, and n are the harmonics, and ν_b is the RF beat frequency. The sum, l+m+n+1 is the mixing order. The signal to noise ratios obtainable with this device are shown in Fig. 1. These are obtained with single frequency

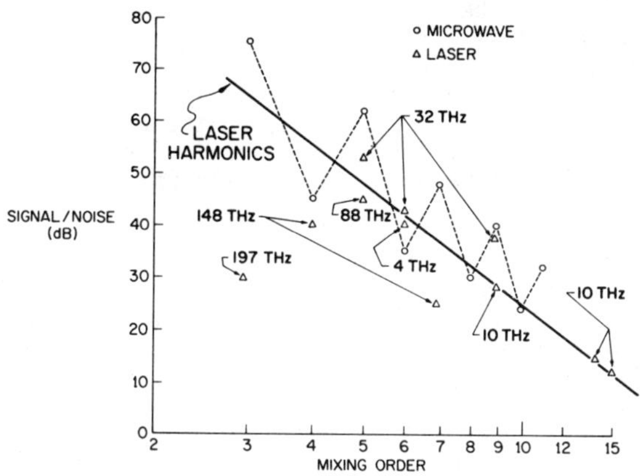

Fig. 1. Signal to noise ratios obtained with tungsten-nickel MIM diodes as a function of mixing order and at various frequencies between 10 and 1000 MHz with a bandwidth of 30 or 100 KHz, a scan width of 200 or 500 KHz/cm, a scan time of 2 ms/cm, and a video filter of 1 KHz.

lasers with outputs of about 100 mw. They were taken with a standard spectrum analyser at between 10 and 1000 MHz with a bandwidth of 30 to 100 kHz, a scan width of 200 to 500 kHz/cm, scan time of 2m sec/cm and a video filter of 1 kHz. Some of the details of microwave and lower frequency laser operation are presented in an earlier paper [25]. The signals at higher frequencies were taken with the larger settings on the spectrum analyser; this is necessary at the higher frequencies since a free-running laser's jitter is approximately proportional to its frequency. This accounts for only a part of the fall off at higher frequency.

The most widely accepted explanation of the diode's operation is that the mechanism is due to tunnelling [26, 27], but, there is some evidence for more than one phenomena occurring in the diodes. The impedance of the diode may be adjusted from about 10 to 10,000 Ω by the microadjustment of the contact pressure. The more sensitive contacts are at higher impedances, and the more stable are at the lower impedances. Usually a compromise at about 200 to 600 ohms is made and the diode is then often stable for several days at a time, and with readjustment of the contact, the whisker can be used for several weeks.

Normally (about 95% of the time), and at frequencies less than 100 THz the laser radiation is rectified in the diode and it drives the tungsten whisker negative. Under some conditions, a bias improves the heterodyne signal; however, it usually is not a significant improvement, and consequently, the diode does not have a dc bias applied.

At frequencies above 100 THz, the rectified polarity usually reverses (ie. the tungsten is driven positive). However, by increasing the dc impedance of the diode a negative driven whisker can sometimes be achieved. For example, at 148 THz, an impedance of about 600 ohms is generally necessary for negative rectification. The negative signal is not necessary for harmonic generation, but somewhat higher sensitivity generally accompanies this signal. Although we have observed antenna patterns at wavelengths as short as 2 μm, generally the antenna becomes less effective due to an increasing resistivity of tungsten at these frequencies. Consequently, we have achieved significantly greater signals with the use of high quality, long working distance microscope objectives to focus the radiation at wavelengths of 2 μm and less. Sharp angular dependence is absent, and the radiation is usually focused at 45° with respect to the antenna with the polarization of the laser radiation lying in the plane determined by the whisker and the laser beam.

The generation of third order or higher harmonic signals is the only satisfactory criterion we have found to test the diodes. For example, at 197 THz, the rectified voltage is generally positive, and the positive signal is not critical with respect to the laser polarization, nor is it extremely critical with respect to the focus. However, to obtain a third order signal, an extremely critical focus must be found which is indicated only by a somewhat sharper square wave signal from a 1 kHz chopped laser beam. The heterodyne signal is over an order magnitude larger with the polarization in the plane of the antenna and laser beam. Thin film diodes have been fabricated [26, 28], however, as yet, they have not been tested for high frequency operation (i.e. with third order mixing).

Chebotayev and co-workers [29] used a very clever scheme to achieve 3 wave frequency synthesis of visible radiation which relies upon the nonlinear properties of neon itself in the Helium-Neon laser. This unique mixing occurs because of some common energy levels in 4 lasing transitions in Ne: the He-Ne laser can oscillate on the 0.6330 μm, 3.39 μm, the 2.39 μm, and the 1.15 μm lines. The sum of all of the IR energies is exactly the energy of the 0.6330 μm line since the top-most energy level of the 3 IR frequencies and the bottom most are common with the 0.6330 μm energy levels. Therefore, when a prism and separate mirrors for each of the 3 IR lines are installed on the laser, the induced polarization from the sum of the IR radiations induces a visible transition, and about 1 μwatt of 6330 radiation emerges (with no 0.6330 μm mirror on the cavity). The 0.6330 μm radiation is thus the sum of all 3 IR frequencies. Of the 3 IR frequencies, only that of the 1.15 μm line has not been reached; thus, once the 1.15 μm frequency is measured, synthesis to the visible will have been accomplished.

Alternatively, it is quite easy to double the frequency of the 1.15 μm laser in a lithium niobate crystal; in fact, we have obtained 100 μw of visible light from 30 mw of 1.15 μm radiation.

The Speed of Light

The advent of lasers and in particular the extension of direct frequency measurement into the infrared region, has been responsible for the 100 fold improvement in the accuracy of the measurement of the speed of light. It has taken c from one of the less accurately known fundamental constants and made it the most accurately known. In fact, there is now the definite possibility of fixing the value of c with a redefinition of the meter [2, 30].

We will summarize all recent measurements of $c = \lambda\nu$ where c is obtained by the measurement of both the wavelength, λ, and the frequency, ν, of a stabilized laser. Table 2 gives the results of various frequency and wavelength measurements yielding values of c; these values are graphed in Fig. 2. Also listed is the value of c recommended by the CCDM [2]: for use

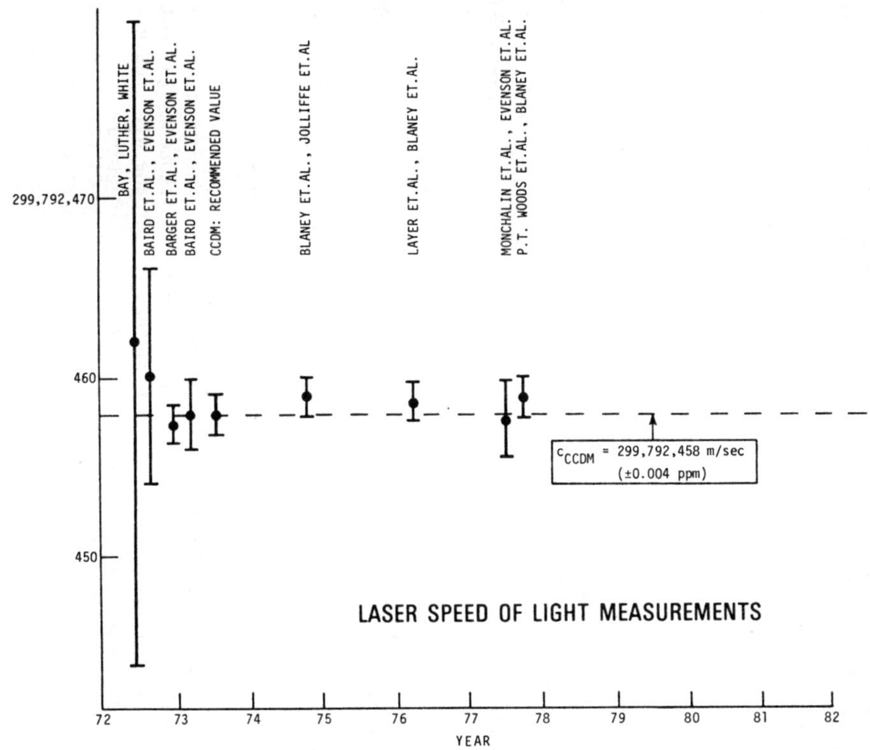

Fig. 2. Values of the speed of light obtained using lasers since 1972.

Table 2. Laser Measurements of $c = \lambda\nu$

Laser	Stab.	$\lambda^{1}_{vac.}$, μm	Ref	ν, THz	Ref	c, m/sec
He-Ne	Ne	0.63299147(1)	30	473.612166(29)THz[2]	30	299,792,462(18)
CO_2	CO_2	nine values (9 to 10 μm)	31	nine frequencies	32, 33	299,792,460(6)
He-Ne	CH_4	3.392231390(14)	34, 35	88.376181627^D(50)[3]	34, 36	299,792,457.4(1.2)
He-Ne	CH_4	3.392231400(20)	37	88.376181627^D(50)[3]	34, 36	299,792,458(2)
CO_2 R(12)	CO_2	9.317246348(44)[4]	38	32.176079482(27)	39	299,792,459.0(1.2)[3]
He-Ne	CH_4	3.392231405(14)[4]	40,1	88.376181608(43)[3]	41	299,792,458.7(1.2)[3]
CO_2 R(14)	CO_2	9.305385613(70)	42	32.217091275(24)	36,43	299,792,457.6(2.2)
CO_2R(12)	CO_2	9.317246340(38)[4,5]	45	32.176079482(27)	39	299,792,458.8(1.2)[3]
		CCDM Value 2, 20				299,792,458(1.2)

1. Average of the center of gravity and peak of the asymetrical line of Krypton length standard.
2. In this frequency determination, the optical frequency was not directly counted.
3. A third determination of the frequency of methane with respect to CO_2 is in excellent agreement with the values here and helps confirm these frequencies [44].
4. The uncertainty in the krypton standard ($\pm 4 \times 10^{-9}$) was added in quadrature to the published uncertainty.
5. This number represents a remeasurement of the wavelength shown in line 5 by the same workers.

in distance measurements where time-of-flight is converted to length; for the fundamental constant; and for use in converting frequency to wavelength. The uncertainty in this value ($\pm 4 \times 10^{-9}$) is the estimated uncertainty in the realization of the definition of the meter itself and results from asymmetries in the line, the line width, and variations from lamp to lamp. This uncertainty has been added to some of the recent determinations of λ (see footnote 3 in Table 2).

All of the values of c except that in row 6 have appeared in the literature; this highly accurate value is independent of the other methane value since both the wavelength and frequency were independently measured. One sees a remarkable convergence of the values of c for the first time in history.

<u>Redefinition of the Meter</u>

As a result of the recommendations made by CCDM, two different definitions of a new length standard must be considered. First, we can continue as before with separate standards for the second and meter, but with the meter defined as the length equal to 1/λ wavelengths in vacuum of the radiation from a stabilized laser instead of from a ^{86}Kr lamp. Either the methane-stabilized He-Ne laser at 3.39 μm (88 THz) or the I_2-stabilized He-Ne laser at 0.633 μm (474 THz) appears to be suitable candidates. The 3.39 μm laser is already a secondary frequency standard in the infrared, and hopefully, direct measurements of the frequency of the 0.633 μm radiation will give the latter laser the same status in the visible. The 3.39 μm laser frequency is presently known to within 6 parts in 10^{10}, and the reproducibility and long term stability have been demonstrated to be better by more than two orders of magnitude. Hence, frequency measurements with improved apparatus in the next year or two are expected to reduce this uncertainty to a few parts in 10^{11}. A new value of the speed of light with improved accuracy would thus be achievable if the standard of length were redefined in terms of the wavelength of this laser.

Alternately, one can consider defining the meter as a specified fraction of the distance light travels in one second in vacuum (that is,one can fix the value of the speed of light). The meter would thus be defined in terms of the second; and, hence, a single unified standard would be used for frequency, time, and length. What at first sounds like a rather radical and new approach to defining the meter is actually nearly one hundred years old. It was first proposed by Kelvin in 1879 [46], and has been recently reemphasized by Townes [47] and Bay [48]. With this definition, the wavelength of all stabilized lasers would be known to the same accuracy with which their frequencies can be measured. Stabilized lasers would thus provide secondary standards of both frequency and length for laboratory measurements, with the accuracy being limited only by the reproducibility, measurability, and long term stability. It should be noted that, prior to the CCDM value, an adopted nominal value for the speed of light was already in use for high-accuracy astronomical measurements [49]; now however, with the adoption of the CCDM value [2] by the IAU, there is only one standard of length in existence. A definition which fixes c for all users would certainly be desirable from a philosophical point of view.

<u>Extension of Direct Frequency Measurements to 197 THz</u>

The frequency of a cw, free running, 8 m long, 15 to 20 mw, HeNe 1.5 μm laser line was directly measured using both third and fourth order mixing in a tungsten nickel point contact diode. The third order frequency synthesis

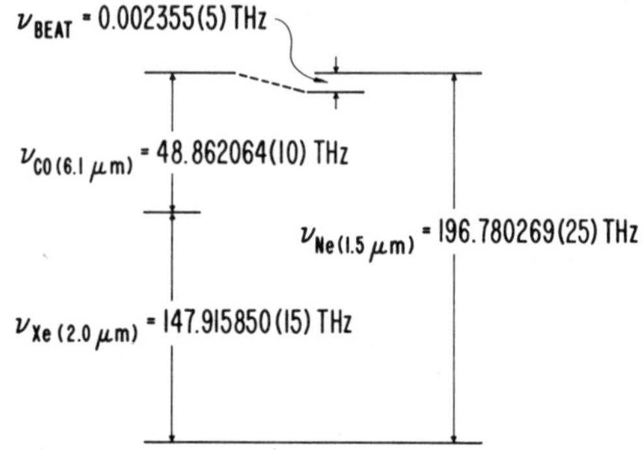

FREQUENCY SYNTHESIS OF 1.5 μm

Fig. 3. Third order synthesis of 197 THz in MIM diode.

scheme is shown in Fig. 3. Forty mw of 148 THz radiation from the Xe laser was focused at 45° with respect to the tungsten antenna on one side of the diode, and the 197 THz was focused at 45° on the other side. Radiation from each of these lasers was isolated from other lines by means of a prism external to the laser. About 100 mw of 49 THz CO laser radiation from the J=14, 19-18 band was focused at about 5° with respect to the antenna. High quality (i.e. expensive) long-working distance microscope objectives were used for the 148 and 197 THz radiations. A 10 db signal to noise beat note at 2.355 GHz was obtained, however a 20 db transmission loss was measured at 2.4 GHz; thus, an estimated signal to noise of 30 db is plotted in figure one. To optimize the signal to noise, the spectrum analyser was set at a band width of 100 kHz, the video filter, at 1 kHz, and the sweep rate, at 5ms/cm.

The frequency obtained by third order mixing was:

$$\nu_{Ne} = 196.780\ 269(25)\ \text{THz} = \nu_{2\mu m} + \nu_{CO} + \nu_{beat}$$

where $\nu_{2\mu m}$ = 147.915 850(15) THz[51], ν_{CO} = 48.862 064(10) THz [52], and ν_{beat} = 0.002355(5) THz. Additional uncertainties resulted from setting the Xe and CO lasers to the tops of their free running gain curves, 10 and 5 MHz respectively. The uncertainties were added in quadrature, giving an estimated uncertainty of 25 MHz.

The 4th order experiment was done with a different CO laser line, the J=20 line from the 18-17 band, ν_{CO} = 48.917 212(10) THz; plus the ν_{kl} = 0.051745(1) THz radiation from a microwave klystron. The result was:

$$\nu_{Ne} = 196.780\ 274(25)\ \text{THz} = \nu_{2\ \mu m} + \nu_{CO} + \nu_{kl} + \nu_{beat}.$$

In this case ν_{beat} = 0.001043(5) THz and the signal to noise was 10 db. The frequencies of the CO lines were calculated from the molecular constants of Todd et al [53] and the two resultant Ne frequencies were in much closer agreement than those using earlier CO molecular constants. Each experiment was repeated 3 times and the results were always within 5 MHz of these average values.

The average of each of these two experiments, thus, gives a value of

$$\nu_{Ne} = 196.780\ 271(25)\ \text{THz},$$

and using the recommended value of c

$$\lambda_{vac} = 1.523\ 4884(2) \times 10^{-6}\text{m}.$$

This number is in agreement with that given in spectral tables [53].

The entire frequency chain connecting this laser with the Cs frequency standard is shown in Fig. 4. The 197 THz radiation should be useful in reaching 260 THz (1.15 μm) via

$$\nu_{1.15} = \nu_{1.5} + \nu'_{CO_2} + \nu''_{CO_2} + \nu_{beat}.$$

However; this experiment has been tried unsuccessfully on two different occasions. Diode-burnout with 1.15 μm radiation seemed to be the main problem. It is not too surprising because the reflectivity of tungsten drops to about 60% at 1 μm. It is hoped that gold coating might decrease the burnout problem and allow the continuation of frequency measurements to the visible. Even if this is not the case, we are now close enought to the visible that 2nd harmonic generation or 2 wave mixing in bulk materials will certainly produce direct frequency measurement in the visible soon.

References

1. H. P. Layer, R. D. Deslattes, and W. G. Schweitzer, Jr., Appl. Optics, 15, 734, (1976).
2. J. Terrien, Metrologia, 10, 9, (1974).
3. David J. Wineland, David W. Allan, David J. Glaze, Helmut W. Hellwig, and Stephen Jarvis, Jr., IEEE Trans. Instr. and Meas. IM25, 453 (1976).
4. J. L. Hall, Conference on Atomic Masses and Fundamental Constants, Paris, (June 1975). (Plenum Press 1976).
5. T. S. Jaseja, A. Javan, and C. H. Townes, Phys. Rev. Letters 10, 165 (1963).
6. C. Freed and A. Javan, Appl. Phys. Lett. 17, 53 (1970).
7. K. Shimoda and A. Javan, J. Appl. Phys. 36, 718 (1965).
8. G. R. Haines, C. E. Dahlstrom, Appl. Phys. Letters, 14, 362 (1969).
9. L. A. Hackel, R. P. Hackel, and S. Ezekiel, Proceedings of the 2nd Frequency Standards and Metrology Symposium, Copper Mountain, Colorado, USA (1976).
10. Jacques Vanier and Robert Larouche, Proceedings of the 2nd Frequency Standards and Metrology Symposium, Copper Mountain, Colorado, USA (1976).
11. Helmut Hellwig, Robert F. C. Vessot, Martin W. Levine, Paul W. Zitzewitz, David W. Allan, and David J. Glaze, IEEE Trans. Instrum. Meas. IM19 200 (1970).
12. David W. Allan, Proceedings of Symposium on 1/f Fluctuations, Tokyo, Japan (1977).
13. Claude Audoin and Jacques Vanier, J. of Physics, E, 9, 697 (1976).
14. B. G. Whitford, K. J. Siemsen, H. D. Riccius, and G. R. Hanes, Optics Communications, 14, 70 (1975).

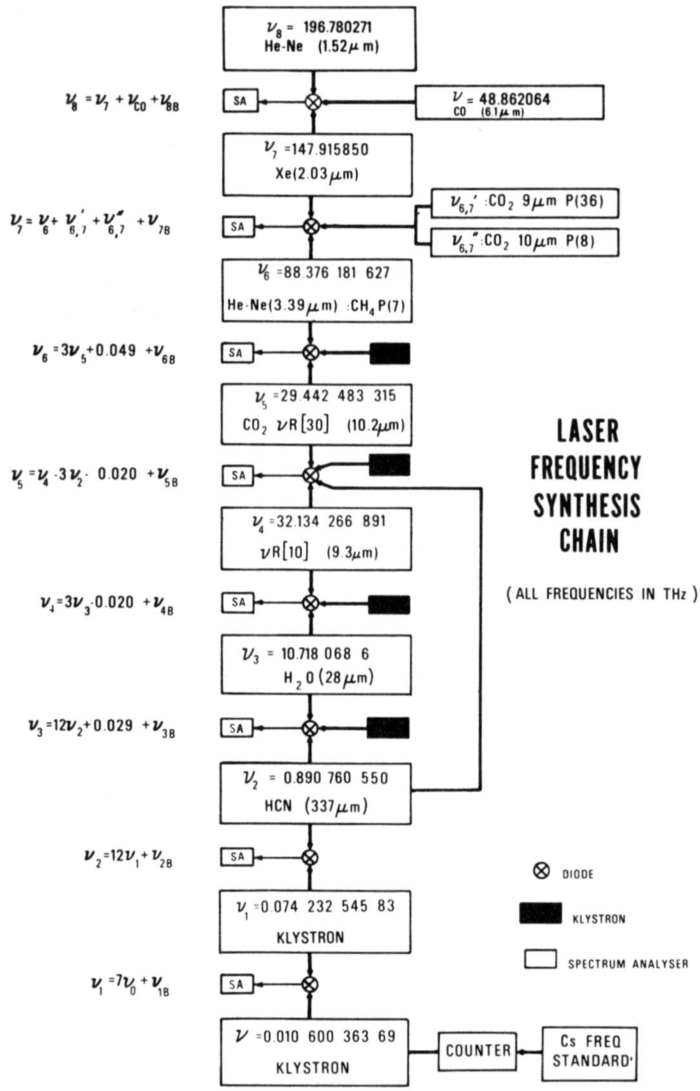

Fig. 4. Laser frequency synthesis chain from Cs (0.009 THz) to 197 THz.

15. Charles Freed and Robert G. O'Donnell, Proceedings of the 2nd Frequency Standards and Metrology Symposium, Copper Mountain, Colorado, USA (1976).
16. Barrie W. Jolliffe, Gunter Kramer, and JeanMarie Chartier, IEEE Trans. on Instr. and Meas. IM25 447 (1976).
17. V. P. Chebotayev, Proceedings of the 2nd Frequency Standards and Metrology Symposium, Copper Mountain, Colorado, USA (1976).
18. S. N. Bagayev, V. P. Chebotavev and L. S. Vasilenko, Conference on Laser Engineering and Applications, Washington, D.C. (June, 1977).
19. P. Cerez and A. Brillet, Metrologia, 13, 29 (1977).
20. Comite Consultatif Pour La Definition Du Metre, 5th Session, Rapport (Bureau International des Poids et Mesures, Sevres, France, 1973).
21. D. J. E. Knight and P. T. Woods, J. Phys. E, 9, 898 (1976).
22. V. Daneu, D. Sokoloff, A. Sanchez, and A. Javan, Appl. Phys. Letters 15, 398 (1969).
23. K. M. Evenson, J. S. Wells, L. M. Matarrese, and L. B. Elwell, Appl. Phys. Lett, 16, 159 (1970).
24. L. M. Matarresse and K. M. Evenson, Appl. Phys. Letters 17, 8, (1970).
25. E. Sakuma and K. M. Evenson, IEEE J. Quantum Electronics QE 10,599, (1974).
26. G. M. Elchinger, A. Sanchez, C. F. Davis, Jr., and A. Javan, J. Appl. Phys. 47, 591 (1976).
27. S. M. Faris, T. Kenneth Gustafson and John C. Wiesner, IEEE J. Quantum Electronics QE9, 737 (1973).
28. M. Heiblum, S. Y. Wang, J. R. Whinnery, and T. K. Gustafson, to be published, J. Appl. Phys.
29. V. M. Klementyev, Yu. A. Matyugin, and V. P. Chebotaev, JETP Lett. 24, 5 (1976).
30. Z. Bay, G. G. Luther, and J. A. White, Phys. Rev. Letters, 29, 189 (1972). 31. K. M. Baird, H. D. Riccius, and K. J. Siemsen, Opt. Commun. 6, 91 (1972).
31. K. M. Baird, H. D. Riccius, and K. J. Siemsen, Opt. Commun. 6, 91 (1972).
32. J. D. Cupp, B. L. Danielson, G. W. Day, L. B. Elwell, K. M. Evenson, D. G. McDonald, L. O. Mullen, F. R. Petersen, A. S. Risley, and J. S. Wells, Conf. on Precision Electromagnetic Measurements, Boulder, Colorado, USA (1972).
33. T. Y. Chang, Opt. Commun. 2, 77, (1970).
34. K. M. Evenson, J. S. Wells, F. R. Petersen, B. L. Danielson, G. W. Day, R. L. Barger, and J. L. Hall, Phys. Rev. Letters, 29, 1346 (1972).
35. R. L. Barger and J. L. Hall, Appl. Phy. Lett. 22, 196 (1973).
36. K. M. Evenson, J. S. Wells, F. R. Petersen, B. L. Danielson, and G. W. Day, Appl. Phy. Lett, 22, 192 (1973).
37. K. M. Baird, D. S. Smith, and W. E. Berger, Opt. Commun. 7, 107 (1973).
38. B. W. Jolliffe, W. R. C. Rowley, K. C. Shotton, A. J. Wallard, and P. T. Woods, Nature, 251, 46 (1974).
39. T. G. Blaney, C. C. Bradey, G. J. Edwards, B. W. Jolliffe, D. J. E. Knight, W. R. C. Rowley, K. C. Shutton, and P. T. Woods, Nature, 251, 46 (1974).
40. W. G. Schweitzer, Jr., E. G. Kessler, Jr., R. O. Deslattes, H. P. Layer, and J. R. Whetstone, Appl. Optics, 12, 2927 (1973).
41. G. J. Edwards, B. W. Jollife, D. J. E. Knight, and P. T. Woods, T. G. Blaney, J. Phys. D. 9, 1323 (1976).
42. J. P. Monchalin, M. J. Kelly, J. E. Thomas, N. A. Kurnit, A. Szöke, and A. Javan, Optics Letters, 1, 5 (1977).
43. F. R. Petersen, D. G. McDonald, J. D. Cupp, and B. L. Danielson, Laser Spectroscopy, Proceedings of the Vail (Colorado) Conference, R. G. Brewer and A. Mooradian, eds., Plenum Press, New York (1973).
44. B. G. Whitford and D. S. Smith, Optics Commun. 20, 280 (1977).
45. P. T. Woods, B. W. Jolliffe, W. R. C. Rowley, K. C. Shotton, and A. J. Wallard, to be published, Applied Optics 1977.

46. W. F. Snyder, IEEE Trans. Instr. and Measurement, IM-22 (1973) 99.
47. C. H. Townes, in Advances in Quantum Electronics, edited by J. R. Singer (Columbia U. P. New York, 1961).
48. Z. Bay, in Proceedings of the Fourth International Conference on Atomic Masses and Fundamental Constants, edited by J. H. Sanders and A. H. Wapstra (Plenum, New York, 1972).
49. P. Bender, Science, 168, 1012 (1970).
50. Resolution #6, Information Bulletin #31 of the International Astronomical Union, (Jun., 1974).
51. D. A. Jennings, F. R. Petersen, and K. M. Evenson, Appl. Phys. Lett. 26, 510 (1975).
52. T. R. Todd, C. M. Clayton, W. B. Telfair, T. K. McCubbin, Jr., and J. Pliva, J. Mol. Spec. 62, 201 (1976).
53. Charlotte E. Moore, Atomic Energy Levels, Vol. I, National Bureau of Standards Circular 467 (U.S. GPO, Washington, D.C. 1958).

LASER SPECTROSCOPY OF THE HYDROGEN MOLECULAR ION HD^+

W.H. Wing

Department of Physics and Optical Sciences, University of Arizona
Tucson, AZ 85721, USA

The hydrogen molecular ion is the simplest bound system that can be called a molecule. Therefore it plays a role in the theory of molecular structure that corresponds to the role of the hydrogen atom in atomic structure theory. The experimental situations, however, little resemble each other, for while the hydrogen atom spectrum is among the most precisely and the earliest studied, that of the hydrogen molecular ion has languished until recently among the "forgotten poor," spectroscopically speaking. In this talk I will summarize briefly the work [1] several colleagues and I are doing in this area. While by comparison with some of the other spectroscopic experiments reported at this conference it is not yet correct to call our results extremely precise, we have at least been able to uplift the hydrogen molecular ion spectrum to "middle-class respectability," and have been able to make a significant test of the first-principles theoretical calculations of its spectrum. Our absolute accuracy, as I will show, now is a few parts in 10^7.

As is well known from elementary texts [2] on quantum mechanics, the three-body problem can be solved analytically with the assumption that two of the bodies (the nuclei) are massive and have negligible kinetic energies compared to the third. With the nuclei fixed in space, the electron's SCHRÖDINGER equation can be separated into 3 ordinary differential equations, which can then be solved by function series or continued-fraction methods. The electron probability distribution in the H_2^+ ground state was first obtained in this way by BURRAU [3], as shown in Fig.1. One then treats

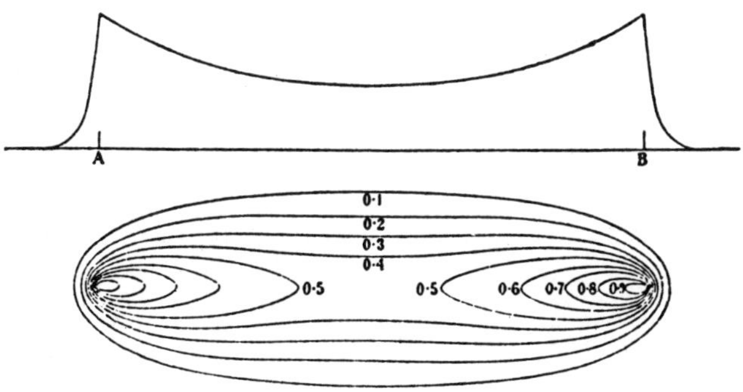

Fig.1 Electron probability distribution in the lowest state of H_2^+, as calculated by Burrau in 1927

the electron's energy eigenvalue as part of the effective central potential in which the nuclei vibrate and rotate, and solves for the nuclear eigenfunctions and eigenenergies numerically. This is the so-called "clamped-nuclei," or lowest BORN-OPPENHEIMER approximation. More accurate treatments lead to the adiabatic approximation and to non adiabatic, or non-relativistically exact, calculations, usually done variationally. When "full reality" is considered, one must include corrections due to special relativity, quantum electrodynamics, finite nuclear size, hyperfine structure, etc. - in short, all the effects that arise in high-accuracy treatments of the hydrogen atom - and in addition a new effect, non-electromagnetic interactions between the nuclei. Table I summarizes typical sizes of some of these for the HD^+ isotope.

Table I(a) HD^+ Energies

Quantity	Energy (MHz)
Dissociation energy	660,000,000
Lowest vibrational spacing	57,000,000
Lowest rotational spacing	1,300,000
Largest hyperfine splitting	1000

Table I(b) Post- Born-Oppenheimer Theoretical Corrections to Lowest Vibrational Spacing in HD^+

Correction	Shift (MHz)
Adiabatic	-15,000
Nonadiabatic	- 4500
Relativistic	+ 1000
Radiative	- 150

$1\ cm^{-1} = 30,000$ MHz

Although most of these effects are very small, ultimately they must be included to bring theory and experiment into accord. This process will provide a stringent test of both experimental techniques and theoretical calculations in a system other than the familiar hydrogen atom. While it is not now clear what the accuracy limits are for either theory or experiment, it is clear that they have not been reached in either direction. Ultimately, of course, the possibility of agreement will be restricted by our knowledge of the fundamental constants. By a slight alteration of emphasis, new constant values can then be obtained. As we will see, this point is already quite close.

The only electronic state of the hydrogen molecular ion that is strongly bound and readily accessible (and the only one that has been observed) is the ground state, $1s\sigma\,^2\Sigma_g^+$. Thus it is logical to study the vibration-rotational spectrum of this level. We have chosen to study the isotope HD^+

because of its large electric-dipole transition moment. In the experiment, HD^+ molecules are processed in three stages, as sketched in Fig.2. The apparatus is diagrammed in Fig.3. The equipment is an optical-frequency

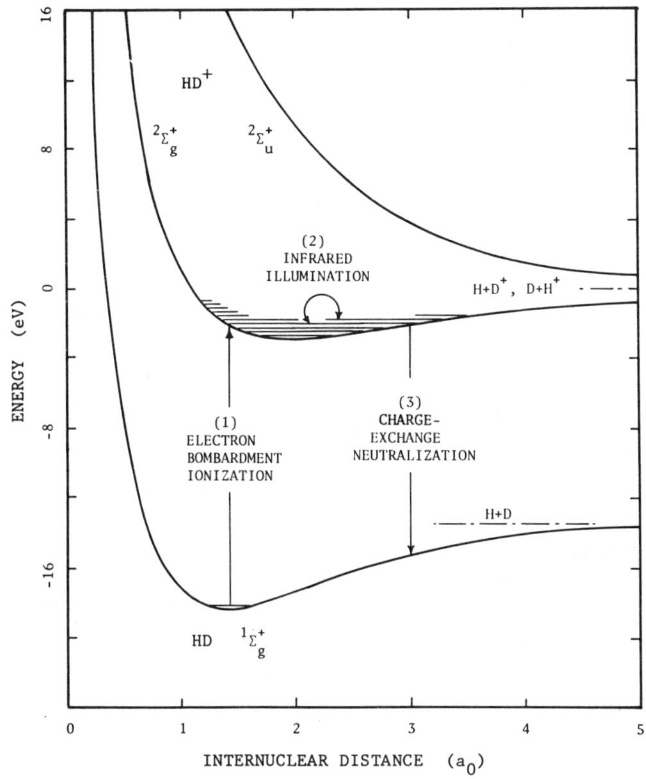

Fig.2 Transition scheme for HD^+ laser resonance experiment

analogue of a molecular beam electric resonance spectrometer, but each part is, of course, quite different. In an electron gun HD^+ ions are formed from HD, and quickly extracted and accelerated before gas-phase reactions can occur. Vibrational populations are proportional to FRANCK-CONDON factors for HD ionization, while rotational populations are determined by the BOLTZMANN distribution at the source gas temperature of about 300°K. Relatively little angular momentum (hence little heat) is added during the ionization process. Next the ion beam is illuminated by a nearly-colinear CO laser beam. The ion beam velocity is adjusted to bring a CO laser line frequency into resonance with a nearby HD^+ transition frequency via the DOPPLER effect. An extra benefit of the fast-beam method is that the velocity spread in the moving beam is reduced by a factor of 20 to 50 from that in the ion source. This is a readily obtainable result of the acceleration kinematics. Thus quite a narrow resonance is seen.

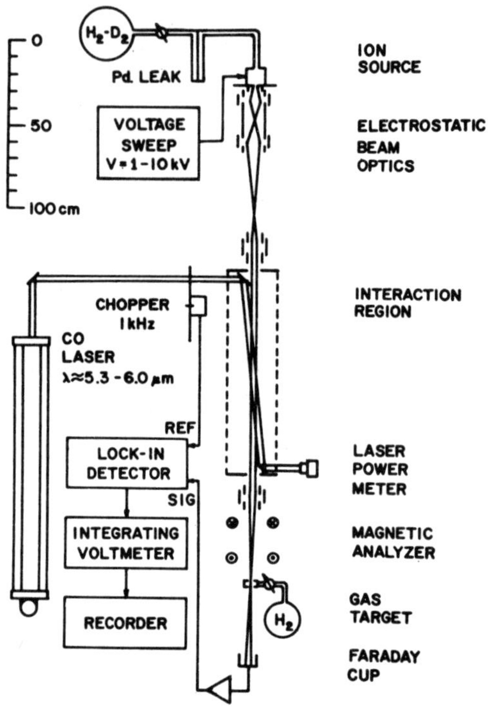

Fig.3 Laser beam-ion beam resonance apparatus

If the populations of the resonating states are unequal, as is generally the case, a net transfer will occur, allowing the possibility of detection. Then the HD^+ frequency is calculated from the CO frequency and the relativistic DOPPLER shift.

Because of extremely low population densities in the ion beam, it seemed impractical to look directly for optical absorption or emission at resonance. Instead we pass the irradiated ion beam into a gas target chamber where a fraction of the ions are neutralized by charge exchange. If ions in the two resonating states have unequal charge-exchange cross sections, it is easy to show that when the laser beam is turned on the charged-particle fraction of the beam will change slightly in intensity.

In our case, the beam current is typically 3×10^{-7} A and the resonance signal typically 3×10^{-6} of that. A 1-kHz laser beam chopper and a lock-in amplifier permits detection of signals of this size, if attention is paid to minimizing the ion beam noise. We have succeeded in keeping it near the shot-noise level. The choice of target gas is not critical; signals have been seen using N_2, Ar, and H_2 targets.

Figure 4 shows a relatively good-quality resonance trace. A frequency

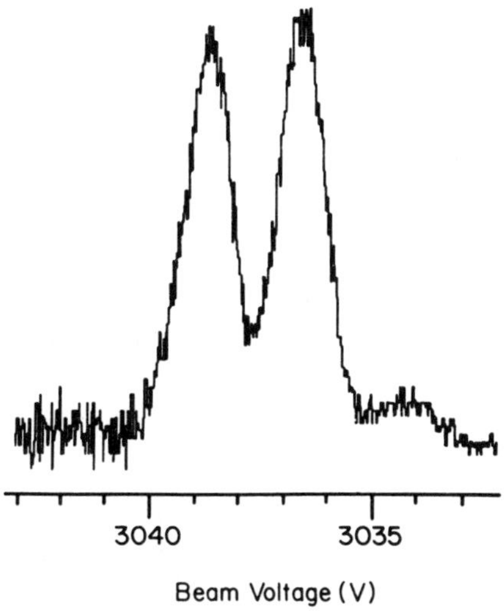

Fig.4 Trace of the (v=1,N=1) - (v=0,N=2) HD^+ resonance. Integrating time, 8 sec. See text for notation

scale can be established from the 30-MHz splitting of the two large peaks. The structure results from hyperfine splitting of the vibration-rotational energy levels. Since four angular momenta are present (electron spin, proton spin, deuteron spin, and molecular rotation) the pattern is quite complicated. More than a dozen individual resonances contribute, most of which are unresolved. The main splitting shown here results primarily from the electron-spin-rotation interaction. However, in other states the appearance is often quite different.

The hyperfine structure has been analyzed theoretically by RAY and CERTAIN, whose theoretically reconstituted resonance curve agrees reasonably well with Fig.4. The differences that exist most likely result from errors in the exact lineshapes and intensities used for the individual resonance components.

More than a dozen different transitions have been found so far, involving vibrational quantum numbers v and rotational quantum numbers N both as high as 4. An accurate comparison with the latest theory by BISHOP and CHEUNG [5] is possible in three cases, as shown in Table II. The experimental

Table II Experiment vs. Theory (Units: cm^{-1})

Transition (v,N) - (v',N')	Preliminary Observed Frequency 1977 (E)	Bishop & Cheung Theory 1977, 515-Term Basis (T)	Difference (E-T)
(1,0) - (0,1)	1869.1339	1869.1351	-.0012
(1,1) - (0,2)	1823.5326	1823.5342	-.0016
(2,1) - (1,0)	1856.7786	1856.7815	-.0029
Estimated Uncertainty	±0.0007	±0.002	

values are the results of a preliminary analysis of data which is more accurate than that used in our original 1976 Physical Review Letter [1]. It can be seen that agreement within 1 part in 10^6 has been achieved, and also that the disagreement that remains increases systematically with v and N.

It is possible to show that a vibrational frequency is approximately proportional to the fundamental constant combination $cR(m_e/M_{12})^{1/2}$ where c = speed of light, R = Rydberg constant, m_e = electron mass, and M_{12} = nuclear reduced mass = $M_1M_2/(M_1+M_2)$. The proton-deuteron mass ratio is known to a few parts in 10^8, while the electron-proton ratio is uncertain to ±4 parts in 10^7. c and R are much better known. Thus, considering the square root, the theoretical vibrational frequency has an uncertainty limit of ±2 parts in 10^7, or .0004 cm^{-1}, from this source. Put another way, the theory is only about a factor of 5, and the experiment a factor of 2, away from a significant redetermination of the electron-proton mass ratio.

A number of benchmarks on the way to high accuracy in both experiment and theory are shown in Fig.5. This figure was taken from a research proposal. What was proposed was a two-quantum sub-DOPPLER modification of the experiment. If it succeeds, it appears likely to put us in an accuracy range in which experimentally all the fundamental constant uncertainties, as well as infrared calibration standards, will be the limits, and theoretically, hitherto unconsidered effects will be significant.

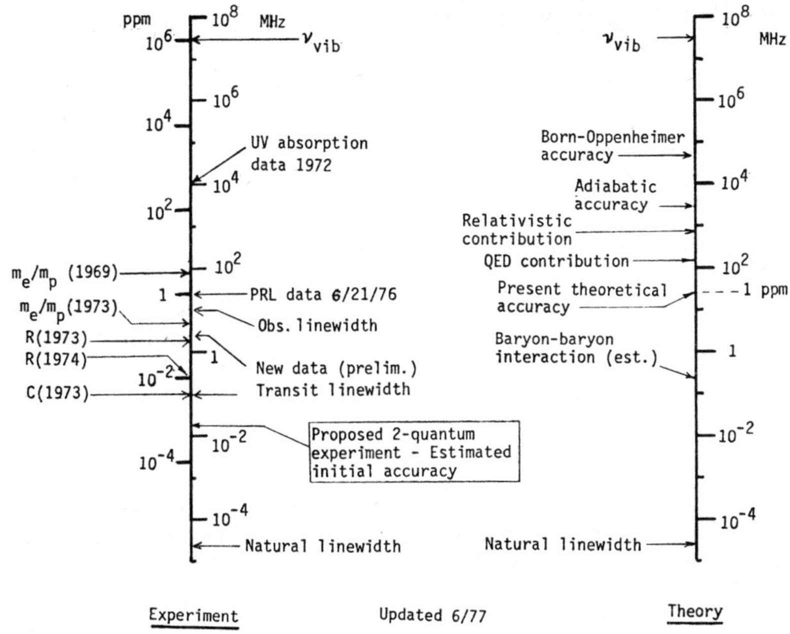

Fig.5 Benchmarks on the way to high experimental and theoretical accuracy for HD^+ vibration-rotational frequencies. (1969)=[6]; (1973)=[7]; (1974)=[8]

References

1. W. H. Wing, G. A. Ruff, W. E. Lamb, Jr., and J. J. Spezeski, Phys. Rev. Lett. 36, 1488 (1976).
2. L. Pauling and E. B. Wilson, Introduction to Quantum Mechanics (McGraw-Hill, New York and London, 1935), Sec. 42.
3. Ø. Burrau, Det. Kgl. Danske Vid. Selskab 7, 1 (1927).
4. R. D. Ray and P. R. Certain, Phys. Rev. Lett. 38, 824 (1977).
5. D. M. Bishop and L. M. Cheung, Phys. Rev. A, August, 1977.
6. B. N. Taylor, W. H. Parker, and D. N. Langenberg, Rev. Mod. Phys. 41, 375 (1969).
7. E. R. Cohen and B. N. Taylor, J. Phys. Chem. Ref. Data 2, 663 (1973).
8. T. W. Hänsch, M. H. Nayfeh, S. A. Lee, S. M. Curry, and I. S. Shalin, Phys. Rev. Lett. 32, 1336 (1974).

II. Multiple Photon Dissociation

SELECTIVE DISSOCIATION OF POLYATOMIC MOLECULES BY TWO INFRARED PULSES

R.V. Ambartzumian, G.N. Makarov, and A.A. Puretzky

Institute of Spectroscopy, Academgorodok, Podolski Rayon
Moscow 142092, USSR

Introduction

Recently, the collisionless multiple photon dissociation of polyatomic molecules by infrared lasers has become one of the most interesting and intriguing problems in laser physics and laser chemistry. The possibility that an isolated molecule, placed in the intense infrared field of a laser, can absorb many photons (enough to dissociate as a result of absorption) in a very short time span, was first suggested in 1971 by Isenor and Richardson [1] based on their observations and studies of the prompt luminescence which was observed in the focal and near focal volume of the lens that focused the IR laser radiation into the absorption cell. Interest in this phenomenon was heated by the observation that the multiple photon dissociation process under certain conditions is isotopically selective [2,3]. In these reports, extremely large enrichment factors of boron and sulfur isotopes were reported. Later, these results were successfully duplicated in many laboratories [4,5].

From a scientific point of view, the multiple photon dissociation process is an extremely intriguing problem. In order to reach the dissociation threshold, a single molecule must absorb at least several tens of infrared photons. If a nonlinear process were responsible for the absorption of several tens of photons simultaneously, one would expect the required laser intensity to be very close to the gas breakdown threshold, namely 10^9-10^{10} W/cm^2. Therefore, it was quite surprising, when it was found [6] that laser intensity of only $\sim$20 MW/cm^2 (pulse energy $\sim$2 J/cm^2) is needed for multiple photon dissociation. Then, the questions which must be answered are (a) what is the physical mechanism through which a single molecule can absorb so many photons of the same frequency with a large probability, (b) how many quanta are really absorbed by a molecule prior to dissociation, (c) how is the absorbed energy distributed in the excited molecule over the vibrational manifold before dissociation, and (d) what are the dynamics of the dissociation process.

Theoretical explanations of multiple photon excitation to high lying vibrational levels were offered in several papers [7,8]. Much more became known about the last stage of dissociation processes from the beautiful molecular beam experiments where the dynamic behavior of the dissociation process was studied and the primary dissociation fragments were identified [9]. In these experiments, it was shown that for molecules which are near the dissociation threshold the absorbed energy is completely randomized prior to the dissociation and the dissociation process can be adequately explained in terms of the simple RRKM theory of unimolecular reactions. However, it is still not known at which vibrational levels this randomization occurs, what are the rates of intramolecular collision free energy transfer, and how do these rates depend on the vibrational excitation level.

The above mentioned papers described the behavior of a molecular system in a monochromatic, single frequency field. Further progress in understanding the physical processes involved in multiple photon absorption and dissociation is connected with the multiple photon dissociation of polyatomic molecules by two or more infrared pulses [10].

Two IR Pulse Dissociation

Careful measurements of absorbed energy made by a direct calorimetry technique showed that the amount of energy that is absorbed by an isolated molecule from the infrared laser field is a smooth function of the incident energy as shown in Fig. 1.

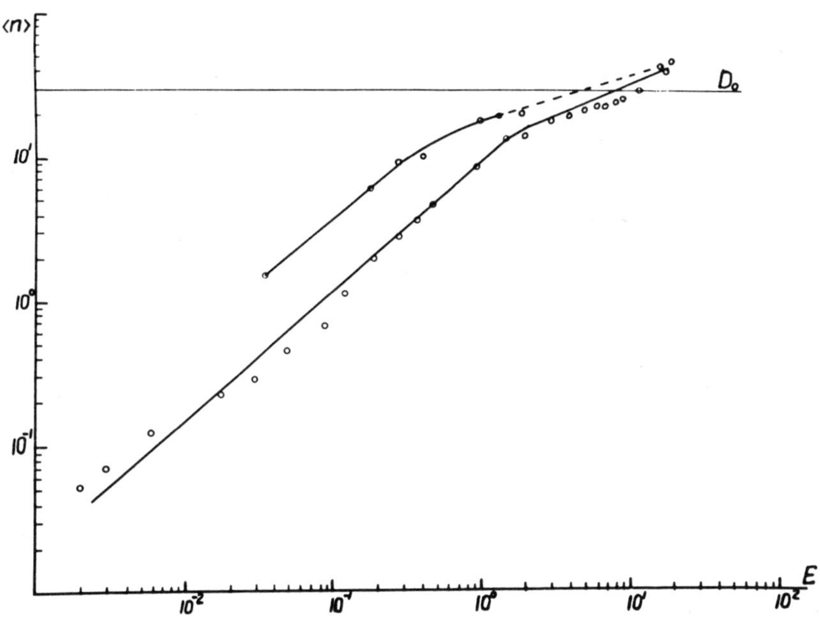

Fig. 1 Dependence of the number of absorbed CO_2 quanta per molecule, <n>, on energy flux for the case of SF_6. The upper curve is for 0.5 nsec pulses [11], the lower curve corresponds to pulse lengths of 100 nsec. The data for the lower curve is taken from Refs. 10,11, and 12. D_0 is the energy required for dissociation of SF_6.

The quantity <n>, which is shown in Fig. 1, is defined by

$$\langle n \rangle = \frac{E_{abs}}{\hbar\omega N_0} \tag{1}$$

where E_{abs} is the energy absorbed by molecules from the beam, N_0 is the total number of molecules which are irradiated, and $\hbar\omega$ is the energy of an IR quantum. Data were usually taken at SF_6 pressures of $\sim$0.1-0.2 torr to minimize effects due to collisions. The upper curve corresponds to a pulse length of 0.5 nsec [11], and the lower curve corresponds to a pulse length $\sim$100 nsec. The data for this curve are taken from [10b,11,and 12].

One can see that the accumulation of absorbed energy is a very smooth function of incident energy without a sharply pronounced threshold. From this data, one can see that there is no anharmonicity bottleneck for the excitation process up to the high vibrational levels. For example, with 100 nsec pulses and $\langle n \rangle = 5$, the intensity of the beam is only $\simeq$5 MW/cm^2, and power broadening cannot compensate for the anharmonicity of the vibrational transitions, even for the levels $v=1 \rightarrow v=2$ in the ν_3 mode of SF_6.

On the other hand, the yield of dissociated SF_6 exhibits very sharp threshold behavior, in the range from 1.4-2.0 J cm^{-2}, depending on the excitation frequency [5,6,13].

The model of excitation and dissociation via multiple photon absorption can be roughly divided into three parts: (a) excitation of several low lying vibrational levels is accomplished through multiquantum processes; (b) above the vibrational levels $v>$ or $\gg 3$, the vibrational level density in polyatomic molecules is high enough that there is no need for radiation to be in resonance with a specific vibrational mode, and further vibrational excitation is performed through the vibrational quasicontinuum; (c) the third state of dissociation is connected with the dissociation process itself. The observed thresholds [5,6,13] were attributed to saturation of certain, though unknown, vibrational levels in the quasicontinuum of vibrational states.

If the proposed model of excitation is valid, then it is obvious that the first stage and the subsequent stages can be performed by two IR pulses at different frequencies. The first pulse, with its frequency in resonance with vibrational transitions of the molecule, excites the molecule to the intermediate vibrational states, where the quasicontinuum starts. The second pulse, which can be off-resonance from the ground state absorption, can deliver energy to the excited molecule which is needed for its dissociation [10].

This technique can be very useful for obtaining (a) spectral data of the vibrational quasicontinuum through which the molecules reach the dissociation threshold; (b) the excitation function to the vibrational states from which the molecules can further absorb energy up to the dissociation limit; (c) information concerning the energy flow rates between the vibrational manifold, (e.g. rates of V-V exchange between distinct molecules or different isotopic species). This type of dissociation also can be extremely useful for the application of the multiple photon dissociation technique to isotope separation. It is very important in those cases when the value of isotope shift, $\Delta\nu_{isotope}$, is comparable or even less than the power broadening, $\Delta\nu_{pb}$, at the threshold value of the laser field required for dissociation of certain types of molecules. And as it will be clear from the data presented below, the two IR pulse dissociation technique makes the isotope separation process easily scalable, permitting the utilization of most of the laser energy.

Experiments

Two line tunable CO_2 TEA lasers were used to study the dissociation by two

IR pulses. The frequencies of both lasers could be tuned independently. The pulse shapes of both lasers were essentially the same (100 nsec spike followed by 400-500 nsec tail). The energy in the tail was roughly 1/3 of the total pulse energy. The dissociation yield was measured by infrared spectrophotometry for SF_6 and OsO_4 molecules, as well as the amplitude of the visible liminescence resulting from OsO_4 photodissociation as described below. The two pulses propogated in opposite directions through the cell (no difference was observed when both beams propogated in the same direction). The beams were unfocused, in order to minimize spatial effects. To obtain 12-15 nsec FWHM pulses, optical breakdown of air was used with further recollimation of the focused beam.

It was found that the peak intensity of visible luminescence in OsO_4 is linearly proportional to the dissociation yield which was measured by IR spectrophotometry. This direct proportionality between dissociation yield measured by the IR technique and the visible luminescence intensity was checked over a wide dynamic range for various combinations of frequencies ν_1 and ν_2. The dependence of the visible luminescence signal showed all of the properties which are intrinsic to unimolecular reactions: independence of pulse risetime on pressure of OsO_4 in the cell (from 10^{-5} to 0.25 torr), pressure dependence, etc.

Though the nature of this luminescence is not yet known, it reflects the unimolecular process connected with the disappearance of molecules. This permitted single pulse measurements of the dissociation yield (by measuring the luminescence intensity) instead of the lengthy irradiation of sample cells and measurements of IR absorption of OsO_4. The method also avoids difficulties connected with the chemistry of the dissociation fragments.

Experimental Results

A. Dissociation yields

Figure 2 shows the dependence of the visible luminescence intensity (dissociation yield) for OsO_4 when dissociation is performed by two IR pulses of different frequencies. The highest dissociation yield corresponds to 1% per pulse. The measurements were performed with different CO_2 excitation and dissociation frequencies and energy dependencies are shown for (1) a fixed amount of energy in the excitation pulse which is in resonance with the ν_3 absorption band of OsO_4, and (2) for a fixed energy at the frequency ν_2, which was off resonance from the ν_3 absorption band. As one can see, no threshold behavior is observed when the dissociation frequency ν_2 was tuned to the red of the ν_3 band. This differs from the situation of SF_6 molecules, where the same threshold behavior was observed as for the single frequency dissociation case. An important difference between the SF_6 and OsO_4 cases is in the relative position of the ν_2 dissociation frequency, which was tuned to the blue of the ν_3 absorption in the case of SF_6.

The observed high yields allowed us to take measurements in directly propogating light beams at relatively low intensities, therefore avoiding spatial effects.

B. Spectral Measurements

Figure 3 shows the luminescence intensity vs. the frequency of the excitation pulse, ν_1, with ν_2 fixed to the red of the OsO_4 absorption as shown.

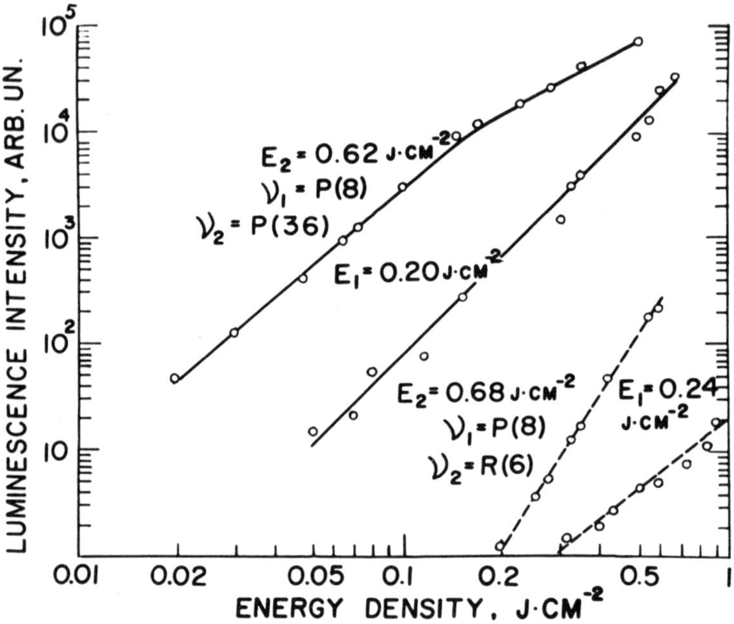

Fig. 2 Luminescence intensities in OsO_4 and dissociation yields per pulse vs. energy flux for different pairs of ν_1 and ν_2. The fixed energies of the pulses are shown on the figure. The highest point corresponds to 1% dissociation yield. OsO_4 pressure was 0.15 torr.

One can easily see that this dependence shows several peaks and minima. It should be noted that the curve shown in Fig. 3 only connects the points where the measurements were taken (line-to-line tuning of the CO_2 laser), and that the real dependence might be more complex. These observed peaks may represent branches of multiquantum transitions corresponding to variations of the J values in the initial and final vibrational states. Also occasional resonances between successive vibrational levels should not be neglected [10]. The peaks are more pronounced in the case shown here than in our previous results [14], when similar measurements were made at higher laser intensities.

A final assignment of the peaks is rather difficult because a substantial contribution to this dependence arises from transitions in the vibrational quasicontinuum which have their own frequency dependence. This kind of dependence is shown in Fig. 4, where the dependence of the luminescence intensity vs. ν_2 (dissociation pulse) was measured. The frequency which initially excited OsO_4 molecules, ν_1, was fixed. The curve shows a sharp increase of the dissociation yield when ν_2 was shifted to the red [10,15]. This result is more or less trivial, because it shows only that vibrationally excited molecules, due to anharmonicity and/or anharmonic splitting, have absorption bands which are wider and shifted to the red compared to the room temperature absorption features of the unexcited molecules [16].

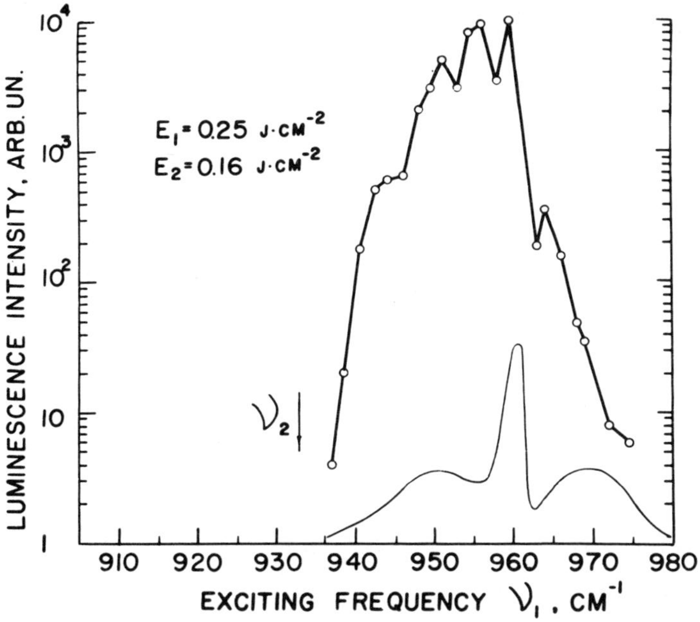

Fig. 3 Dependence of the luminescence intensity on the frequency of the excitation pulse, ν_1, with ν_2 fixed. The energies in both pulses were kept constant. OsO_4 pressure was 0.15 torr. The delay between ν_1 and ν_2 was 500 nsec.

C. Kinetic Studies

A serious question is whether the spectrum shown in Fig. 4 represents transitions from vibrationally heated molecules (energy is randomized over vibrational manifold) or whether it represents transitions from the vibrationally excited mode which is preferentially excited. For these two cases the spectrum of transitions to the dissociative state should be different because the various vibrational modes have different anharmonicity, etc.

It is believed at this time that randomization occurs on a picosecond time scale for SF_6 and that randomization occurs from $v \sim 5$-6 of the ν_3 mode (E. Yablonovitch and N. Bloembergen, this volume). However, some experimental results on the spectroscopy of high lying vibrational levels (overtones) contradict this point of view. For example, the $0 \rightarrow 3\nu_3$ absorption of SF_6 consists of only Doppler broadened rotational transitions (A. Paine, this volume). Also, intracavity absorption spectroscopy of NH_3 and C_2H_4 in the 6000 Å region show sharp, Doppler broadened rotational lines. These results show that the randomization time from the levels corresponding to the excitation energy is rather long. Of course, for different molecules the randomization time for a particular energy level will vary.

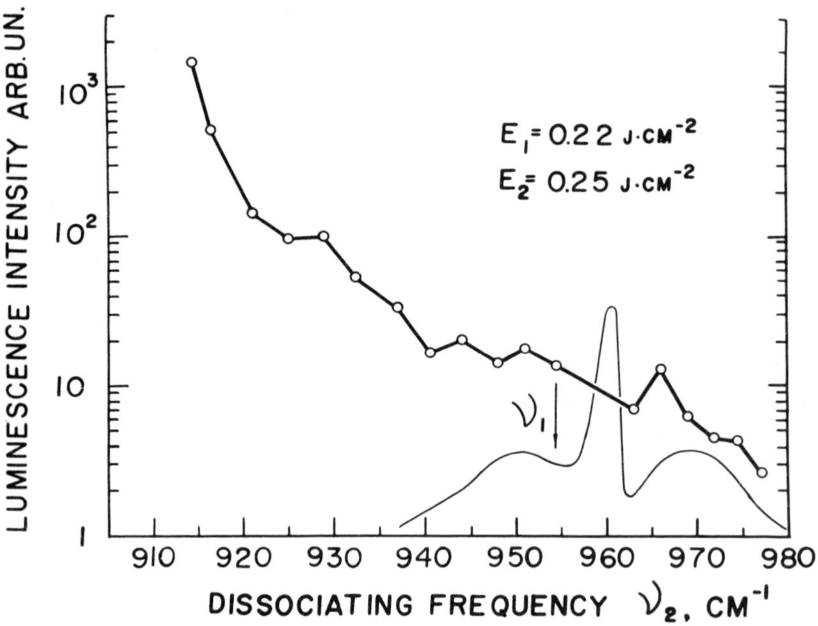

Fig. 4 Dependence of the luminescence intensity of the frequency of the dissociation pulse, ν_2, with ν_1 fixed. The signal level in the absence of the ν_2 pulse was at least 10 times less than the lowest observed value. OsO_4 pressure was 0.15 torr. The delay between ν_1 and ν_2 was 500 nsec.

The two IR pulse dissociation technique permits the measurement of this effect. In our experiments, we have used a very short excitation pulse at ν_1 (12-15 nsec FWHM), and delayed the dissociation pulse at ν_2 (100 nsec FWHM). The observed dependence of luminescence intensity on the delay between these two pulses is shown in Fig. 5. It was found that in the region 0-1 μsec delay, the decay rate is approximately independent of OsO_4 pressure while the long tail which follows is pressure dependent. One can also see that the slope of the pressure independent part varies slightly with pump intensity. At the lowest gas pressure, the pressure independent part takes a goose-head shape. The only explanation of such behavior that we can offer at this time is that the dissociation cross section, for the case when a particular mode is excited, differs from that when the vibrational energy is randomized over the vibrational manifold. As discussed above, the observed pressure independent decay rate is the randomization rate. It is impossible to say which vibrational levels are pumped by the pulse at ν_1.

D. Osmium Isotope Separation

The possibility of achieving dissociation with low intensity fields at ν_1 and ν_2 enabled isotope selective dissociation of OsO_4 by two infrared pulses to be accomplished when the sum of both pulse intensities is significantly lower than the single frequency threshold value ($\sim 3J/cm^2$)[18].

The experiments on isotope separation were done with OsO_4 of natural abundance. The pressure of OsO_4 in the cell was 0.3 torr, with 2.0 torr of OCS added as a scavenger. Observation of fast reverse reactions forced us to keep the energy of both pulses $\sim$1 J/cm^2 in order to dissociate OsO_4 on a time scale comparable to that of the reverse reaction. In each experiment the burnout level of OsO_4 was 90%. The enrichment was measured in the residual OsO_4. The results of mass spectrometer analyses are given in Table 1. The results of dissociation by an intense single freqeuncy laser is also shown there.

Comparison of the results of selective dissociation of OsO_4 with the laser intensity near threshold (single frequency case [19]) with the results described here shows that for the two infrared pulse technique the selectivity increases at least by a factor of 5-6. Estimates show that an additional decrease of intensity at ν_1, and the use of a laser with a higher repetition rate (to discriminate against reverse reactions) can further increase the selectivity of dissociation. We would like to emphasize that this experiment was performed with directly propagating beams (see also Fig. 2), and that this makes the isotope separation process quite scalable. It is possible that by using three or more frequencies, the selectivity of the excitation and dissociation can be increased by an order of magnitude.

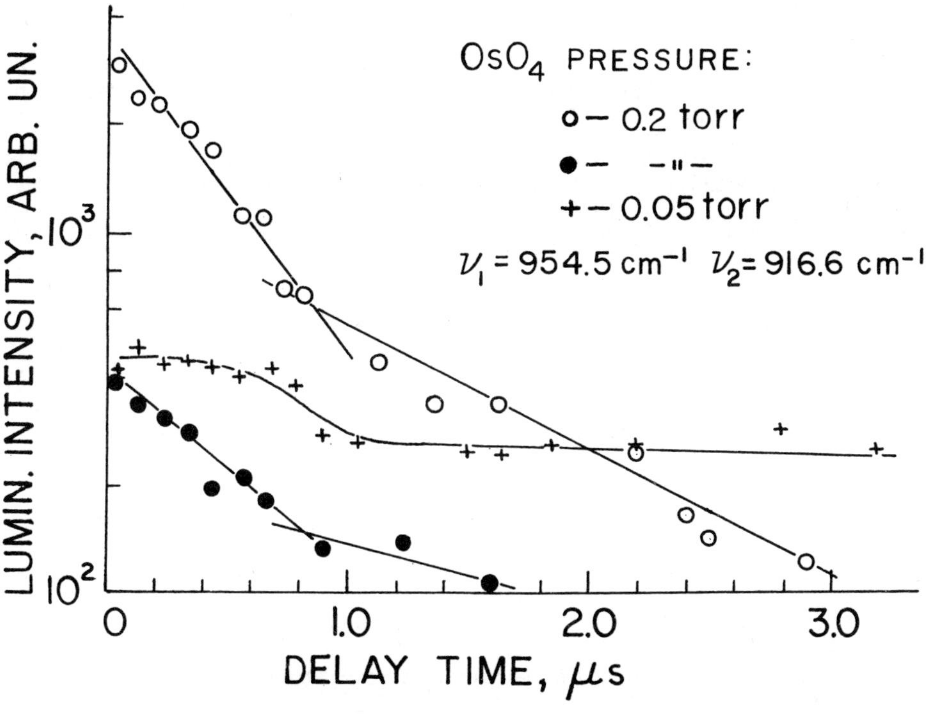

Fig. 5 Dependence of the luminescence intensity on the delay between the pulses at ν_1 and ν_2. The frequencies and gas pressure are shown on the figure. The two upper curves correspond to an energy flux of 0.025 J/cm^2, the lowest curve corresponds to an energy flux of 0.013 J/cm^2.

Table 1 Mass-spectrometer analyses

Laser frequencies		Enrichment Factor $\dot{K}$(±0.02)				
ν_1	ν_2	$^{192}Os/^{190}Os$	$^{192}Os/^{189}Os$	$^{192}Os/^{188}Os$	$^{192}Os/^{187}Os$	$^{192}Os/^{186}Os$
P(6)	P(20)	1.11	1.13	1.24	1.48	1.58
P(6)	P(12)	1.04	1.04	1.08	1.04	1.02
P(20)	P(20)	1.02	1.02	1.08	1.00	1.01
P(4)	P(20)	1.08	1.10	1.17	—	—
P(4)-one frequency case I∿10J/cm^2		1.01	1.01	1.01	1.01	1.01

Acknowledgment

The authors would like to thank V. S. Letokhov, Yu. A. Gorokhov, and N. P. Farzinov for many useful discussions.

References

1. N. R. Isenor, V. Merchant, R. S. Hallsworth, and M. C. Richardson, Can. Journ. Phys. 51, 1281 (1973).
2. R. V. Ambartzumian, V. S. Letokhov, E. A. Ryabov, and N. V. Chekalin, JETP Lett. 20, 515 (1974).
3. R. V. Ambartzumian, Y. A. Gorokhov, V. S. Letokhov, and G. N. Makarov, JETP Lett. 21, 171 (1975).
4. J. L. Lyman, R. J. Jensen, J. Rink, C. P. Robinson, and S. D. Rockwood, Appl. Phys. Lett. 27, 87 (1975).
5. M. C. Gower, K. W. Billman, Optics Comm. 20, 123 (1977).
6. R. V. Ambartzumian, Y. A. Gorokhov, V. S. Letokhov, G. N. Makarov, A. A. Puretzky, JETP Lett. 22, 177 (1975).
7. D. M. Larsen, C. Cantrell, and N. Bloembergen, in "Tunable Lasers and Applications," Proceedings of Nordfjord Conference, edited by A. Moordian, T. Jaeger and P. Stokseth, (Springer-Verlag, 1976) and references therein.
8. V. S. Letokhov, A. A. Makorov, Optics Comm. 17, 250 (1976).
9. E. R. Grant, P. A. Schulz, Aa. S. Sudlo, M. J. Coggiola, Y. R. Shen, and Y. T. Lee, Phys. Rev. Lett. 38, 17 (1977), and paper presented to the International Conference on Multiphoton Processes Proceedings, University of Rochester, Rochester, N.Y., June 6-9, 1977.
10. R. V. Ambartzumian, Y. A. Gorokhov, V. S. Letokhov, G. N. Makarov, N. P. Furzikov, A. A. Puretzky, a) JETP Lett. 23, 194 (1976), b) Optics Comm. 18, 517 (1976).
11. J. G. Black, E. Yablonovitch, N. Bloembergen and S. Mukamel, Phys. Rev. Lett 38, 1131 (1977).
12. R. V. Ambartzumian, Y. A. Gorokhov, V. S. Letokhov and G. N. Makarov, JETP 69, 1956 (1975) (Russian).
13. P. Colodner, C. Winterfield and E. Yablonovitch, Optics Comm. 20, 119 (1977).
14. R. V. Ambartzumian, Y. A. Gorokhov, N. P. Furzikov, V. S. Letokhov, G. N. Makarov and A. A. Puretzky, Optics Letters 1, 22 (1977).

15. R. V. Ambartzumian, Y. A. Gorokhov, N. P. Furzikov, G. N. Makarov and A. A. Puretzky, Kvantovay Electionica N7, 1977.
16. J. L. Lyman and A. V. Nowak, Journ. Quant. Specta. Radiative Transfer, 15, 945 (1975).
17. E. Borik, V. N. Koloshnikov, E. Antonov and V. Mironenko, Proceedings of XVIII All Union Spectroscopy Conference, Gorkiy, 1977.
18. R. V. Ambartzumian, N. P. Furzikov, Yu. A. Gorokhov, G. N. Makarov, and A. A. Puretzky, Chem. Phys. Lett. 45, 231 (1977).
19. R. V. Ambartzumian, Y. A. Gorokhov, V. S. Letokhov, and G. N. Makarov, JETP Lett. 22, 43 (1975).

COLLISIONLESS MULTIPHOTON DISSOCIATION OF SF_6: A STATISTICAL THERMODYNAMIC PROCESS

N. Bloembergen and E. Yablonovitch

Division of Applied Sciences, Harvard University
Cambridge, MA 02138, USA

1. Introduction

The interaction of an SF_6 molecule with strong infrared CO_2 laser pulses has been the subject of intense theoretical and experimental investigation in many laboratories. Several review papers [1,2] have appeared, summarizing the state of the art in 1976, but new results are still being published in abundance [3-8]. In this paper we shall restrict ourselves to results where the effect of collisions is clearly eliminated, i.e., data obtained with molecular beams [3,4], or with very short laser pulses in vessels at sufficiently low pressure [5,6]. The problem, how one isolated SF_6 molecule bathed in a sea of infrared photons, absorbs more than thirty quanta necessary for dissociation, may be discussed with the aid of the schematic diagram of vibrational energy levels shown in Fig.1. Three regions may be distinguished. The molecule is excited from the ground vibrational state in a coherent multiphoton process, in which three to six quanta are involved [9]. This process is frequency - and therefore isotope - selective. The data in Fig.2 show that the dissociation rate, or dissociation probability per pulse, is nearly independent of pulse duration for constant pulse energy fluence. The dissociation process does not exhibit the high power dependence on the peak intensity, as a true multiphoton transition would have. The transitions in region I are therefore not the dissociation rate limiting step.

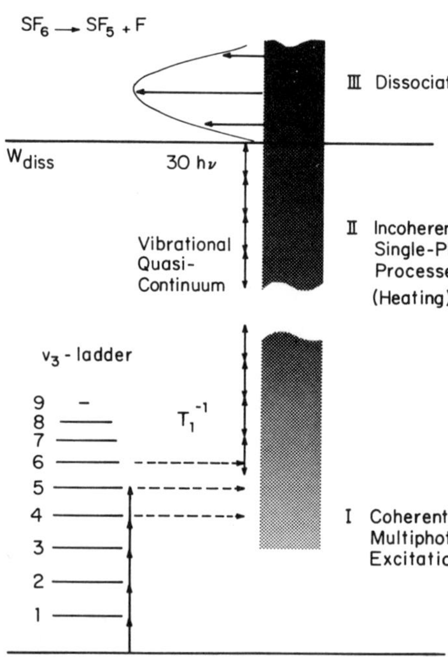

Fig.1 A schematic energy level diagram for the infrared excitation and dissociation of the SF_6 molecule

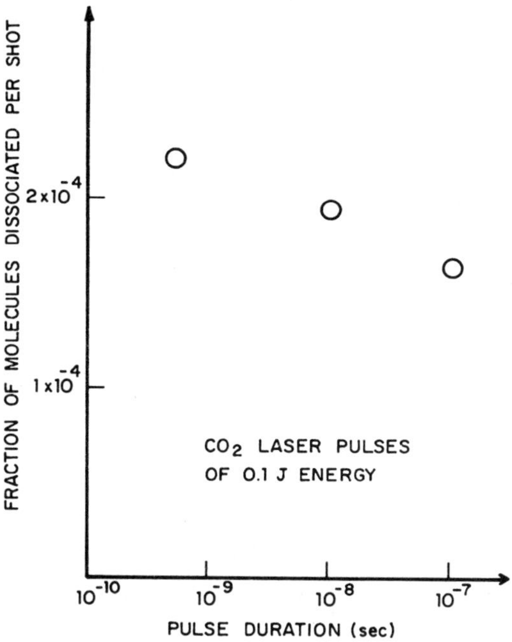

Fig.2 The dependence of dissociation rate on laser pulse duration, for constant pulse energy (from KOLODNER et al. [5])

The dissociation rate is a very steeply rising function of the energy fluence, as shown in Fig.3. Because the experimental detectivity sets a lower limit of about 10^{-4} for the dissociation probability, one may talk about an energy fluence threshold of about 1 Joule/cm^2. About one order of magnitude above this threshold, the dissociation probability per pulse approaches its saturation value of unity. For a pulse duration of 0.5×10^{-9} sec, the intensity at the dissociation threshold is about 2×10^9 watts/cm^2. At this intensity the effective Rabi frequency, $\hbar^{-1}|\mu||E| \sim 10$ cm^{-1}, is comparable to the width of the rotational band and to the anharmonic shift for a transition to the vibrational level $v_3 = 5$. Under these conditions the question of compensation of anharmonicity by rotational energy or by Coriolis splittings of degenerate vibrational states loses its importance [10]. Every molecule, regardless of its initial rotational-vibrational state, will acquire an energy excitation of 5000 cm^{-1} or more in a short fraction of the pulse duration. For longer pulses, at the same energy fluence, only a fraction of molecules will be able to reach the quasi-continuum of vibrational states.

In this paper emphasis is placed on this second region with energy excitation of the vibrational manifold. In the next section evidence is presented that the molecule undergoes a sequence of one-photon absorption and stimulated emission processes. Energy is rapidly exchanged with the other vibrational degrees of freedom through higher order anharmonic coupling terms. An effective vibrational temperature may be introduced.

This same temperature may be used in the description of the dissociation in regime 3. The molecules in the statistical distribution with an energy exceeding the dissociation limit W_{diss} will dissociate at a rate, given by the standard statistical kinetic theories of unimolecular reaction rates. It will be shown in section 3 that the rates so calculated are consistent with the experimental data shown in Fig.3.

The conditions under which high mode-selective vibrational excitation could be achieved are discussed in a concluding section.

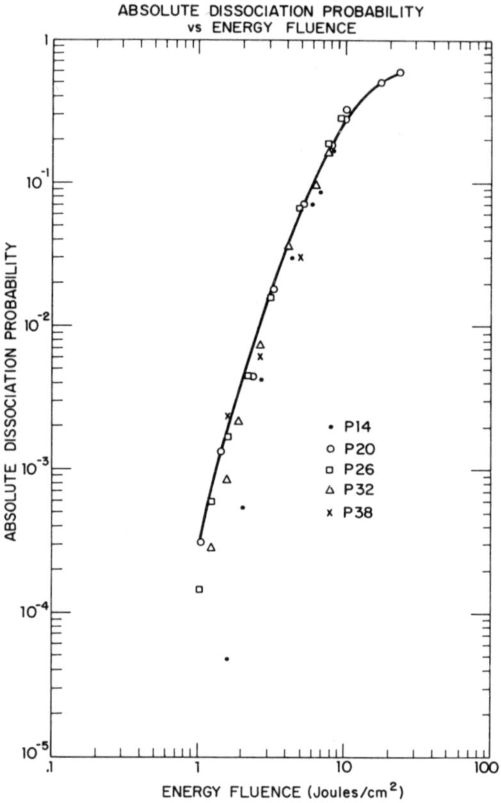

Fig.3 Experimental dissociation probability of SF_6 as a function of energy fluence, for different CO_2 laser line frequencies. The P(14) line is at a higher frequency, where the absorption cross section in the quasi-continuum is smaller

2. Vibrational Heating of Polyatomic Molecules

Below the dissociation threshold the energy absorbed by the molecules can be observed by direct attenuation of the transmitted pulse. For this method the pressure in the cell has to be fairly high in order to obtain sufficient optical density. Only with the shortest pulses will the collisionless condition be satisfied. For lower densities and pulse energy fluencies, the amount of energy absorbed by the molecules can be determined by detecting the thermal pressure wave with a condenser microphone. This signal develops after the laser pulse has passed, in a time of the order of the vibrational-translational relaxation time. This acousto-optic detection method requires a simple collimated beam geometry and accurate calibration [6,11]. Then the average number of infrared photons absorbed per molecule can be determined with an accuracy of about twenty percent. The results of such a determination are shown in Fig.4. The rate of energy absorption may be written as

$$\hbar\omega \frac{d\langle n_v\rangle}{dt} = \sigma(\langle n_v\rangle, \omega) I(t) \tag{1}$$

where $\sigma(\langle n_v\rangle,\omega)$ is the absorption cross section at an average excitation $\langle n_v\rangle \hbar\omega$ and frequency ω, while $I_\omega(t)$ is the intensity in the pulse. Integration over the pulse duration yields

$$\sigma = \frac{d}{d\mathscr{J}} (\langle n_v\rangle \hbar\omega) \tag{2}$$

where

$$\mathscr{J} = \int I(t)dt \tag{3}$$

is the energy fluence in Joules/cm^2. The derived values of σ are indicated by the dashed line in Fig.4. If it is assumed that the absorbed energy is

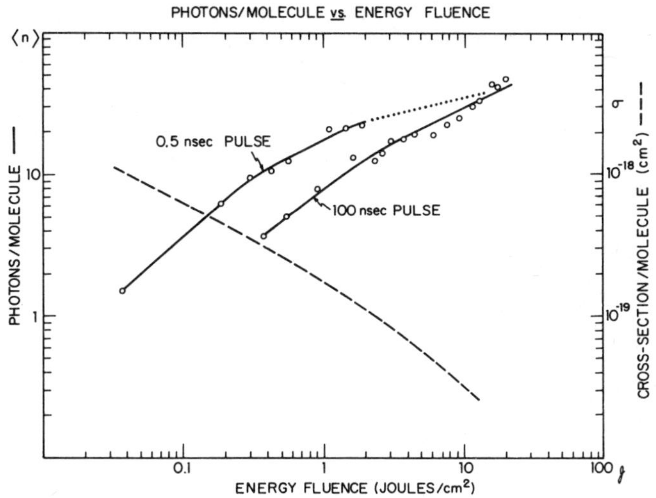

Fig.4 The vibrational excitation per molecule as a function of laser energy fluence. The dashed line (right hand) scale gives the absorption cross section derived from the short pulse data (from BLACK et al.[6])

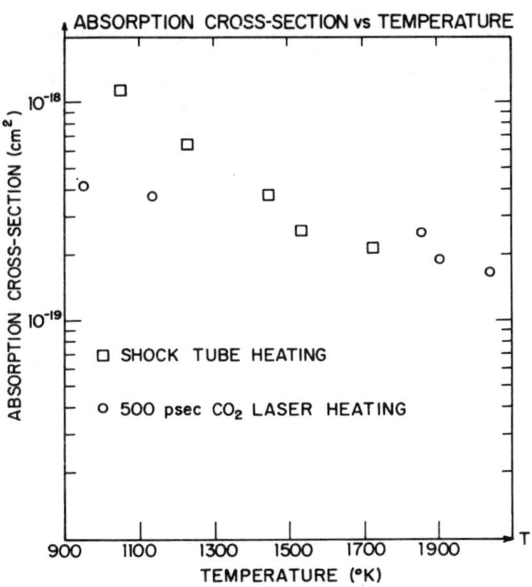

Fig.5 The absorption cross section on the P(20) line as a function of vibrational temperature

distributed rapidly, in a time T_1 short compared to the laser pulse duration, over all 15 vibrational degrees of freedom, a vibrational temperature may be defined by

$$T_{vib} = \frac{\langle n_v \rangle \hbar\omega}{15\ k} \qquad (4)$$

Thus $\sigma(\langle n_v \rangle)$ may be compared with $\sigma(T)$ derived from absorption measurements in shock tubes [12]. This comparison is shown in Fig.5 and reveals that the assumption of thermodynamic equilibrium is consistent with the data for the shortest pulses. The difference for longer pulses must be attributed to the fact that not all molecules can reach the quasi-continuum. The peak intensity is not high enough in the 200 nanosecond pulses, so that only molecules in a certain range of rotational states are excited through region I and can take part in the subsequent heating. The cross section $\sigma(\omega)$ decreases with increasing vibrational temperature, because the absorption maximum shifts toward longer wavelengths and the ν_3-oscillator strength is spread over a wider frequency range. Shock tube data of NOVAK and LYMAN [12] show that for $T = 10^{3}$°K the absorption maximum is at 930 cm^{-1} and the width (FWHM) is about 25 cm^{-1}. The absorption cross section at 940 cm^{-1} therefore decreases with increasing vibrational excitation. The oscillator strength is not distributed uniformly in the quasi-continuum of vibrational states. The down-shifted and broadened ν_3-band is still clearly distinguishable and sits on top of a combination band absorption background with a smaller cross section of about 10^{-20} cm^2. Vibrational heating by irradiation in this background has been demonstrated. It leads to dissociation with a higher energy fluence threshold.

3. Dissociation

It must be emphasized that the heating of the vibrational manifold is the net result of a random sequence of absorption and stimulated emission processes. Some molecules have more than the average energy, and the simplest assumption of true thermodynamic equilibrium is that they are distributed in energy according to the tail of a Maxwell-Boltzmann distribution at a temperature T_{vib}. In this connection it is important to observe that at the dissociation "threshold" of 1.4 Joule/cm^2, the average excitation corresponds to about $\langle n_v \rangle = 20$, which is two-thirds the energy required for dissociation. Some molecules in the tail of the energy distribution may have $n_v > 30$ and are capable of dissociation. The kinetic rate theory for unimolecular reaction is fully developed [13]. It should apply particularly well to the present situation, where a system has been prepared with a high vibrational temperature, without the intervention of any intermolecular collisions. The rotational and translational degrees of freedom remain cold. A molecule with a vibrational energy larger than W_{diss} will probe the vibrational phase space and eventually concentrate its energy in the dissociation coordinate and dissociate.

The probability for a molecule to dissociate would be simply the probability to have an energy larger than W_{diss}, if an infinite time were available. Actually deactivation processes will take place with a characteristic time of about 10^{-5} or 10^{-6} sec due to wall collision and intermolecular collisions with cold molecules. Molecules which exceed W_{diss} by a very small amount (one to three photons) may not have sufficient time to dissociate. Most dissociating molecules will have vibrational excitations between three to nine photons above W_{diss}. The majority of these molecules will dissociate after the short laser pulse, but before a deactivating collision. This fraction of the molecules is called the absolute dissociation probability and is plotted as a function of temperature in Fig.6. Denote the probability for a molecule to have an excitation energy $n\hbar\omega$ at temperature T by

$$P_n(T) = \alpha^n (1 - \alpha)^s \frac{(n - s - 1)!}{n!(s - 1)!} \tag{5}$$

with $\alpha = \exp(-\hbar\omega/kT)$. The dissociation probability is then given by

$$P_{diss} = \sum_{n>30} \frac{k_n}{k_n + k_d} P_n(T) \tag{6}$$

Here $k_d \sim 10^5$ or 10^6 is the deactivation rate, and the dissociation rate for a molecule with vibrational energy $n\hbar\omega$ is given in Kassel's theory [13] by

$$k_n = \omega \frac{n!(n-m+s-1)!}{(n-m)!(n+s-1)!} \tag{7}$$

Here ω is the vibrational or laser frequency, $s = 15$ is the number of vibrational degrees of freedom, $m \approx 30$ is the minimum number of photons needed for dissociation.

The dissociation probability calculated from (5-7) is plotted in Fig.6 as a function of temperature for $m = 25$ and $m = 30$. The results are insensitive to the precise choice of k_d, or to the substitution of the more elaborate RRKM theory for Kassel's expression [13]. These curves may be compared directly with the experimental data in Fig.3. One obtains for each value of the energy fluence an effective temperature T. This relationship may, in turn, be compared with that derived from the calorimetric experiment. The calorimetric data are replotted in Fig.7. An almost perfect agreement is obtained between the temperatures derived from the dissociation and heating experiments for the sub-nanosecond pulses with $W_{diss} = 30\ \hbar\omega$.

The molecular beam data [3,4] also fully support the statistical thermodynamic picture. They show that for short laser pulses the reaction products are exclusively SF_5 and F. The translational kinetic energy and the angular distribution of the reaction products is consistent with absolute reaction rates of about $10^6\ sec^{-1}$ for an excitation of four photons above W_{diss} and of about $10^8\ sec^{-1}$ for eight photons excess over W_{diss}. It is very difficult to excite a molecule more than $10^4\ cm^{-1}$ in excess of W_{diss}, because it finds the dissociation channel for SF_5 and F so rapidly.

The SF_5 dissociation product retains a considerable fraction of the excess energy as vibrational excitation. On the average, it may have about 5 $\hbar\omega$ of vibrational energy and is therefore ready to absorb additional infrared quanta from the laser pulse in its own quasi-continuum. This explains why secondary dissociation reaction yielding SF_4 (and SF_3) occurs in longer laser pulses [3,4].

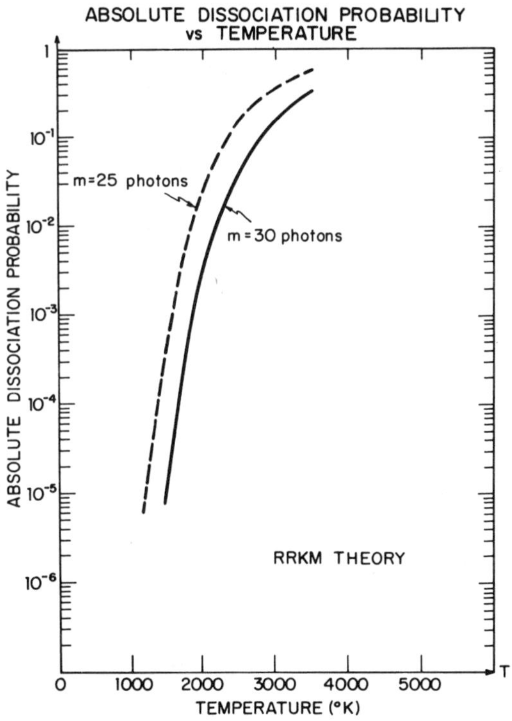

Fig.6 The dissociation probability, defined by (5-7), as a function of vibrational temperature

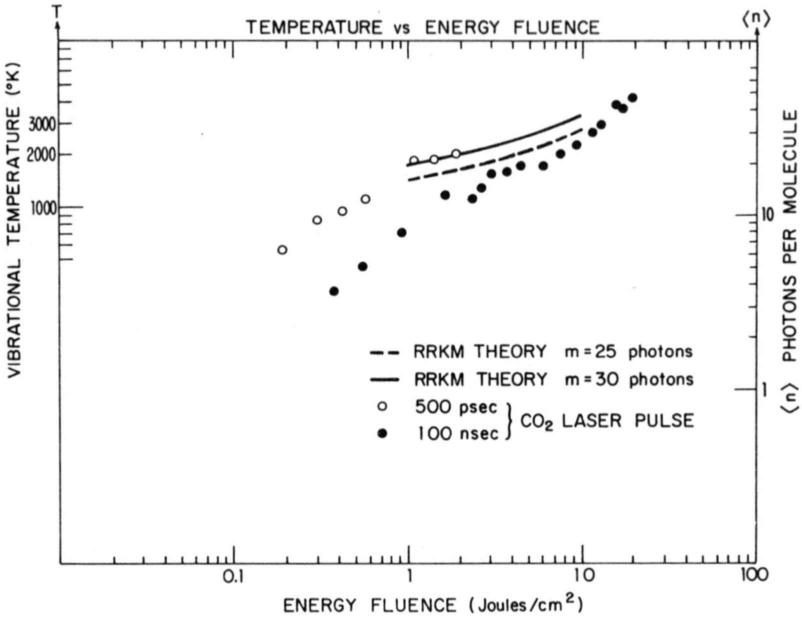

Fig.7 Vibrational excitation, derived from absorption data (dots and circles), is compared with the vibrational temperature derived from dissociation data (drawn and dashed curves)

4. Discussion

The data on SF_6 are consistent with the statistical picture of a canonical ensemble, in which the vibrational energy is distributed over the ergodic surface in vibrational phase space on a picosecond time scale. It would appear to be very difficult to keep a large amount of energy concentrated in one particular vibrational mode or one molecular bond. This dims the prospects for selective laser photo-chemistry to a considerable extent.

The other degrees of vibrational freedom in a polyatomic molecule serve as a thermal reservoir, to which a particular mode is coupled by anharmonic forces. The analogy with relaxation between magnetic spin systems has been worked out further by YABLONOVITCH [14]. He introduces relaxation times T_1 and T_2 for intramolecular vibrational transfer, and shows that $\hbar^{-1}|\mu||E|T_{1,2} > 1$ is required for nonequilibrium excitation. In practice this will require extremely fast pulse rise times.

It is possible that molecules with a particular vibrational frequency or symmetry exist, in which one mode is better isolated from the remaining vibrational manifold than is the case for SF_6. The question also remains open what happens in smaller molecules, particularly in linear triatomic ones. Further experimentation on high vibrational excitation of different molecular species by infrared radiation at several frequencies is clearly desirable. Rapid intramolecular vibrational thermalization is, however, expected to be the rule rather than the exception in polyatomic molecules.

References

1. See papers by R.V. Ambartzumian, K.L. Kompa, and N.Bloembergen, C.D. Cantrell, D.M. Larsen: In *Tunable Lasers and Applications* Proc. Nordfjord Conf. (1976), in Optical Sciences, Vol. 3, ed. by A. Mooradian, T. Jaeger, P. Stokseth (Springer-Verlag, Berlin, Heidelberg, New York, 1976)
2. (a) V.S. Letokhov: Physics Today May, 23 (1977) (b) V.S. Letokhov, C.B. Moore: Sov. J. Quant. El. 6, 259 (1976)
3. M.J. Coggiola, P.A. Schulz, Y.T. Lee, Y.R. Shen: Phys. Rev. Lett. 38, 17 (1977)
4. E.R. Grant, M.J. Coggiola, Y.T. Lee, P.A. Schulz, Y.R. Shen: Chem. Phys. Lett. to be published (1977)
5. P. Kolodner, C. Winterfeld, E. Yablonovitch: Opt. Commun. 20, 119 (1977)
6. J.G. Black, E. Yablonovitch, N.Bloembergen, S. Mukamel: Phys. Rev. Lett. 38, 1131 (1977)
7. H. Stafast, W.E. Schmid, K.L. Kompa: Opt. Commun. 21, 121 (1977)
8. J.L. Lyman, J.W. Hudson, S.M. Freund: Opt. Commun. 21, 112 (1977)
9. (a) V.S. Letokhov, A.A. Makarov: Opt. Commun. 17, 250 (1976) (b) D.M. Larsen, N. Bloembergen: Opt. Commun. 17, 254 (1976) (c) M.F.T. Loy, D. Grischkowsky: to be published
10. N. Bloembergen: Opt. Commun. 15, 250 (1976)
11. V.N. Bagratashvili, I.N. Knyazev, V.S. Letokhov, V.V. Lobko: Opt. Commun. 18, 525 (1976)
12. A.V. Novak, J.L. Lyman: J. Quant. Spectrosc. Radiat. Transfer 15, 945 (1975)
13. P.J. Robinson, K.A. Holbrook: *Unimolecular Reactions* (Wiley, New York, 1972)
14. E. Yablonovitch: submitted for publication in Optics Letters

MULTIPHOTON DISSOCIATION OF POLYATOMIC MOLECULES STUDIED WITH A MOLECULAR BEAM

E.R. Grant, P.A. Schulz, Aa.S. Sudbo,
M.J. Coggiola, Y.T. Lee, and Y.R. Shen

Materials and Molecular Research Division, Lawrence Berkeley Laboratory
University of California, Berkeley, CA 94720, USA

Collisionless multiphoton dissociation (MPD) of polyatomic molecules is now a well-known subject in quantum electronics. The process has been shown to be isotopically selective and rather efficient. It has also been suggested as a potential method for exciting mode-control unimolecular reactions for chemical synthesis. Since there already exist a number of extensive review articles on the subject [1], we shall not discuss here in any detail what has already been established, but shall limit ourselves to the most recent progress in our understanding of the subject.

Among the many problems of MPD, the following are most important:

1. How can a molecule absorb several tens of infrared photons from a moderately intense laser field with a high probability? In other words, what is the physical mechanism responsible for such an efficient multiphoton excitation?
2. Is the multiphoton excitation of a molecule mode-selective or non-selective, or is the laser energy deposition into the molecule randomized among all vibrational modes?
3. For each molecule dissociated, how many photons (or how much laser energy) does it absorb? How does the molecular structure limit the laser energy deposition?
4. What is the dynamics of multiphoton dissociation? Is the dissociation always dominated by the lowest dissociation channel? How does the molecular structure affect the dynamics of dissociation?

The first question has already had a qualitative but reasonable answer [1]. A polyatomic molecule has discrete states at low energies, but the density of states increases very rapidly with increase of energy and soon forms a quasi-continuum. It is believed that a moderately strong laser field can selectively excite the molecule over the discrete states via a near-resonant multiphoton transition and then through the quasi-continuum via resonant stepwise transitions to and beyond the dissociation threshold. This explanation is strongly supported by the results of the two-laser experiments of AMBARTZUMIAN et al [2].

The other questions, however, have not yet received satisfactory answers. The main difficulty of the usual experiments on MPD of molecules in a gas cell is that molecular collisions during and after the laser pulse excitation and the chemical reactions following the collisionless unimolecular dissociation often make the experimental results very confusing and sometimes even inconsistent. Then in these usual experiments, study of dissociation dynamics is also impossible. It is clear that in order to be able to understand a collisionless process, one must first eliminate molecular collisions in the experimental investigation. The best way to achieve this is to use a molecular beam. With an appropriately designed molecular beam apparatus, the dynamics of dissociation can also be studied [3-5]. In this paper, we shall describe and discuss the preliminary experimental results on MPD of polyatomic molecules obtained from our recent crossed laser and molecular beam experiments. We show that with the help of a phenomenological model for multiphoton excitation and a statistical model for molecular dissociation, we can essentially answer all those important questions posted above.

Our experimental arrangement has been described elsewhere [3]. Briefly, a Tachisto CO_2 TEA laser was used to produce a laser beam which crossed with a molecular beam at the collision center in a molecular beam apparatus. The dissociation fragments from the collision center were detected and analyzed by a mass spectrometer rotatable around the collision center. A gated counting system attached to the mass spectrometer was used to obtain time-of-flight spectra of the fragments. Thus, both the angular distributions and the velocity distributions of the fragments could be readily obtained. From these results together with the measured velocity distribution of the primary beam, we could then deduce by deconvolution the kinetic energy distribution of the fragments.

We have so far studied MPD of three different polyatomic molecules: SF_6, CF_3Br, and $CFC\ell_3$. In all three cases, we found from mass spectroscopy that MPD occurred through the lowest dissociation channel

$$SF_6 + nh\nu \longrightarrow SF_5^* + F$$
$$CF_3Br + nh\nu \longrightarrow CF_3^* + Br$$
$$CFC\ell_3 + nh\nu \longrightarrow CFC\ell_2^* + C\ell$$

where * denotes internal energization of the dissociation products. The case of SF_6 turned out to be much more complicated than the others. First, the fragmentation pattern of SF_5^* in the ionization chamber of the mass spectrometer was not known and had to be established. Then, we realized that at higher laser energies SF_5^* produced during the laser pulse could absorb more photons and undergo a secondary dissociation $SF_5^* + n'h\nu \longrightarrow SF_4^* + F$ [4]. Why we have found no similar secondary dissociation in MPD of CF_3Br and $CFC\ell_3$ is of course a rather interesting question. We shall see later that the question can be answered by the statistical theory of molecular dissociation with its explicit dependence on the molecular structure.

As was mentioned earlier, we can deduce, from the measured angular and velocity distributions of the fragments, the kinetic-energy (or recoil-energy) distribution $g(\mathcal{E})$ for the fragments. We show in Fig. 1 an example of SF_6 obtained with ~ 10 J/cm^2 of laser excitation. One can readily draw several conclusions from the results in Fig. 1. First, the average recoil energy of the fragments is only 2.5 KCal/mole (~ 0.11 eV), suggesting that if a molecule absorbs more than one photon above the dissociation threshold, then a large fraction of this excess energy must be retained by the SF_5 fragment in its internal degrees of freedom. Second, $g(\mathcal{E})$ strongly peaks at zero kinetic energy, indicating that there is little energy barrier for dissociation of SF_6 and the laser energy deposition in the molecule before dissociation must be randomized in many accessible states.

We believe that because of the very strong coupling among the vibrational

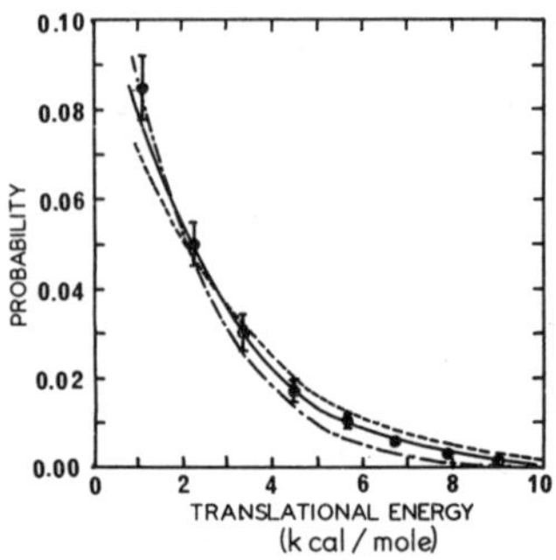

Fig.1 Fragment recoil energy distribution for $SF_6 \to SF_5 + F$. Experimental data points are denoted by the dots. Curves are calculated from the RRKM theory assuming a molecular excitation of $E = E_0 + nh\nu$ with n = 7 (— · —), n = 9 (——), and n = 11 (----) where E_0 is the dissociation threshold energy and $h\nu$ is the CO_2 laser photon energy.

modes of a highly excited polyatomic molecule, the excitation energy deposited in the molecule is likely to be randomized in all vibrational degrees of freedom. Then, the well-known RRKM statistical theory for unimolecular dissociation [6] which assumes complete energy randomization can be used to calculate $g(\mathcal{E})$. We may assume that the molecules are initially excited to an energy $E - E_0 = nh\nu$ above the dissociation level where E_0 is the dissociation threshold energy, $h\nu$ is the CO_2 photon energy, and n is an integer. Knowing the molecular structure, we can then calculate $g_{nh\nu}(\mathcal{E})$. In Fig. 1, three theoretical curves with n = 7, 9, and 11 are shown; the n = 9 curve is in fair agreement with the experimental results. Actually, because of the statistical nature of the laser excitation process, there should be a significant spread of populations in different n levels before dissociation with n = 9 being the average. This will be seen more clearly later in our model calculation. The RRKM calculation also predicts a laser energy dependence of $g(\mathcal{E})$ which agrees well with our experimental results.

From the good agreement between theory and experiment, we can then conclude that (1) the laser energy deposited in SF_6 before dissociation is completely randomized in all vibrational modes, (2) mode-controlled dissociation of SF_6 does not occur with nanosecond pulse excitation, and (3) each SF_6 molecule absorbs on the average 36 to 40 CO_2 laser photons before dissociation, assuming that absorption of 29 photons is needed to reach the dissociation threshold. The RRKM calculation also yields a dissociation rate corresponding to each specific level of excitation. As shown in Fig. 2, the dissociation rate in-

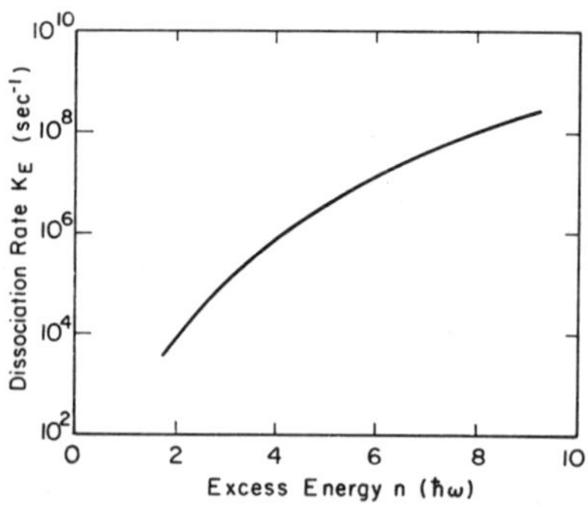

Fig.2 Dissociation rate of SF_6 calculated from the RRKM theory as a function of level of excitation $nh\nu = E - E_0$.

creases very rapidly with n, and for n = 7 to 11, it varies from 2×10^9 to 2×10^7 sec^{-1}. This explains why on average, the dissociation seems to have initiated from the n = 9 level of excitation. For $n < 7$, the up-excitation rate is much larger than the dissociation rate so that only a small fraction of molecules will dissociate. For $n > 11$, the reverse is true and only a small fraction of molecules can be excited to higher levels.

We have also obtained preliminary results of recoil energy distributions of fragments from MPD of CF_3Br and $CFC\ell_3$ under ~ 10 J/cm^2 laser excitation. The RRKM calculations for these cases show the same qualitative behavior as in the case of SF_6. In particular, the excitation energy in these molecules seems to be completely randomized in all degrees of freedom before dissociation, and only a small fraction of the excess energy appears in the form of recoil energy after dissociation. However, the observed recoil energy distributions in the cases of CF_3Br and $CFC\ell_3$ correspond respectively to an average excitation energy of 1-3 and 3-5 CO_2 laser photons beyond the dissociation level. This average excess energy seems to be quite different for different molecules, but actually, it corresponds to a dissociation rate from 10^7 to 10^9 sec^{-1} which is the same for all three molecules we have investigated. Clearly, the balance between the up-excitation rate and the dissociation rate is responsible for what we have observed. The dissociation thresholds for SF_6, $CFC\ell_3$, and CF_3Br are 76, 77, and 65 KCal/mole respectively. At a given energy above the dissociation threshold, the statistical rate for unimolecular dissociation is proportional to the ratio of the density of states of the critical configuration for dissociation to that of the energized molecule [6]. This ratio is smaller for larger or/and heavier molecules.

The difference in the excess energies in different molecules explains why SF_6 can undergo stepwise secondary dissociation while the others cannot. In all cases, a major portion of the excess energy appears as internal energy of the fragment after dissociation. Thus in the case of SF_6, the dissociation product SF_5^* has an average internal energy of $6h\nu$ —— $10h\nu$, and must have already been excited to the quasi-continuum states. It can therefore easily absorb more photons to and beyond its dissociation threshold as long as the laser field is present. Consequently, stepwise dissociation of SF_6 can be expected if the exciting laser pulse is sufficiently long and contains enough energy. This is however not true for the other molecules. Because of the lower excess energies, the fragments CF_3^* and $CFC\ell_2^*$ are not quite in the quasi-continuum states. As a result, they cannot resonantly absorb more photons and hence the secondary dissociation process becomes less probable. From these results, we can then predict that for MPD of polyatomic molecules, the larger and heavier molecules with a large excess energy before dissociation will most likely undergo stepwise dissociation.

To help our understanding of MPD, we have developed a simple phenomenological model which we believe is realistic enough to exhibit at least the qualitative behavior of the multiphoton excitation and dissociation process. We assume in the model that the molecular system can be described by a set of evenly spaced energy levels with the corresponding densities of states being the degeneracy factors. In this respect, we have neglected the possible initial multiphoton transition step or steps to reach the quasi-continuum by jumping over the discrete states. This is probably a good approximation as long as the laser intensity is much larger than the threshold intensity for overcoming the discrete state barrier, e.g., ~ 30 KW/cm^2 for SF_6 [7]. We then assume that the transitions between levels are incoherent and the populations N_j of all levels are governed by the following set of rate equations.

$$\frac{d}{dt} N_i = C^a_{i-1} N_{i-1} + C^e_i N_{i+1} - (C^a_i + C^e_{i-1}) N_i \tag{1}$$

for levels below the dissociation threshold E_0, and

$$\frac{d}{dt} N_m = C^a_{m-1} N_{m-1} + C^e_m N_{m+1} - (C^a_m + C^e_{m-1}) N_m - k_m N_m \tag{2}$$

for levels above the dissociation threshold. In the above equations, k_m is the dissociation rate of molecules in the mth level calculated from the RRKM theory, and C^a_j and C^e_j are respectively the absorption rate from level j to j + 1 and the emission rate from j to j - 1. For one-photon transitions, we have

$$C^a_j = \sigma_j I$$
$$C^e_j = (g_j/g_{j+1}) C^a_j \tag{3}$$

where σ_j is the absorption cross-section, I is the laser intensity, and g_j is the degeneracy factor of level j.

We can solve the above set of equations numerically for a given molecule with σ_j and I(t) specified. Our results for SF_6 are shown in Figs. 3 and 4.

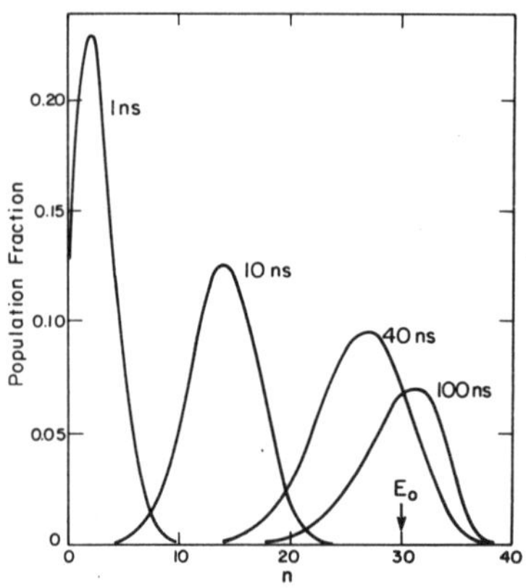

Fig.3 Calculated excited state population distributions of SF_6 for various times during a square laser pulse (200 MW/cm^2, 100 nsec.).

We used in the calculations a square laser pulse of 200 MW/cm^2 with a 100 - nsec duration, and an absorption cross-section

$$\sigma_j = \exp[- .02936\ j - 42.93]\ \text{cm}^2.$$

This relation for σ_j was chosen so that our numerical results yield both the observed dependence of average number of photons absorbed per molecule on laser fluence [8] and the observed dependence of dissociation yield on laser energy fluence [9].

Figure 3 shows that as time goes on, the laser excitation effectively drives the population distribution up to higher levels; the average number of photons absorbed per molecule of course increases correspondingly. Only after ~ 30 nsec, does the high-energy tail of the population distribution start to have a

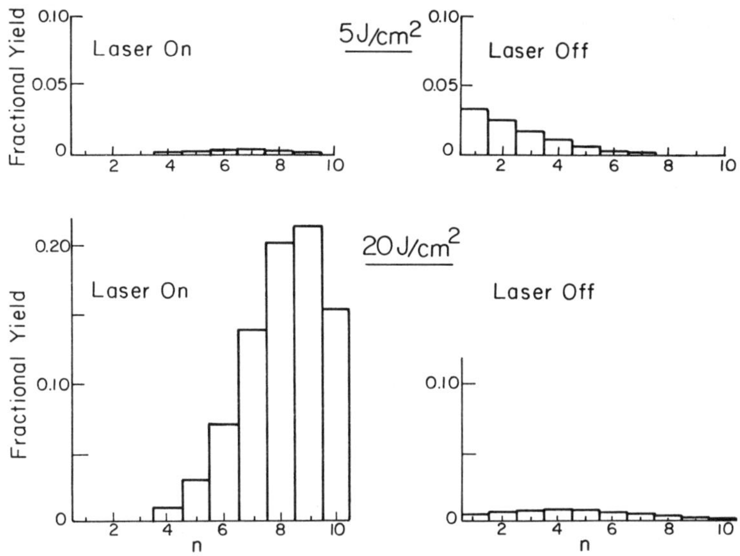

Fig.4 Calculated dissociation yields during and after the 100 - nsec laser pulse for two laser energy fluences: 5 J/cm^2 and 20 J/cm^2

significant portion above the dissociation threshold (assuming E_0 = 30 $h\nu$). Dissociation then occurs. The laser excitation continues to drive the population distribution further up, but the action is soon limited by the very high dissociation rates at higher energy levels which deplete the populations effectively. This is seen by the more abrupt cutoff on the high-energy side of the 100 - nsec population distribution curve.

Knowing the populations above the dissociation threshold, we can then calculate the dissociation yield as shown in Fig. 4 for two different laser energy fluences. We have calculated separately the yield during the laser pulse and the yield after the pulse is off. The total yield is of course the sum of the two. It is seen that with 20 J/cm^2 of laser excitation, already a large fraction of the molecules is dissociated during the laser pulse. Then, the fragments produced during the pulse can absorb more photons and undergo secondary dissociation if the laser pulse is sufficiently long and intense as we have experimentally observed.

Recently, BLOEMBERGEN et al have also concluded from their optoacoustic measurements that MPD is a statistical process [8,10]. They used the quantum Kassal theory to interpret their results. In their model, they assumed that the laser multiphoton excitation of a molecule is equivalent to a heating process. The population distribution is then governed by the thermal Boltzmann distribution characterized by an effective temperature T. To find T, they assumed that the classical equipartition relation $skT = \langle n \rangle h\nu$ holds, where s is the total number of vibrational degrees of freedom (s = 15 for SF_6) and $\langle n \rangle$ is the average number of photons absorbed per molecule which can be obtained from the optoacoustic measurement. They also assumed that the thermal distribution is not affected by dissociation. This limits the validity of their calculations to cases with low dissociation yield. Now, it is not obvious a priori that the above assumptions are correct. In particular, we wonder whether laser excitation will indeed yield a thermal distribution with an effective temperature $T = \langle n \rangle h\nu/sk$. Using our more realistic model calculations, we can now answer this question directly.

Figure 5 shows a population distribution created by laser excitation with an average number of photons absorbed per SF_6 molecule $\langle n \rangle = 20$. Two thermal distribution curves are also shown for comparison. Clearly, the one at T = 1800° K calculated from $T = \langle n \rangle h\nu/sk$ with s = 15 is very different from the laser-excited distribution. The other at T = 2200° K (corresponding to a reduced number of vibrational degrees of freedom $s' \cong 12$) has the same average excitation energy as that of the laser-excited distribution, but the thermal distribution curve is appreciably broader and has a longer high-energy tail.

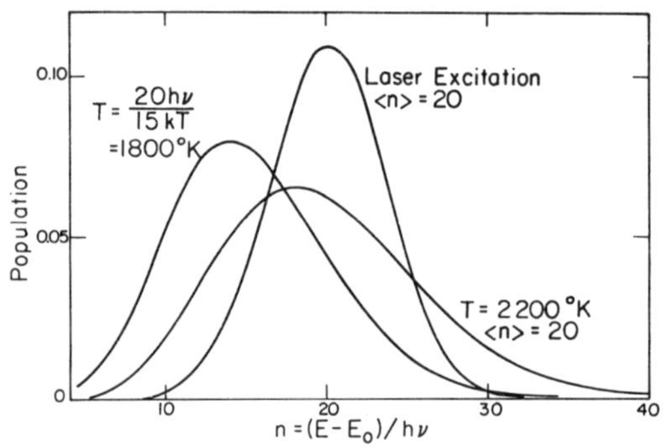

Fig.5 Comparison between population distributions obtained by laser excitation and by thermal heating. See the text.

Thus, we can conclude that a thermal distribution is only a rather crude approximation of the real distribution created by laser excitation, and the corresponding "temperature" is appreciably larger than the one calculated from $T = \langle n \rangle h\nu/sk$. The discrepancy in "temperature" comes in mainly because in reality different vibrational modes have different frequencies and the inequality $h\nu_i \ll kT$ necessary for the validity of equipartition theorem does not hold for all modes. However, if we consider only the population distribution near and above the dissociation threshold, then the T = 1800° K thermal distribution agrees better with the laser-excited distribution. In other words, the T = 1800° K curve gives a fair prediction of the dissociation yield. This is probably the reason why BLOEMBERGEN et al found that thermal distributions with $T = \langle n \rangle h\nu/sk$ seem to describe the observed dissociation yield near threshold fairly well.

We summarize here the most important results we have obtained from our studies. First, in MPD, the laser energy deposition into a molecule is quickly randomized among all vibrational degrees of freedom, suggesting that mode-controlled dissociation of molecules is not possible at least in the cases we have studied. Second, the number of excess photons absorbed per molecule above the dissociation threshold varies with molecules, ranging from 1-3 for CF_3Br, 3-5 for $CFC\ell_3$, and 7-11 for SF_6. The dissociation rate however ranges from 10^7 to

10^9 sec^{-1} for all these molecules and is the limiting mechanism for excitation to higher levels. Third, the primary dissociation of a molecule usually occurs through the lowest dissociation channel. A larger or heavier molecule such as SF_6 is more likely to undergo a secondary dissociation process. Fourth, during dissociation, only a small fraction of the excess energy appears as the recoil energy of the fragments; the rest is retained by the fragments in their internal degrees of freedom. We have developed a model calculation which exhibits the qualitative behavior of multiphoton excitation and dissociation and strongly corroborates our physical interpretations. In addition, we have shown that the population distribution obtained from laser multiphoton excitation is appreciably different from that resulting from thermal heating.

This work was supported by the U.S. Energy Research and Development Administration.

References

1. V. S. Letokhov, Physics Today 30, 23 (1977); V. S. Letokhov and C. B. Moore, Sov. J. Quant. Electronics 6, 259 (1976); R. V. Ambartzumian, Yu A. Gorokhov, V. S. Letokhov, G. N. Makarov, E. A. Ryabov, and N. V. Chekalin, Sov. J. Quant. Electronics 6, 437 (1976); See also papers on the subject in "Tunable Lasers and Applications," Proc. of the Nordfjord Conference, edited by A. Mooradian, T. Jaeger, and P. Stokseth (Springer-Verlag, Berlin, 1976) and references therein.
2. R. V. Ambartzumian, N. P. Furzikov, Yu A. Gorokhov, V. S. Letokhov, G. N. Makarov, and A. A. Puretzky, Optics Comm. 18, 517 (1976); V. S. Letokhov, Proc. International Conference on Multiphoton Processes (Rochester, N.Y., June 1977)(to be published); R. V. Ambartzumian, paper presented at this conference.
3. M. J. Coggiola, P. A. Schulz, Y. T. Lee, and Y. R. Shen, Phys. Rev. Lett. 38, 17 (1977).
4. E. R. Grant, M. J. Coggiola, Y. T. Lee, P. A. Schulz, and Y. R. Shen, Chem. Phys. Lett. (to be published).
5. E. R. Grant, P. A. Schulz, Aa. S. Sudbo, M. J. Coggiola, Y. R. Shen, and Y. T. Lee, Proc. International Conference on Multiphoton Processes (Rochester, N.Y., June 1977)(to be published).
6. See, for example, P. J. Robinson and K. A. Holbrook, "Unimolecular Reactions" (J. Wiley, New York, 1972).
7. M. C. Gower and K. W. Billman, Opt. Comm. 20, 123 (1977); M. C. Gower and K. W. Billman, Appl. Phys. Lett. 30, 514 (1977).
8. J. G. Black, E. Yablonovitch, N. Bloembergen, and S. Mukamel, Phys. Rev. Lett. 38, 1131 (1977).
9. E. R. Grant, P. A. Schulz, Aa. S. Sudbo, M. J. Coggiola, Y. T. Lee, and Y. R. Shen (unpublished).
10. N. Bloembergen and E. Yablonovitch, paper given in this conference; J. G. Black, E. Yablonovitch, N. Bloembergen, and S. Mukamel, Proc. International Conference on Multiphoton Processes (Rochester, N.Y., June 1977)(to be published).

LASER DIODE SPECTRA OF SPHERICAL-TOP MOLECULES

R.S. McDowell

University of California, Los Alamos Scientific Laboratory
Los Alamos, NM 87545, USA

Introduction

About ten years ago the tunable semiconductor diode laser was developed [1], which provided an infrared spectral source with a linewidth of ~50 kHz, or about 2×10^{-6} cm^{-1} [2]. The potential usefulness of such a device for high-resolution absorption spectroscopy was immediately recognized [3], but progress remained slow because of the limited tuning ranges of the early diodes, and because many spectra recorded at this resolution reveal an almost embarrassing wealth of detail, the analysis of which requires a fairly sophisticated approach.

These problems are particularly severe in the vibration-rotation spectra of spherical-top molecules (i.e., molecules of tetrahedral or octahedral symmetry), which exhibit a complex splitting of the individual rotational manifolds due to tensor perturbations. The theory of this band structure was developed in the early 1960's by MORET-BAILLY [4] and HECHT [5] and was extensively applied to the infrared absorption spectra of methane and its analogues. In such molecules the total angular momentum states of interest are limited to $J < 15$; there was no need to consider larger angular momenta because the bands of heavier molecules that exhibit high-J transitions simply could not be adequately resolved before the development of the diode laser.

Since such molecules as SF_6 and OsO_4 have recently been used as saturable absorbers, as media for self-induced transparency and other experiments in non-linear optics, and as subjects for laser isotope separation, there has been increased interest in the analysis of these complex vibration-rotation bands. At Los Alamos we have over the last few years obtained tunable diode laser spectra of CH_4, CF_4, OsO_4, SF_6, and UF_6, and have undertaken an extensive program for the analysis of such data. As examples of the methods employed and their applicability, this talk will concentrate on the SF_6 and CF_4 molecules.

Sulfur Hexafluoride

Doppler-limited spectra of the ν_3 stretching fundamental of SF_6 in the regions within ±1 GHz of the various 10.4-μm CO_2 laser lines were obtained by HINKLEY in 1970 [3,6]. At about the same time sub-Doppler saturation spectra were reported within ±50 MHz of the laser lines [7-9]. These revealed a complex vibration-rotation structure with a line density of 10^4 per cm^{-1} in some regions, but were not complete enough to allow the individual transitions to be assigned. On the other hand, a study of this band using the 0.07-cm^{-1} resolution of a "grille" spectrometer [10] showed only the overall PQR structure and a few hot-band Q branches.

Our attack on this problem began in 1975 when ALDRIDGE et al. [11] were able to record much of the band in a series of overlapping scans. In conjunction with this experimental work, CANTRELL and GALBRAITH [12] derived the correct nuclear-spin statistical weights for octahedral XY_6 molecules, which are 2,10,8,6,6 for sublevels of A_1,A_2,E,F_1,F_2 symmetry if the Y-nuclei have spins I = 1/2. These advances allowed an assignment of much of the rotational structure in the P and R branches [11]. From this beginning it proved possible to assign completely Hinkley's earlier heterodyne-calibrated spectra near CO_2 P(14), P(18), and P(20) [13]; the SF_6 transitions in these three regions are R(28-30), P(32-33), and P(55-60), respectively.

SF_6 absorbs most strongly the P(16) CO_2 laser line at 947.742 cm^{-1}, but this falls in the SF_6 Q branch where the overlapping of different rotational manifolds makes the analysis particularly difficult. Despite the complexity of the spectrum in this region, the assignments have recently been made [14]. The structure of the Q branch can be most easily understood by considering the expression for the line frequencies, which in dominant approximation (with off-diagonal terms neglected) is [4,15]

$$\nu(J,p) = m + vJ(J+1) + wJ^2(J+1)^2 + \cdots + [-2g + uJ(J+1) + \cdots]\bar{F}^{(4)}. \quad (1)$$

Here p designates the sublevel and $\bar{F}^{(4)}$ is the product of the eigenvalue of the octahedrally-invariant fourth-rank tensor operator times a J-dependent term; $\bar{F}^{(4)}$ has recently been tabulated for all (J,p) levels up to J = 100 by KROHN [16]. By retaining only the most important terms in (1), we can write for the displacement of any Q-branch transition from the band origin

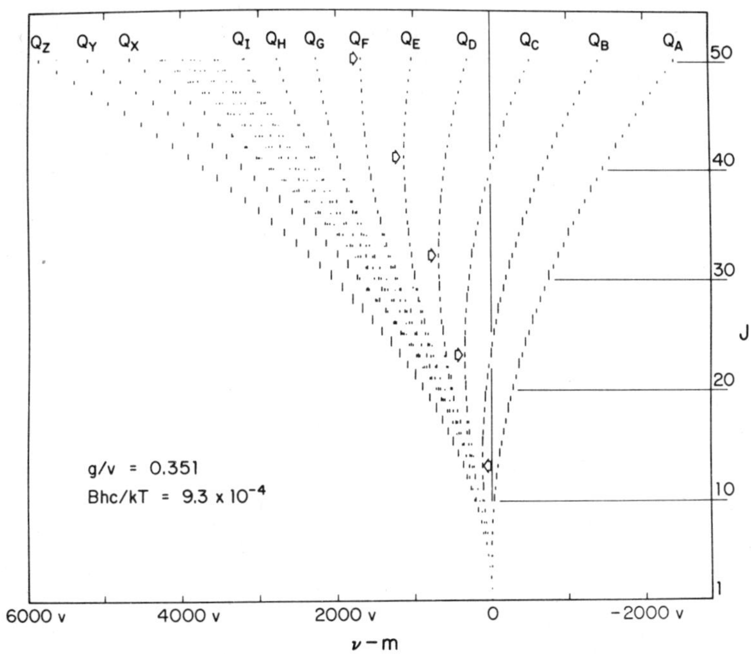

Fig.1 The patterns exhibited by the manifolds J = 1 to J = 50 in the Q branch of SF_6

$$\nu(J,p) - m = v[J(J+1) - 2(g/v)\bar{F}^{(4)}]. \qquad (2)$$

A plot of this equation for the first 50 manifolds is shown in Fig.1. The parameters chosen correspond to ν_3 of SF_6 and the relative intensities of the transitions are correct for a temperature of 140 K. The lines tend to group into sub-branches [17,18] that are here labelled $Q_A, Q_B, \ldots, Q_Z$, where Q_A and Q_Z correspond to the largest positive and negative values, respectively, of $\bar{F}^{(4)}$ for each manifold. Note that some of the sub-branches form bandheads, indicated by arrows.

With the aid of such a plot, the entire Q branch can be assigned; the results are shown in Fig.2. Most of these assigned "lines" represent the grouping of from two to four separate transitions. Some 3000 such individual Q(J,p) transitions contribute to the observed Q-branch structure at 117 K, and perhaps twice that number at room temperature.

A detailed scan of the SF_6 Q branch near 947.74 cm^{-1} is shown in Fig.3. Here the temperature was 300 K and (unlike Fig.2) the Q_H bandhead is prominent. The moderately strong transition very near CO_2 P(16) is assigned as $Q_Z(38) = Q(38)\ F_1{}^0{+}E^0{+}F_2{}^0$, the lowest-frequency group in the Q(38) manifold. That this "line" is indeed a triplet was confirmed by saturation spectroscopy,

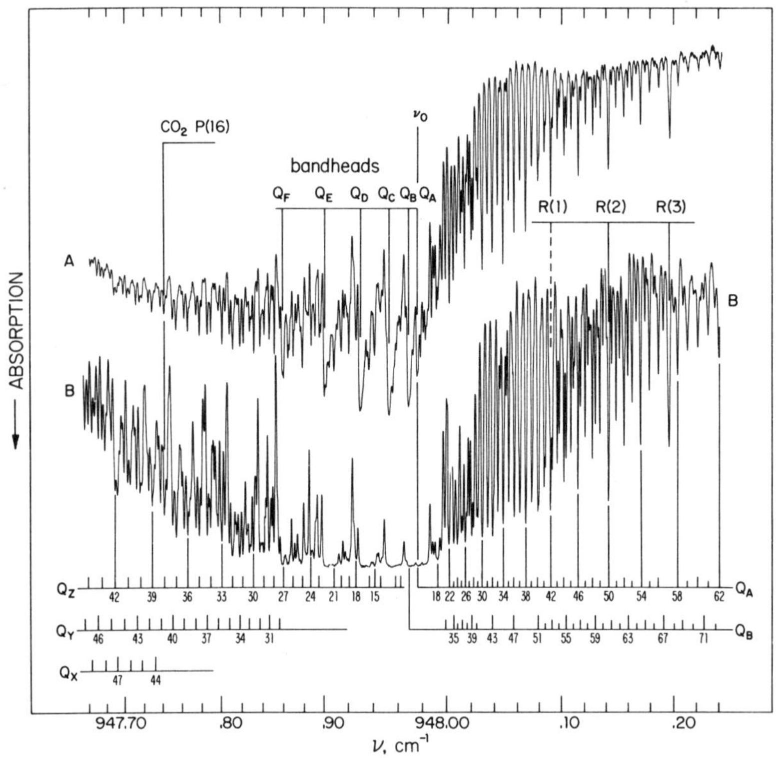

Fig.2 The ν_3 Q branch of $^{32}SF_6$, recorded at (A) 117 K and (B) 126 K using a tunable semiconductor diode laser

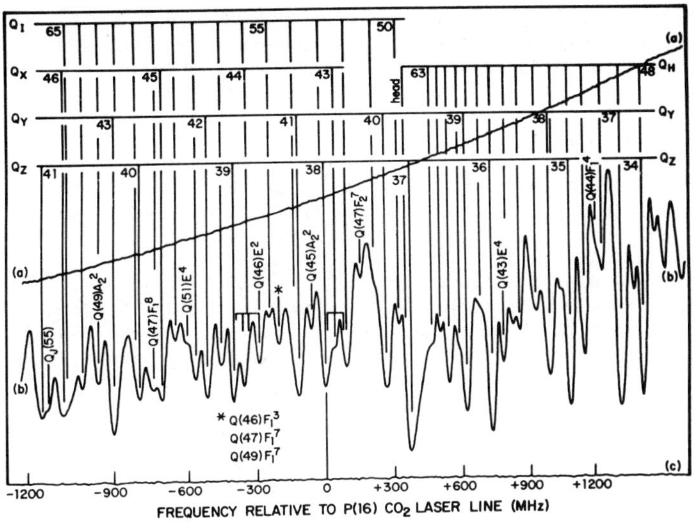

Fig.3 Detail of the SF_6 Q-branch absorption between 947.70 and 947.80 cm^{-1}

which revealed the expected 6:8:6 relative intensities and an overall F_1-F_2 separation of 1.02 MHz [8], compared with 1.08 MHz calculated using the value of g obtained from the P- and R-branch analysis [13]. The assignments have also been confirmed by the double-resonance experiments of MOULTON et al. [19], in which pumping of the SF_6 Q branch causes perturbations in the corresponding P-branch lines due to depopulation of their ground states.

Observed and calculated line positions near CO_2 P(16) are summarized in Table 1, which includes estimates of the transition strengths according to

Table 1 SF_6 lines within ±60 MHz of CO_2 P(16)

Identification	Peak strength at 273 K [20] $[cm^{-1}torr^{-1}]$	Detuning [MHz] calc.	obs.[9]
Q(53) $F_2{}^6$	0.29	-29.6	-31.1
Q(53) $F_1{}^6$	0.29	-17.3	-17.3
Q(38) $F_2{}^0$	0.41	-7.9	
Q(38) E^0	0.55	-7.4	-7.8
Q(38) $F_1{}^0$	0.41	-6.9	
Q(43) $F_1{}^8$	0.38	+8.5	+8.8
Q(45) $F_2{}^7$	0.36	+14.2	+12.0
Q(48) $F_2{}^5$	0.34	+36.3	+35.0
Q(43) E^5	0.51	+37.5	

FOX [20]. Of these lines, only the frequency of the Q(38) triplet was used in the data set from which the molecular parameters were determined (this data set included also lines from the P and R branches and other portions of the Q branch). Despite this, the rms deviation between the calculated frequencies and those reported by GOLDBERG and YUSEK [9] is only 1.4 MHz, which illustrates the predictive capability of the present analysis.

However, these molecular parameters, while good, are already obsolete, for L. Henry of Université Pierre et Marie Curie (Paris) has recently obtained saturation spectra in which many of the SF_6 frequencies are determined to within ±30 kHz, or ±1 part in 10^9, and even this uncertainty is mainly that of the CO_2 laser frequencies themselves. We are now carrying out a least-squares adjustment of the frequency expressions (with full diagonalization of the Hamiltonian) to these data, and the results will be published shortly as a collaborative effort by the Paris, Dijon, and Los Alamos groups. The band origin, m, should have a precision comparable to that of the line frequencies, or rather better than that of the currently accepted value of the velocity of light! This is the first time that such precision in the molecular constants has been achieved for any polyatomic molecule, and illustrates the potential of laser spectroscopy.

Table 2 summarizes the SF_6 assignments near the P(12) to P(20) CO_2 laser lines [that at P(12) is based on an unpublished collaboration with P. F. Moulton of M.I.T. Lincoln Laboratories]. The frequency offsets in column 3 are those determined by Henry, and may differ slightly from previously published values (cf. Table 1). Table 2 also includes a summary of pulse breakup as observed in self-induced transparency (SIT) experiments [21,22]. It will be noted that ideal pulse breakup depends more upon the amount of hot-band absorption than on the nature of the ground-state SF_6 transitions. This supports the conclusion of GIBBS, McCALL, and SALAMO [23] that pulse reshaping is associated with *isolated* transitions (i.e., those free of hot-band interference) that need not be nondegenerate.

Table 2 Summary of assignments and observed pulse breakup in SIT

CO_2 line	Nearest SF_6 transition	Detuning [MHz]	Hot band absorption [$cm^{-1}torr^{-1}$]	Observed pulse breakup
P(12)	R(66) $A_2^0+F_2^0+F_1^0+A_1^0$	-23.0	~0.03	Near ideal; most evident at -20 MHz [21]
P(14)	R(28) A_2^0	+17.6	0.14	Near ideal; most evident at +15 MHz [21]
P(16)	Q(38) $F_1^0+E^0+F_2^0$	-7.2	0.23	Poor [21]
P(18)	P(33) A_2^1	+6.7	300 K: 0.32 195 K: 0.14	None Appreciable [22]
P(20)	P(59) A_2^3	+27.8	0.25	None

Carbon Tetrafluoride

As an example of a tetrahedral molecule we will consider the ν_4 bending fundamental of CF_4 at 631 cm^{-1}. Fig.4 shows a portion of the P branch at high J. Because the manifolds in this band are separated by about 0.5 cm^{-1} (com-

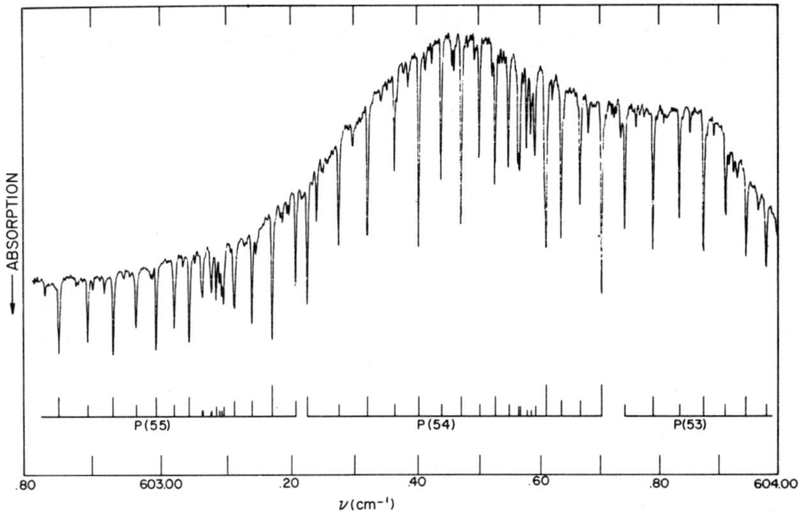

Fig.4 Spectrum of CF_4 between 602.8 and 604.0 cm^{-1} at 296 K and 2.1 torr

pared with 0.056 cm^{-1} in SF_6), overlap is much less of a problem, and in fact P(54) in Fig.4 is the highest-J non-overlapped manifold that we have observed to date in any spherical top. From such data on the P(54) to R(46) transitions, the tensor splitting constants have been determined for the species $^{12}CF_4$, $^{13}CF_4$, and $^{14}CF_4$ [24]; g for $^{12}CF_4$ is $-(2.668 \pm 0.009) \times 10^{-5}$ cm^{-1}, close to its value for SF_6.

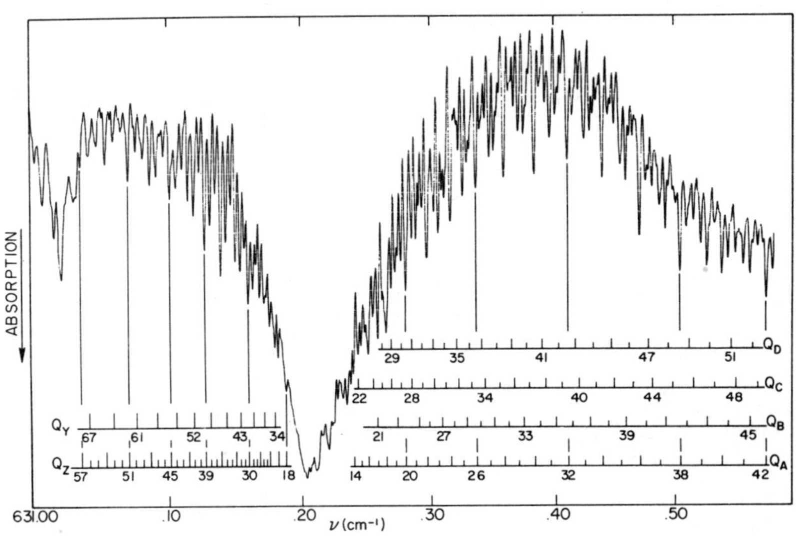

Fig.5 The $^{12}CF_4$ ν_4 Q branch, taken at 296 K and ~0.2 torr

The Q branch of ν_4 has also been resolved and assigned, Fig.5. Here the molecular constants are such ($g/v \approx -0.51$) that the bandheads observed in SF_6 (Fig.2) do not occur. Instead, many lines coincide near the band origin and the sub-branches fan out from this position. A full analysis of this fundamental is in progress and will be reported shortly [24].

Acknowledgements

The spectra reported here were recorded in the Applied Photochemistry Division at Los Alamos, and thanks are due J. P. Aldridge, H. Flicker, N. G. Nereson, and M. J. Reisfeld. The analyses would not have been possible without the support of the Theoretical Division, particularly by H. W. Galbraith, C. D. Cantrell, and B. J. Krohn. From outside Los Alamos, E. D. Hinkley, P. F. Moulton, and L. Henry have been most generous in sharing unpublished data with us. This research was performed under the auspices of the U. S. Energy Research and Development Administration.

References

1. J. F. Butler, T. C. Harman: Appl. Phys. Lett. 12, 347 (1968), and references cited therein
2. E. D. Hinkley, C. Freed: Phys. Rev. Lett. 23, 277 (1969)
3. E. D. Hinkley: Appl. Phys. Lett. 16, 351 (1970)
4. J. Moret-Bailly: Cahiers Phys. 15, 237 (1961); J. Mol. Spectr. 15, 344 (1965)
5. K. T. Hecht: J. Mol. Spectr. 5, 355, 390 (1960)
6. E. D. Hinkley, P. L. Kelley: Science 171, 635 (1971)
7. P. Rabinowitz, R. Keller, J. T. LaTourrette: Appl. Phys. Lett. 14, 376 (1969)
8. P. Rabinowitz: Thesis (Polytechnic Inst. of Brooklyn, Brooklyn, N. Y., 1970); Polytechnic Inst. of Brooklyn Dept. of Electrophysics Tech. Report PIBEP-70-065 (1970)
9. M. W. Goldberg, R. Yusek: Appl. Phys. Lett. 17, 349 (1970)
10. H. Brunet, M. Perez: J. Mol. Spectr. 29, 472 (1969)
11. J. P. Aldridge, H. Filip, H. Flicker, R. F. Holland, R. S. McDowell, N. G. Nereson, K. Fox: J. Mol. Spectr. 58, 165 (1975)
12. C. D. Cantrell, H. W. Galbraith: J. Mol. Spectr. 58, 158 (1975)
13. R. S. McDowell, H. W. Galbraith, B. J. Krohn, C. D. Cantrell, E. D. Hinkley: Opt. Comm. 17, 178 (1976)
14. R. S. McDowell, H. W. Galbraith, C. D. Cantrell, N. G. Nereson, E. D. Hinkley: J. Mol. Spectr. (in press)
15. B. Bobin, K. Fox: J. Phys. (Paris) 34, 571 (1973)
16. B. J. Krohn: J. Mol. Spectr. (in press); Los Alamos Scientific Laboratory Report LA-6554-MS (1976)
17. W. G. Harter, C. W. Patterson: J. Chem. Phys. 66, 4872 (1977)
18. C. W. Patterson, W. G. Harter: J. Chem. Phys. 66, 4886 (1977)
19. P. F. Moulton, D. M. Larsen, J. N. Walpole, A. Mooradian: Opt. Lett. (in press), and other unpublished work
20. K. Fox: Opt. Comm. 19, 397 (1976)
21. A. Zembrod, Th. Gruhl: Phys. Rev. Lett. 27, 287 (1971)
22. G. B. Hocker, C. L. Tang: Phys. Rev. 184, 356 (1969)
23. H. M. Gibbs, S. L. McCall, G. J. Salamo: Phys. Rev. A 12, 1032 (1975)
24. R. S. McDowell, H. W. Galbraith, M. J. Reisfeld, J. P. Aldridge: To be published

COHERENT VERSUS STOCHASTIC THEORIES OF COLLISIONLESS MULTIPLE-PHOTON EXCITATION OF POLYATOMIC MOLECULES

C.D. Cantrell

Theoretical Division, University of California
Los Alamos Scientific Laboratory
Los Alamos, NM 87545, USA

1. Introduction

At the present time there is no fundamental theory which gives a satisfactory quantitative-or even qualitative-account of all the observed features of the phenomena of collisionless multiple-photon excitation (CMPE) and dissociation (CMPD) of polyatomic molecules. Such relatively simple, experimentally observable characteristics as the dependence of the unimolecular rate of dissociation upon the intensity, energy per pulse, and frequency of the laser(s) used, are the result of many details of molecular structure. These details include (i) the structure of the vibrational and rotational energy levels of the overtone states (in SF_6, the states $2\nu_3$, $3\nu_3$, etc.) of the vibrational mode pumped by the laser; (ii) the collisionless transfer of energy from the pumped mode to other vibrational modes; (iii) the processes by which dissociation occurs. The models or theories of CMPE which have been published so far include realistic treatments of (i) or (ii), but not both. The purpose of this report is to describe a theoretical framework into which both (i) and (ii) fit naturally, and in which the transition from coherent laser pumping of one mode to incoherent heating of the entire molecule occurs in a continuous manner. The framework encompasses both the rate-equation approach of RRKM theory [1-3] and the multilevel Schrodinger equations which are characteristic of coherent excitation [4-7]. The parameters which occur in the theory described here can, in principle, be evaluated using a model for the molecular vibrational force field. This theory therefore includes a first-principles derivation of the pheonomenological models of GOODMAN et al. [8] and HODGKINSON and BRIGGS [9], which employ equations similar to those derived here. This report, like previous work [7,10], will concentrate on the ν_3 mode of SF_6, although the methods are applicable to all spherical-top molecules.

It is customary and convenient to divide the vibrational energy levels of SF_6 into three regions: Region I, where laser excitation proceeds coherently via sharp, discrete vibration-rotation levels; Region II, where resonant absorption of laser photons is possible over a substantial range of frequencies near the small-amplitude frequency ν_o of the pumped mode; and Region III, above the first dissociation threshold of the molecule. There is, of course, no sharp dividing line between Region I and Region II; and the processes which broaden the absorption spectrum of the molecule in Region II are still a matter of debate. It has been shown that the structure of the vibrational energy levels of a spherical-top molecule can allow for many successive nearly resonant transitions, even in Region I [10,11]. However, the most popular view of Region II is that many successive resonant transitions are enabled by the broadening and shifting of the energy levels of the pumped (degenerate) mode which arise from the interaction between the pumped mode and the other vibrational degrees of freedom of the molecule. It is probably the case that coherent excitation plays a major role at lower energies in Region II, while at higher energies the rate of transfer of energy to other modes becomes larger than the rate of excitation by the laser, enabling one to speak correctly if incoherent excitation [4]. In the past two years estimates of the level of excitation of SF_6 at which Region II (i.e. the vibrational quasi-continuum) begins have crept upwards from v=3 [12] to v=5 at the present time. These uncertainties in the details of the application of the concept of the vibrational quasicontinuum make a unified, first-principles approach imperative.

2. The Reduced Density Matrix

Of the six normal modes of an octahedral spherical-top molecule, only two (ν_3 and ν_4) are able to absorb energy from the electromagnetic field by dipole-allowed processes. This statement remains true regardless of the amplitude of vibrational displacement, since it is based only on the symmetry of the vibrational potential-energy surface in the electronic ground state, and upon the symmetry properties of the normal modes. For large vibrational excitation, the normal modes no longer diagonalize the vibrational Hamiltonian, owing to nonlinear intermode couplings. Conceptually the interaction of the SF_6 molecule with a laser field near the frequency of the ν_3 mode can be viewed as the problem of a system (the ν_3 mode) interacting with a coherent pump (the laser field and with an unobserved reservoir (the remaining vibrational degrees of freedom). Such a relationship between a system, a coherent driving field and a reservoir is familiar from the theory of spin relaxation in nuclear magnetic resonance [13,14]. In that case the well-known Bloch equations may be derived using the reduced density matrix [15] for the nuclear-spin degrees of freedom. We recall that the states $|s\rangle$ of a system and the states $|r\rangle$ of a reservoir may be used to construct a basis of states $|s\rangle \otimes |r\rangle$ for the combined system composed of (system + reservoir), and that the density matrix ρ for the combined system will have the matrix elements $\rho_{sr;s'r'}$. For SF_6, the states $|s\rangle \otimes |r\rangle$ are of the form $|v_1v_2v_3v_4v_5v_6\rangle$, where v_i indicates the number of quanta in the mode v_i. The <u>reduced density matrix</u> for the system, $\rho^{(s)}$, has the matrix elements

$$\rho^{(s)}_{ss'} = \sum_r \rho_{sr;s'r} = (tr_R\,\rho)_{ss'} \quad . \qquad (1)$$

For SF_6, the sum on $|r\rangle$ in (1) would run over all v_i ($i \neq 3$).

Usually the diagonal elements of the density matrix are interpreted as populations. From (1) we learn that a diagonal element of the reduced density matrix, $\rho^{(s)}_{ss}$, is the sum of the populations in all states of the combined system which include the state $|s\rangle$. For SF_6, Eq. (1) implies that the diagonal reduced density matrix element $\rho^{(3)}_{vv}$ for v quanta of the ν_3 mode is the sum of the populations in all states $|v_1v_2vv_4v_5v_6\rangle$ which have $v\nu_3$ quanta. Thus a rate equation describing a flow of "population" from v to v' in terms of the reduced density matrix elements $\rho^{(3)}_{vv}$ and $\rho^{(3)}_{v'v'}$ is, in fact, describing the flow of summed populations from all states of the form $|v_1v_2vv_4v_5v_6\rangle$ to all states of the form $|v_1v_2v'v_4v_5v_6\rangle$. By virtue of its definition, the reduced density matrix for the ν_3 mode $\rho^{(3)}$, can thus describe "hot-band" transitions, as well as transitions between states in which only ν_3 quanta are excited. This capacity is important if, as has been suggested in [1-3], all the normal modes and not merely the ν_3 mode are collisionlessly excited during laser pumping.

3. Reduced-Density-Matrix Equation of Motion

Equations of motion for reduced density matrices have been studied by many authors. We shall draw on the work of LAX [16-17] and LOUISELL [18]. Using their methods, it is straightforward but tedious to show that in the Markov approximation [18] the equation of motion for the reduced density matrix $\rho^{(3)}$ of the ν_3 mode of SF_6 is the following:

$$\frac{\partial\rho^{(3)}_{ss'}}{\partial t} = \left(\frac{\partial\rho^{(3)}_{ss'}}{\partial t}\right)_{\text{laser}} + \left(\frac{\partial\rho^{(3)}_{ss'}}{\partial t}\right)_{\text{reservoir}} \tag{2}$$

$$\left(\frac{\partial\rho^{(3)}_{ss'}}{\partial t}\right)_{\text{laser}} = -\frac{i}{\hbar}\sum_{s''}\left\{H^{(3F)}_{ss}(t)\,\rho^{(3)}_{s''s'} - \rho^{(3)}_{ss'}\,H^{(3F)}_{s''s'}(t)\right\} \tag{3}$$

$$\left(\frac{\partial\rho^{(3)}_{ss'}}{\partial t}\right)_{\text{reservoir}} = -\frac{1}{2}(\gamma_{ss}+\gamma_{s's'})(1-\delta_{ss'})\,\rho^{(3)}_{ss'} - \delta_{ss'}\sum_{s''}\left[\rho^{(3)}_{ss}w(s\to s'') - \rho^{(3)}_{s''s}w(s''\to s)\right] \tag{4}$$

In (2) - (4), the reduced density matrix $\rho^{(3)}$ and the interaction Hamiltonian $H^{(3F)}$ between the ν_3 mode and the laser field are in an interaction picture based on the Hamiltonian $H^{(3)}$ for the ν_3 mode alone. We may conceptually separate the contributions to $H^{(3)}$ into two parts: that which contains only operators on the ν_3 mode and at least one other normal mode. The calculation of the part of $H^{(3)}$ which refers only to the ν_3 mode has been discussed in [7] and [10]; that discussion will not be repeated here. The part of $H^{(3)}$ which contains both ν_3-mode and other-mode operators, denoted $H^{(3)}{}_{\nu_3\text{-res}}$, may conveniently be chosen as [16]

$$\left(H^{(3)}\right)_{\nu_3\text{-res}} = \mathrm{tr}_R\left(\rho^{(R)}H^{(3R)}\right) . \tag{5}$$

In (5), $\rho^{(R)}$ is the reduced density matrix for the reservoir, defined as in (1); and $H^{(3R)}$ is the interaction Hamiltonian between the ν_3 mode and the reservoir. The remaining symbols in (4) will be defined below.

Physically, (5) represents the shift of the energy levels of the ν_3 mode due to the vibrational excitation of other modes. To see this, let us temporarily take for $H^{(3R)}$ the familiar form [19]

$$H^{(3R)} = \sum_{j\neq 3} X_{3j}\left(v_j+\frac{dj}{2}\right)\left(v_3+\frac{3}{2}\right) \tag{6}$$

where d_j is the degeneracy of the mode ν_j, and v_j is the number of quanta in ν_j. Then

$$\left(H^{(3)}\right)_{\nu_3\text{-res}} = \sum_{j\neq 3} X_{3j}\left(\langle v_j\rangle+\frac{dj}{2}\right)\left(v_3+\frac{3}{2}\right) \tag{7}$$

is the average shift of the energy levels with v_3 quanta of the ν_3 mode, owing to the excitation of a number of quanta $\langle v_j \rangle$ (where $j \neq 3$) in each of the other vibrational modes. Eq. (7) predicts that the average frequency of a transition in which one additional ν_3 quantum is excited will be shifted by the amount

$$\Delta\nu_{\nu_3\text{-res}} = \sum_{j \neq 3} \langle v_j \rangle \, X_{3j} \tag{8}$$

Since the anharmonicities X_{3j} are believed [20] to be all negative for SF_6, (8) predicts a shift of the peak of molecular absorbtion of laser light towards lower frequencies as molecular excitation increases. If we estimate the anharmonicities X_{3j} as in [21], we find that at a molecular vibrational temperature, T=2000 K the shift in the peak of absorption should be $\Delta\nu_{\nu_3\text{-res}} = 16\ cm^{-1}$, relative to the transition frequency from $v_3=0$ to $v_3=1$ (all other $v_j=0$, $j \neq 3$), namely $\nu_0=948\ cm^{-1}$. At a vibrational temperature T=3000 K, the shift is $\Delta\nu_{\nu_3\text{-res}} = 25\ cm^{-1}$. These examples are consistent both with the vibrational temperatures proposed in [1,2], and with the order of magnitude of the shift with respect to ν_0 of the laser frequency at which the molecular dissociation rate is observed to peak [22].

The shifts predicted above are also in rough agreement with the shift observed in transient absorption measurements on SF_6 [23]. Complete data are not yet published for SF_6, so it is not yet possible to say whether the predicted shift (8) does in fact substantiate the hypotheses advanced in [1,2]. A similar red shift as a function of vibrational temperature will, of course, occur in other spherical-top molecules such as SiF_4 [24] and OsO_4 [25].

Unfortunately the experimental demonstration of a red shift of the peak of molecular absorption as a function of the number of laser photons previously absorbed cannot be unambiguously ascribed to the phenomenon of general vibrational heating just described. The center of gravity of the manifolds into which the overtone states $v\nu_3$ are split, as discussed in [7,10,11], may tend towards the red, causing a shift of the mean absorption frequency towards the red as the number of ν_3 quanta increases. The approximate analytic solution for the ν_3 overtone levels presented in [26] may be used to show that the center of gravity of the manifold of vibrational levels (into which the overtone state $v\nu_3$ is split) lies approximately at the frequency

$$\overline{E}_v = v\nu_3 + v^2(X_{33}+\tfrac{2}{3}T_{33}) - v(X_{33}+4T_{33}) \tag{9}$$

above the vibrational ground state. The frequency of a transition from $\overline{E}_v$ to $\overline{E}_{v+1}$, which is approximately equal to the frequency of maximum absorption strength, is

$$\overline{E}_{v+1} - \overline{E}_v = \nu_3 + 2v\, X_{33} + \frac{2}{3}\,(2v-5)\, T_{33}. \tag{10}$$

For particular combinations of the vibrational anharmonic parameters X_{33} and T_{33}, which are defined in [7], [10] and [26], the transition frequency (10) will be somewhat lower than ν_3, as is apparently observed [23]. For the current best estimate [27] of the parameters X_{33} and T_{33}, (10) predicts a red shift of 184 cm^{-1} for v=30, far larger than is observed near the dissociation limit of SF_6.

In this section an equation of motion for the reduced density operator of the ν_3 mode has been written down, and some of the most important effects of the interaction between the ν_3 mode and the reservoir have been described. In the following section a qualitative discussion of the collisionless transfer of excitation from the ν_3 mode to other modes will be presented.

4. Collisionless Intermode Vibrational Coupling

At high excitation the normal modes of a molecule no longer diagonalize the vibrational Hamiltonian. An attempt at describing the actual state of the molecule in terms of the small-amplitude normal modes must therefore confront the terms in the vibrational Hamiltonian which cause energy to be transferred from one stongly pumped mode to other modes. Physically this transfer of energy leads to a shift and a spreading of the infrared absorption cross section of the molecule (as a function of frequency) as the molecule absorbs more and more laser photons. In the previous section the shift was discussed briefly. In this section the spreading of the cross section as a function of frequency will be discussed, following a brief discussion of intermode energy transfer under collisionless conditions.

Following LAX [16], we shall assume that the interaction between the ν_3 mode and the remaining modes of the SF_6 molecule (the reservoir) can be written in the separable form

$$H^{(3R)} = \hbar \sum_i (Q_i F_i + Q_i^+ F_i^+). \tag{11}$$

Eq. (11) leads to the following forms for the quantities γ_{ss}, $w(s \to s'')$, and $\Delta\omega_{ss''}$ which appear in (4):

$$\gamma_{ss} \equiv \gamma_{ss}^+ + \gamma_{ss}^-$$

$$= \sum_{s''} \sum_i \left\{ |(Q_i)_{ss''}|^2 \gamma_i^+(\omega_{ss''}) + |(Q_i)_{s''s}|^2 \gamma_i(\omega_{s''s}) \right\} \tag{12}$$

$$w(s \to s'') = \sum_i \left\{ |(Q_i)_{ss''}|^2 \gamma_i^+(\omega_{ss''}) + |(Q_i)_{s''s}|^2 \gamma_i^-(\omega_s s) \right\} \tag{13}$$

$$\Delta\omega_{ss'} = \hbar^{-1} [\Delta H_{ss}^{(3)} - \Delta H_{s's'}^{(3)}] \tag{14}$$

$$\Delta H_{ss}^{(3)} = \hbar \sum_{s''} \sum_i \left\{ |(Q_i)_{ss''}|^2 \Delta_i^+(\omega_{ss''}) - |(Q_i)_{s''s}|^2 \Delta_i^- (\omega_{s''s}) \right\} \tag{15}$$

$$\frac{1}{2} \gamma_i^{+(-)}(\omega) + i\, \Delta_i^{+(-)}(\omega) = \sum_{r,r'} \rho_{rr}^{(R)} |\langle r|(F_i^{(+)}|r'\rangle)|^2 \int_0^{\Delta t} e^{i(\omega +(-) \omega_{rr'})\tau} d\tau \tag{16}$$

In (16), the symbols which occur in parentheses are to be used together as a group. It has been assumed that the reservoir density matrix is diagonal in a basis of energy eigenstates. Also, the time Δt which occurs in (16) is the time which is used to form the finite-difference approximation $\Delta\rho^{(3)}/\Delta t$ from which (2) is derived. The exact value of Δt makes little quantitative difference so long as certain inequalities are obeyed [18]. The point of writing down (12)-(16) is to make clear how a model for the vibrational force field which explicitly gives the coupling $H^{(3R)}$ in (11) may be used to calculate the parameters which appear in (2) - (4).

Physically the terms in the reduced-density-matrix equation of motion may be recognized as follows: Eq. (3) is the coherent driving of the ν_3 mode by the laser field, described in exact quantum-mechanical terms. Eq. (4) contains three physically different contributions. The first line of (4) describes the decay of coherence (represented by the off-diagonal elements of $\rho^{(3)}$) with a rate which may be calculated from (12). The second line of (4) is a familiar stochastic rate equation for the change of the summed populations $\rho^{(3)}_{ss}$, as a result of the transfer of energy to the reservoir. The third line of (4) is a second-order shift of the transition frequency from $|s\rangle$ to $|s'\rangle$ (within the ν_3 mode).

From these comments it is clear that the parameters which appear in the phenomenological models [1-3,8,9] of CMPE can in principle be estimated from a model of the vibrational force field of the molecule. Further, it is clear that the populations of the states of the ν_3 mode which occur in these models are in fact summed populations as described in Section 2, and are not the populations of single vibrational states. Finally, we may use (4) and (13) to give an improved definition of the quasicontinuum (Region II): the beginning of the quasicontinuum occurs at the ν_3 state $|s\rangle$ whose width γ_{ss} is greater than the smallest anharmonic defect of a transition from $|s\rangle$ to a state with one more ν_3 quantum. It is evident from the comments already made on the physical interpretation of the populations $\rho^{(3)}_{ss}$ that the ν_3 state $|s\rangle$ at which the quasicontinuum begins, depends upon the state of the reservoir, i.e. depends upon the degree of excitation of the other vibrational modes. This point is discussed in somewhat more detail elsewhere [28].

5. Acknowledgements

It is a pleasure to thank J. R. Ackerhalt, R. V. Ambartzumian, N. Bloembergen, J. H. Eberly and H. W. Galbraith for stimulating discussions.

References

[1] M. J. Coggiola, P. A. Schulz, Y. T. Lee and Y. R. Shen, Phys. Rev. Lett. 38, 17 (1977).
[2] J. G. Black, E. Yablonovitch, N. Bloembergen and S. Mukamel, Phys. Rev. Lett. 38, 1131 (1977).
[3] J. L. Lyman, to be published.
[4] J. R. Ackerhalt and J. H. Eberly, Phys. Rev. A 14, 1705 (1976).
[5] D. M. Larsen and N. Bloembergen, Optics Commun. 17, 254 (1976).
[6] D. M. Larsen, Optic Commun. 19, 404 (1976).
[7] C. D. Cantrell and H. W. Galbraith, Optics Commun. 21, 374 (1977).
[8] M. F. Goodman, J. Stone, and D. A. Dows, J. Chem. Phys. 65, 5052 (1976); J. Stone, M. F. Goodman and D. A. Dows, J. Chem. Phys. 65, 5062 (1976).
[9] D. P. Hodgkinson and J. S. Briggs, Chem. Phys. Lett. 43, 451 (1976).

[10] C. D. Cantrell and H. W. Galbraith, Optics Commun. 18, 513 (1976).
[11] V. M. Akulin, S. S. Alimpiev, N. V. Karlov, and B. G. Sartakov, Zh. Eksper. i Teor. Fiz. 72, 88 (1977).
[12] N. Bloembergen, Optics Commun. 15, 416 (1975).
[13] N. Bloembergen, Nuclear Magnetic Relaxation (New York, W. A. Benjamin, Inc., 1961).
[14] C. P. Slichter, Principles of Magnetic Resonance (New York, Harper & Row Publishers, 1963).
[15] U. Fano, Rev. Mod. Phys. 29, 74 (1957).
[16] M. Lax, J. Phys. Chem. Solids 25, 487 (1964).
[17] M. Lax, Phys. Rev. 145, 110 (1966).
[18] W. H. Louisell, Quantum Statistical Properties of Radiation (New York, John Wiley & Sons Inc., 1973).
[19] G. Herzberg, Molecular Spectra and Molecular Structure. II. Infrared and Raman Spectra of Polyatomic Molecules (New York, Van Nostrand Reinhold Company, 1945).
[20] R. S. McDowell, J. P. Aldridge and R. F. Holland, J. Phys. Chem. 80, 1203 (1976).
[21] A. V. Nowak and J. L. Lyman, J. Quant. Spectrosc. Rad. Transfer 15, 945 (1975).
[22] R. V. Ambartzumian, Yu. A. Gorokhov, V. S. Letokhov, G. N. Makarov, and A. A. Puretsky, ZhETF Pis. Red. 23, 26 (1976).
[23] A. B. Petersen, J. Tiee and C. Wittig, Optics Commun. 17, 259 (1976).
[24] V. M. Akulin, S. S. Alimpiev, N. V. Karlov, A. M. Prokhorov, B. G. Sartakov and E. M. Khokhlov, ZhETF Pis. Red. 25, 428 (1977).
[25] R. V. Ambartzumian, Third International Conference on Laser Spectroscopy, 1977 (to be published).
[26] H. W. Galbraith and C. D. Cantrell, Proc. Orbis Scientiae 1977 (in press).
[27] C. C. Jensen, W. B. Person, B. J. Krohn and J. Overend, Optics Commun. 20, 275 (1977).
[28] C. D. Cantrell and J. R. Ackerhalt, Proc. International Conference on Multiphoton Processes, 1977 (to be published).

MULTIPHOTON DISSOCIATION OF POLYATOMIC MOLECULES: QUANTUM OR CLASSICAL?[1]

W.E. Lamb, Jr.

Optical Sciences Center, University of Arizona
Tucson, AZ 85721, USA

This talk deals with a theoretical study of the interaction of polyatomic molecules with an intense laser beam. The main interest in the problem comes from the experimental observation of isotopically selective multiphoton laser induced dissociation of molecules such as NH_3, SF_6, etc. under illumination by intense infrared laser radiation. The following discussion will treat the specific case of SF_6 as an example. When SF_6 is illuminated by CO_2 laser radiation of a suitable frequency, an appreciable change in the dissociation rate of SF_6 is found when the S^{32} nucleus is replaced by S^{34}. This phenomenon clearly has great potential for laser isotope separation technology. Other very important applications may be found in laser induced photochemistry.

I will first describe the main steps in an "ideal" theoretical description of laser induced photodissociation of a polyatomic molecule. In practice, however, this program is far too complicated to carry out. After making comments on the most promising current suggestion to deal with the complexities of a proper treatment, I will describe an alternative classical method which I believe is likely to give more reliable results than any but the most thorough quantum mechanical treatment.

It is characteristic of research on laser isotope separation that there are many experimental parameters to be varied: molecule under study, laser frequency, pulse duration, power, coherence and other characteristics of the radiation, buffer gas, effect of collisions, etc. I expect that the development of a realistic theoretical treatment could provide a very useful supplement to the very extensive (and expensive) experimental program on laser isotope separation.

There is little reason to doubt that, in principle, non-relativistic quantum mechanics should be able to deal very well with the molecular problem. If necessary, nuclear spin could be described by Pauli two component wave functions. The effect of molecular symmetry on allowed rotational states would follow from the symmetry imposed on the overall wave function. Hyperfine and magnetic interactions and relativistic corrections are probably not important, but could be considered if desired.

The laser field which dissociates the molecule can, in any process of interest for isotope separation, be very well represented as a classical electromagnetic field which interacts with the electrostatic charge distri-

[1] Research supported in part by the U.S. Energy and Development Administration, (Office of Advanced Isotope Separation).

bution of the molecule. The simplest case to treat would be that of a monochromatic laser, but it would easily be possible to treat multimode or noise modulated laser input. Some form of time dependent frequency variation could also be well worth considering since the active molecular "resonant frequencies" are strongly dependent on the molecular state, and will change with time during the break-up of the molecule.

The concept of a "photon" has no real utility when one deals with a classical electromagnetic field. Hence the adjective "multiphoton" used in the first paragraph really means only that one must use high order perturbation theory in coupling the classical laser field to the molecule.

The SF_6 molecule is a system of seven nuclei and seventy electrons. The Hamiltonian operator for the molecular system in the presence of the laser field is complicated but well defined. The corresponding Schrödinger equation would determine the time evolution of the wave function $\Psi(t)$ from its initial value $\Psi(0)$ at $t = 0$. Unfortunately, at this level of description the problem is hopelessly complicated. It becomes more manageable only when the adiabatic approximation is made. This exploits the fact that the nuclear masses are very much larger than the electron mass, (or that the electrons move more rapidly than the nuclei).

In the Born-Oppenheimer (adiabatic) approximation, one first finds, with more or less accuracy, the ground state electronic wave function for the seventy electrons moving in the field of the seven nuclei fixed in an arbitrary configuration. The electronic energy then acts as a potential energy for the motion of the nuclei. In all of this discussion, I assume that the electrons remain in their ground state. This prevents the formation of ions, Landau-Zener crossings and other complications too great for the present approach.

The total wave function of the molecule is a product of the above electronic wave function and a wave function for the motion of the seven nuclei. The latter wave function can be approximately factored into wave functions for the rotation, vibration and translation. This procedure works quite well for the lower rotational and vibrational states, despite the many complications such as centrifugal distortion, Coriolis interactions, Fermi resonances, tunneling, anharmonic corrections to the fifteen normal modes of vibration and their strong mixing with each other, rotation-vibration interaction, etc. However, in the photo-dissociation problem we are vitally concerned with strongly vibrating and rotating molecules. The eigenfunction basis provided by the fifteen normal modes of vibration is a clumsy one for the representation of the dissociation of the molecule. It would take a very high order perturbation theory and the consideration of an enormous number of states to give an adequate treatment of the dissociative process. Even at this level of description, the problem is still hopelessly complicated.

In order to deal with the interaction of the molecule with the time dependent electric field of the laser one would have to calculate the charge density distribution of the molecule as a function of the nuclear coordinates. A very crude approximation would be to assign fictitious charges: $-Q_i$, $i = 1,2,\cdots,6$ for the six fluorine atoms and $\Sigma_i Q_i$ to the sulphur atom, with the Q_i determined from data on the intensities of the vibrational spectrum of the molecule. The values of Q_i would have to be allowed to vary with atomic configuration in some plausible way as the dissociation process proceeds.

Bloembergen has suggested that the dissociation process should be thought of as occurring in two stages. The state of the molecule is treated by conventional quantum mechanics in the first stage. At something like the fourth or fifth vibrational level of excitation, he exploits the high density of molecular states. The molecule is considered to "leak into a quasi-continuum" of bound vibrational and rotational states. The transition to this quasi-continuum is described in what seems to me to be a rather casual manner by use of rate constants. The upward diffusion through the quasi-continuum until the true continuum and dissociation is finally reached seems to be even less based on fundamental quantum mechanical theory.

There is a close analogy between the problem of breaking up of a molecule by a laser field, and a science fiction project for breaking up the solar system by a strong gravitational wave. Because of the large masses involved and large "quantum numbers", no one would try to deal with the latter problem using quantum mechanics. The classical Newtonian equations of motion would be used to describe the changing orbits of the planets. On the other hand, almost all physicists would agree that quantum dynamics should be used to describe molecular structure. However, as I have indicated above, this would involve a very difficult calculation. Besides this practical consideration, moreover, there are some other good reasons for considering the classical approach more carefully. The border line between classical and quantum mechanics is not well understood. One usually thinks that classical mechanics applies in the "limit of large quantum numbers." Large vibrational and rotational quantum numbers are certainly involved in the molecular dissociation problem. The large number of degrees of freedom, and the closely spaced molecular energy levels found even at moderate excitation may also favor a classical approach. Because of their large mass differences it is certainly a better approximation to treat the nuclear motion classically than that of the electrons. The Born-Oppenheimer approximation quite naturally provides such a separation of classical and quantum regions.

I will now describe the proposed treatment of the molecular dissociation problem. As in the above outlined quantum calculation, one could start with the best available eigen-energy for the electronic motion in the field of fixed nuclei and the corresponding charge density distribution. These would give the potential energy of the atoms and a perturbation term for the laser interaction. If desired, one could instead choose the potential energy function empirically so that it would lead to the correct equilibrium configuration of the molecule and the correct frequencies of the fifteen normal modes of vibration. Other severe tests would be that the dissociation energy for removal of a fluorine atom was obtained, as well as the correct equilibrium configuration of the SF_5 fragment molecule. Then one could write down the classical Newtonian equations of motion for the seven atoms of SF_6 under the action of the interatomic forces and the laser field. These would be a coupled system of twenty one second order differential equations. With elimination of the center of mass coordinates, the number reduces to eighteen.

One would then begin integration of the classical equations of motion from some initial configuration of the molecule SF_6. Several possibilities exist. One could start in the equilibrium configuration with all atoms at rest, and find out whether, according to some reasonable criterion, the molecule was dissociated by the laser field in a specified time.

Admittedly, classical mechanics is not a complete substitute for quantum mechanics. Here is a partial list of shortcomings of the classical treatment. (1) The zero point oscillations and quantum fluctuations are ignored. (2) In general, a quantum mechanical wave packet does not follow the corresponding classical trajectory. Also the packet usually spreads, but less so for large masses than for small ones. (3) Tunneling through potential barriers does not occur. (4) No rotational states are excluded by symmetry conditions imposed on the wave function. I now comment briefly on these four points.

(1) A plausible way to allow for zero point oscillation would be to start the molecule off in a number of configurations slightly different from the equilibrium position, and to accumulate statistics on the dissociation process. The initial configurations could be distributed in accord with the square of the ground state wave function in order to simulate the effects of zero point vibrations.

(2) Ehrenfest's theorem of quantum mechanics can be used to show that the centroid of a wave packet for a harmonic oscillator under the action of a laser field follows exactly a classical trajectory for the same problem. This result is only valid for an anharmonic oscillator to the extent that the anharmonic part of the force does not vary across the wave packet. If the variation is small, an approximation, based on a Taylor's expansion, shows that the motion of the centroid is still classical, but for a slightly modified Hooke's law restoring force. This slightly shifts the resonance frequency of the oscillator. Unfortunately, this shift depends on the (variable and unknown) spreading of the wave packet. It seems that a better method is needed. The early stages of the molecular excitation involve mostly the nearly resonant interaction of the CO_2 laser radiation with the three-fold degenerate ν_3 modes. I have oversimplified the problem by considering a single degree of freedom forced anharmonic oscillator (Duffing problem). I integrated the Heisenberg matrix equations of motion for the Duffing oscillator. I found that for realistic anharmonicities and laser intensities the frequency dependence of energy absorption was very similar to that given by a classical calculation.

(3) In many molecules the tunneling frequency is low compared to the time scale of the laser induced disruption. In any case, the classical treatment will use whatever multi-minimum potential is used to describe the tunneling phenomenon.

(4) It seems to me that a few missing rotational states (out of so many) would be the least of our worries.

One of the big advantages of the classical approach is that one is always working in an inertial frame of reference. It is not necessary to transform to rotating axes fixed in the molecule. As a result, no centrifugal distortions, Coriolis forces or vibration-rotation interactions have to be considered. Once a suitable interatomic potential has been chosen, it is not necessary to consider the normal modes of vibration or the "radiationless transitions" between them which arise from the anharmonic forces. All of these things are automatically contained in the solution of the Newtonian equations of motion.

Obviously one must worry about the extent to which the classical treatment should be used. My present belief is that one will get a good approximation by starting from the equilibrium configuration, but this can easily be tested in the following manner. We can follow Bloembergen's approach with a quantum treatment for the first few vibrational levels. We can then use the wave function of the molecule to get a starting probability distribution of initial configurations for a classical treatment of the later stages of the dissociation process. I have already mentioned the possible use of this kind of recipe for dealing with the zero point oscillations.

Another very great advantage of the suggested method is that the vague concepts of "leakage into the quasi-continuum" and diffusion into the continuum are avoided entirely. It is not necessary to introduce phenomenological rate and diffusion constants. It is also very easy to similate the effect of collisions in a classical treatment by subjecting one or more of the atoms of the molecule to random impulsive forces.

Calculations based on this program are now underway.

III. New Sub-Doppler Interaction Techniques

PROGRESS IN UNDERSTANDING SUB-DOPPLER LINE SHAPES

C.J. Bordé

Laboratoire de Physique des Lasers, Université Paris-Nord
Avenue J.-B. Clément, 93430 Villetaneuse, France

The application of sub-Doppler spectroscopic methods to optical frequency standards has stimulated many efforts to improve our knowledge of the line shapes in order to provide realistic estimates of the accuracy. In this paper we shall consider four directions of recent progress :

1- the recoil structure of the saturation resonances
2- transit effects on the line shapes and especially curvature-induced shifts
3- the influence of weak elastic collisions in saturation spectroscopy
4- the use of RAMSEY interference fringes in Doppler-free two-photon and saturation spectroscopies.

I - LIGHT SHIFT AND POLARIZATION DEPENDENCE OF THE RECOIL PEAKS

Owing to the atomic recoil during the interaction with light, every saturation resonance is usually a doublet (with the exception of crossover resonances). The origin of this doubling for a two-level system is illustrated in figure 1.

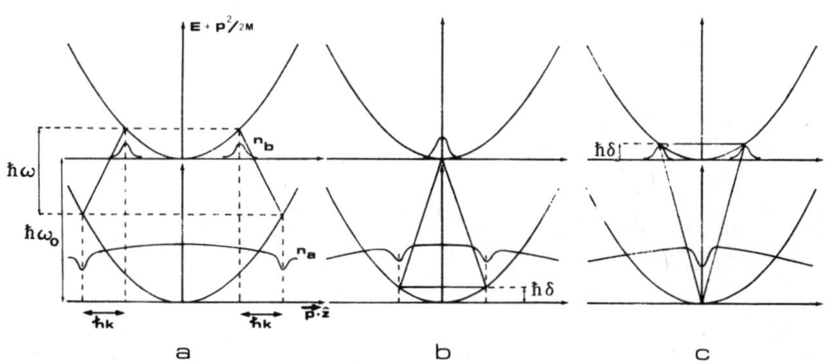

Figure 1

In a non-relativistic approximation, the curve representing the total energy versus the momentum component along the optical $\hat{z}$-axis is a parabola in each level. Because of momentum conservation, molecules interacting with a monochromatic field of a given direction have to belong to two different velocity classes in the lower and in the upper levels. The momentum change is precisely equal to the light quantum momentum $\pm\hbar k$ (where the sign depends upon the wave direction with respect to the $\hat{z}$-axis). There are two situations where one of the two waves can produce a change in the absorption of the other. The first (figure 1b) is obtained when molecules in the upper level can interact with both waves. The corresponding energy exchanged with the light can be read directly on the vertical axis. From the equation of the parabola it is easily seen that this energy is smaller than the Bohr energy E_b-E_a by the amount $(\hbar k)^2/2M$. Similarly there is a second resonance obtained when molecules in the lower level can interact with both waves (figure 1c) and for which the energy exchanged is bigger than the Bohr energy by the same quantity. In first approximation each of these peaks has an intensity proportional to the lifetime of the velocity group in the corresponding level. This doublet structure was demonstrated with CH_4 at 3.39 µm, first in 1973 through a careful line shape study [1] . The 2.16 kHz splitting was then clearly resolved with large aperture parabolic optics to reduce the half-width below one kilohertz [2]. In fact Planck's constant measured by this method has a tendency to be too small and even to decrease with the light intensity ! This effect is now understood as a light shift of each recoil component towards the other, in a theory which treats one field to all orders and the probe field to first order [3]. In this approximation only four energy-momentum classes are coupled together by the light as illustrated on figure 2.

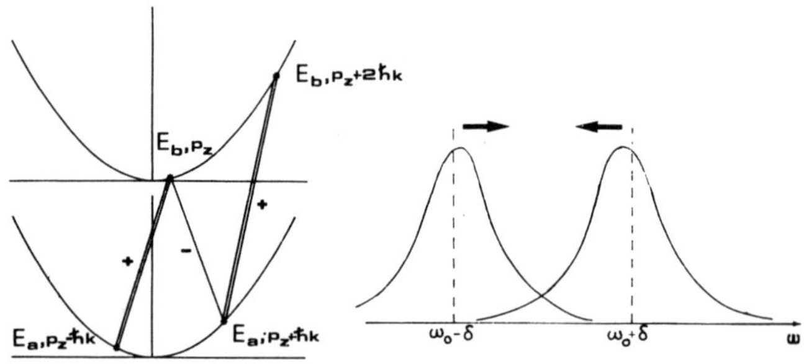

Figure 2

For plane waves this leads to a generalization of the BAKLANOV-CHEBOTAEV line shape :

$$\mathrm{Re} \sum_{\alpha=a,b} \left[\gamma_{ba} + \Gamma - 2i\,(\omega-\omega_o + \varepsilon_\alpha \delta)\right]^{-1} \left[\gamma_\alpha^{-1} - D_\alpha^{-1} \left\{ \left(\gamma_{ba} + \Gamma - 2i\,(\omega-\omega_o + \varepsilon_\alpha\delta)\right)^{-1} \right.\right.$$

$$\left[\left(\gamma_{\beta\neq\alpha} + 2\Gamma - 2i\,(\omega-\omega_o + \varepsilon_\alpha\delta)\right)^{-1} + \left(\gamma_\alpha + 2\Gamma - 2i\,(\omega-\omega_o + 3\varepsilon_\alpha\delta)\right)^{-1}\right] (\Omega^+)^2 \gamma_\alpha^{-1}$$

$$\left.\left. + \tfrac{1}{2}\left(\sqrt{1+S} - 1\right) \left[\gamma_{\beta\neq\alpha} + 2\Gamma - 2i\,(\omega-\omega_o + \varepsilon_\alpha\delta)\right]^{-1} \right\}\right]$$

with $D_\alpha = 1 + (\Omega^+)^2 \left[\left(\gamma_\alpha + 2\Gamma - 2i\,(\omega-\omega_o + 3\varepsilon_\alpha\delta)\right)^{-1} + \left(\gamma_{\beta\neq\alpha} + 2\Gamma - 2i\,(\omega-\omega_o + \varepsilon_\alpha\delta)\right)^{-1}\right]$

$$\left[\left(\gamma_{ba} + \Gamma - 2i\,(\omega-\omega_o + \varepsilon_\alpha\delta)\right)^{-1} + \left(\gamma_{ba} + 3\Gamma - 2i\,(\omega-\omega_o + 3\varepsilon_\alpha\delta)\right)^{-1}\right]$$

where $\beta = a,b$, $\hbar\omega_o = E_b - E_a$, $\varepsilon_a = -1$, $\varepsilon_b = +1$,

$\gamma_\alpha, \gamma_{ba}$ are the relaxation constants for the population of level α and for the off-diagonal element of the density matrix,

$\Omega^+ = \mu_{ba} E_o^+ /2\hbar$ is the Rabi pulsation for the matrix element μ_{ba} of the dipole operator and the amplitude E_o^+ of the saturating field

$S = 2\,(\Omega^+)^2\,(\gamma_a^{-1} + \gamma_b^{-1})/\gamma_{ba}$ is the saturation parameter,

and finally $\Gamma = \gamma_{ba}\sqrt{1+S}$, and $\delta = \hbar\omega_o^2 / 2Mc^2$

Besides the main contributions coming from population effects, we find contributions from higher order coherent processes, which are not symmetric with respect to the centers of each peak. These extra-contributions build up the inner side of each peak and thus lead to an apparent pulling of the peaks towards each other. In the limit of well resolved peaks the shift is given in first approximation by $(\Omega^+)^2/8\delta$ in circular frequency units. This light shift is a problem for optical frequency standards such as CH_4 at 3.39 µm and Ca at 6573 Å which are based on saturation lines for which the recoil doubling is not very large compared to the line-width. Other possible dependences of the line center with the light intensity come through the light shift caused by a third coupled level, differential saturation of any unresolved structure (recoil or hyperfine) [1] or differential saturation among transverse velocity classes changing the effective second-order Doppler shift [4]. Another interesting feature about the recoil doublet is the dependence of the relative intensities of the two components upon the polarization properties of the laser beams [5]. As an example let us consider the simple case of a J = o ↔ J = 1 transition such as the $^1S_0 \rightarrow {}^3P_1$ transition of calcium at 6573 Å. With opposite circular polarizations of the counter-propagating beams (retro-reflected circularly polarized light) and the quantization axis chosen along the propagation axis, both beams induce σ^+ (or σ^-) transitions between M sublevels as illustrated on figure 3a.

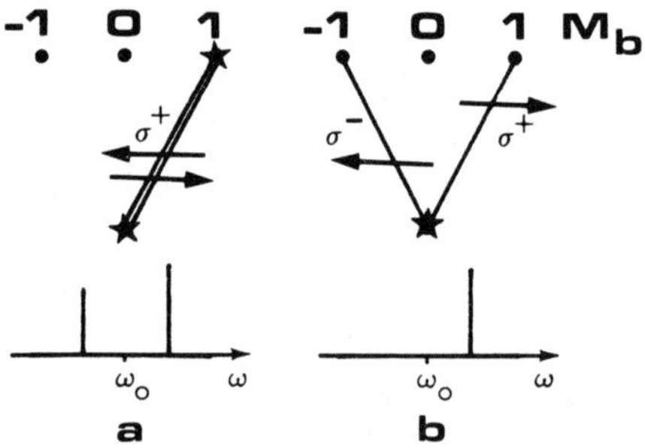

Figure 3

Both levels contribute symmetrically and, if the relaxation constants are approximately equal, the two recoil peaks have equal intensities. An identical conlusion is reached with parallel linear polarizations and is applicable to the CH_4 experiments. With identical circular polarizations, the two counter-propagating beams induce respectively σ^+ and σ^- transitions and only the crossover resonance of the J = o level is obtained. If the J = o level is the lower one, as on figure 3b, only the higher frequency recoil peak can occur. The same result is obtained for perpendicular linear polarizations. This property may be used to avoid the accuracy problem brought by the recoil structure. Closed-form formulae obtained by summing the corresponding products of Clebsch-Gordan coefficients are given in reference [5]. As another example, in the case of the F = 6 ↔ F = 7 transition of CH_4 at 3.39 µm and identical circular polarizations, the lower frequency recoil peak is 2.06 times bigger than the higher frequency one.

II - <u>INFLUENCE OF TRANSIT EFFECTS ON THE LINE SHAPES - DENSITY MATRIX DIAGRAMS</u>

In the high resolution limit the line shape is dominated by transit effects across the laser beam. It becomes essential to introduce in the equations the exact time-dependent perturbation corresponding to the shape of the pulse seen by the molecule in its flight across the beam. We have taken two approaches to this problem [4]. The first one is purely numerical : the coupled density matrix equations are solved with a computer and a numerical integration is performed on axial and transverse velocities. This method is especially useful for strong field problems (saturation effects) or arbitrary spatial dependence of the laser fields [6]. The second approach is a third order perturbation method which leads to analytical results [4].

During these studies density matrix diagrams have been a most useful tool [4,7]. It seems therefore appropriate to give first a brief introduction to these diagrams. The simplest process that one can consider is a single quantum transition of a molecule from level a to level b. The corresponding diagram is shown on figure 4a.

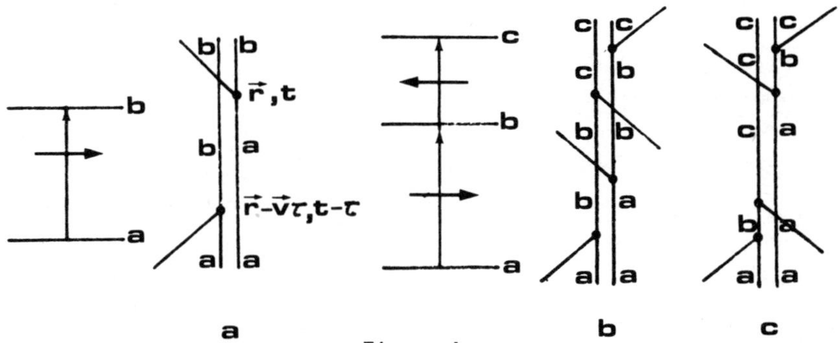

Figure 4

The density matrix elements for the molecule are represented by a vertical double bar along the time axis. To each segment corresponds a subscript of the matrix element ρ_{aa}, ρ_{ba}, ρ_{bb}. Each interaction with the field is represented by a vertex between two segments. The field is represented by a lateral line coming from the right or from the left according to its direction with respect to the $\hat{z}$-axis. The field is separated in the exp $(-i\omega t)$ part (line going upwards) and in the exp $(i\omega t)$ part (line going downwards). At each vertex the rotating wave approximation leaves a single possibility between these two : on the first column the line has to go upwards when the subscript is changed from that of a lower energy level to a higher energy one (for example a to b on figure 4a) the line goes down when the corresponding energy decreases (b to a) and these rules are reversed on the second column. To each time interval τ appearing between two vertices and corresponding to the density matrix element $\rho_{\alpha\beta}$ we associate the propagator exp- $(i\omega_{\alpha\beta} + \gamma_{\alpha\beta})\tau$. The diagram is terminated at space-time point $(\vec{r},t)$ by a last vertex. If the molecules have the velocity $\vec{v}$ the space-time coordinates of the previous vertices are determined by the various time intervals : $(\vec{r}-\vec{v}\tau,\ t-\tau)$, $(\vec{r}-\vec{v}\tau-\vec{v}\tau',\ t-\tau-\tau')$... The matrix element of the interaction hamiltonian for each vertex is taken at the corresponding space-time point : $V_{\alpha'\alpha}(\vec{r}-\vec{v}\tau-\vec{v}\tau'...,\ t-\tau-\tau'...)/i\hbar$ for a change from α to α' on the first column, $-V_{\beta\beta'}/i\hbar$ for a change from β to β' on the second column. Only the part of the field corresponding to the line orientation is kept in the interaction hamiltonian. One has then to perform the integration over all time intervals $\tau, \tau' \ldots$ (from 0 to $+\infty$) of the products of all the quantities appearing on the diagram, times the initial equilibrium population. The diagram then gives the rate of change of the last density matrix element (usually a population) at space-time point $(\vec{r},t)$. To get a fluorescence signal or the absorbed power, final integrations over space and the velocity distribution need to be performed. The mathematical justification of these diagrams may be found in the Appendix A of reference [4] and a detailed example of application to Doppler-free two-photon spectroscopy is given in reference [7]. In the linear absorption diagram of figure 4a, the equilibrium population of the lower level ρ_{aa} is turned by a first interaction into the off-diagonal element ρ_{ba} which in turn will give rise to a change in the upper state population ρ_{bb} at the second vertex ; a second contribution comes from the complex conjugate diagram involving ρ_{ab}. For a plane wave the application of the preceding rules leads to the familiar VOIGT profile. As a second example, let us consider the three-level system of figure 4b interacting with two fields having opposite directions. If we put two of the previous diagrams in sequence, we get one of the diagrams representing stepwise excitation from level a to level b to level c. If we exchange the order of the two middle vertices, instead of going through the intermediate level population ρ_{bb}, the corresponding process goes through the off-diagonal element ρ_{ca} which has a resonant behaviour for the frequency ω_{ca} (figure 4c). The corresponding diagram is thus one of those describing Doppler-free two-photon spectroscopy [7]. If we finally turn to saturated absorption spectroscopy for a two-level system, the only two fourth-order diagrams of interest are represented on figure 5.

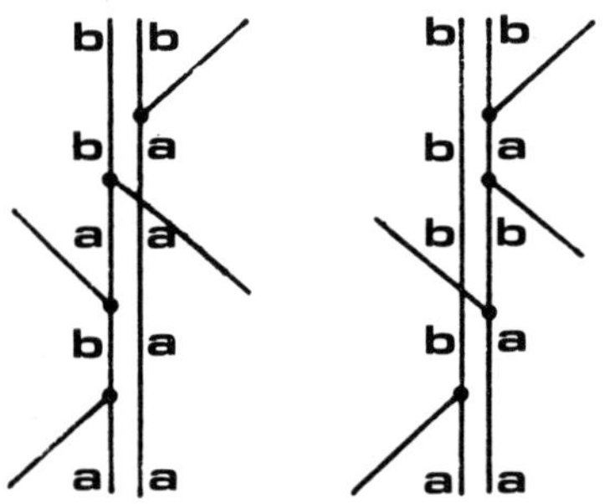

Figure 5

In a first process the lower state population is turned into an optical coherence which in turn gives a population hole in the lower state population. This hole is then probed by the opposite direction wave as described by the two next interaction vertices. This diagram represents the higher frequency recoil peak and there is a similar diagram for the low frequency one, going through the upper state population change. For Gaussian laser beams, all the integrations but one may be performed analytically and the Fourier transform of the line shape for each recoil peak is obtained [4] :

$$\mathrm{Im}\int_0^{+\infty} d\tau\ \gamma_\alpha Y^{-1}\Big[\exp(X-iY)E_1(X-iY)-\exp(X+iY)E_1(X+iY)\Big]\ \exp^{-}\Big\{2\big[i(\omega-\omega_0+\varepsilon_\alpha\delta)-\gamma_{ba}\big]\tau\Big\}$$

where E_1 is the exponential integral function of the following arguments :
$X = \gamma_\alpha\ (B/A)\tau,\quad Y = \gamma_\alpha\left(1+ Du^2\tau^2 - i\omega_0(u^2/c^2)\tau\right)^{1/2}/u\sqrt{A}$
A,B,D, are coefficients easily calculable from the beam geometry. The second order Doppler shift is included in Y. One may take advantage of having a Fourier transform to include the effect of the laser frequency modulation by multiplying the integrand by the proper Bessel function [4,8] or the broadening due to the laser frequency jitter with the help of the convolution theorem. This line shape formula describes fairly well the transit-time broadening and especially the line-width collapse at low pressures due to the anomalous contribution of slow molecules. The following interesting formula for the half-width was derived in [4] for this low pressure regime :
$\Delta\ \xi = \sqrt{\eta}\,2^{-1/4}\exp\left(\frac{3G}{\pi} - \frac{\Gamma}{2}\right)$ where the half-width and the relaxation constant are in units of the average reciprocal transit-time across the beam waist radius $\Delta\ \xi = \Delta\omega\ w_0/u$, $\eta = \gamma w_0/u$; G and Γ are respectively Catalan's and Euler's constants. The reduction in second-order Doppler shift in this regime is also well described by our line shape. Another new important result was brought by this theory : owing to the curvature of the wave-fronts, the coefficient B is usually complex and introduces a shift and an asymmetry in the line shape. A quick but rough physical explanation for this shift is the following : let us assume that we have a diverging saturation beam and a matched converging probe beam. The molecules entering on either side of this system first see a blue-shifted saturation beam because of the outwards component of its wave-vector. Part of this information (population hole or peak created) is carried across the beam by the molecular motion and probed at a later time by the probe beam which is also blue-looking for the molecules leaving the field region. The net result is a red shift of the resonance. This shift is exactly reversed if the roles of both beams are reversed. This effect is a very serious problem for the accuracy of optical frequency standards based on saturation spectroscopy since shifts of several hundred kilohertz have been observed. A quantitative demonstration of the effect has been performed with methane at 3.39 µm [8] and fortunately the agreement with

the calculated line shape is so good that one can feel confident enough in the present understanding of the physics to suggest conditions where this shift is very small and to calculate an upper limit of its value. The Fourier transform of the line shape has also been derived for Doppler-free two photon spectroscopy with Gaussian beams [7] :

$$\mathrm{Re}\int_0^{+\infty} d\tau\ (1+\beta u^2\tau^2-i\omega_{ca}\tau u^2/2c^2)^{-1}\ (1-i\omega_{ca}\tau u^2/2c^2)^{-1/2}\ \exp\left\{\left[i(2\omega-\omega_{ca})-\gamma_{ca}\right]\tau\right\}$$

where β is a real parameter (that can be derived from the beam geometry) and thus no curvature-induced shift can occur in this case (for matched beams β is simply w^{-2}). The integration leads to simple formulae for the line shape when either the transit time broadening or the second-order Doppler shift may be neglected [7] .

III - INFLUENCE OF WEAK ELASTIC COLLISIONS ON THE LINE SHAPE

An important piece of physics is missing from our previous saturation line shape : there is a pressure range where weak elastic collisions may play

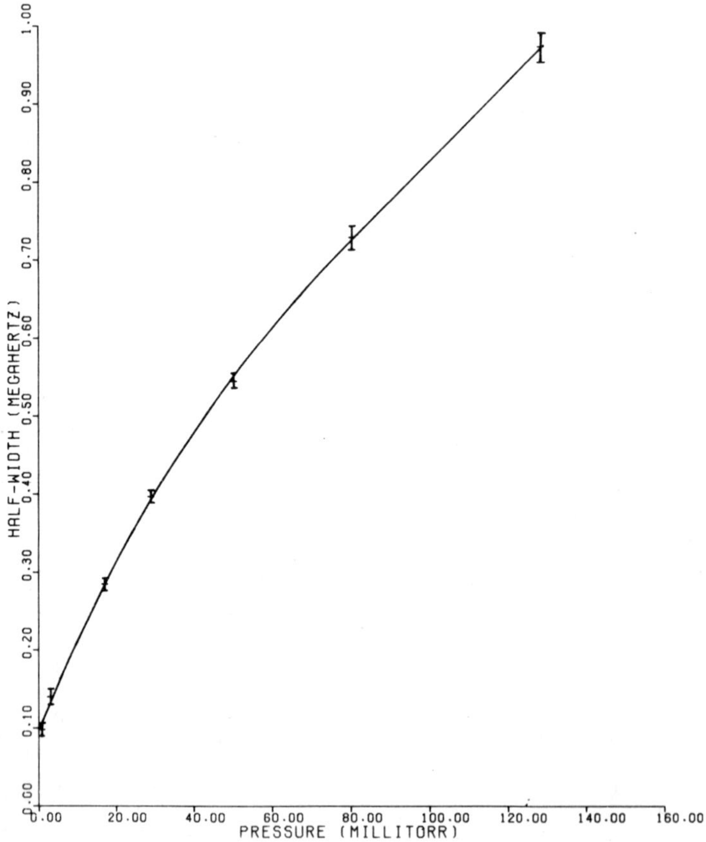

Figure 6

a major role in the broadening of saturation resonances. There has been an increasing set of experimental evidences for this, either from transient experiments [9] or directly from the anomalous behaviour of the broadening rate versus pressure [10,11]. As an example the figure 6 shows the width versus pressure of one of the iodine saturation peaks in coincidence with the Argon laser at 5145 Å [11].

The non-linearity in this curve may be explained and even fitted by introducing weak elastic collisions in the theory. Other observations of this kind have been reported with molecules in the infrared : CH_4, CO_2, NH_3 [10]. We have been able to show, either by a fully diagrammatic approach or by taking the Fourier transform with respect to v_z [12], that the effect of these collisions may be simply introduced in the previous saturation line shape formulae by the following replacement rules :
$\gamma_\alpha \rightarrow \gamma_\alpha - \widetilde{W}_\alpha(\tau)$, $\gamma_{ba} \rightarrow \gamma_{ba} - \Phi(\tau)$ where $\widetilde{W}_\alpha$ and Φ are obtained from the collision kernels W_α for the populations and W_{ba} for the off-diagonal density matrix element [12] by :

$$\widetilde{W}_\alpha(\tau) = \int_{-\infty}^{+\infty} dv_z W_\alpha(v_z) \exp(ikv_z\tau), \qquad \Phi(\tau) = \int_{-\infty}^{+\infty} dv_z W_{ba}(v_z) \operatorname{sinc}(kv_z\tau)$$

Of course γ_α and γ_{ba} now include the velocity-changing departure rates as well as dephasing and quenching contributions. The collision kernels can now be calculated from the knowledge of the intermolecular potential. These functions are very strongly peaked around the initial velocity with a width of the order of 10^{-2} that of the Maxwell distribution. They will therefore play a major role in the line shape only when the line-width is of the order of 1% of the Doppler width. At very high resolution they give only a very broad and flat contribution and the width comes only from the transit time and from the total decay rate of the optical dipole. At intermediate resolution the collision kernels are responsible for an increase in pressure broadening. At low resolution they mostly act like delta-functions and compensate the departure decay rates coming from elastic collisions. At very high pressures the wings and the slightly asymmetric character of the kernels manifest themselves in the atomic case through a broad background, whereas in the molecular case, rotational relaxation dilutes this contribution from strong or multiple weak collisions over many levels. The saturated absorption line shape has thus a much richer content than what could be anticipated in the early days of saturation spectroscopy and some further interactions between theory and experiments are still necessary to get the full picture.

IV - RAMSEY INTERFERENCE FRINGES IN OPTICAL SPECTROSCOPY

A very elegant way out of the limitation imposed by transit-time broadening was proposed by N.F. RAMSEY in 1950 [13]. The proper way to extend this technique to sub-Doppler spectroscopy was discussed by Y.V. BAKLANOV and coll. very recently [14] and beautifully demonstrated by J. BERGQUIST and coll. [15] in the case of saturation resonances of CH_4 and Ne. The idea is to replace the large beam by two or more smaller field regions. To illustrate the principle of the method it is usual to describe the motion of the pseudo-spin but one gets an even simpler picture by considering the first-order perturbation term described by the diagram on figure 7.

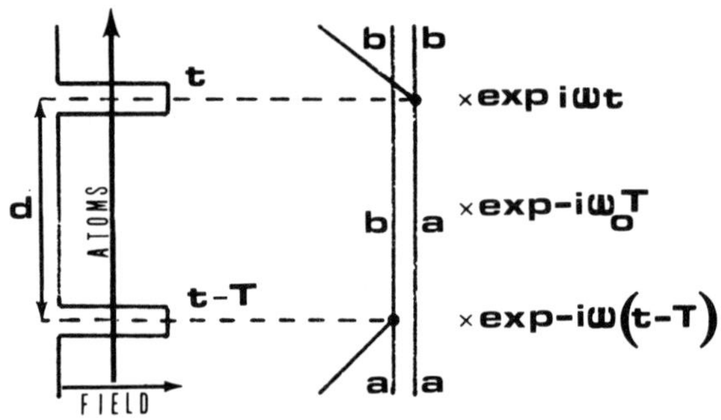

Figure 7

In the first field region a first interaction creates a dipole whose complex representation is proportional to the off-diagonal density matrix element ρ_{ba}. In the region free of field this dipole precesses at the Bohr frequency ω_0 and therefore accumulates a phase angle $\omega_0 T$ appearing in the corresponding propagator, where T is the time of flight across the distance d with velocity v. In the same time interval the field vector accumulates the phase angle ωT. The projection of the dipole on the second field is therefore multiplied by the factor exp i $(\omega-\omega_0)T$. If we add the complex conjugate diagram contribution we find that the power absorbed in the second zone is a broad function corresponding to the precise shape of each field region times cos $(\omega-\omega_0)d/v$ and that the line shape therefore exhibits fringes with an oscillation frequency proportional to the distance between the two regions of field. The integration over a velocity distribution will tend to destroy the side fringes and leave only the central one. Any fixed phase difference between the fields in both regions will appear in the cosine and displace the fringe system so that great care has to be taken to avoid this source of error in microwave spectroscopy.

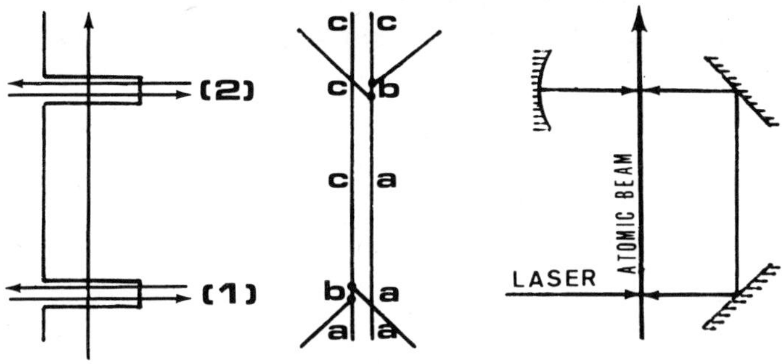

Figure 8

Figure 8 illustrates the extension of RAMSEY's idea to Doppler-free two-photon spectroscopy. This time we have at least two interactions in each field region, one with each of the two counter-propagating fields and it is the off-diagonal element ρ_{ca} which freely precesses between the two field regions. The phase factor that appears is exp $\left[i(\varphi_2^- + \varphi_2^+ - \varphi_1^- - \varphi_1^+) \right]$ where each φ comes frome one of the four fields. This total phase cancels out in a folded standing wave and this is precisely the geometry that was proposed by Y. BAKLANOV in [14] . The precise line shape with the relaxation and the Gaussian structure of the laser beams taken into account is derived in [16] where it is also shown that the beams need to be carefully matched to avoid shifts. In this case if $\Delta = 2\omega - \omega_{ca}$ is the detuning, β a real parameter depending upon the beam geometry (β is w^{-2} for beams having the same waist radius w) the fringes are proportionate to : $\exp(-\Delta^2/4\beta v_r^2) \cos\left[\Delta\,(2\beta v_x d - \gamma_{ca})/2\beta v_r^2\right]$ where d is the distance between the beams in the x direction and $v_r^2 = v_x^2 + v_y^2$.

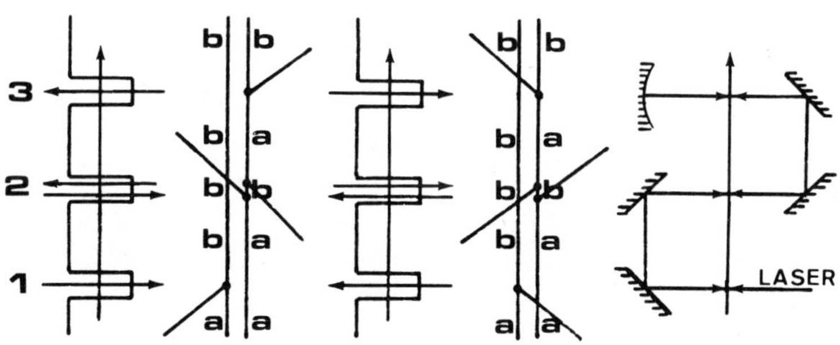

Figure 9

In saturation spectroscopy it is essential to add a third field region to the two others (figure 9). The reason for this third region lies in the fact that saturation spectroscopy is a two-step process : first the selection process of a velocity group, second the probing process of this selected group. Roughly speaking, to each of these two steps corresponds a response function and it is the convolution over v_z of these two functions that gives the final shape of the resonance. We must therefore repeat the RAMSEY operation twice : once to create a system of fringes on the population holes or peaks and a second time to probe this population change. If there were only two field zones, only one of these two steps would beneficiate from the narrowing, and the convolution with the second response function would average out the fringes. If we look at the time sequence on the fourth-order density matrix diagrams for one of the recoil peaks, we have a first interaction with one of the two waves in the first field zone giving a first-order dipole, which then freely precesses in the first dark zone and is turned into a population change with fringes in the v_z space. This population change interacts with the second wave in the central zone, gives a second dipole that precesses freely in the second dark zone to end up as energy absorbed for the last wave. Again there is a phase factor involving the fixed phases of the four fields : $\exp\left[i(\varphi_3^- - \varphi_2^- + \varphi_2^+ - \varphi_1^+) \right]$ which can be a source of shifts. Again the way out is the use of a folded standing wave. In this case we add the contribution of the second diagram of figure 9 where the first interaction is with the other choice in the first zone and the phases are such that the phase factor is complex conjugate of the first one. The asymmetric part of the two contributions versus frequency cancels out if the intensities of the two waves are identical (reflexion coefficent equal to one). This ideal situation of two waves of equal intensities was already required to cancel the curvature-induced shift [4,8] and we find here an analogous phenomenon. Some care has therefore to be taken to avoid these shifts. For small beam separations the cat's eye geometry proposed by J.C. BERGQUIST et al. [15] is even better than the simple folded standing wave since the total phase is identically zero. The line shape with relaxation and Gaussian beams is also derived in [16] . It is shown that the fringes disappear very quickly when the lengths of the dark zones differ from each other. In the special case of equal distances d and equal relaxation constants $\gamma_\alpha \equiv \gamma_{ba}$ it is found that the fringes are proportional to :

$$\exp\left[-(\omega-\omega_o)^2 w_o{}^2/v_r^2 \right] \cos\left[(\omega-\omega_o)\ (2v_x d/w_o{}^2 - \gamma_{ba})\ w_o{}^2/v_r^2 \right]$$

The strong-field numerical program [6] solving the coupled density matrix equations has also been found especially useful to study RAMSEY fringes.

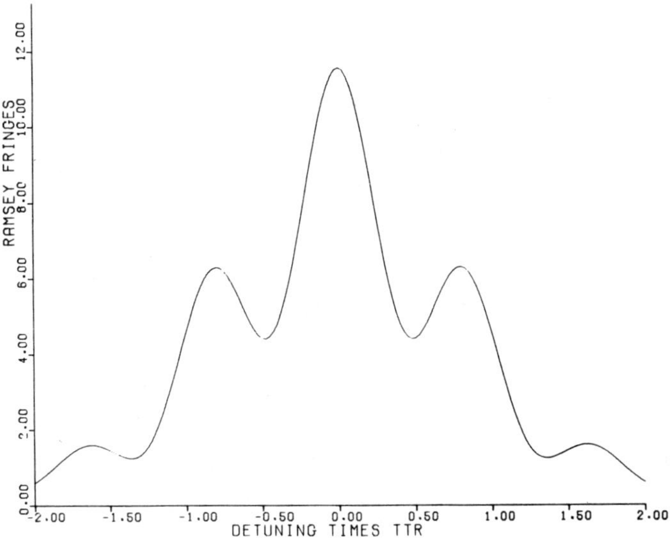

Figure 10

Figure 10 shows an example of application to three Gaussian beams separated by four times their beam waist redius w_o. The relaxation constants were all chosen equal to 0.1 and the two Rabi frequencies $\Omega^{\pm}$ equal to 0.01 in units of TTR = w_o/v_x. The same program gave asymmetric fringes for unequal beam intensities and a total phase different from zero. Some further confrontations between theory and experiments should bring us enough knowledge of these systematic effects to fully benefit from the increase of resolving power resulting from this beautiful technique.

Aknowledgment

I have shared the concern in the accuracy problem in high precision spectroscopy with Dr. John HALL for a number of years and many results of this paper are either a direct output or a consequence of our friendly collaboration.

REFERENCES

1. C.J. Bordé and J.L. Hall, in Laser Spectroscopy, edited by R.G. Brewer and A. Mooradian (Plenum, New York, 1974), pp. 125-142.

2. J.L. Hall and C.J. Bordé, Bull. Am. Phys. Soc. Series II, 19, 1196 (1974); J.L. Hall, C.J. Bordé and K. Uehara, Phys. Rev. Letters 37, 1339-1342 (1976)

3. C.J. Bordé, C.R. Acad. Sc. Paris 283 B, 181-184 (1976)

4. C.J. Bordé, J.L. Hall, C.V. Kunasz and D.G. Hummer, Phys. Rev. A 14, 236-263 (1976)

5. J. Bordé and C.J. Bordé, C.R. Acad. Sc. Paris, to be published.

6. C.J. Bordé and J.L. Hall, Eighth International Quantum Electronics Conference, San Francisco (1974) ; C.J. Bordé, J.L. Hall and C.J. Kunasz, to be published

7. C.J. Bordé, C.R. Acad. Sc. Paris 282 B, 341-344 (1976)

8. J.L. Hall and C.J. Bordé, Appl. Phys. Letters 29, 788-790 (1976)

9. T.W. Hänsch, I.S. Shahin and A.L. Schawlow, Phys. Rev. Letters 27, 707 (1970)

10. S.N. Bagaev, E.V. Baklanov and V.P. Chebotaev, JETP Letters 16, 9-12 (1972)
T.W. Meyer, C.K. Rhodes and H.A. Haus, Phys. Rev. 12, 1993-2008 (1975)
L.S. Vasilenko, V.P. Kochanov and V.P. Chebotaev, Optics Comm. 20, 409-411 (1977)
A.T. Mattick, N.A. Kurnit and A. Javan, Chem. Phys. Letters 38, 176-180 (1976)

11. G. Camy, B. Decomps and C.J. Bordé, to be published

12. C.J. Bordé, S. Avrillier and M. Gorlicki, Journal de physique Lettres (July 1977)

13. N.F. Ramsey, Phys. Rev. 78, 695-699 (1950)

14. Y.V. Baklanov, B.Y. Dubetsky and V.P. Chebotaev, Applied Physics, 9, 171-173 (1976)
Y.V. Baklanov, V.P. Chebotaev and B.Y. Dubetsky, Appl. Phys. 11, 201 (1976)

15. J.C. Bergquist, S.A. Lee and J.L. Hall, Phys. Rev. Letters 38, 159-162 (1977)

16. C.J. Bordé, C.R. Acad. Sc. Paris 284B, 101-104 (1977)

OPTICAL RAMSEY FRINGES IN TWO-PHOTON SPECTROSCOPY

M.M. Salour

Department of Physics, Harvard University
Cambridge, MA 02138, USA

Interaction of atoms and molecules with intense monochromatic laser standing waves can give rise to very narrow Doppler-free two-photon resonances. This new high resolution spectroscopic method has recently been extensively applied to the study of several atomic and molecular optical transitions [1]. The discovery and accurate measurement of these narrow, frequency-stable resonance lines in the emission and absorption spectra of substances are significant because advances in this area will not only provide direct clues to the innermost processes of matter, but may also lead to novel applications in many areas of science.

The ultimate resolution achievable by the method of Doppler-free two-photon spectroscopy is limited in principle only by the natural linewidth of the excited state; in practice, however, the observed linewidth has thus far been limited by the laser. Thus in order to take full advantage of the elimination of the Doppler width, the spectral linewidth of the laser light must be as small as possible, which explains the motivation for using cw lasers. But, on the other hand, two-photon resonances require an appreciable intensity, and it is well known that pulsed dye lasers deliver higher powers and, furthermore, cover a broader spectral range. Recently, by using pulsed dye lasers pumped with an N_2 laser, it has been possible to observe two-photon resonances in several highly excited states of various atomic vapors which could not have been reached by cw excitation [2]. The resolution was limited by the spectral width $1/\tau$ of the pulse (τ: duration of each pulse), thus making it impossible to resolve many closely spaced two-photon resonance lines.

We have recently demonstrated [3] that, by exciting atoms with two time-delayed coherent laser pulses, one can obtain interference fringes in the profile of the Doppler-free two-photon resonances with a splitting $1/2T$ (T: delay between the two pulses) much smaller than the spectral width $1/\tau$ of the laser pulse (τ: duration of each pulse). This technique combined the advantages of pulsed dye lasers (power, spectral range) with the high resolution usually associated with a cw excitation, and could be considered as an extension, to two-photon Doppler-free resonances in the optical range, of the well-known Ramsey method of using two separated RF or microwave fields in atomic beam experiments [4].

Such an idea was first suggested in a slightly different context by BAKLANOV, CHEBOTAYEV AND DUBETSKII [5], who proposed using two spatially separated cw light standing waves. These authors considered two-photon

Doppler-free resonances between two very sharp levels, such as ground states or metastable states (an important example being the $1S_{1/2}$-$2S_{1/2}$ transition of H). In such a case, the use of a single laser beam leads to linewidths which are generally limited by the inverse of the transit time of atoms through the laser beam. Rather than increasing this time by expanding the laser beam diameter (with a consequent loss of intensity), BAKLANOV *et al.* proposed using two spatially separated beams to obtain structures, in the profile of the resonance, having a width determined by the time of flight between the two beams. The experiment described below deals with short-lived atomic states (lifetime ~ 5×10^{-8} sec), so that the transit time through the laser beam (~ 10^{-7} sec) plays no role in the problem. Consequently, we use two time-delayed short pulses instead of two spatially separated beams [6].

The interference structure appearing in the resonance profile can be simply understood in the following way. Consider first a single pulse of the form $E(t)e^{-i\omega t}$, where the envelope function $E(t)$ has the shape represented in Fig. 1a (τ: duration of the pulse). The Fourier analysis of such a pulse leads to a frequency spectrum spread over an interval $1/\tau$

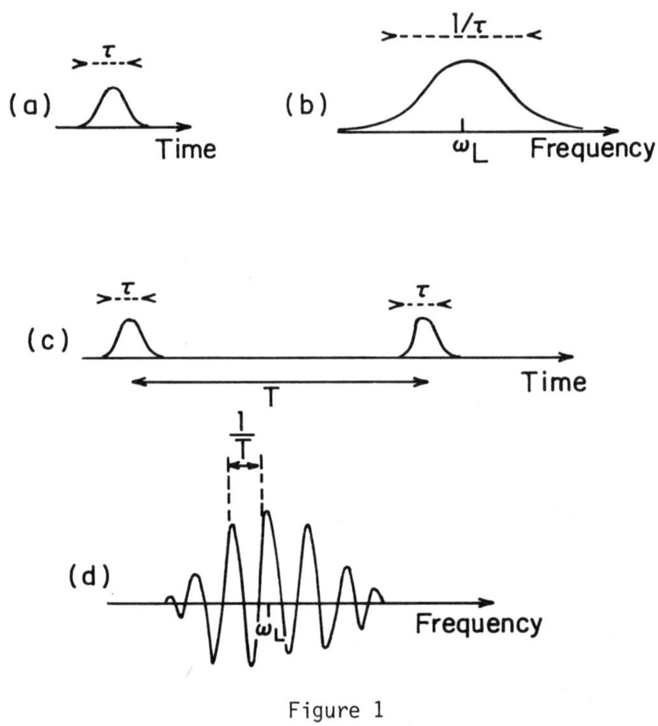

Figure 1

around ω (Fig. 1b). This non-monochromatic excitation acting upon the atom leads to a Doppler-free two-photon resonance having a width $1/\tau$. Suppose now that $E(t)$ has the shape represented in Fig. 1c (two pulses of duration τ separated by a delay T). Just as the diffraction pattern through two spatially separated slits exhibits interference fringes within the diffraction profile corresponding to a single slit, the Fourier transform of the electric field corresponding to Fig. 1c exhibits interference fringes within the spectrum profile of a single pulse (Fig. 1d), with a frequency splitting of 1/T between two fringes.

Two important requirements must be fulfilled in order to obtain such interference fringes in the profile of the two-photon resonance. First, each pulse must be reflected against a mirror placed near the atomic cell in order to expose the atoms to a pulsed standing wave and in this way to suppress any dephasing factor due to the motion of the atoms. The probability amplitude for absorbing two counterpropagating photons is proportional to $e^{i(\omega t-kz)}e^{i(\omega t+kz)} = e^{2i\omega t}$ and does not depend on the spatial position z of the atoms. Second, the phase difference between the two pulses must remain constant during the entire experiment. Significant phase fluctuations between the two pulses will wash out the interference fringes, since any phase variation produces a shift of the whole interference structure within the diffraction background. To avoid such fluctuations, the experiment must be done not with two independent pulses, but with two time-delayed pulses having a constant phase difference during the entire experiment. Furthermore, these two pulses have to be Fourier limited, that is, their coherence times must be not shorter than their duration.

From a more physical point of view, one can think of a two level system, $|g\rangle$ and $|e\rangle$. After the first pulse the atomic state is a coherent superposition of the two levels. Atoms in this superposition freely precess and, depending on the point in time at which the delayed pulse arrives (while the excited atom is still freely precessing), one can see either constructive or destructive interference with the atomic precession.

Note again that it is important that the initial and the delayed pulses be phase-locked. Otherwise, when the delayed pulse arrives, the random phase difference between initial and delayed pulse will give rise randomly to constructive or destructive interference, and when averaged the interference signal will be washed out.

Another interpretation of this interference can be shown diagrammatically:

Figure 2 represents two quantum mechanical amplitudes whose sum represents the total amplitude, of order E^2 (each photon contributes a factor of E), for reaching the excited state $|e\rangle$ from the ground state $|g\rangle$. The probability of reaching the $|e\rangle$ state is then the absolute square of this sum, and the cross term appearing in this absolute square represents the interference signal.

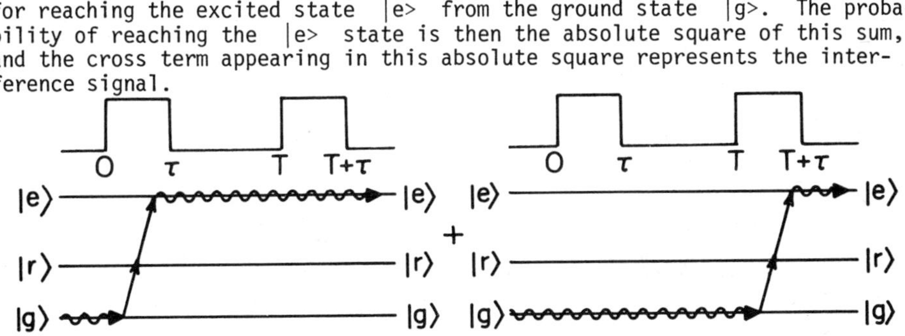

Figure 2

Higher order processes can also become important, especially at higher laser intensities. Figure 3 illustrates additional amplitudes of order E^4 which contribute to the interference signal. These diagrams have the physical interpretation of ac Stark shift of the $|g\rangle$ and $|e\rangle$ states via mixing with the intermediate state $|r\rangle$. These processes would result in linear power shift of the interference signal.

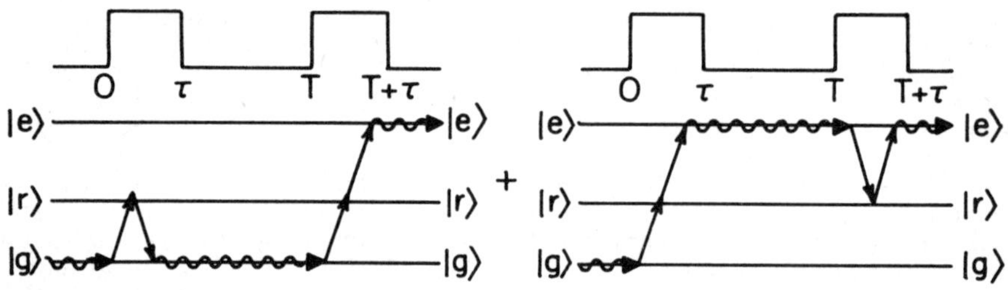

Figure 3

We have discussed elsewhere [3] two possible methods for obtaining such a sequence of two phase-locked and Fourier limited pulses, by utilizing a frequency stabilized cw dye laser oscillator with synchronized pulse amplifiers pumped by a high power pulsed laser.

Figure 4-I shows the four well known two-photon resonances of the 3^2S-4^2D transition of Na, observed in a reference cell with a single pulse. Figure 4-II shows the same four resonances observed in the sample cell when excited

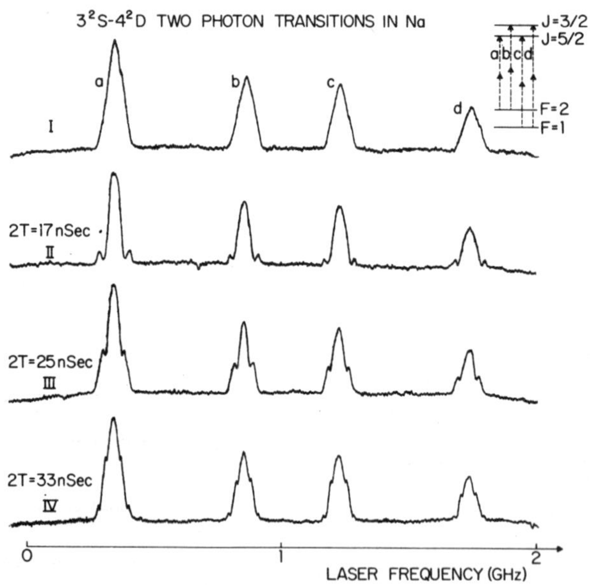

Figure 4

by the two time-delayed coherent pulses with $2T \simeq 17$ nsec. An interference structure clearly appears on each resonance. We have verified that the splitting between the fringes is inversely proportional to the effective delay time T_{eff} ($T_{eff} = 2T$) between the two coherent pulses. Figures 4-III and 4-IV show the same experimental traces as those of Fig. 4-II except for effective delays of $T_{eff} = 2T = 25$ and 33 nsec., respectively. Note that in these cases (Figs. 4-III and 4-IV) the contrast of the fringes is progressively less than that of Fig. 4-II, where a shorter delay is used. We have also verified that, with the delay line unlocked, the interference fringes disappear (when ω is varied).

Clearly, increasing the delay time T between the two pulses causes both the fringe spacing and the fringe contrast to decrease; consequently, one obtains better resolution, but a smaller signal-to-noise ratio. This can be understood by noting that when $T > 1/\Gamma$, a smaller number of atoms excited by the first pulse can live for a sufficiently long time to experience the second pulse; since it is the interference due to the second pulse that provides the high resolution, only the still-excited atoms contribute to the signal (i.e., only those atoms that remain in the excited state for a time at least equal to T contribute to the interference effect, and consequently if T is large, the fringes will have a very poor contrast).

Going back to the discussion of Figs. 2 and 3, the sum of the two quantum mechanical amplitudes A_1 and A_2 represents the total amplitude $A = A_1+A_2$ for reaching the excited state $|e\rangle$ from the ground state $|g\rangle$. When one introduces a phase shift $e^{i\delta}$ on the second pulse only, then the total amplitude is $A_1 + e^{i\delta}A_2$, and the probability of reaching the $|e\rangle$ state is

$$|A|^2 = |A_1 + e^{i\delta}A_2|^2 = \underbrace{|A_1|^2 + |A_2|^2}_{\substack{\text{diffraction}\\ \text{terms}}} + \underbrace{2A_1A_2 \cos\delta}_{\substack{\text{interference}\\ \text{term}}}$$

In order to eliminate the diffraction background appearing in each of the four resonances of Figs. 4-II, 4-III, and 4-IV, one could induce a 90° phase shift (so that $\delta = 180°$ for the two-photon excitation) between every other pair of coherent pulses into the sample cell. By subtracting the resulting fluorescence from pairs of coherent pulses with and without the phase shift:

$$|A|^2_{\delta=0} - |A|^2_{\delta=\pi} = 4A_1A_2$$

one can eliminate the diffraction background and thus isolate the interference fringes. Note that in this way not only does one get rid of the diffraction background, but also one doubles the interference signal automatically. Such a technique seems promising for situations where the contrast of the interference fringes is poor; for example, when T is chosen much longer than the radiative lifetime in order to get a fringe splitting smaller than the natural linewidth. A detailed study of this scheme has appeared elsewhere [7].

Figure 5-II shows the experimental results for a delay of 25 nsec. Above it, Fig. 5-I corresponds to the signal obtained when atoms are excited by a single pulse in the reference sodium cell.

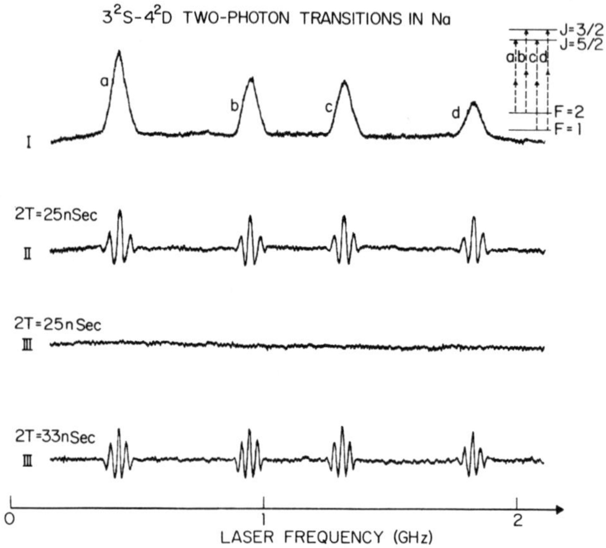

Figure 5

Figure 5-IV shows similar experimental results for a delay of 33 nsec. Note that in this case the fringe spacings are much smaller than in the case of T_{eff} = 25 nsec delay. It was verified that the fringe spacing was proportional to T_{eff}^{-1}. In comparison with the results of Fig. 4 note the enormous gain in contrast associated with the traces of Figs. 5-II and 5-IV. It was also verified that no signal was observed when zero phase shift was introduced between alternate pulses. Fig. 5-III shows this result.

In conclusion, we have introduced a technique which could mark a logical beginning of a stimulating and exciting new generation of optical experiments, in which through the use of delayed pulses, pulse-length broadening can be overcome. Thus, pulsed dye lasers, the only spectroscopic tool presently available in many spectral regions, can be used for ultra high resolution spectroscopy of atomic and molecular systems. Our technique can be generalized. Instead of two pulses, one could use a sequence of N equally spaced pulses (obtained, for example, by sending an initial pulse into a confocal resonator [8]), and thus submit atoms to a series of N equally spaced time-delayed and phase-locked pulsed standing waves. An argument parallel to the discussion of Fig. 1 would show that the optical analogue of such a system is a grating with N lines; the total length of the grating corresponds to the total duration NT of the pulse train. In such a case, the frequency spectrum experienced by the atom would be a series of narrow peaks, N times narrower in width and N^2 times stronger in intensity, separated by a frequency interval T_{eff}^{-1}.

Acknowledgements

The work reported here represents contributions by Professor C. Cohen-Tannoudji and expert technical assistance of L. Donaldson, S. S. Mertz,

R. W. Stanley, and D. A. Van Baak. The work was supported in part by the Joint Services Electronics Program.

References

1. L.S. Vasilenko, V.P. Chebotayev and A.V. Shishaev, Pis'ma Zh. Eksp. Teor. Fiz. *12*, 161 (1970) [JETP Lett. *12*, 113 (1970)]; B. Cagnac, G. Grynberg, and F. Biraben, J. Phys. (Paris) *34*, 56 (1973) and Phys. Rev. Lett. *32*, 645 (1974); M.D. Levenson and N. Bloembergen, Phys. Rev. Lett. *32*, 645 (1974); T.W. Hänsch, K. Harvey, G. Meisel, and A.L. Schawlow, Opt. Comm. *11*, 50 (1974); M.M. Salour, Opt. Comm. *18*, 377 (1976).

2. M.M. Salour, Opt. Comm. *18*, 377 (1976); M.M. Salour (to be published).

3. M.M. Salour, Bull. Am. Phys. Soc. *21*, 1245 (1976); M.M. Salour and C. Cohen-Tannoudji, Phys. Rev. Lett. *38*, 757 (1977).

4. N.F. Ramsey, Phys. Rev. *76*, 996 (1949); *Molecular Beams* (New York: Oxford University Press, 1956); P.B. Kramer, S.R. Lundeen, B.O. Clark, and F.M. Pipkin, Phys. Rev. Lett. *32*, 635 (1974).

5. Ye. V. Baklanov, B. Ya. Dubetskii, and V.P. Chebotayev, Appl. Phys. *9*, 171 (1976); Ye. V. Baklanov, B. Ya. Dubetskii, and V.P. Chebotayev, Appl. Phys. *11*, 201 (1976); B. Ya. Dubetskii, Kvantovayev Elektron. (Moscow) *3*, 1258 (1976) [Sov. J. Quantum Electron. *6*, 682 (1976)]. Note that the method of using spatially separated laser fields has recently been applied to saturated absorption spectroscopy by J. C. Bergquist, S.A. Lee, and J.L. Hall, Phys. Rev. Lett. *38*, 159 (1977).

6. Note that the method of using two coherent time-delayed electromagnetic pulses has been already used in the microwave domain. See, for example, C.O. Alley, in *Quantum Electronics, A Symposium* (New York: Columbia University Press, 1960), p. 146; M. Arditi and T.R. Carver, IEEE Trans. Instrumentation and Measurement, IM-13(2), (3), (1964), p. 146.

7. M.M. Salour, Appl. Phys. Lett. (to be published, September 1977).

8. R. Teets, J. Eckstein, and T.W. Hänsch, Phys. Rev. Lett. *38*, 760 (1977).

RAMSEY FRINGES IN SATURATION SPECTROSCOPY

J.C. Bergquist, S.A. Lee, and J.L. Hall*

Joint Institute for Laboratory Astrophysics
National Bureau of Standards and University of Colorado
Boulder, CO 80309, USA

Provided one has been able to develop/acquire frequency and intensity stabilized sources of coherent radiation, there remain two main problems in the pursuit of narrow lines: transit time broadening and Doppler broadening. The first-order Doppler effect may be essentially eliminated by various techniques: saturated absorption, two-photon spectroscopy, highly collimated beams, ... Transit time (interaction time) broadening is more difficult to eliminate. Still, it has recently proved possible to resolve the radiative recoil-induced doublets in the three main hyperfine components of methane at 3.39 µm. This high resolution (2 parts in 10^{11}), derived from an external absorption cell with a 30 cm aperture, nevertheless remained 2 orders of magnitude removed from the natural lifetime line width. Larger cells and associated optics being cumbersome, one is induced to find alternative schemes to reduce transit time broadening and so approach natural line width resolution.

An attractive solution to this transit time problem in saturated absorption was recently studied by Ye.V. Baklanov and his coworkers [1]. A beam of quantum absorbers could be sent transversely through three consecutive standing wave light fields equally separated in space. The resulting absorption/emission line shape would be an interference fringe pattern. This is the optical analog of the well-known Ramsey fringes [2] which are routinely utilized in rf spectroscopy of atomic and molecular beams. In these interference methods the spectral resolution is limited by the travel time between radiation zones rather than by the transit time through each zone.

The need for three radiation zones with saturated absorption at optical wavelengths should be explained, especially when one remembers that two zones sufficed in the microwave case. (Also only two interaction regions are needed with two-photon spectroscopy at optical wavelengths, cf. papers by Salour *et al.* and Hänsch *et al.* in this volume.) In the rf and microwave region, Ramsey used the interference resulting from the interaction of a quantum absorber with two separated radiation fields. After being initially prepared by the first interaction region, the atomic polarization phase evolves during the dark interzone interval at a rate which, in general, differs from that of the driving fields. The effect of the second radiation field depends on the phase of that radiation relative to the phase of the quantum system's oscillation, thus bringing a frequency tuning sensitivity dependent on the

---

*Staff Member, Quantum Physics Division, National Bureau of Standards.

interzone transit time. However, the immediate synthesis of Ramsey's idea and saturated absorption in the optical domain meets with difficulties related to the shortness of the wavelength in comparison with the dimensions of the interaction region. Even in a collimated atomic beam there will be a spread of the residual Doppler velocity projections on the direction of light wave propagation.

Consider a beam of quantum absorbers crossing a collimated laser beam of waist size w_0. Near perpendicular incidence there is a narrow angular slice, $\delta\theta \approx \lambda/3w_0$, within which the absorbers experience no progressive phase shift. That is, for absorbers within this slice, the transit-time broadening exceeds the residual Doppler broadening. Adding a second optical interaction zone a distance L downstream does not lead to strong Ramsey fringes since the angular slice defined by the first interaction maps into a large extension, $\Delta Z = L \cdot \delta\theta = L\lambda/3w_0$, at the second beam. The condition for increased resolution by the Ramsey interference, $L/w_0 >> 1$, is just the same condition that the dipoles originating at one spatial position in light zone 1 will be spread out several wavelengths along beam 2. This results in a spatial averaging of the Ramsey effect since the quantum systems with the *same* interzone transit time phase evolution experience *different* phases of the second driving field dependent on their spatial entry position into the second zone.

Baklanov and his colleagues [1] have drawn attention to the spatial-modulation aspects of these interactions and have introduced the idea of a third equally spaced interaction zone as a method to recover the Ramsey fringes. However, as a way to see the physical effects -- and especially to appreciate the phase relationships -- it is more convenient to assume we have four interaction zones. Basically, in saturation spectroscopy, the narrow resonances arise from four conceptually-separate time-ordered interaction processes: lower state population $P_L \to$ dipole, dipole $\to$ upper state population, P_U, $P_U \to$ dipole, and dipole $\to P_L$ and P_U. Each process will be proportional to the electric field at the interaction point, and the system response will therefore acquire a phase associated with the electric field. Thus the atomic system will sequentially interact at four space time points with four running wave fields, which may be given in the laboratory frame as

$$\varepsilon_1 = E_1\, e^{i(\omega t_1 - kz_1 + \phi_1)}$$

$$\varepsilon_2 = E_2\, e^{-i(\omega t_2 - kz_2 + \phi_2)}$$

$$\varepsilon_3 = E_3\, e^{+i(\omega t_3 + kz_3 + \phi_3)}$$

$$\varepsilon_4 = E_4\, e^{-i(\omega t_4 + kz_4 + \phi_4)}$$

The space-time points z_i, t_i are related by the absorber's free-flight trajectory. In the first interzone darkness the dipole prepared by the first interaction will precess at its own natural frequency ω_0 and decay with a dipole decay rate γ_{ab}. In the second interzone space the excited state population will only decay, with the population decay rate γ_b. In the third interzone space the system again carries a dipole moment, and precesses and decays as before. Assuming the interzone distances to be L, aL, bL respectively leads to the following expression for the total phase of the Ramsey signal:

$$e^{i\phi_{total}} = \underbrace{e^{-ikv_z \frac{L}{v}(b-1)}}_{\text{Doppler Phase}} \; \underbrace{e^{-i(\omega-\omega_0)\frac{L}{v}(b+1)}}_{\text{Detuning Phase}} \; \underbrace{e^{i(\phi_1-\phi_2+\phi_3-\phi_4)}}_{\text{Cavity Phase}}$$

$$\times \underbrace{e^{-\gamma_{ab}\frac{L}{v}(b+1)}}_{\text{Dipole Decay}} \; \underbrace{e^{-\gamma_b \frac{L}{v} a}}_{\text{Population Decay}} \; . \qquad (1)$$

For typical atomic systems of interest in connection with Ramsey fringes the lower level may be long-lived, so the dipole decay rate, γ_{ab}, and the upper level population decay rate, γ_b, are related by $\gamma_{ab} = (1/2)\gamma_b$. Considering that even a collimated beam contains a distribution of v_z values, one can conclude that the Ramsey fringes are observable only for the case $b = 1$, that is, when the dipole free precession distances are equal. One can also show that the fringes arise only for the case of two separated interactions with parallel-running beams followed by two separated interactions with oppositely-running beams.

It is clear that the spectral resolution is not enhanced by the second dark space, although $a \neq 0$ is sometimes a useful condition experimentally. (In the three-beam geometry two interactions occur in the central region.) One may make connection with the usual saturation spectroscopy description by noting that the high resolution provided by Ramsey interference is necessary to define a narrow Bennett hole in the saturation process and is also necessary in the probe interrogation process. Thus two (equal) dark spaces are necessary.

Such saturated absorption optical Ramsey fringes were observed recently in our laboratory [3]. The experimental setup is shown in Fig.1. A fast ($v/c \approx 10^{-3}$) monovelocity ($\Delta v/v \approx 10^{-4}$) beam of metastable $1s_5$ ^{20}Ne atoms is efficiently produced by charge exchange of a 5-50 keV Ne^+ beam focused through a Na oven. This metastable atomic beam sequentially interacts with three spatially-separated, standing-wave light beams from a single mode frequency stabilized dye laser. The 588 nm Ne $1s_5 \rightarrow 2p_2$ transition is excited, and the fluorescence emission of the $2p_2 \rightarrow 1s_2$ at 660 nm is detected with excellent signal-to-noise ratio with an appropriately filtered photomultiplier. ^{20}Ne, having zero nuclear spin, is free of hyperfine structure, thus allowing a clear interpretation of our results.

We recall that the fringes are produced by the transport of phase information between the separated radiation zones by the atom's freely precessing induced dipole polarization. It is necessary then that the dipole "lives" between zones. In our experiments, the common spatial separation of the light fields was ~5 mm while the dipole decay length for a typical beam energy of 20 keV was ~16 mm.

The experimentally observed fluorescent profiles are shown in Fig.2. Curve a shows most of the beam Doppler profile, the saturated absorption dip, and the fringes due to the atom's interaction with three equally spaced

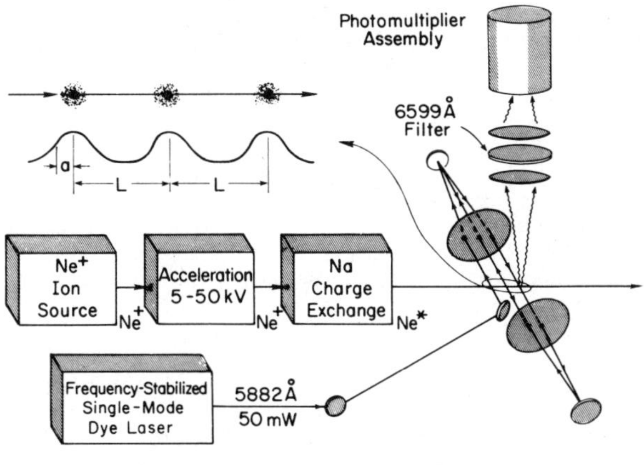

Fig.1 Schematic of experiment. The three standing-wave interaction regions are formed by two well-corrected cat's-eye retroreflectors. Typical values are the following: $i(Ne^+) = 3\ \mu A$, $V = 20$ kV, and the laser power is 50 mW. Fluorescence signals $\sim 10^8$ photons/sec reach the multiplier through the f/2 collection optics and filter.

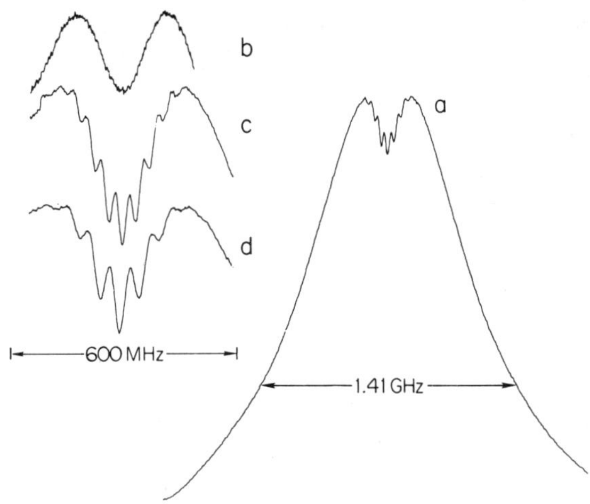

Fig.2 Fluorescence signals. Curve a, most of beam Doppler profile (full width at half-maximum of 1.41 GHz) showing saturation dip and Ramsey fringes; $V = 19.5$ kV. Fringe contrast $\simeq 3.8\%$. Curve b, saturation dip observed with two separated laser beams, $V = 19.5$ kV. Curves c and d, Ramsey fringes and saturation dips at $V = 35$ kV. Curve c, three beams; and curve d, four beams.

standing-wave radiation zones. In the inset, we compare the signals of three different light beam geometries. Curve b is the fluorescent profile produced when the Ne beam interacts with only two standing waves. Consistent with the theoretical ideas discussed above, no fringes result due to the interaction with two separated radiation fields. Curve c shows the emission profile when the atomic beam interacts with three equally spaced standing waves: one easily observes the optical Ramsey fringes. Their expected form is essentially a cosine multiplied by the saturated absorption envelope [1,4]. Adjacent fringe separations, $\Delta\nu$, were measured as a function of atomic beam velocity, v, and interzone separation, L. The expected relation $\Delta\omega = 2\pi\Delta\nu = \pi v/L$, $\Delta\nu = v/2L$, was verified to within the experimental precision of $\leq 10\%$. We note that the periodic phase shift between adjacent fringes will be developed by a smaller frequency interval if the interzone transit time is increased (larger L and/or smaller v).

The fringes produced with four equally spaced light beams are shown in curve d. The interzone separation was 2/3 that of the three light beam cases with the result that the fringe frequency separation is 3/2 the previous value (the atomic beam velocity remained the same).

Symmetric fringes result when the spatial phases in the three zones are such that they appear to be samples of a large planar wave front. This condition is intrinsically provided by the opposition of two correctly focused cat's-eye retroreflectors, as may be verified by applying Fermat's principle to the optical system illustrated in Fig.1. In our experiment the cat's-eye retroreflectors were correctly focused using an auxiliary interferometer. As may be seen in Fig.2, the fringes were symmetric, as expected.

It is interesting to consider shifting the phase of the inner zone relative to the outer ones, for example, by defocusing the cat's-eye retroreflector (i.e., we want the spatial phases to appear to be samples of a large curved wavefront). For a phase shift of $0 < \phi < \pi$ in the central zone we would expect to obtain a Ramsey pattern which is asymmetric and shifted in frequency, analogous to the asymmetric line shapes obtained with curved wavefronts in the cell experiments of Ref. 5. However in the present experiments we employ standing waves in each zone, and so do not distinguish in the first two interactions which running beam is playing the role analogous to the saturating beam in the cell experiments. Thus we expect to have *two* overlapping Ramsey contributions of essentially equal size where one corresponds to a blue-shifted Ramsey fringe system, the other to red-shifted fringes. Their sum, as detected by fluorescence from the third zone and beyond will show essentially no asymmetry or shift in the remaining Ramsey structure. In the idealized case of standing waves, composed of equal intensity counter running beams, the *fringe pattern will remain symmetrical* for all values of the spatial phase difference, only the *intensity* of the Ramsey pattern will have a cosine dependence on this phase. In particular, for a phase shift of 180° the fringes will be inverted. This cosine behavior was first obtained by Baklanov *et al.* [1].

It is attractive to consider use of an electro-optic phase modulator to switch the cavity phase between 0° and 180°. The fringe signal recovered by a lock-in amplifier may decrease -- even to zero -- but will not essentially shift in frequency with variations of the additive phase delay induced into one zone (for example, by temperature variations of the electro-optic modulator). This property will clearly be of fundamental utility in the use of optical Ramsey techniques for precision spectroscopy.

In exploring the use of defocusing the cat's eye to provide a stable phase offset between the beams, we found the three-beam configuration to be particularly phase stable: no significant Ramsey fringe shape changes occurred over any reasonable focus range. With a slight lateral offset of one cat's eye, one can obtain four standing-wave radiation zones. This configuration was found suitable for providing usable phase shifts with reasonable defocusing. The fringe signals in this case can be shown to arise from the above-described three-zone interaction sequences in radiation zones 1,2,3 and 2,3,4. Another term of about the same resolution is contributed by one interaction in each zone, with an excited state population prepared in zone 2 and traveling as a population to zone 3. Figure 3a shows the four-zone Ramsey fringes with the cat's eye properly focused. For some defocus distance of the cat's eye, the phase of the central zone is appropriate to give essential cancellation of the Ramsey fringe signals, as may be seen in Fig.3b. With the four-zone geometry, the cancellation condition is somewhat complicated and occurs for a phase shift $\lesssim 90°$ where the contributions of the 1,2,3 and 2,3,4 fringes just cancel the 1,2,3,4 fringes. Finally, with further defocusing, the central fringe becomes inverted as shown in Fig.3c. Since the phase offsets for the two types of fringes are not the same (differing by a factor of 2), it is reasonable that a significant loss of the fringe contrast occurs.

We plan to explore several phase shift/fringe modulation techniques to further elucidate the behavior of the optical Ramsey effect, especially in connection with bias-free line center algorithms for precision spectroscopy and optical frequency standards applications.

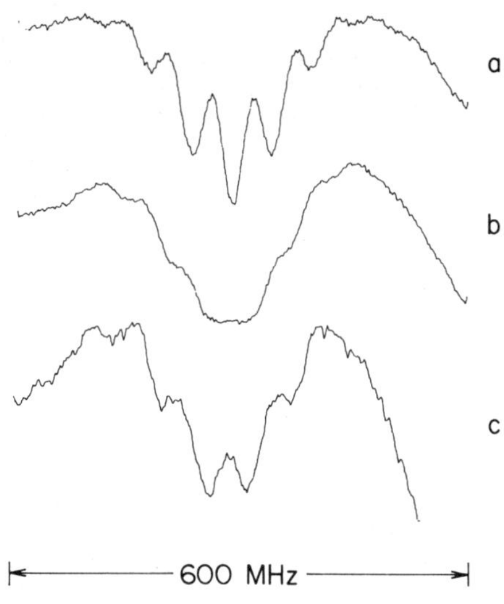

Fig.3 Four-zone Ramsey fringes with different cavity phase shifts. Top: Correct focus, negligible phase shift. Middle: Somewhat defocused, $0 < \phi \lesssim \pi/2$. The two systems of fringes essentially cancel. Lower: Increased defocusing, $\pi/2 < \phi \lesssim \pi$. The fringes are inverted. See text for discussion of contrast.

There are two distinct, important advantages of the nonlinear separated interaction Ramsey method when compared to other nonlinear techniques. First, most of the quantum absorber's phase evolution occurs in the absence of the driving field, therefore power broadening is minimized. Secondly, we note that three-zone coherent interaction occurs for all the absorbers in the angular slice $\delta\theta \sim \lambda/3w_0$ allowed by the transit time frequency uncertainty of one zone. If the radiation field filled the entire effective aperture 2L, only those particles in the angular interval $\delta\theta' \sim \lambda/3L$ would contribute. Thus, we can achieve significant line narrowing without loss of signal-to-noise ratio.

To summarize, we have observed optical "Ramsey" fringes with spatially separated light beams and nonlinear absorption. Symmetry of the fringes is controlled by the phase relationships between the light beams. The fringes can be highly symmetric if one uses interferometric quality cat's-eye retro-reflectors. With the distinct advantages of high contrast and a very sharp spectral feature (ultimately only lifetime limited), this method offers the possibility of significant improvement in optical spectroscopy and frequency metrology. An immediate candidate which should show high resolution with good signal-to-noise ratio is the calcium intercombination line ${}^1S_0 \rightarrow {}^3P_1$ at 657 nm with a natural line width limit of 400 Hz. Finally, we note that these fringes have been computer synthesized with the high intensity theory [6] (program SHAPE) for both the case of separated Gaussian beams and for the case of three zones induced on a single Gaussian beam by using an annular aperture. The aperture-induced fringes were first observed by one of us (J.C.B.) in an external CH_4 cell at 3.39 μm.

This work was supported in part by the National Bureau of Standards program of research on improved measurement techniques for application to basic standards, and in part by the National Science Foundation under grant no. PHY76-04761 through the University of Colorado.

References

1. Ye.V. Baklanov, B. Ya. Dubetsky and V. P. Chebotayev, Appl. Phys. 9, 171 (1976); 11, 201 (1976).
2. N. F. Ramsey, Phys. Rev. 78, 695 (1950).
3. J. C. Bergquist, S. A. Lee and J. L. Hall, Phys. Rev. Lett. 38, 159 (1977).
4. Bordé has given analytic expressions for the line shape in the case of three separated Gaussian beams. See C. J. Bordé, C. R. Acad Sci. (Paris) 284B, 101 (1977).
5. J. L. Hall and C. J. Bordé, Appl. Phys. Lett. 29, 788 (1976).
6. C. J. Bordé, C. V. Kunasz and J. L. Hall, in preparation.

MULTIPLE COHERENT INTERACTIONS

T.W. Hänsch

Department of Physics, Stanford University
Stanford, CA 94305, USA

The observation of narrow optical Ramsey fringes in two-photon excitation of atoms with separate short light pulses, as discussed by M. SALOUR in a preceeding paper, can greatly enhance the potential of Doppler-free laser spectroscopy. It can, in particular, extend the accessible wavelength range into the ultraviolet and vacuum ultraviolet, where there are presently only pulsed tunable lasers available which have by necessity relatively large bandwidths.

At Stanford we began about one year ago to explore the merits of coherent two-photon excitation with more than two light pulses (1), and it soon became apparent that such multi-pulse spectroscopy can offer important advantages in resolution and signal strength over two-pulse experiments.

To better understand the principles and advantages of such an approach, let us briefly recall the origin of two-pulse Ramsey fringes. As illustrated in Fig. 1, top, a single short laser pulse can, at best, produce a broad spectrum whose transform-limited width is about equal to the inverse pulse length. Two-photon excitation with two identical separate pulses produces a sinusoidal spectral interference pattern, as shown at the center of Fig. 1. The first pulse leaves the atoms in a coherent superposition of upper and lower states, and causes atomic oscillations at the two-photon resonance frequency. The effect of the second pulse depends then on the relative phase between these atomic oscillations and the light field. The atoms are either further excited, or they can return to the ground state by stimulated two-photon emission. If the pulse delay is varied in proportion to the laser wavelength, the spacing between the resulting fringes (in frequency) is equal to the inverese delay time between the pulses, while the envelope corresponds to the spectrum of a single pulse.

The width of these two-pulse Ramsey fringes can in principle be arbitrarily narrow, down to the natural atomic linewidth, if only the two light pulses are sufficiently far apart in time. But the number of fringes within the pulse bandwidth increases accordingly, and can easily become confusingly large. More seriously, the sinusoidal fringe pattern makes it difficult, if not impossible, to resolve closely spaced line components, even if the number of fringes remains small.

It is easy to recognize a solution to these problems, if we observe the similarity between the spectrum of such a two-pulse excitation experiment and the well known spatial interference pattern that is obtained in the diffraction of light from a double slit. Based on this analogy we should expect that coherent excitation with a whole train of equidistant identical light pulses will produce a spectrum that resembles the interference pattern from an array of slits or a diffraction grating. As indicated in Fig. 1, bottom, the fringe spacing should remain the same for a given pulse separation, but the fringes should condense into narrow spectral lines, which correspond to the diffraction orders of a grating. The fringe width should now be determined by the length of the entire pulse train, as long as the transverse relaxation of the atomic oscillators remains negligible.

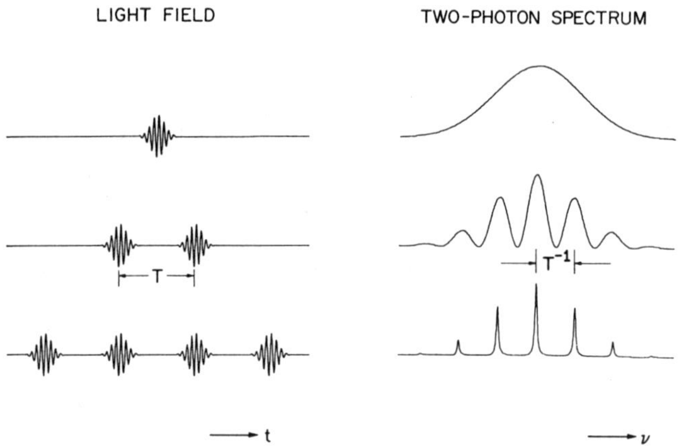

Fig. 1 Short light pulses (left) and resulting Doppler-free two-photon spectra (right). One single pulse produces a broad, transform-limited spectrum (top). Two successive pulses give rise to sinusoidal optical Ramsey fringes (center). A train of multiple identical pulses can produce narrow spectral lines (bottom).

Such multi-pulse excitation is obviously analogous to molecular beam spectroscopy with multiple, spatially separated rf field regions, and it is interesting to note the N. RAMSEY, twenty years ago, has already considered up to four separated oscillatory fields (2). But unlike in this latter case, the complexity and expense of an apparatus for optical two-photon excitation does not have to grow with increasing number of pulses.

The signal magnitude produced by two or more light pulses can be readily calculated by second order perturbation theory, as long as the total two-photon excitation probability remains small (3). Alternatively, we can describe an atom in a coherent superposition of ground state and excited state by a 3-dimensional Feynman vector (4), whose vertical component gives the difference of the level populations or the excitation probability. Its horizontal components give the atomic oscillation amplitude and phase, and can be made to stand still in the absence of light fields by going into a rotating coordinate system. Just as in the familiar case of single-photon excitation of a two-level atom, the two-photon excitation by a light pulse can be described as a finite rotation of this Feynman vector around some external "torque" vector. (This is true even for pulse which is chirping in frequency or otherwise not transform limited, although the net rotation will then be the result of a more complicated detailed angular motion.) In order to predict the excitation by a train of identical light pulses, we need only look at the corresponding sequence of rotations. The relative angle between subsequent rotation vectors, as seen in the rotating frame, is simply determined by the delay time between the pulses. At exact resonance, all rotations add constructively, and the total excitation probability, to lowest order, is proportional to the square of the number of light pulses in the train. For a given pulse energy, coherent multi-pulse excitation can hence lead to a dramatic signal enhancement. Such an enhancement is often extremely welcome in two-photon spectroscopy, because it can permit one to avoid tight focusing of the laser light with its associated problems of transit time broadening and light shifts.

There are several simple ways of generating a proper train of light pulses experimentally. One attractive possibility is the use of a mode-locked cw dye laser. To control the exact pulse roundtrip time inside the laser resonator, one can employ similar stabilization schemes as are commonly used for single-frequency dye lasers. The accessible wavelength range of such devices is unfortunately still limited, but the possible high pulse peak power can permit an efficient wavelength extension by nonlinear frequency mixing in crystals or gases. We are presently exploring the merits of this approach.

Another, more general scheme, has been employed in a recent experiment by R. TEETS and J. ECKSTEIN in our laboratory (3). Here, the pulse train is generated by injecting a single 7 nsec long light pulse from a nitrogen-pumped dye laser into a simple optical resonator, formed by two mirrors. The light enters the resonator through one of the partially transmitting mirrors. The gas sample is placed near one end mirror, where the atoms see a pulsed standing wave field once during each pulse roundtrip. The roundtrip time is controlled by changing the resonator length with a piezotranslator. As long as the laser bandwidth is large compared to the free spectral range of the resonator, the fringes can be observed by simply keeping the laser frequency fixed, and only tuning the length of the passive resonator. The Na 3s-5s transition was studied in this initial experiment, and the two-photon excitation was observed by monitoring the 3p-3s UV fluorescence light. Fringes of very high contrast and of a few MHz width, much below the 300 MHz laser line width, were easily observed with a 2 m long confocal resonator and about 6 effective pulse roundtrips.

This simple setup immediately suggests an alternative interpretation of the multi-pulse Ramsey fringes. We can ascribe these fringes simply to the discrete axial modes of the passive resonator. This resonator filters narrow spectral lines out of the broadband spectrum of the laser pulse. We could also say that modes within the laser bandwidth are shock-excited by the incoming laser pulse. A fringe maximum is observed if the resonator is tuned so that its modes can excite the two-photon resonance. In this description, multi-pulse two-photon spectroscopy does not seem much different from conventional Fabry-Perot spectroscopy, and one may question whether we are entitled at all to use the term "Ramsey fringes" to describe the observed narrow spectral lines.

It is instructive to look at the two-pulse experiment of M. SALOUR and C. COHEN-TANNOUDJI (4) from the same point of view. The optical delay line, which produces the second pulse, can also be considered as a passive optical filter. Similar to a Michelson interferometer, it filters a comb of narrow spectral bands out of the original broad pulse spectrum, and the observed Ramsey fringes can again be simply ascribed to this spectral filtering. Because of the uncertainty principle, such filtering can, of course, not be performed without changing the pulse shape in time. And the temporal effect of the considered comb-filter is simply the generation of a second, delayed pulse.

Such a description in terms of passive spectral filtering indicates that the exact central frequency of the laser pulse, if it can be defined at all, plays no important role in such experiments. In practice it can nonetheless be highly desirable to employ a cw oscillator with pulsed amplifier and to lock the oscillator frequency to the passive filter or resonator, as in the experiment of SALOUR, if for no other reason than to monitor and control the exact position of the filter transmission bands.

Have we now succeeded in explaining a quantum interference effect without resorting to quantum mechanics? A description simply in terms of passive spectral filtering would certainly seem adequate for weak single-photon excitation of atoms at rest. The Schrödinger equation provides the atoms simply with a means to take the Fourier transform of the incoming pulse train and to analyze the filtered spectrum. For two-photon excitation, however, the situation is not quite as simple. Here, we can satisfy the resonance condition by tuning one of the cavity modes to half the transition frequency. The atom can then be excited by absorbing two photons from this mode. But it can also absorb one photon from the next higher mode and one from the next lower one and so forth. As a consequence, all the modes of the resonator contribute simultaneously to the resonant excitation. The same is also true, if half the transition frequency falls exactly halfway in between two resonator modes. In order to predict the signal magnitude, we hence need to know the relative phases of all oscillating modes. And it is only these phases that imply that, in the experiment of TEETS and ECKSTEIN, for instance, the atomic sample has to be placed near one resonator mirror. The description in the time domain, which seems certainly more intriguing, can then actually be simpler as well.

Nonetheless we can use our spectral filter arguments to explain the predicted strong resonant signal enhancement in multi-pulse excitation. If we send a single light pulse through an optical resonator, consisting of two lossless mirrors of equal reflectivity R, then the filtered, transmitted light (a damped pulse train) will exhibit a comb-like spectrum of narrow lines. The light inside the resonator has obviously the same spectrum, but its intensity is higher by a factor 1/R, i.e. the spectral energy density in the line maxima is increased over that of the original pulse. But such a passive interferometer still wastes a large portion of the incoming light by reflection. One might say that this reflection simply discards the unwanted light outside the narrow transmission bands of the resonator. But we have to remember that the dominant loss occurs during the initial injection and that the reflected pulse has exactly the same shape as the "good" light that is transmitted into the resonator. It is hence possible to avoid any energy loss by injecting the entire pulse actively into the resonator, for instance with the help of a fast electrooptic or acoustooptic light switch. The signal magnitude can then be many orders of magnitude higher than for single-pulse excitation outside the resonator. The active pulse injection is in fact changing the spectral energy distribution of the available light, and is condensing a broadband spectrum into narrow channels.

Regardless of its interpretation, coherent two-photon spectroscopy with multiple light pulses promises to become a valuable tool for high resolution spectroscopy of atoms and molecules. We expect, for instance, that it should be quite feasible to observe the hydrogen 1S-2S two-photon transition at 2430 Angstroms in this way with a resolution of about 50 kHz, close to the limit set by the transverse relativistic Doppler effect at room temperature. Such a resolution would be more than sufficient for several important new precision measurements. Moreover, the possible strong signal enhancement may make multi-pulse two-photon excitation attractive for the effective excitation of high atomic and molecular levels, for applications such as selective photochemistry.

* Work supported by the National Science Foundation under Grant No. 14789 and by the U.S. Office of Naval Research under Contract No. N00014-75-C-0841.

REFERENCES

1. T. W. Hänsch, in "Tunable Lasers and Applications," A. Mooradian et al., eds., Springer Series in Optical Sciences, Vol. 3, Berlin, Heidelberg, New York 1976, p. 326

2. N. F. Ramsey, Phys. Rev. 109, 822 (1958)

3. R. Teets, J. Eckstein, and T. W. Hänsch, Phys. Rev. Letters 38, 760 (1977)

4. R. P. Feynman, F. L. Vernon, Jr., and R. W. Hellwarth, J. Appl. Phys. 28, 49 (1957)

5. M. M. Salour and C. Cohen-Tannoudji, Phys. Rev. Letters 38, 757 (1977)

DOPPLER-FREE LASER-INDUCED DICHROISM AND BIREFRINGENCE

C. Delsart and J.-C. Keller
Laboratoire Aimé Cotton, C.N.R.S. II, Bât. 505
91405 Orsay, France

1. Introduction

There has been recently several papers devoted to laser-induced polarization effects in atomic vapours and to the corresponding applications to spectroscopy [1 - 5]. It is a matter of fact that such laser-induced anisotropies can be observed both for absorption and dispersion, i. e. both for the real and the imaginary part of the induced susceptibility. In the following paper we present an experimental investigation of the Doppler-free dichroism and birefringence induced by a linearly-polarized single-mode laser beam in two- and three-level systems.

Let us consider a Doppler-broadened and degenerate atomic system illuminated by a linearly-polarized and resonant laser beam (pump beam) and also by a weak probe beam propagating along the same direction. The probe can be resonant either with the same atomic transition as the pump (two-level system) or with a transition sharing a common level (three-level system). The linearly-polarized pump beam induces an anisotropic macroscopic polarization ; this polarization is supposed not to be modified by the weak probe beam (linear response to the probe). When the two beams are interacting with atoms belonging to the same longitudinal velocity group, the probe beam experiences an anisotropic optical medium . This results in Doppler-free absorption-coefficient and refractive-index anisotropy (i. e. respectively dichroism and birefringence) that can be observed when the probe-beam frequency is tuned across the atomic line.

Let $\alpha_{\parallel}$ (resp. $\alpha_{\perp}$) be the absorption coefficient per unit length and $n_{\parallel}$ (resp. $n_{\perp}$) be the refractive index experienced by the probe beam with linear polarization parallel (resp. perpendicular) to the pump-beam polarization direction π . The change of the probe beam polarization after crossing the interaction region is analyzed by a linear analyzer rotated through an angle β from the reference direction π . In the particular case of a circularly-polarized probe beam the light intensity $I(\beta)$ transmitted through the analyzer is given by

$$I_c(\beta) - I_1 = \frac{-I}{2}\left\{\Delta\alpha \frac{\ell}{2} \cos 2\beta + \frac{\omega}{c} \Delta n \, \ell \sin 2\beta\right\} \qquad (1)$$

$$I_1 = \frac{I}{2}\left\{1 - \frac{\alpha_{\parallel}+\alpha_{\perp}}{2} \ell\right\} \quad ; \quad \Delta\alpha = \alpha_{\parallel}-\alpha_{\perp} \quad ; \quad \Delta n = n_{\parallel}-n_{\perp}$$

I is the total incident light intensity and ℓ is the absorption-path length. To obtain (1), an optically thin medium has been assumed ($\alpha_{\parallel}\ell$, $\alpha_{\perp}\ell \ll 1$ and $\Delta n\,\ell/\lambda \ll 1$). Under our assumptions, pure dichroism and pure birefringence signals are given respectively by $\Delta_1 = I_c(0°) - I_c(90°)$ and $\Delta_2 = I_c(45°) - I_c(-45°)$ (maximum birefringence signal).

In [3] the method has been discussed in term of interferences of polarized light and a more general expression of $I(\beta)$ has been given for the case of an elliptically-polarized incident probe beam. Let us point out that in the case of a linearly-polarized probe beam the method is not sensitive to birefringence.

2. Laser Induced Dichroism and Birefringence in a Two-Level System.

The experimental arrangement (Fig. 1) is the usual one for saturated absorption experiments except that, after the interaction region, the probe beam polarization is analyzed by means of a birefringent prism (Wollaston prism). The Wollaston prism allows us to simultaneously analyze the polarization of the probe beam along two perpendicular directions β and $(\beta + \pi/2)$, and to get a spatial separation of the corresponding beams. The two beams are sent on two identical photodiodes and the corresponding output signals enter a differential amplifier. With a suitable orientation of the axes of the birefringent prism and a correct balance of the signals delivered by the two photodetectors, one can record directly the quantities Δ_1 and Δ_2. The signal from one of the detectors gives $I(\beta)$; $I(0°)$ and $I(90°)$ represents respectively the usual saturated absorption resonance curve $\alpha_{\parallel}$ and $\alpha_{\perp}$.

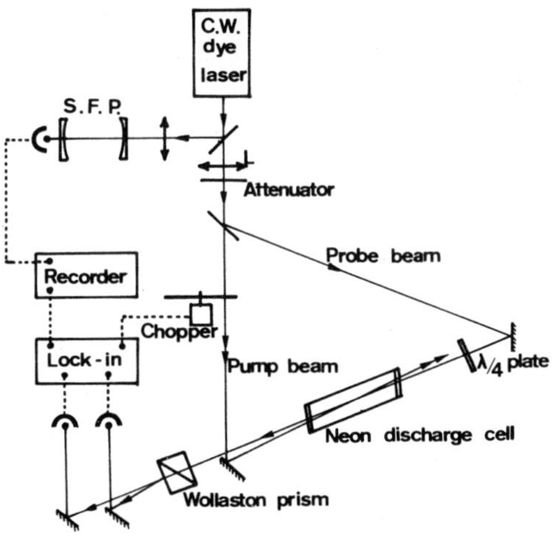

Fig. 1 Experimental arrangement for observation of Doppler-free laser-induced dichroism and birefringence in a two level system.

The experiments have been performed with various Ne I transitions in a neon discharge cell (discharge current : 5 mA ; neon pressure : 0.2 Torr ; cell diameter : 1.5 cm ; discharge length : 5 cm). On Fig. 2 the recorded spectrum $I_c(\beta)$ clearly shows a superposition of absorption and dispersion resonance curves as expected from (1).

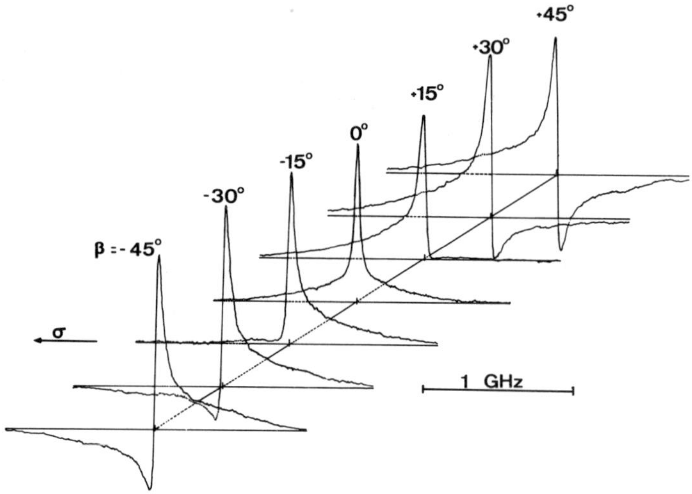

Fig. 2 $\lambda = 6074$ Å ($1s_4 \rightarrow 2p_3$) neon transition ; recorded spectrum $I_c(\beta)$ for various directions of the linear analyzer.

For most of the $1s \rightarrow 2p$ Ne I yellow lines the saturated-absorption resonance shows, in addition to the narrow peak, a large and rather intense background due to collisional redistribution of the atomic velocities (Fig. 3a). Under our experimental conditions it appears that the observed background is independent of the probe-beam polarization ; in other words the collisions which redistribute the velocities of the pumped atoms over the Doppler profile drastically reduce the initial laser-induced anisotropy . Consequently, the broad structure is not present on the dichroism and birefringence resonances curves (Fig. 3b and 3c).

As they allow the elimination of the large and broad background which can appear in saturated-absorption experiments, our Doppler-free resonances can be helpful , on one hand , to improve the resolution in the case of closely-spaced structures and, on the other hand, for the investigation of saturated-absorption profiles.

3. Laser-Induced Dichroism and Birefringence in a Three-Level System

For observation of laser-induced dichroism and birefringence in three-level systems the experimental arrangement is very similar to that of Fig. 1 , except that : i) the probe beam is generated by a second CW single-mode dye laser ; ii) the two beams are now co-propagating (a small angle between the two beams allows spatial separation).

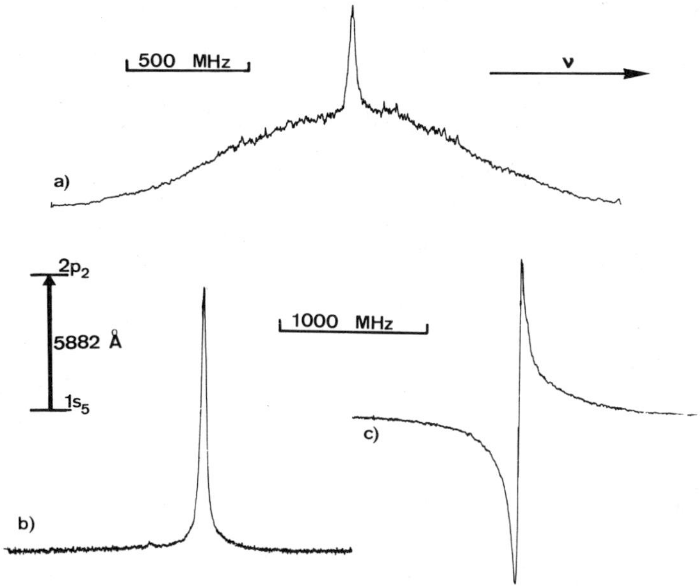

Fig. 3 $\lambda = 5882$ Å ($1s_5 \rightarrow 2p_2$) neon transition ; a) Saturated absorption spectrum ; b) Laser-induced dichroism spectrum : Δ_1 ; c) Laser-induced birefringence spectrum : Δ_2 (the amplitude of the curve is about $\Delta n_{max} \simeq 4 \times 10^{-8}$).

We deal with three-level systems in folded arrays involving $1s \rightarrow 2p$ transitions of neon. The pump-beam frequency is fixed and is very close to the atomic frequency ω_1 ; the probe-beam frequency Ω_2 can be tuned over the Doppler profile of the atomic transition ω_2 ; the two atomic transitions have a common level. A narrow-resonance occurs when the two beams are interacting with the same velocity group, i. e. for $v = (\Omega_1 - \omega_1)/k_1 = (\Omega_2 - \omega_2)/k_2$ (laser-induced line-narrowing effect).

We have investigated mainly two three-level systems of neon :

$$2p_2 \xleftarrow[\text{(probe)}]{5882\ \text{Å}} 1s_5 \xrightarrow[\text{(pump)}]{5945\ \text{Å}} 2p_4 \quad \text{and} \quad 1s_4 \xrightarrow[\text{(pump)}]{6096\ \text{Å}} 2p_4 \xleftarrow[\text{(probe)}]{5945\ \text{Å}} 1s_5 \ . \quad [5]$$

In the first case the resonance curve exhibits a narrow peak superimposed on a large background. Owing to the fact that the selected longitudinal velocity v can be non-zero whereas the velocity redistribution corresponding to the background is centered at $v = 0$, the narrow peak position can now be moved relatively to that of the background by changing the pump-laser detuning [5]. Otherwise the birefringence and dichroism spectrum are found to be very similar to that shown on Fig. 2, i. e. they have no broad background [5] . The discussion of the previous section on the possible applications and advantages of the method also holds for the case of three-level systems.

4. Simultaneous Observation of Laser-Induced Dichroism and Birefringence

According to (1), when the linear analyzer is rotated in its plane at frequency Ω , a modulation of $I_c(\beta)$ at frequency 2Ω appears if the sample exhibits some dichroism and birefringence. The amplitude of the modulation is dependent both upon dichroism $(\Delta\alpha)$ and upon birefringence (Δn) but the two contributions are in quadrature ; using phase sensitive detection it is possible to separate them. The experimental arrangement is shown on Fig. 4a and examples of simultaneously recorded dichroism and birefringence spectra are given on Fig. 4b, Fig. 5 and Fig. 6.

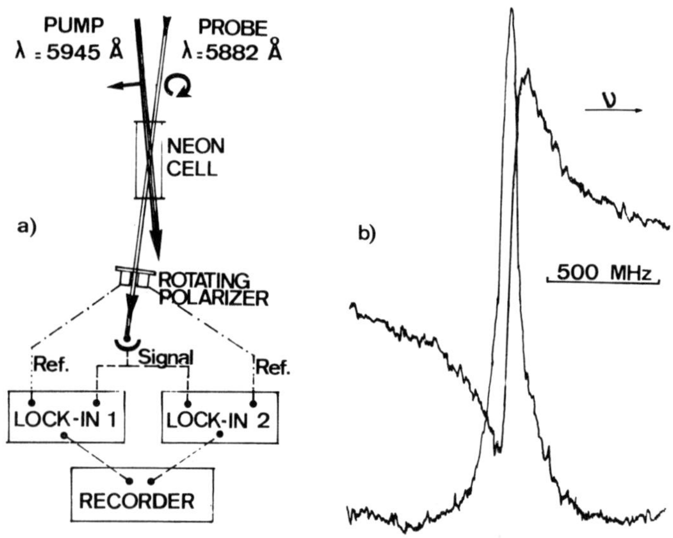

Fig. 4 a) Experimental arrangement for the simultaneous observation of Doppler-free laser-induced dichroism and birefringence ; b) Simultaneously obtained dichroism and birefringence resonance curves for the three-level system $2p_2 \leftarrow 1s_5 \rightarrow 2p_4$ of neon.

According to our initial assumptions, the dichroism and the birefringence resonance curves are expected to be related to one another through the Kramers-Kronig relation. For instance, for a Lorentzian-shaped dichroism curve $\Psi = \Delta\alpha\ \ell/2 = \frac{A}{1+x^2}$, $x = \frac{\omega-\omega_0}{\Gamma}$, the dephasing angle is given, according to the Kramers-Kronig relation, by $\varphi = \frac{\omega}{c}\Delta n\ \ell = -x\ \Psi = \frac{-Ax}{1+x^2}$. After some improvements in the experiment we hope it would be possible, using curves similar to that of Fig. 5, to check whether the expected relation between dichroism and birefringence, as observed in our experiments, holds or not.

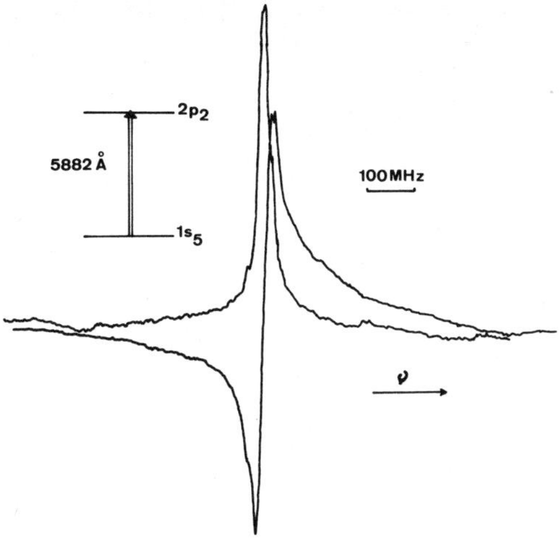

Fig. 5 λ = 5882 Å ($1s_5 \rightarrow 2p_2$) neon transition ; simultaneously recorded dichroism and birefringence spectra (circularly-polarized probe beam and rotating polarizer on use).

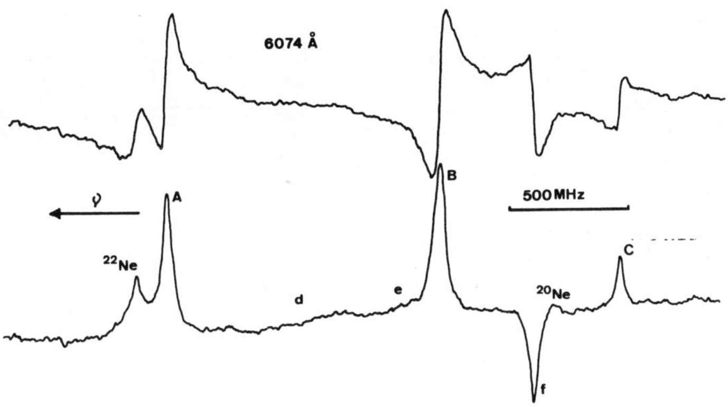

Fig. 6 λ = 6074 Å ($1s_4 \rightarrow 2p_3$) neon transition ; birefringence and dichroism spectrum recorded simultaneously with a mixture of ^{20}Ne , ^{21}Ne (nucleon spin : 3/2) and ^{22}Ne .

References

[1] P. F. Liao and G. C. Bjorklund, Phys. Rev. Letters 36, 584 (1976).
[2] C. Wieman and T. W. Hänsch, Phys. Rev. Letters 36, 1170 (1976).
[3] J.-C. Keller and C. Delsart, Optics Commun. 20, 42 (1977).
[4] V. P. Kaftandjian, Thesis, University of Provence (1977).
[5] C. Delsart and J.-C. Keller, to be published.

IV. Highly Excited States, Ionization, and High Intensity Interactions

IDENTIFICATION OF RYDBERG STATES IN THE ATOMIC LANTHANIDES AND ACTINIDES*

J.A. Paisner, R.W. Solarz, and E.F. Worden

Lawrence Livermore Laboratory, P.O. Box 808, Livermore, CA 94550, USA

and

J.G. Conway

Lawrence Berkeley Laboratory, Berkeley, CA 94720, USA

1. Introduction

An accurate determination of an atomic ionization potential usually requires an observation of a Rydberg progression or the determination of a photoionization threshold. In the lanthanides and the actinides direct measurements of limit values using conventional techniques are prohibitively difficult. For example, a uv absorption spectrum is usually hard to interpret due to the existence of many high (low) lying states of low (high) lying valence configurations. Transitions to these high lying valence states are stronger and more numerous than transitions to Rydberg states. As a rule, Rydberg series are hopelessly obscured in the absorption spectrum. Similarly, uv photoionization spectra are complicated by transitions from many low lying metastable levels that have significant populations at the temperatures at which the atomic species have reasonable vapor pressures. Usually, photoionization occurs at wavelengths significantly longer than that necessary to ionize ground state atoms. It is nearly an impossible task to separate the photoionization spectrum originating on the atomic ground state from that on the metastable states. While long Rydberg series in 3 lanthanide elements have been identified using conventional methods [1,2], more sensitive and flexible methods are required to systematically study these heavy multi-electron atoms.

About one year ago we reported the detection of several Rydberg series in atomic uranium [3]. This was the first observation of Rydberg progressions in any actinide. In those experiments high lying states were accessed by time resolved stepwise excitation using pulsed dye lasers tuned to resonant transitions. Atoms excited to levels within 1000 cm^{-1} of the ionization limit were then photoionized by 10.6 μ radiation from a pulsed CO_2 laser. By delaying the infrared ionizing pulse, and thus discriminating against short lived valence states, longer lived Rydberg states having $n \geqslant 14$ were observed. Series convergences yielded a very precise value for the uranium

*Work performed under the auspices of the U.S. Energy Research and Development Administration under Contract No. W-7405-Eng-48.

ionization limit (6.1941(5) eV or 49958(4) cm^{-1}). We would now like to report the study of Rydberg spectra and ionization thresholds of ten lanthanides using several variations of this time-resolved resonant multistep technique. The ionization limits for the lanthanides determined in this way show a systematic dependence on atomic number. We offer a simple physical model to explain these results.

2. Experimental Methods

The experimental apparatus is shown schematically in Fig. 1. Briefly, the apparatus is a crossed beam spectrometer in which the atomic beam is irradiated and eventually ionized by the output of two or three pulsed dye lasers tuned to resonant transitions. The oven consists of a resistively heated tungsten crucible operated at temperatures to give a vapor pressure of about 10^{-3} torr of the atomic species under study. The vapor issues through a slit into an interaction chamber where the number density is 10^9 - 10^{11} atoms/cm^3. The chamber is electrically biased to suppress thermal ions from the oven and to efficiently focus photoion product into the detection system. The detector is a channeltron particle multiplier contained within a quadrupole mass analyzer. The excitation sources are N_2 laser pumped dye lasers which provide 5-10 nsec pulses having 0.5-1.0 cm^{-1} spectral line widths. The separate N_2 lasers are fired by a common master trigger. The master control unit also provides a pulse to trigger the gate of a boxcar which in turn integrates the signal from the particle multiplier. Wavelength calibrations are provided by emission lines from either a uranium or thorium microwave electrodeless discharge lamp.

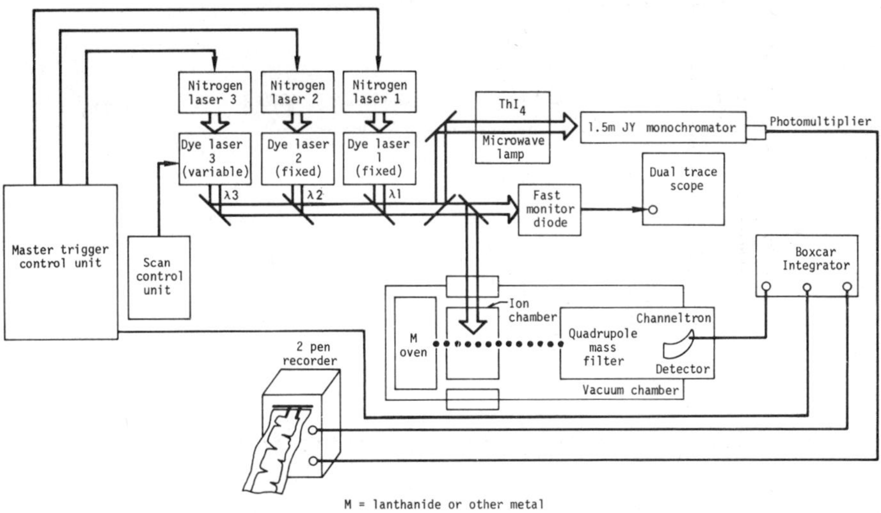

Figure 1. Laser spectroscopy apparatus.

The experimental procedure is to first identify a threshold for photoionization in the atom under study to an accuracy of ten to twenty wavenumbers using a two step photoionization process from several parent levels. Once this threshold value is well established, the photoionizing laser is next tuned through regions in the ionization spectrum where one would expect to observe autoionizing Rydberg states which converge to well known low lying excited states of the ion. The two lasers are generally delayed in time by about 10 nanoseconds to provide an unambiguous excitation sequence. That is, the first laser is held at wavelength λ_1 to bring the atom to an excited state near 3 eV. The second laser, which is scanned in wavelength λ_2, ionizes the atom when the combined energy of the two laser frequencies exceeds the ionization energy of the atom. If necessary, three step photoionization studies are also performed. As before, a small time delay between the output of the lasers is introduced to provide an unambiguous excitation sequence of λ_1 then λ_2 and finally λ_3. A threshold for photoionization is determined to check the two laser two photon results. This is followed by a search for autoionizing Rydberg states. All experiments consist of monitoring the photoion production as a function of the continuously scanned final wavelength.

Figure 2 illustrates an autoionizing Rydberg series in Dy. We have also observed autoionizing progressions in Nd, Sm, Eu and Er. In some cases autoionizing series are obscured by the large density of non-Rydberg autoionizing states. This was the situation in our original work on U. In those cases we employed time resolved ionization techniques. The atom is brought to a state just below the ionization limit, and is subsequently ionized by a pulsed electric field[4] or by collisions with other atoms in the beam[5]. Bound Rydberg states are preferentially ionized and detected because they are long lived and highly polarizable. Figure 3 shows a bound Rydberg series in Ce converging to the ground state of the ion obtained using the time delayed (~2 μsec) field ionization method. We have also observed bound Rydberg series in Gd and Tb in this way. Collisional ionization was used to detect bound Rydberg series in Ho, as illustrated in Fig. 4.

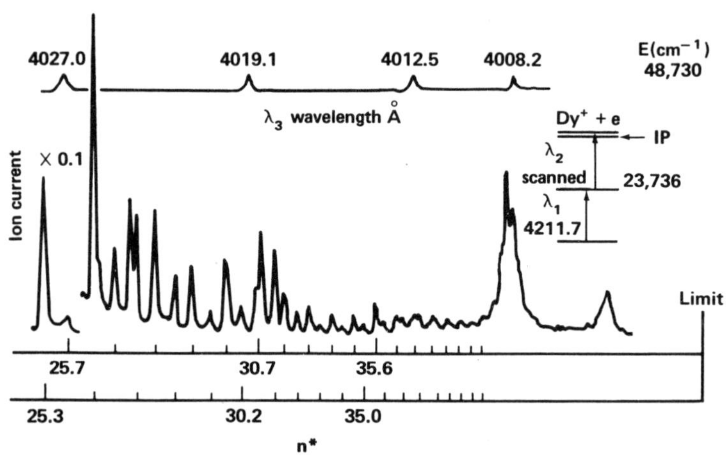

Figure 2. Dysprosium autoionizing Rydberg series converging to the $^4I_{15/2}$ limit 828.3 cm^{-1} above the $^6I_{17/2}$ ground state of the ion. This is a double series converging to the same limit.

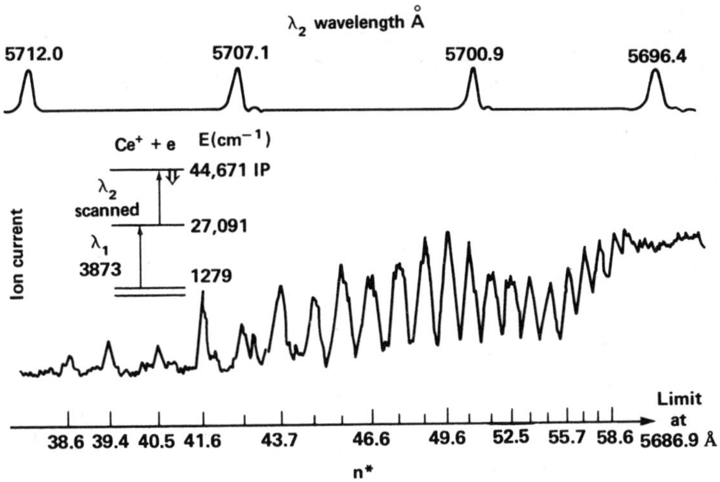

Figure 3. Cerium series converging to the $^4H^o_{7/2}$ ground state obtained by pulsed field ionization of the bound Rydberg levels. The pulsed field of 100 V/cm was delayed 4 μs from the populating laser pulse.

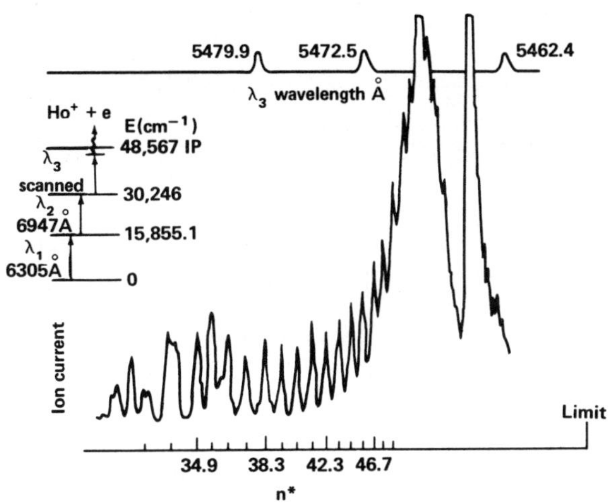

Figure 4. Collisionally ionized Rydberg series of holmium converging to the $^5I^o_8$ ground state of the ion. The excitation sequence is shown with collisional ionization indicated by the wavy arrow.

3. Results

We have observed Rydberg progressions in U, Ce, Nd, Sm, Eu, Gd, Tb, Dy, Ho and Er. With the exception of Pr and Pm, series have been observed in all the lanthanides since Rydberg spectra in Eu, Tm and Yb have been reported in the literature [1,2]. The criterion we used in determining an ionization limit is to choose that value which yields the smoothest and most constant value of the quantum defect as a function of n for the high Rydberg states. This procedure ignores the effects of perturbations in the spectrum which can cause sudden deviations in the quantum defect from state to state. For very high principal quantum number states, however, such deviations are unlikely to cause an error larger than a few wavenumbers. In those cases where the effects of perturbations are obviously present, that is when smooth progressions over an extended range are not observed, we have adopted larger error bars in our quoted values for the ionization limits. Figure 5 shows

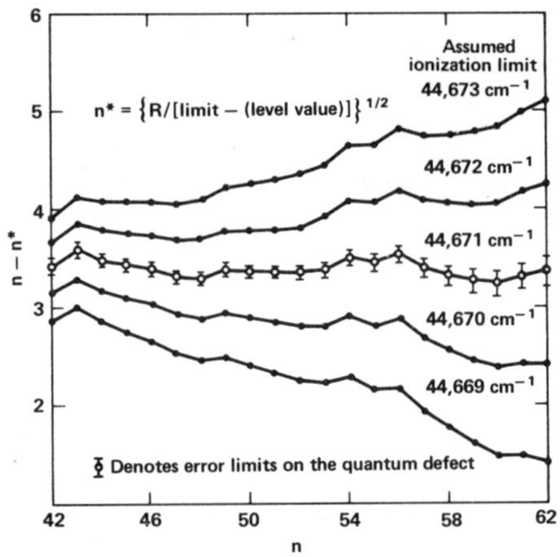

Figure 5. Variation in quantum defect (n - n*) vs n with change in assumed limit for the Cerium Rydberg series shown in Fig. 3.

the variations in quantum defect (n-n*) for a change in the assumed limit value for a field ionized bound series in Ce. The value of n is not necessarily the principle quantum number but is a running integer chosen close to n* in order to evaluate the variation in quantum defect. The sensitive dependence of the slope of the quantum defect curve on the assumed ionization limit is quite evident, and illustrates the usefulness of this procedure in extracting a series convergence limit. For Pr, we derived a value for the ionization limit from the observed sharp onsets of photoionization.

Table 1 lists the lanthanide ionization limits that we have measured. The typical uncertainties are .0005 eV for Rydberg convergences and .005 eV for photoionization thresholds. The ionization limit listed for Pm was obtained using the interpolation procedure described in the next section. Our measured limit for Eu overlaps the earlier more accurate determination of Smith and Tomkins[2]. The values for Tm and Yb are taken from Camus[1]. The best previous values for the other lanthanides are those of Reader and Sugar[6] who used a spectroscopic extrapolation procedure. For several cases our direct measurements are outside of their .008-.050 eV error bars. The electron impact values of Ackermann et al.[7] have quoted uncertainties of 0.1 eV.

Table 1. LANTHANIDE FIRST IONIZATION POTENTIALS

			Laser Spectroscopy LLL	
Element	Electron Impact Ref. 7	Spectroscopic Extrapolation Ref. 6	Photoionization Threshold	Rydberg Convergence
Ce	5.44	5.47(5)	5.537(4)	5.5387(4)
Pr	5.37	5.42(3)	5.464(6)	---
Nd	5.49	5.489(20)	5.523(3)	5.5250(6)
Pm	---	5.554(20)	---	[5.582(10)]
Sm	5.58	5.631(20)	5.639(3)	5.6437(10)
Eu	5.68	5.67045(2)[a]	5.666(3)	5.6704(3)
Gd	6.24	6.141(20)	---	6.1502(6)
Tb	5.84	5.852(20)	---	5.8639(6)
Dy	5.90	5.927(8)	5.936(3)	5.9390(6)
Ho	5.99	6.018(20)	6.017(3)	6.0216(6)
Er	5.93	6.101(20)	6.104(3)	6.1077(6)
Tm	6.11	6.18436(20)[b]	---	---
Yb	6.21	6.25394(20)[b]	---	---

( ) = error in last digit

[ ] = extrapolated value

[a]Taken from reference 2

[b]Taken from reference 1

We have used the conversion 1 eV = 8065.479 cm^{-1}.

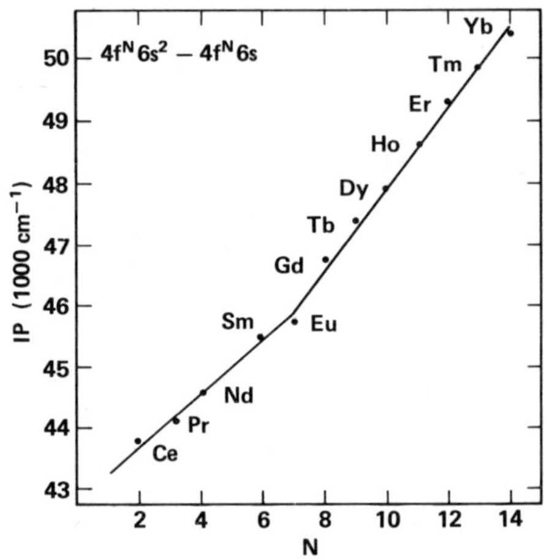

$$IP = \begin{cases} 441\,N + 42822\ cm^{-1} & \text{for } N \leq 7 \\ 661\,N + 41280\ cm^{-1} & \text{for } N \geq 7 \end{cases}$$

Figure 6. Normalized ionization potentials of the lanthanides plotted as a function of number of f electrons. Only the Ce and Gd points required normalization to the $4f^N6s^2 - 4f^N6s$ process; see text.

4. Discussion

With the exception of Ce and Gd, the ionization process is a removal of an s electron from the $4f^N6s^2$ (N = 2 to 14) configuration, and the ionization potential is the difference between the energy of the lowest level of the $4f^N6s^2$ configuration of the neutral and the lowest level of the $4f^N6s$ configuration of the ion. For Ce(N = 2) and Gd(N = 8) the ground configurations for the neutral are $4f^{N-1}5d6s^2$. The ground ionic configuration of Ce is $4f^{N-1}5d^2$ and that for Gd is $4f^{N-1}5d6s$. However, the positions of the lowest lying levels of the $4f^N6s^2$ and $4f^N6s$ configurations are known very accurately from spectroscopic studies[8]. We can therefore normalize our results so that all the ionization limits correspond to the removal of an s electron from the lowest level of $4f^N6s^2$. The resulting data is plotted in Fig. 6. The values for Tm and Yb are taken from reference 1. The ionization limits normalized in this way clearly show a piecewise linear variation with a change of slope at the half shell (N = 7). This behavior is not specific to the lanthanides as illustrated in Figs. 7 and 8 in which the normalized ionizations limits for the first short period ($2p^N2s^2 - 2p^N2s$) and the iron transition series ($3d^N4s^2 - 3d^N4s$) are plotted as a function of N. In all cases the limit corresponding to the difference in energy between the lowest levels of the $n\ell^Nn's^2$ configuration of the neutral and $n\ell^Nn's$ of the ion has the same characteristic piecewise linear dependence on N with a slope break at the half-shell. To our knowledge this is the first time that a systematic dependence of ionization limit on atomic number within a period has been clearly shown.

This behavior for the normalized limit values can be explained simply in terms of the Pauli exclusion principle and Hund's rule. Consider the ionization process $n\ell^Nn's^2 - n\ell^Nn's$. In the absence of correlation between the

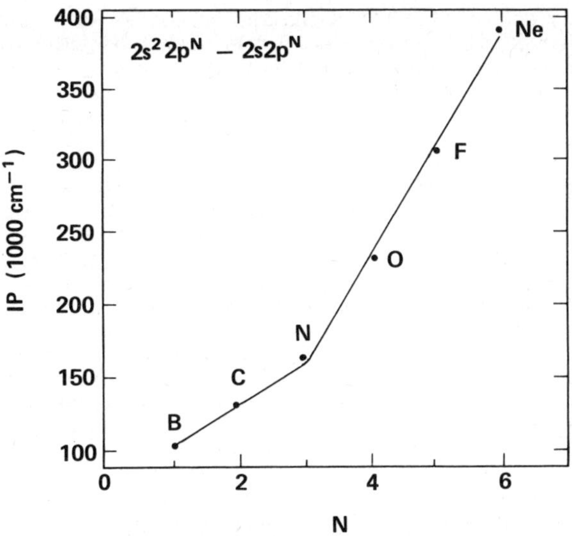

$$IP = \begin{cases} 27260N + 77771\ cm^{-1} & N \leq 3 \\ 75426N - 66727\ cm^{-1} & N \geq 3 \end{cases}$$

Figure 7. Normalized ionization potentials for the 2p series. In each case a correction for the ion from $2s^2 2p^{N-1}$ to $2s2p^N$ was required to normalize the ionization process to $2s^2 2p^N$ - $2s2p^N$.

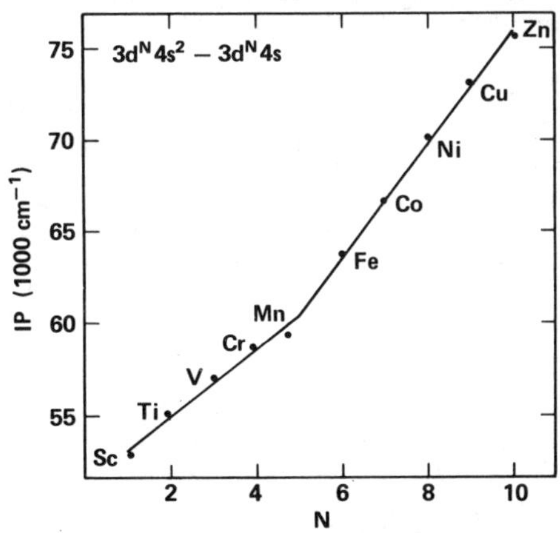

$$IP = \begin{cases} 1824N + 51333\ cm^{-1} \\ 3117N + 44867\ cm^{-1} \end{cases}$$

Figure 8. Normalized ionization potentials for the 3d series. Corrections were necessary for V, Cr, Co, Ni and Cu to normalize the ionization process to $3d^N 4s^2$ - $3d^N 4s$.

electrons motions and positions the ionization energy we associate with this configuration change will increase uniformly with N because of incomplete electrostatic shielding. However, the exclusion principle constrains the electrons motions to be correlated through its requirement that the multi-electron wave function be anti-symmetrized. Hund's rule dictates that the state which has the highest spin multiplicity lies lowest in energy. As one proceeds to the half filled shell in the ion ($n\ell^N n's$) the s electron has its spin aligned along the parallel spins of the n ℓ-orbital electrons. This reduces the electrostatic repulsion energy in the ground ionic state as a function of atomic number because the s electron is forced to spatially avoid in a pairwise way each of the N ℓ-orbital electrons. As a result this additional interaction will depress, monotonically with N, the ionization energy that would exist in the absence of exchange. In fact, this is the origin of Hund's rule. However, after the shell has been half filled, the ℓ-orbital electrons start pairing and the exchange effect reverses. Every unit change in N lifts the constraints on electron overlap and thus reduces the lowering of the ionization energy. Paired s electrons make no contribution to this symmetry dependent energy; only the ion with its single s electron is affected by the spin alignment of the partially filled shell.

Under the assumption that the ℓ-and s-orbitals do not change as the ℓ shell is progressively filled, Slater-Condon theory[9] predicts this piecewise linear behavior with a break of slope at the half-shell. In fact, the slope change is identically the exchange integral $G_\ell(\ell s)$ of the ion. Fortunately, these conditions are approximately satisfied by the lanthanides. The least squares fit to our experimental data (solid line figure 6) yields a slope difference of 220(23) cm^{-1} which is in excellent agreement with the average value of $G_3(fs)$ = 210(10) determined from the parameterized fitting of energy levels in the lanthanides[10]. This demonstrates that the exchange effect is indeed responsible for the slope break. Unfortunately, the Slater-Condon approach is not very useful in describing the behavior of the other series because in these cases the Slater parameters vary drastically across the periods.

Kathy Rajnak and Bruce Shore of our laboratory have extended the analysis by performing ab initio Hartree-Fock (HF) calculations[11]. In all cases (Figs. 6, 7 and 8), the HF results show the same piecewise linear behavior with a slope break at the half-shell. In fact, the HF calculations reproduce the slope differences for the short period and transition metal series, as well as the lanthanides, quite accurately. Although the HF calculations fail to predict the absolute ionization limits, they can be shifted to fit the experimental data because they have the correct dependence on N. In this way we hope to accurately predict the ionization potentials in the actinides based on our knowledge of the limit in U.

The Rydberg series we have reported were taken with a resolution (0.5-1.0 cm^{-1}) sufficient to extract a good value of the ionization limit, but insufficient to study spectral perturbations. Thus the errors in our extrapolated ionization limits reflect in part the variations in the quantum defects. In order to understand the nature and the magnitude of perturbations on Rydberg states in very heavy multi-electron atoms we have initiated an experimental program to measure absolute level energies of Rydberg states in atomic uranium to 0.03 cm^{-1} precision. Our preliminary results indicate that we may be able to apply multichannel quantum defect theory (MQDT) to uranium[12]. Indeed, it would be marked achievement if MQDT, which has been successful in explaining periodic perturbations and predicting level positions in light atoms[13], can provide a similar understanding of configuration interactions in the heaviest atomic systems.

5. Acknowledgements

We would like to acknowledge valuable discussions with B. Shore, K. Rajnak, J. Sugar, K.T. Lu and L. Carlson as well as the expert technical assistance of C. May and S. Johnson.

References

1. P. Camus, Thesis, Univ. of Paris, Orsay, (1971) Unpublished.
   P. Camus and F. Tomkings, J. Phys. 30, 545 (1969).

2. G. Smith and F.S. Tomkins, Proc. R. Soc. Lond. A 342, 149 (1975) and G. Smith and F.S. Tomkins, Phil. Trans. R. Soc. Lond. 283, 345 (1976).

3. R.W. Solarz, C.A. May, L.R. Carlson, E.F. Worden, S.A. Johnson, J.A. Paisner, and L.F. Radziemski, Phys. Rev. A 14, 1129 (1976).

4. T.W. Ducas, M.F. Littman, R.R. Freeman and D. Kleppner, Phys. Rev. Letters 35, 366 (1975).

5. W.P. West, G.W. Foltz, F.B. Dunning, C.J. Latimer and R.F. Stebbings, Phys. Rev. Letters 36, 854 (1976).

6. J. Reader and J. Sugar, J. Opt. Soc. Am. 56, 1189 (1966) and W.C. Martin, L. Hagan, J. Reader and J. Sugar, J. Phys. Chem. Ref. Data 3, 771 (1974).

7. R.J. Ackermann, E.G. Rauh and R.J. Thorn, J. Chem. Phys. 65, 1027 (1976).

8. J. Blaise, P. Camus and J.F. Wyart, "Rare Earths Elements" in Gmelin Handbuch der Anorganischen Chemie System 39-B4 (Springer-Verlag, Berlin, 1976) pp. 124ff.

9. E.U. Condon and G.H. Shortley, The Theory of Atomic Spectra (Cambridge University Press, London, 1967) pp. 158ff.

10. B.G. Wybourne, Spectroscopic Properties of Rare Earths (John Wiley and Sons, Inc., New York, 1965) p. 55-56.

11. K. Rajnak and B. Shore, to be published.

12. For a brief review of MQDT see K.T. Lu and U. Fano, Phys. Rev. A 2, 81 (1970).

13. J. Wynne, J. Armstrong and P. Esherick, paper in this Conference.

MULTIPHOTON IONIZATION SPECTROSCOPY OF THE ALKALINE EARTHS

P. Esherick[1], J.J. Wynne, and J.A. Armstrong

IBM Thomas J. Watson Research Center, P.O. Box 218
Yorktown Heights, NY 10598, USA

1. Introduction

Many aspects of the electronic structure of many-electron atoms are still not well understood. In many-electron atoms, the Coulomb repulsion and spin-orbit coupling effects between electrons strongly influence the nature of the electronic states. The one-electron picture, where each electron is viewed as moving in a well-defined orbit, independent of the motion of the other electrons, breaks down when these effects are strong enough. This has hindered efforts to identify and classify the states of many-electron atoms with one-electron labels. We have shown how multiphoton ionization spectroscopy provides us with new spectroscopic data allowing us to identify previously unknown states of alkaline earth atoms and how the Coulomb repulsion of the two outer electrons may be taken into account via a model based on multichannel quantum defect theory (MQDT).

Many laser spectroscopists have been studying the alkali metal atoms, where the energies of the bound states are known to better than 1 cm^{-1} through the use of conventional, pre-laser absorption and emission spectroscopy. The laser spectroscopists have been looking at these atoms with high resolution, multiphoton techniques to study fine and hyperfine structure. Conventional survey spectroscopy of the alkali atoms has been successful in identifying the states because these are essentially one-electron atoms. They behave as one-electron atoms because they have an outer electron "orbiting around" a closed electronic shell. The bound states of the one-electron atoms fall into series where the energy of each state is given by

$$E = I - R/(n^*)^2 \qquad (1)$$

where R is the Rydberg constant and I is the ionization limit. For hydrogen, n^* is an integer, n, and for the alkali metal atoms

$$n^* = n - \delta \qquad (2)$$

[1] Current address: Sandia Laboratories, Albuquerque, NM 87115

where δ, the so-called quantum defect, is found to be nearly constant for states of the same orbital angular momentum, ℓ.

The contrast between the alkali metal atoms and the alkaline earth atoms Ca, Sr and Ba is brought into focus by considering the series of even-parity, J=2 states of Ca, converging on the first ionization limit, that have been labeled 4snd $^1D_2$. Of these, only the 4s3d, 4s4d, 4s5d, and 4s6d states were correctly identified in the literature before our multiphoton ionization spectroscopic studies. The 4s7d $^1D_2$ had been incorrectly identified by emission spectroscopy [1], and we found the correct position more than 100 cm^{-1} away.[2] Furthermore, this state is more than 1200 cm^{-1} higher than the 4s6d $^1D_2$. Were we to look for this state using high resolution lasers by scanning and recording on chart paper at the rate of 10 MHz/cm starting from the 4s6d state, we would have had to use 30 km of chart paper before we reached the 4s7d state! It would still be 3 km away from the expected position based on the results of emission spectroscopy! This points to the need for doing survey spectroscopy on atoms other than the alkali metals.

In this paper, we shall describe the experimental technique of multiphoton ionization spectroscopy and show how we have used it to study series of even- and odd-parity states in Ca, Sr and Ba. We shall indicate how to use MQDT to analyze the data, describe some interesting trends in the results and point out some future directions for studies in these atoms.

2. Experimental Techniques

In the technique of multiphoton ionization spectroscopy, two or more photons excite atoms from the ground state to an excited state. If the excited state is bound, there are several processes which may ionize it, including photo-ionization, collisional ionization, or chemi-ionization. Whatever the mechanism, strong ionization signals occur when an atomic vapor is irradiated by lasers tuned to multiple photon resonances. The simple set-up shown in Fig. 1 may be used to detect ionization. An ionization probe consisting of a thin wire is inserted into a pipe containing the atomic vapor. For Ca, Sr and Ba, the pipe is heated to $\sim$ 700 $^{\circ}$C. If the probe is negatively biased relative to the walls of the pipe, thermionic emission from the probe

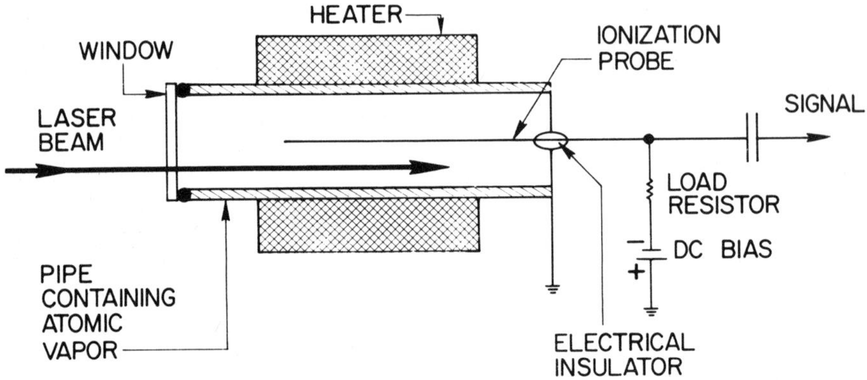

Figure 1 Experimental set-up for detecting multiphoton ionization in a hot atomic vapor.

will lead to space-charge limited current. Ions produced by the laser excitation partially neutralize the space charge, thereby allowing an increased electron current to flow.[3] One may then detect ionization by observing the pulsed current, synchronous with the pulsed lasers, which flows through the external load resistor. Figure 2 is a typical spectrum obtained when Ca vapor is irradiated with a single, repetitively pulsed nitrogen-laser-pumped dye laser. The observed resonances have been successfully identified[2] in terms of two-photon excitations to $^1S_0$ and $^1D_2$ states. Bias voltages of only ∿ 1 V are needed to produce this high signal-to-noise ratio. Although the simple ionization probe shown in Fig. 1 serves as an excellent ion detector, the electric field near the probe is not uniform. To remove the uncertainty as to the magnitude of the applied field, we have adopted a parallel plate configuration. (An alternative is to irradiate the atoms in a field free region and let them drift out of this region into the region where they can be collected by the electric field around the probes[4].)

Variations of this basic experiment may be performed. For example, we have used three dye lasers, simultaneously pumped by the same nitrogen laser, to stepwise excite Ca, Sr and Ba, from the $^1S_0$ ground state to high-lying odd parity states which are primarily triplet in character. These states are not observed in absorption spectra from the ground state because of the spin-forbidden nature of the transitions. The specific three laser

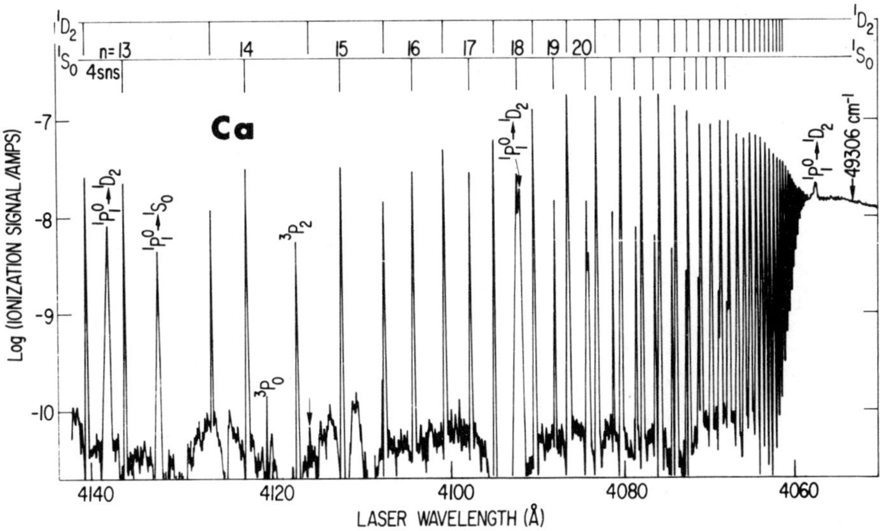

Figure 2 Multiphoton ionization spectrum of even-parity states of Ca converging on the ionization limit at 49306 cm^{-1}. Experimental conditions: laser polarization, linear; laser intensity, 10^7 W/cm^2; Ca pressure, 0.1 Torr; buffer gas pressure, 10 Torr; voltage on probe, 1 V.

technique we have used, which will be described below, overcomes this barrier.

3. Experimental Results: Even-Parity States

The spectrum shown in Fig. 2 is readily separated into two distinct series. By tuning the laser frequency ν to lower energies, we verified that a peak was observed when 2ν was tuned to the energies of the highest previously known $^1S_0$ and $^1D_2$ states of Ca, namely the 4s12s $^1S_0$ and the 4s6d $^1D_2$ states. We also observed that the signal strengths of the peaks varied as the square of the input laser power. We thus concluded that we were seeing two photon excitation from the $^1S_0$ ground state to series of 4sns $^1S_0$ and 4snd $^1D_2$ states converging on the 4s $^2S_{1/2}$ ionization limit of Ca.

In order to get accurate energy values for each new state, each spectrum was individually calibrated using the experimental set-up shown in Fig. 3. Accurate calibration of the laser wavelength was obtained by recording, simultaneously with the ionization signal, the absorption of the section harmonic of the

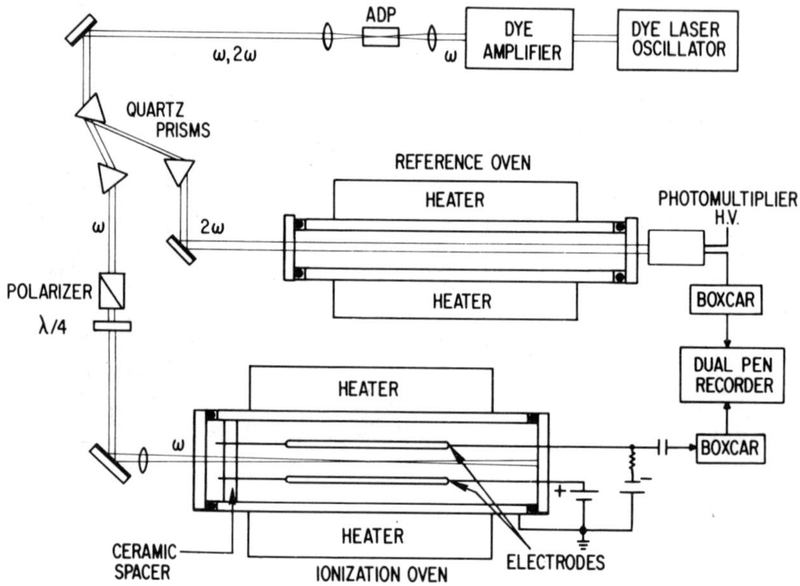

Figure 3 Experimental configuration for measuring ionization spectrum and calibrating it with an absorption spectrum.

laser from the ground state to the well-known $^1P_1^o$ states of the atom under study. A typical spectrum obtained for Sr[5] using these techniques is shown in Fig. 4. The lower trace is the log of the ionization signal showing the two-photon excitation spectrum, while the upper trace is the one-photon transmission spectrum of the second-harmonic of the laser. Using data from several scans and averaging, state energies were determined to an accuracy of better than 0.2 cm^{-1} in almost every case. Detailed results and descriptions of the experimental procedure are presented in [2] and [5].

In the spectrum shown in Fig. 4, we have used circularly polarized laser excitation to eliminate the $^1S_0 \rightarrow {}^1S_0$ excitations, thus isolating the J=2 spectrum from the combined J=0 and J=2 spectrum that is observed with linearly polarized excitation. This is easily understood with reference to Fig. 5. It is important to recognize that these selection rules apply only if the intermediate state is polarized, i.e., when it is a virtual state. As such, it is excited only when the laser field is present. In contrast, the intermediate state can be a real

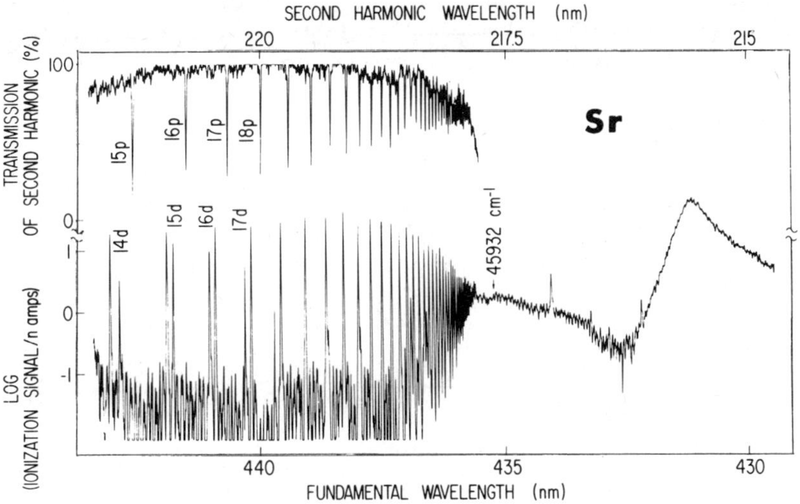

Figure 4 Multiphoton ionization spectrum of even-parity, J=2 states obtained with circularly polarized laser excitation of approximately 0.2 Torr Sr (680^{0}C), with 20 Torr Ne buffer gas. Upper scan is the absorption, in a reference oven, of the second harmonic of the laser by the known $^1P_1^o$ Rydberg series.

state when the laser is tuned directly on resonance (Fig. 5c). In the presence of collisions, this populated real state dephases. In this case, the J=1 intermediate state is unpolarized with the substates M=+1,0 and -1 all equally populated and mutually dephased. The peaks in Fig. 2 are labeled $^1S_0$ and $^1D_2$ on the basis of the determination of J. This method of J selection is vital if one is to sort out a spectrum of states which do not follow the simple Rydberg formula (1).

The two-photon spectrum of J=2 states of Sr, shown in Fig. 4, illustrates several interesting features that can occur in the spectra of many-electron atoms. In addition to the expected transitions from the $^1S_0$ ground state to the 5snd $^1D_2$ states, transitions are also observed to 5snd $^3D_2$ states with unusually strong intensity, particularly for the 5s15d and 5s16d levels. The clue to understanding these anomalous intensities is found by noting that for the 5snd levels with n≤15, the weaker $^3D_2$ state lies higher in energy than the strong $^1D_2$, whereas for

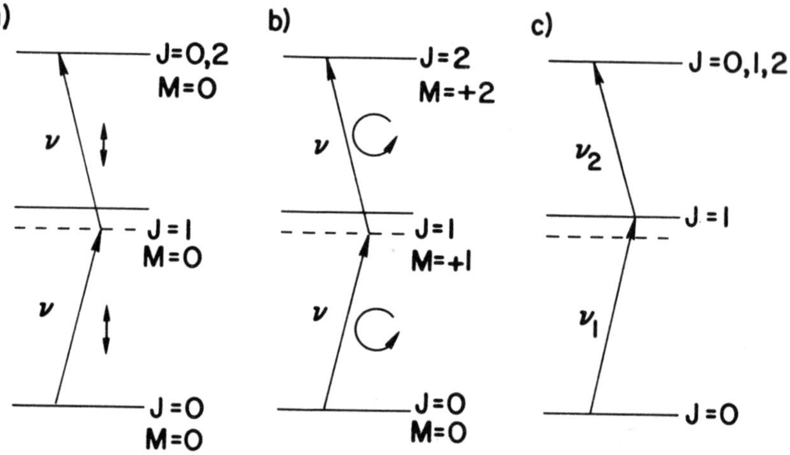

Figure 5 Allowed two-photon transitions from a J=0 initial state. For a direct two-photon transition to a state at energy 2ν, the final state may have a) J=0 or 2 when the laser beam at frequency ν is linearly polarized, but b) only J=2 when the laser is circularly polarized. In case c) one laser at ν_1 excites the intermediate J=1 state on resonance, collisions dephase this state, and the second laser at ν_2 excites final states with J=0,1 or 2 at the energy $\nu_1 + \nu_2$.

$n \geq 16$, this order is reversed. What has happened is that the two series have "crossed" and, at the so-called "crossing", the $^1D_2$ and $^3D_2$ states are heavily mixed, thus giving the unusually strong intensity to the 5s15d and 5s16d $^3D_2$ levels.

The most prominent single feature in Fig. 4 is the presence of a strong autoionizing resonance $\sim$ 450 cm^{-1} above the ionization limit. This resonance has been tentatively assigned[5] to the "$(4d)^2$" doubly-excited configuration. The asymmetric lineshape is typical of autoionizing resonances, and is caused by interference between the transition dipole matrix elements going to the $4d^2$ resonance and the $5s\varepsilon d$ continuim.

4. Experimental Results: Odd-Parity States

The literature shows that the msnp $^3P^o$ states of Ca, Sr, and Ba (m=4,5, and 6 respectively) were correctly known only to n=7. We have observed three-step transitions from the $ms^2\ ^1S_0$ ground state to these states by using a three laser excitation technique. One laser excites atoms from the ground state to the lowest energy ms(m+1)p $^3P^o_1$ state (the intercombination line transition).

A second laser then excites these atoms to the ms(m+1)s $^3S_1$ state. To obtain a spectrum, a third laser is tuned and one measures the ionization signal that occurs when the atoms make transitions from the populated $^3S_1$ level to higher energy msnp $^3P^o$ states. In this way, we have observed the $^3P^o$ states of Ca and Sr and n>70.[6] Additional signals were obtained from transitions from the populated $^3S_1$ level to the msnp $^1P^o_1$ levels whose energies are known accurately from absorption spectroscopy. Signals were also obtained from known two-photon transitions in the same atomic species (i.e., Ca or Sr) or in impurity atoms (e.g., Na, Ba). Furthermore, the third laser was monitored by beam-splitting off part of the beam and measuring its transmission through an air-spaced Fabry-Perot interferometer. The interferometer was calibrated from the ionization signals obtained on known transitions in Ca, Sr, Ba and Na. The calibrated interferometer then provided a means of calibrating the third laser frequency.

The data for Ba are complicated by both configuration mixing and strong spin-orbit effects. Whereas the $^3P^o$ states of Ca and Sr are only perturbed by one interloper (the 3d4p and 4d5p respectively), in Ba the 5d6p, 5d7p, 5d8p and 5d4f configurations all affect the spectrum. Furthermore, the high n $^1P^o_1$ states produce stronger ionization signals than the high n $^3P^o$ states when excited from the 6s7s $^3S_1$ levels. These problems have been resolved by complementing the excitation spectrum from the 6s7s $^3S_1$ with the excitation spectrum from the 6s7s $^1S_0$ intermediate state, and by taking one photon absorption spectra of Ba from the $6s^2$ $^1S_0$ ground state. The result is that the J=1 spectrum of Ba is now well understood and the 6snp $^3P^o_2$ states have been observed and identified to n=50. The detailed results will be published in the near future.

5. Analysis of Results

Once the energies of the states, as well as their parity and J are determined, one calculates the effective quantum number from (1) and looks for trends toward constant quantum defect, δ. If δ is reasonably constant, one says that the states form a relatively unperturbed Rydberg series. On the other hand, if δ is changing from one state to the next, there is strong evidence for configurational mixing and spin-orbit effects. For the $^3P^o$ states of Ca and Sr, the quantum defects are relatively constant. Each spectrum shows evidence for a single low-lying perturber, the 3d4p and 4d5p configuration in Ca and Sr respectively which mixes into the 4snp and 5snp series, respectively, over a limited number of states. Similarly, for the even-parity $^1S_0$ states of Ca and Sr, the data shows evidence for a single configuration, $4p^2$ or $5p^2$, perturbing the 4sns or 5sns series, respectively. Conventional second-order stationary state perturbation theory is perfectly adequate to describe this situation. In contrast, the $^1D_2$ and $^3D_2$ spectra of Ca and Sr shown continuously changing δ. These spectra cannot be adequately treated by perturbation theory. Similarly, although the 6snp $^3P^o_1$ states in Ba have a relatively constant quantum defect, the several perturbers present cause the $^1P^o_1$ spectrum to have a continuously changing δ, and perturbation theory is again inadequate for explaining the situation.

These spectra can be successfully analyzed by MQDT. This theory allows one to treat series of states as channels instead of as individual states. A channel describes a set of states, both bound and unbound, without regard to the specific energy of the highly excited electron. Specification of the angular momentum of the outer electron and the core, along with a description of their coupling, completes the description of the channel. For example, the designation 4snd $^1D_2$ describes a channel which contains bound states of even-parity and J=2, where the core is in the 4s state, the outer electron has ℓ=2, and the electrons are coupled to form a $^1D_2$ state. This channel also contains the continuum states designated by 4sεd $^1D_2$, where ε designates the kinetic energy of the unbound outer electron. Within its range of applicability, MQDT determines the energy levels and wavefunctions of the interacting Rydberg series in terms of interactions between a small number of channels. The interaction is expressed in terms of a small number of physically meaningful parameters.

This is not the place for an exposition of MQDT in detail. A more complete description can be found in [2] and [7]. But a graphical presentation of MQDT in terms of Lu-Fano plots[8] will help the reader to better understand what can be learned from our data. In such a plot, the effective quantum number, $\nu_i = n_i^*$ (mod 1) of a state, calculated from (1) as if it were part of a series converging on an ionization limit I_i, is plotted against $\nu_j = n_j^*$ (mod 1) as if the series converged on a different limit I_j. The limits are chosen as those to which the principal and perturbing series converge. As an example, the 5snp $^3P_1^o$ series of Sr is perturbed by 4d5p $^3P_1^o$. Thus an appropriate Lu-Fano diagram plots ν_{5s} vs. ν_{4d} as shown in Fig. 6. Because unperturbed Rydberg series are periodic in the ν_i, it turns out to

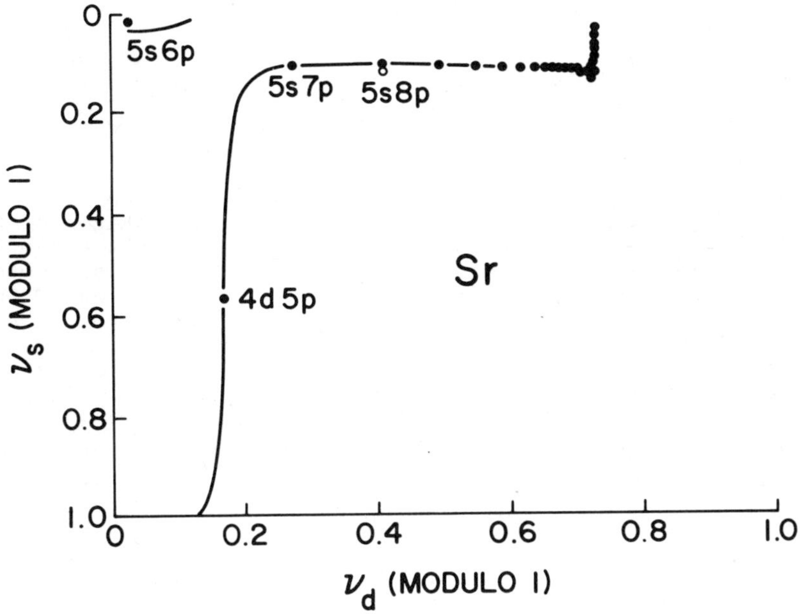

Figure 6 Lu-Fano plot of the $^3P^o$ states of Sr. The solid circles are data points. The open circle corresponds to an incorrect assignment of the 4s8p $^3P_1^o$ in the literature. The solid line is the result of a two channel MQDT fit to the data.[6]

be necessary to plot only the non-integral parts of the ν_i.

Figure 6 is a typical Lu-Fano plot for a Rydberg series which only weakly interacts with a single perturber. Note how the isolated 4d5p $^3P_1^o$ perturber lies far off the nearly horizontal line that can be drawn through the almost constant quantum defects of the 5snp $^3P_1^o$ states.

By contrast, consider the Lu-Fano plots of our $^1D_2$ and $^3D_2$

data in Ca and Sr, shown in Figs. 7 and 8 respectively. The data is plotted as points. In these figures, we see that the

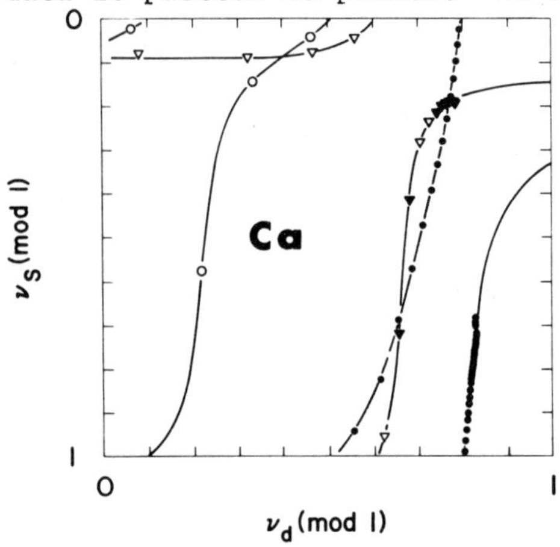

Figure 7 Lu-Fano plot of all J=2 states of Ca. The symbols are o for previously observed $^1D_2$ states, • for new $^1D_2$ states, ∇ for previously observed $^3D_2$ states, and ▼ for $^3D_2$ states observed in our spectra.

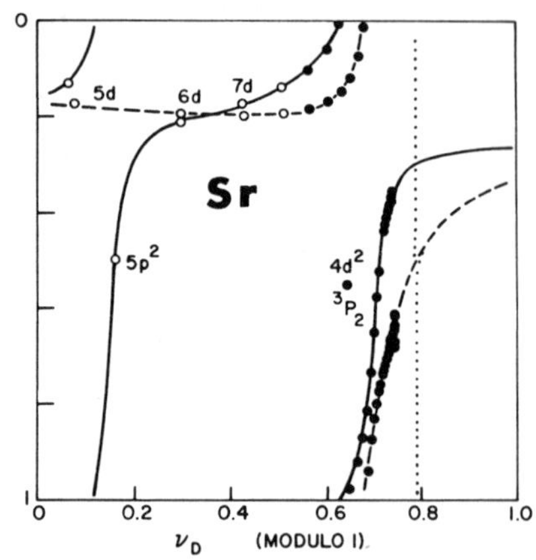

Figure 8 Lu-Fano plot of the J=2 states of Sr. Symbols are o for previously observed states and ● for the new data.

value for ν_s (the effective quantum number assuming the series converges on the Ca 4s or Sr 5s $^2S_{1/2}$ limits) is continuously changing. In Ca (Fig. 7) this is due to the strong configuration interactions amongst the $^1D_2$ states, plotted as circles, with the $4p^2$, 3d5s and $3d^2$ configurations interacting with the 4snd series. The $^3D_2$ series, plotted as triangles, are slightly less complicated since there is only one bound $^3D_2$ perturber, the 3d5s configuration. The solid lines in Fig. 7 are the result of an MQDT fit to the data.[2]

Having found the values for the MQDT parameters for Ca which give the solid curve in Fig. 7, one may calculate the admixture of each contributing configuration into the real states. The results show that whereas the "$4p^2$" configuration is almost entirely in the state labeled $4p^2$ in Fig. 7, the "$3d^2$" and "3d5s" configurations are spread out over dozens of states. (The labels are enclosed in quotation marks because the perturbers are not necessarily pure configurations but may be mixed amongst themselves.) In fact, no individual bound $^1D_2$ state contains more than 3% of the "$3d^2$" configuration, nor more than 13% of the "3d5s" configuration. A more detailed discussion of these results may be found in [2].

The Lu-Fano diagram (Fig. 8) for the even-parity J=2 states of Sr shows that the 5snd $^1D_2$ and $^3D_2$ series are perturbed by $5p^2$ $^1D_2$ and 4d6s $^1D_2$ and 4d6s $^3D_2$ configurations. There is no evidence for a bound $4d^2$ $^1D_2$ perturber analogous to the bound $3d^2$ $^1D_2$ perturber in Ca. This leads one to identify the autoionizing resonance appearing in the Sr spectrum (Fig. 4) as due to the $4d^2$ $^1D_2$ configuration.

Note in comparing Figs. 7 and 8 that in Ca the $^1D_2$ and $^3D_2$ curves cross at several points, whereas in Sr there is a distinct avoided crossing of the curves near the points corresponding to the 5s15d levels. This is the region of the spectrum (Fig. 4) where the 5snd $^3D_2$ states appeared with unusually strong intensities. By fitting MQDT curves through the plotted data points, it is possible to quantitatively determine the extent of interaction between the $^1D_2$ and $^3D_2$ channels due to the breakdown in Sr of L-S coupling. This was done in [5] and the results were found to closely agree with the intensity variations observed for the $^1D_2$ and $^3D_2$ series in the 5s14d to 5s18d region.

6. Discussion

The ease with which new spectroscopic data can be acquired using multiphoton techniques opens up numerous possibilities for study. In the alkaline earths, there are some important unanswered questions about the $^1S_0$ series. In Ca and Sr, the $4p^2$ and $5p^2$ configurations, respectively, make their presence known in the bound state region. But one would also expect to find evidence of the $3d^2$ and $4d^2$ configurations. No such evidence has been found in our studies in the bound state region. These techniques should be extended to the autoionization region to study even-parity states and look for these missing configurations. A preliminary study of the even-parity (J=0,1 and 2) autoionization spectrum of Ca has already been reported.[9]

Our studies in Ca, Sr and Ba show that the configuration interactions are much weaker in the triplet spectra than in the singlet spectra. This is consistent with the tendency of the two electrons to avoid one another in triplet wave functions.

We have seen enormous Stark shifts on the high-lying Rydberg states, but a systematic study remains to be done. The strong Stark effects are accompanyed by easily observed, DC electric field induced second harmonic generation (SHG). Further studies should be done on the Zeeman effect, two-photon pumped optical parametric oscillation, the ionization mechanism, resonantly enhanced optical sum mixing into the vacuum ultraviolet and the autoionization spectrum in these atoms.

Acknowledgements

We thank Mr. L. H. Manganaro for his technical assistance. The U. S. Army Research Office supported this work.

References

1. G. Risberg, Ark. Fys. 37, 231 (1968).

2. J. A. Armstrong, P. Esherick and J. J. Wynne, Phys. Rev. A 15, 180 (1977).

3. D. Popescu, M. L. Pascu, C. B. Collins, B. W. Johnson and I. Popescu, Phys. Rev. A 8, 1666 (1973), and references cited therein.

4. K. C. Harvey and B. P. Stoicheff, Phys. Rev. Letters 38, 537 (1977).

5. P. Esherick, Phys. Rev. A 15, 1920 (1977).

6. P. Esherick, J. J. Wynne and J. A. Armstrong, Optics Letters 1, 19 (July, 1977).

7. U. Fano, J. Opt. Soc. Am. 65, 979 (1975), and references cited therein.

8. K. T. Lu and U. Fano, Phys. Rev. A 2, 81 (1970).

9. J. J. Wynne, J. A. Armstrong and P. Esherick, Bull. Am. Phys. Soc. 22, 64 (1977).

MEASUREMENT AND CALCULATION OF EXCITED ALKALI HYPERFINE AND STARK PARAMETERS

S. Svanberg

Department of Physics, Chalmers University of Technology
402 20 Göteborg, Sweden

1. Introduction

The group of alkali atoms in the periodic chart of the elements has for a long time been a popular playground for atomic physicists. This is due to the relative ease with which these atoms can be handled both theoretically and experimentally. The activities in this field have lately entered a booming phase with the development of new experimental and theoretical methods of great power. In the development of new laser-spectroscopic methods the alkali atoms have played an important part. Especially sodium has become famous as the atom of choice, and quite a few experimentalists have got their color-vision normalized observing the glare from dye-laser beams tuned to the yellow sodium lines! Laser techniques have allowed the exploration of large domains of unknown territory in the spectra of the alkali atoms, including the very highly excited Rydberg states. A lot of information on the basic energy-level structures, fine- and hyperfine structures, isotope shifts, radiative properties and various aspects of the Stark interaction including field ionization phenomena has been obtained.

The basic energy level scheme for an alkali atom is shown in Fig. 1, where rubidium has been chosen as an example. With a single outer electron, sequences of S, P, D, F... doublets are obtained. For quite some time more detailed investigations were limited to the lower P-states which could be readily excited in single-step electric dipole transitions. With the utilization of cascade excitations, HAPPER and coworkers could study low-lying non-P states using decoupling [1] or rf spectroscopy techniques [2]. Using step-wise excitations [3] or two-photon absorption [4], employing laser sources, it is now possible to reach almost any level in the diagram.

The structure of one of the energy levels in Fig. 1 can in the presence of parallel magnetic and electric fields B and E be described by the Hamiltonian:

$$H = H_0 + \underbrace{f_{fs}(\delta W, l, J)}_{\text{fine struct.int}} + \underbrace{a\,\underline{I}\cdot\underline{J}}_{\text{magnetic dipole int.}} + \underbrace{b f_q(\underline{I},\underline{J})}_{\text{el.quadrupole int.}} + \underbrace{\mu_B g_J B J_z - \mu_B g_I' B I_z}_{\text{interaction with ext. magnetic field}} +$$

$$\underbrace{-\frac{1}{2}E^2\left[\alpha_0 + \alpha_2\frac{3J_z^2 - J(J+1)}{J(2J-1)}\right]}_{\text{interaction with ext. el. field}} \qquad (1)$$

H_0 accounts for the basic electronic energy, which actually is slightly different for isotopes of the same element due to mass and volume effects. This

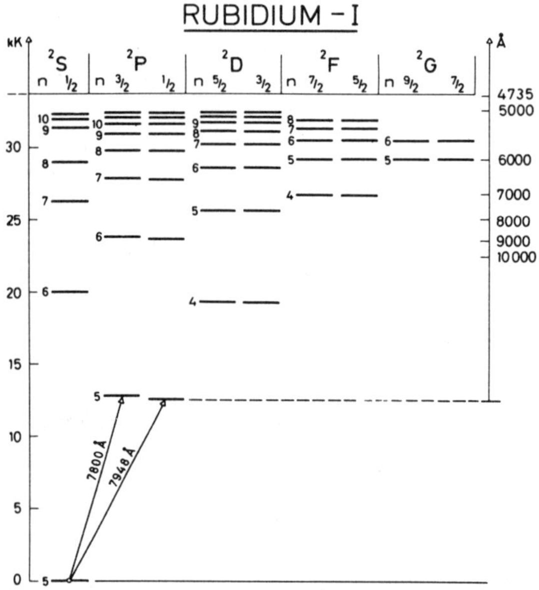

Fig. 1. Energy level diagram for Rb.

gives rise to isotope shifts. f_{fs} is the spin-orbit energy term, resulting in the fine structure splitting δW. The next two terms describe the magnetic dipole- and electric quadrupole hyperfine interactions, respectively. The characteristic constants a and b are products of electronic and nuclear quantities; in the first case a product of the magnetic dipole moment of the nucleus and the electronic magnetic field at the nucleus, in the second case a product of the electric quadrupole moment of the nucleus and the electric field gradient at the nucleus. If the nuclear properties are known the hyperfine interaction can thus be used to probe the electronic shell in a sensitive way. The interaction with the external magnetic field consists of one part due to the magnetism of the electronic shell and one part due to the nuclear magnetism. Finally, the Stark interaction, quadratic in the electric field E, consists of two parts, one scalar part shifting all sublevels equally much and one tensorial part sensitive to the J_z^2 value. This Hamiltonian is applicable when the fine structure splitting is much larger than the hyperfine splittings and the interactions due to the external fields.

The goal for spectroscopic measurements on alkali atoms is often to determine the characteristic constants in (1), i.e. δW, a, b, g_J, α_0, α_2. In this presentation we will mainly consider the hyperfine interaction constants a and b and the Stark parameters α_0 and α_2. We will here for practical reasons limit the discussion to work performed by the atomic spectroscopy group in Göteborg with brief reference to other work.

2. High Resolution Spectroscopy

Often the energies described by the different terms in the Hamiltonian (1) are small compared to the normal Doppler energy spread and thus high-resolution methods are required for accurate experimental studies. We can then consider the traditional resonance methods:

Atomic-beam magnetic resonance (Rabi 1938)
Optical pumping (Kastler 1950)
Optical double resonance (Kastler, Brossel, Bitter, 1949, 1952)
Level-crossing (Hanle 1924, Colgrove et al. 1959)
Anti-crossing (Eck et al. 1963)

These methods have a resolution only limited by the Heisenberg uncertainty relation, i.e. the natural radiation width. The optical methods were originally developed for use with conventional light-sources, but now all the methods mentioned above have been applied in conjunction with single- or multi-step laser excitations, greatly extending the applicability.

Several pure laser-spectroscopic methods have been developed more recently. We can here mention

Saturation (polarization) spectroscopy
2-photon spectroscopy
Collimated atomic beam spectroscopy
Quantum-beat spectroscopy

For these methods the Doppler broadening can also be virtually eliminated. Again the natural radiation width is the basic resolution limitation. For the quantum-beat method, which has certain resemblances with the resonance methods, this is the only line broadening effect. The other methods are also sensitive to laser frequency jitter and transit time effects. However, these methods are the only ones which allow a determination of net, overall displacements, i.e. isotope shifts and scalar Stark shifts. The other interactions in the Hamiltonian (1) give rise to splittings, not changing the center of gravity of the structure. In such cases the resonance methods are readily applicable.

3. Hyperfine Structures

3.1. Resonance Experiments

During the last few years a large number of alkali hyperfine structures have been studied using step-wise laser excitation in conjunction with resonance techniques. The level-crossing and double-resonance methods have been extensively used at Columbia Radiation Laboratory and by the Göteborg group. For the heavy alkali atoms K, Rb and Cs, which have first resonance lines in the infra-red, a powerful rf lamp for populating the first P state by the strong D-lines is very useful. In a second step where a CW dye laser is used, atoms are transferred to a highly excited S or D state. P and F states are then also populated in the cascade decays. Hyperfine structures for about 60 states of K, Rb and Cs have been measured with these techniques [5,6]. These types of measurements are illustrated in Figs. 2 and 3. In the first of these figures level-crossing signals for the 8 and 9 $^2D_{3/2}$ states of ^{87}Rb are shown. From the positions of the three $\Delta m=2$ crossings the hyperfine interaction constants a and b can be accurately determined. In Fig. 3 rf-resonance signals in the Paschen-Back region of the 7 and 8 $^2S_{1/2}$ states of ^{39}K are shown. Four signals are observed corresponding to the nuclear spin I=3/2. From the equally spaced signals the magnetic dipole constant a is deduced, and from the center of gravity for the signals the electronic Landé-factor g_J is inferred. The width of level-crossing or rf resonance signals in principle yields the natural radiative lifetime.

For various reasons hyperfine structure investigations for the light alkali atoms are more difficult than for the heavy ones. We have recently used two CW dye lasers for such cases. The experimental set-up for an investigation on Na is shown in Fig. 4. The hyperfine structure of the 6,7 and 8 $^2S_{1/2}$ states [7] and the 4 and 5 $^2P_{1/2}$ states [8] was measured. As an example the transition scheme and an rf resonance signal for the 6 $^2S_{1/2}$ state is shown in Fig.5.

Clearly, small fine-structure separations can be measured in very much the

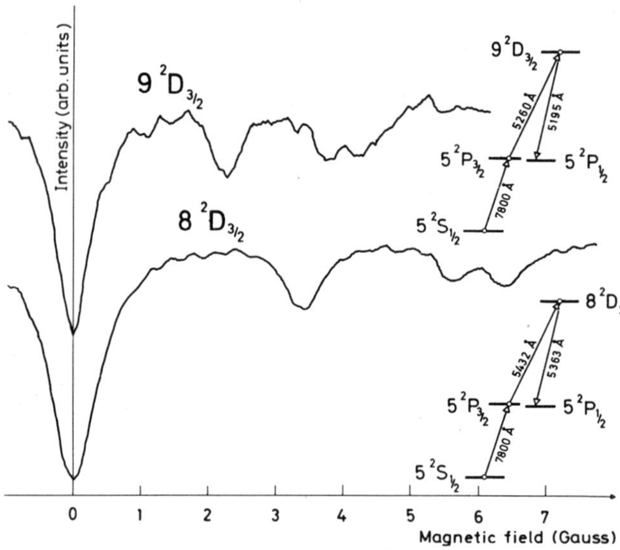

Fig. 2. Level-crossing signals for $^2D_{3/2}$ states in ^{87}Rb.

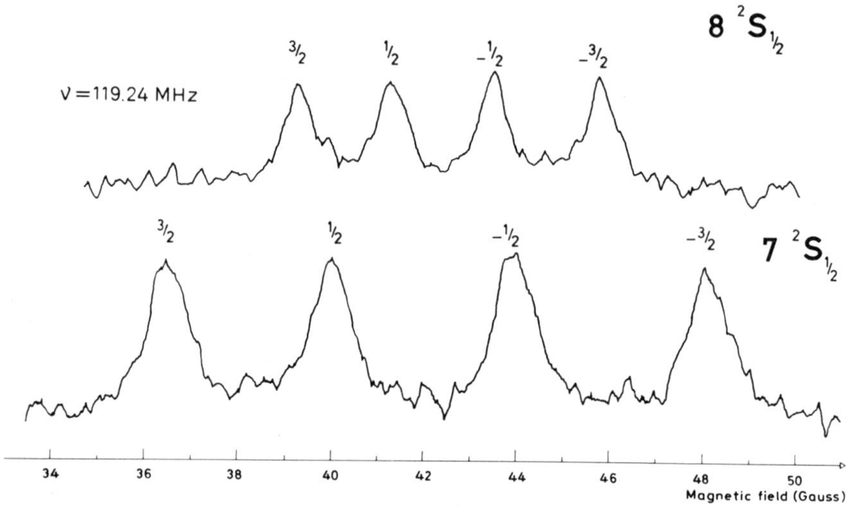

Fig. 3. Rf resonance signals for $^2S_{1/2}$ states in ^{39}K.

same way as hyperfine structures. Thus we have recently measured the fine-structure splitting in the inverted 4-9 ^{2}D states of Na with high precision [9] using the level-crossing method. The 4 ^{2}D state of Li was studied with the same method [10]. Contrary to the sodium states this state was found to be hydrogenic.

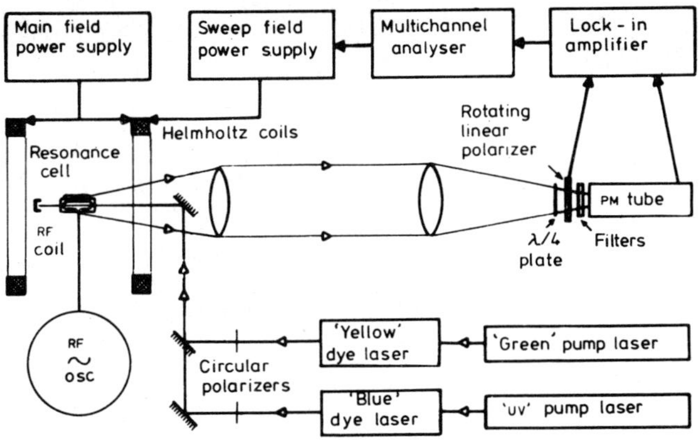

Fig. 4. Experimental set-up for rf resonance experiments.

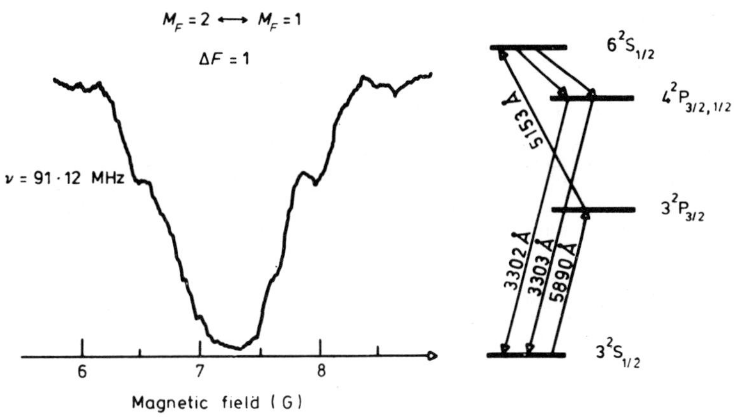

Fig. 5. Rf resonance curve for the 6 $^2S_{1/2}$ state of ^{23}Na.

3.2. Laser Spectroscopic Experiments

Out of the pure laser-spectroscopic methods the quantum-beat technique does not, like the resonance methods, require a narrow bandwidth laser source. Instead short pulses are a prerequisite for abrupt excitations of the atoms, which will exhibit beats superimposed on the exponential decay. The beat frequencies correspond to fine structure-, hyperfine structure- or Zeeman intervals. Hyperfine structure was first measured with quantum beats by HAROCHE, PAISNER and SCHAWLOW, studying the second excited P state of Cs [11]. Lately SERIES and coworkers have used this method to measure the hyperfine structure of several $^2D_{3/2}$ states in Cs [12]. We have also used quantum-beat spectroscopy for studying hyperfine structure and Zeeman effects. An example is shown in Fig. 6, where beats in the 8 $^2P_{3/2}$ state of Cs are shown. As the

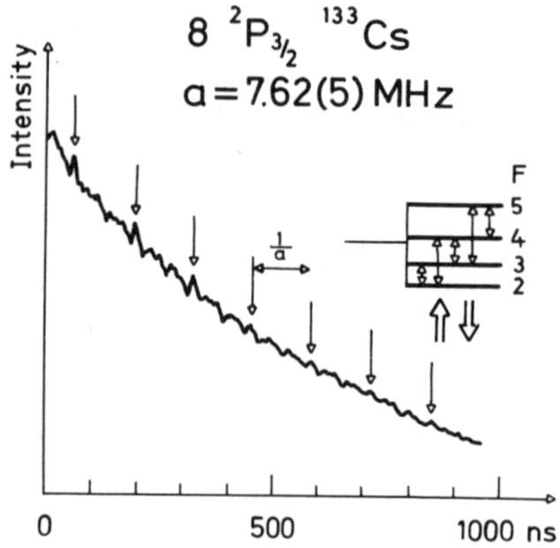

Fig. 6. Hyperfine beats for the $8\ ^2P_{3/2}$ state of ^{133}Cs.

Landé interval rule is very well obeyed in this case the beat structure is particularly simple and a Fourier transformation is not needed to deduce the dipole interaction constant a.

The experimental arrangements used in quantum-beat studies can be readily used for the more simple task of measuring natural radiative lifetimes. We have thus used the delayed-coincidence technique for determining the lifetimes of several excited alkali states [13].

In measurements of very small splittings it is in general difficult to achieve the same accuracy with laser-spectroscopic methods using narrow-band lasers as with resonance methods. This is due to the additional line-broadening effects present in the former methods, and in the difficulty in accurately measuring small optical shifts. However, with perfected equipment the same accuracy should be attainable.A particularly straight-forward type of laser spectroscopy is the use of a well collimated atomic beam in conjunction with a narrow-band laser beam incident at right angles. With CW dye lasers one is restricted to the region above about 4200Å. Pulsed lasers can be used to get access to shorter wavelengths. Combinations of intracavity and extra-cavity etalons have been used by HÄNSCH and others to narrow down the line-width. We have used a particularly simple approach for high-resolution measurements in the blue and near UV regions [14]. A single piezo-electrically-scanned extra-cavity etalon is used, filtering out narrow transmission maxima out of the more or less structureless, 10 GHz broad laser pulse. When the filter is scanned over an absorption line pattern due to hyperfine structure and/or isotope shifts, a repeating fluorescence light pattern, in general with several overlapping orders, is obtained. This enables a very high precision and no additional frequency-marker interferometer is needed. In Fig. 7 a signal curve due to hyperfine structure and isotope shifts in the $5\ ^2S_{1/2} - 6\ ^2P_{1/2}$ 4216Å line of ^{85}Rb and ^{87}Rb is shown.

3.3. Comparison with Theory

Using various methods hyperfine structure of about 90 states of alkali atoms have been measured and thus an extensive material for comparison with theory is available. In a hydrogen-like picture of the alkali atom a simple relation between fine- and hyperfine splittings is predicted. Plotting these splittings versus binding energy on log-log scales, straight lines of the same slope are expected. This is found to be the case for several S and P sequences. In Fig. 8 such a plot is shown for the D sequence of ^{133}Cs. Straight lines are obtained in this case. However, it is well-known that the fine structure for several D sequences is strongly perturbed and inverted which would deeply affect the relation of hyperfine structure to fine structure. This is not the case for the particular sequence plotted, however, the figure still contains a large anomaly. It turns out that the hyperfine structure for all the $^2D_{5/2}$ states is inverted. i.e. a is negative, which is totally inexplicable in a hydrogen-like model. It is necessary to use a more refined model where several radial parameters are used to describe the hyperfine interactions. For D

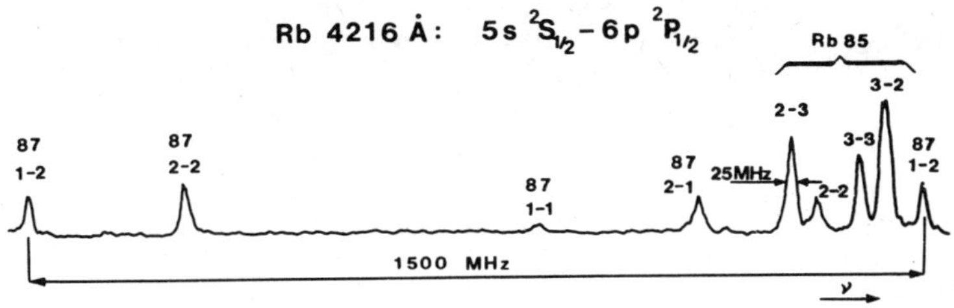

Fig. 7. Fluorescence spectrum from a collimated Rb beam.

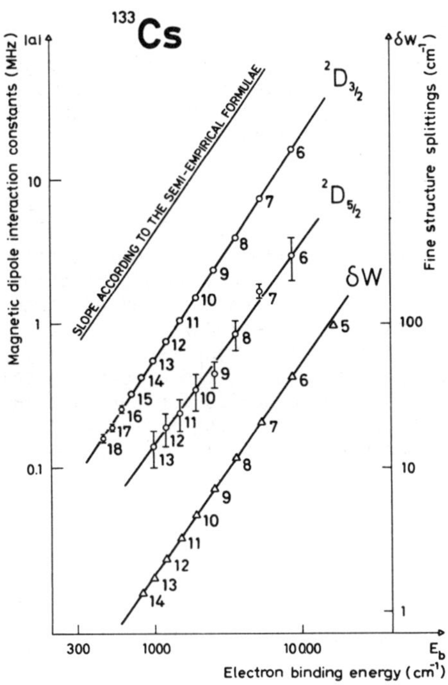

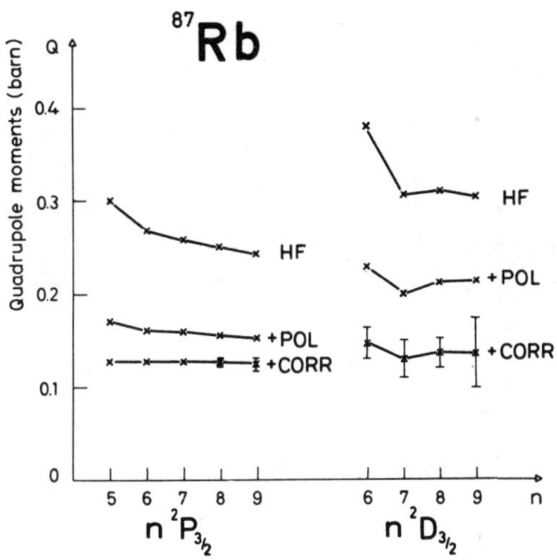

Fig. 8. Plot for Cs ^{2}D states. Fig. 9. Evaluation of Q for ^{87}Rb.

states the following expressions are then obtained [15]:

$$a(^2D_{3/2}) = k\{\tfrac{6}{5}\langle r^{-3}\rangle_l + \tfrac{2}{5}\langle r^{-3}\rangle_{sd} - \tfrac{1}{5}\langle r^{-3}\rangle_c\} \qquad k = \frac{\mu_o}{4\pi}\,2\mu_B^{\,2}\,g_I, \tag{2a}$$

$$a(^2D_{5/2}) = k\{\tfrac{4}{5}\langle r^{-3}\rangle_l - \tfrac{4}{35}\langle r^{-3}\rangle_{sd} + \tfrac{1}{5}\langle r^{-3}\rangle_c\} \tag{2b}$$

$$b(^2D_J) = \frac{e^2}{4\pi\varepsilon_o}\,\frac{2J-1}{2J+2}\,Q\langle r^{-3}\rangle_q \tag{2c}$$

Here $\langle r^{-3}\rangle_l$, $\langle r^{-3}\rangle_{sd}$, $\langle r^{-3}\rangle_c$ and $\langle r^{-3}\rangle_q$ are the orbital, spin-dipole, contact and quadrupole radial parameter, respectively. Q is the nuclear quadrupole moment; the other symbols have their usual meaning. By putting $<r^{-3}>_l = <r^{-3}>_{sd} = <r^{-3}>_q = <r^{-3}>$ and $<r^{-3}>_c = 0$ the above formulae reduce to the hydrogen-like ones. By using e.g. many-body perturbation techniques it is possible to calculate individual values for the different radial parameters, considering polarization effects (single electron excitations) and correlation effects (multiple excitations). A large part of the development of adequate techniques for calculating alkali hyperfine structure has been done by I. LINDGREN and coworkers in Göteborg [16]. For the alkali D states core polarization effects are very strong yielding a large value for $<r^{-3}>_c$. The other radial parameters turn out to be widely different. With the new techniques it is possible to clearly explain the anomalous D states magnetic hyperfine structures [17,5]. In special cases where the calculations have been performed to high order, an agreement with experiment within a few percent is obtained [18].

Also for the quadrupole interaction it is necessary to perform a detailed calculation to obtain a well-defined value for the quadrupole moment as deduced from P and D state data [19,5]. In Fig. 9 the evaluation of Q for ^{87}Rb is shown. When both polarization and correlation effects are taken into account a value of 0.13 barn is consistently obtained.

4. Stark Interactions

The application of an electric field over an atom introduces a polarization in the electronic shell, i.e., a change in the electric charge distribution. Quantum-mechanically it can be seen as a coupling between the studied state and all those other states combining with the state by electric dipole transitions. The theoretical calculation of the polarizability constants α_0 and α_2 in (1) thus involves the evaluation of matrix elements of the electric dipole operator, which are the same matrix elements as those needed for the calculation of oscillator strengths and atomic lifetimes. Thus, the measurement of Stark parameters from a theoretical point of view, yields similar information as the measurement of radiative properties.

4.1. Stark-Effect Measurements

As mentioned earlier the scalar Stark interaction cannot be observed with resonance methods. On the other hand the tensor interaction can be conveniently studied with such methods. A particularly simple approach is to "weigh" the unknown Stark interaction against the well-known Zeeman interaction in a level-crossing experiment in parallel electric and magnetic fields. If the Zeeman level degeneracy at zero magnetic field is broken by the application of a constant electric field, level-crossings are formed when the magnetic field is increased as shown in Fig. 10, where a $^2D_{5/2}$ state with negligible hyperfine structure is considered. In Fig. 11 this is illustrated for the 15 $^2D_{5/2}$ state of Cs. In the presence of hyperfine structure the level diagram is much more complicated, but still similar measurements can be performed. It is interesting to note, that the observed signal curves are sensitive to the sign of the dipole interaction constant. Thus the inversion of several $^2D_{5/2}$ states has been established using this method [20]. The tensor polarizability constant of more than 20 highly excited D states have been measured in level-crossing experiments of this kind [5]. Further states have been investigated by HAROCHE and coworkers [21] using quantum-beat spectroscopy and by GALLAGHER and coworkers employing rf resonance techniques [22].

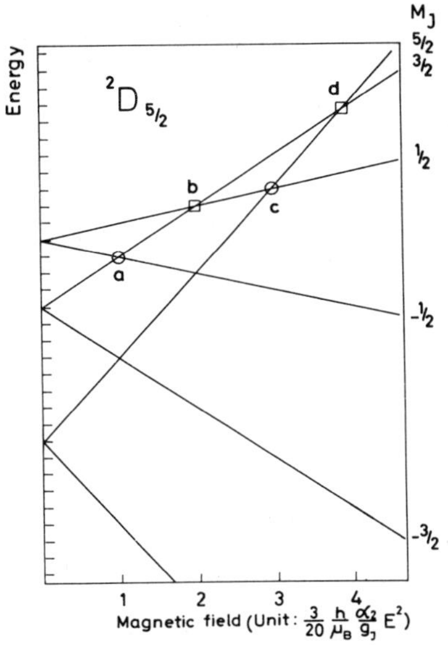

Fig. 10. Formation of $^2D_{5/2}$ level crossing signals.

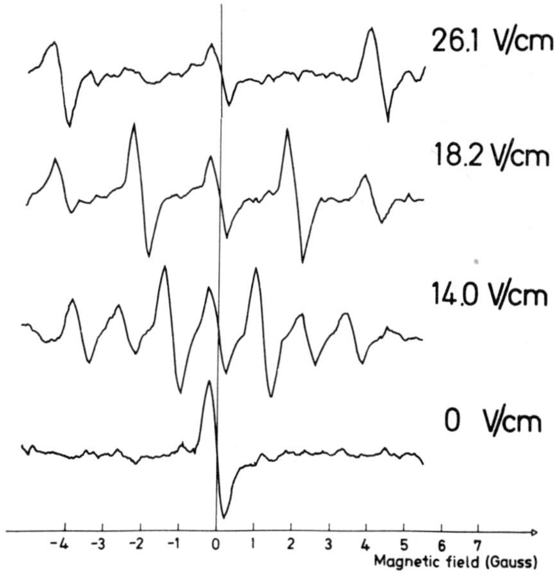

Fig. 11. Experimental curves for 15 $^2D_{5/2}$ Cs corresponding to Fig. 10.

For a measurement of also the scalar polarizability constant α_0 it is very convenient to use a narrow-band tunable laser, sweeping across the Stark-shifted sublevel structure. However, the Stark effect also provides a very nice possibility to sweep the energy level structure across a fixed laser frequency. This allows the use of a quite simple narrow band laser. In Fig. 12 the application of this technique is illustrated in a fluorescence experiment on the 6 $^2P_{3/2}$ - 10 $^2D_{5/2}$ transition in Cs. A collimated atomic beam, passing between Stark plates, is illuminated at right angles by a single-mode dye laser beam, acting in the second step of a two-step excitation from the ground state. The levels of the strongly polarizable D state are tuned to resonance with the laser frequency, once set slightly above, once slightly below the field free resonance value. The scalar polarizability of several S and D states has been measured using this technique [23].With a continuously tunable single mode laser now available we are completing a systematic survey of the alkali Stark structures. The laser is scanned across the Stark-shifted components. A curve for the 8 $^2D_{3/2}$ state of Cs is shown in Fig. 13 [24].

Very similar results can be obtained with two-photon spectroscopy. This technique was first used at Stanford University [25]. Lately, SCHAWLOW and coworkers have used two counter-propagating beams of different frequencies to resonantly enhance the two-photon absorption cross section in Stark-effect measurements on several S and D states in Na [26].

4.2. Comparison with Theory

Using different methods about 50 alkali states have been studied with respect to the Stark interaction. This data can be used for a thourogh comparison with experiment. The theoretical expressions for the polarizability constants are

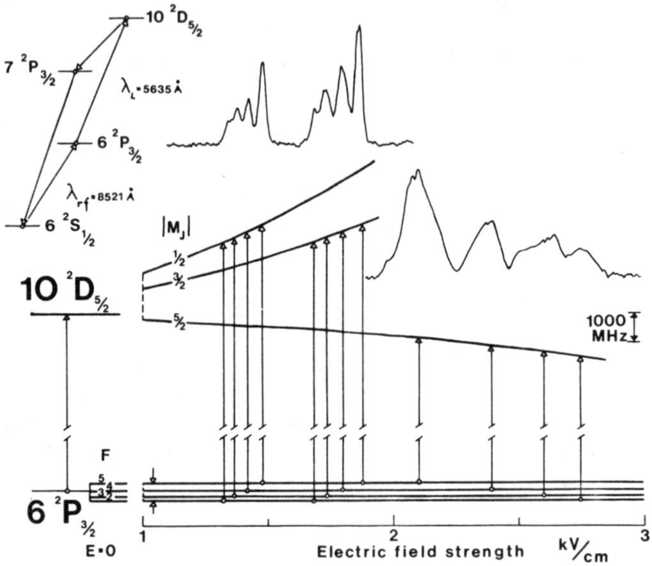

Fig. 12. Stark-tuning spectroscopy for the 5635Å Cs line.

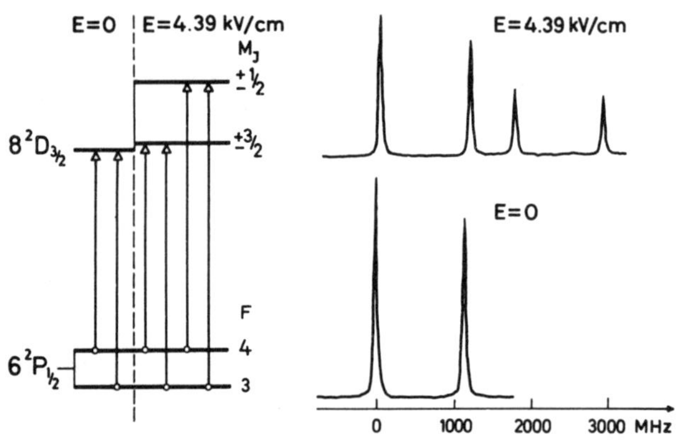

Fig. 13. Dye laser scans for the 6010Å Cs line.

$$\alpha_0 = -\frac{2}{3}\sum_{J'} \frac{|(J\|\underline{p}\|J')|^2}{(2J+1)[W(J)-W(J')]} \tag{3a}$$

$$\alpha_2 = 2\left[\frac{10J(2J-1)}{3(2J+3)(J+1)(2J+1)}\right]^{1/2} \sum_{J'} \frac{|(J\|\underline{p}\|J')|^2}{W(J)-W(J')} \times (-1)^{J+J'+1}\begin{Bmatrix} J & J' & 1 \\ 1 & 2 & J \end{Bmatrix} \tag{3b}$$

Here (J‖p‖J') are reduced matrix elements of the dipole operator between the studied state J and the perturbing states J'. W(J)-W(J') is the energy difference and { } is a 6-j symbol. For an alkali atom the matrix elements can be calculated using e.g. the Coulomb approximation. We have performed extensive calculations of this kind [20,5,23]. This semiempirical method in general gives agreement with experiment within 10 per cent. The influence of the continuum has then not been explicitly taken into account. In a simplified theory where the J-dependence of the radial functions is dropped and spin-orbit perturbations are neglected, simple relations between the α constants for the doublet states are obtained. These have sometimes been used for evaluating experimental data. In Fig. 14 the situation is illuminated for the case of a light and a heavy alkali atom. The dashed lines indicate the expected simplified ratios. From the experiments it is clear that for Na the approximation is valid, whereas it is clearly not for Cs. The deviations observed for the latter atom are closely related to the well-known intensity ratio anomalies for spectral lines of the heavy alkali atoms, where the spin-orbit interaction is strong. Detailed calculations to describe this effect as well as the influence of the continuum would be highly desirable.

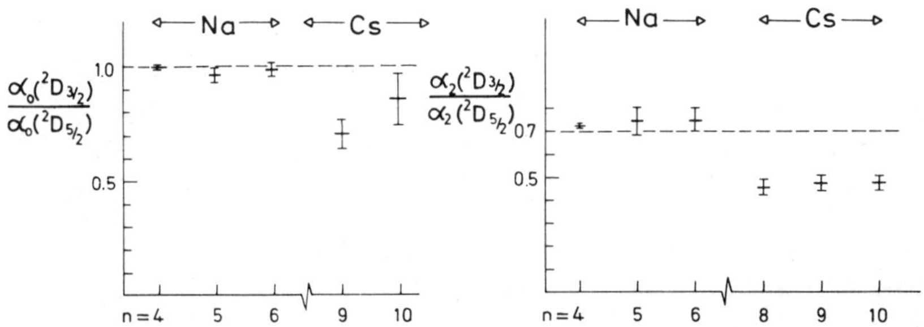

Fig. 14. Plots of Stark parameter ratios for ^2D states of Na and Cs.

Acknowledgements

The author would like to acknowledge a most stimulating cooperation with a number of very able present and former graduate students: G. Belin, K. Fredriksson, P. Grundevik, M. Gustavsson, L. Holmgren, H. Lundberg and A-M. Mårtensson. Furthermore, he is very grateful to Prof. I. Lindgren and Doc. A. Rosén for advice, support and fruitful discussions. This work was financially supported by the Swedish Natural Science Research Council.

References

1. R. Gupta. S. Chang and W. Happer, Phys. Rev. A6, 529 (1972).

2. R. Gupta, S. Chang, C. Tai and W. Happer, Phys. Rev. Lett. 29, 695 (1972); R. Gupta, W. Happer, L.K. Lam and S. Svanberg, Phys. Rev. A8, 2792 (1973).

3. S. Svanberg, P. Tsekeris and W. Happer, Phys. Rev. Lett. 30, 817 (1973); S. Svanberg and P. Tsekeris, Phys. Rev. A11, 1125 (1975).

4. F. Biraben, B. Cagnac and G. Grynberg, Phys. Rev. Lett. 32, 643 (1974); M.D. Levenson and N. Bloembergen, Phys. Rev. Lett. 32, 645 (1974); T.W. Hänsch, K. Harvey, G. Meisel and A.L. Schawlow, Opt. Commun. 11, 50 (1974).

5. G. Belin, L. Holmgren, I. Lindgren and S. Svanberg, Phys. Scr. 12, 287 (1975); G. Belin, L. Holmgren and S. Svanberg, Phys. Scr. 13, 351 (1976), ibid. 14, 39 (1976).

6. P. Tsekeris and R. Gupta, Phys. Rev. A11, 455 (1975); J. Farley, P. Tsekeris and R. Gupta, Phys. Rev. A15, 1530 (1977).

7. H. Lundberg, A-M. Mårtensson and S. Svanberg, J. Phys. B (London), in press.

8. P. Grundevik and H. Lundberg, to be published.

9. K. Fredriksson and S. Svanberg, J. Phys. B (London) 9, 1237 (1976).

10. K. Fredriksson, H. Lundberg and S. Svanberg, Z. Physik, in press.

11. S. Haroche, J. Paisner and A.L. Schawlow, Phys. Rev. Lett. 30, 948 (1973).

12. J.S. Deech, R. Luypaert, L.R. Pendrill and G.W. Series, J. Phys. B (London) 10, L137 (1977).

13. H. Lundberg and S. Svanberg, Phys. Lett. 56A, 31 (1976); M. Gustavsson, H. Lundberg and S. Svanberg, Phys. Lett. A, in press.

14. P. Grundevik, M. Gustavsson, A. Rosén and S. Svanberg, Z. Physik, in press.

15. I. Lindgren and A. Rosén, Case Studies in Atomic Physics 4, 97 (1974).

16. S. Garpman, I. Lindgren, J. Lindgren and J. Morrison, Phys. Rev. A11, 758 (1975); Z. Physik A276, 167 (1976).

17. I. Lindgren, J. Lindgren and A-M. Mårtensson, Z. Physik A279, 113 (1976).

18. I. Lindgren, J. Lindgren and A-M. Mårtensson, Phys. Rev. A, in press; J. Lindgren, Thesis Göteborg 1976 (unpublished).

19. I. Lindgren, Atomic Physics 4, p. 747, Plenum Press, New York, 1975.

20. W. Hogervorst and S. Svanberg, Phys. Scr. 12, 67 (1975).

21. C. Fabre and S. Haroche, Opt. Commun. 15, 254 (1975).

22. T.F. Gallagher, L.M. Humphrey, R.M. Hill, W.E. Cooke and S.A. Edelstein, Phys. Rev. A15, 1937 (1977).

23. K. Fredriksson and S. Svanberg, Z. Physik A281, 189 (1977).

24. K. Fredriksson, L. Nilsson and S. Svanberg, to appear.

25. K.C. Harvey, R.T. Hawkins, G. Meisel and A.L. Schawlow, Phys. Rev. Lett. 34, 1073 (1975).

26. R.T. Hawkins, W.T. Hill, F.V. Kowalski, A.L. Schawlow and S. Svanberg, Phys. Rev. A15, 967 (1977).

SPONTANEOUS RAMAN EFFECT IN INTENSE LASER FIELDS

C. Cohen-Tannoudji and S. Reynaud

Ecole Normale Supérieure and Collège de France
24, rue Lhomond, 75231 Paris Cedex 05, France

1. Introduction

In this paper, we discuss, from a theoretical point of view, how spontaneous Raman effect is modified at very high laser intensities.

Up to now, intense field effects have been mainly investigated, both theoretically [1] and experimentally [2], in the simple case of 2-level systems. Observation of Raman processes requires systems having at least three levels, one in the upper state, two in the lower state. At first sight, one could think that going from two to three levels complicates very much the algebra, since we have now eight Bloch's equations for the density matrix elements instead of three. In fact, this is not true, and we would like to show in this paper how it is possible to understand the modifications of Raman effect with practically no new calculations, even if the presence of two sublevels in the lower state leads, through optical pumping effects, to results qualitatively different from those of the 2-level case.

Let's first give some notations (Fig. 1). We will call e the upper atomic state, g and g' the two sublevels of the lower state separated by a splitting S, ω_0 and ω'_0 the frequencies of the two transitions e-g and e-g', Γ the natural width of e, equal to the sum of the two spontaneous transition rates γ and γ' from e to g and from e to g'.

We consider a beam of such atoms irradiated at right angles by a single mode laser so that one gets rid of Doppler effect. The laser has a frequency ω_L and a polarization such that it can excite both transitions e-g and e-g'. The coupling of the laser with these two transitions is characterized by the two Rabi frequencies $\omega_1 = \vec{E}_L.\vec{d}_{eg}$, $\omega'_1 = \vec{E}_L.\vec{d}_{eg'}$ equal to the product of the laser electric field $\vec{E}_L$ by the dipole moments $\vec{d}_{eg}$ and $\vec{d}_{eg'}$ associated respectively with e-g and e-g'.

The theoretical problem we are interested in concerns the scattered light emitted perpendicularly to the laser and atomic beams. How does the spectrum of this scattered light change when the laser intensity I_L is progressively increased from very low to very high values ?

We will first briefly recall the well known lowest order results concerning Raman effect [3]. Then, some higher order processes which become important at higher intensities will be discussed. We will also consider the case where the laser is tuned in resonance with one transition, saturating this transition but not the second one. Finally, we will discuss the very high intensity limit where the laser is so intense that it can saturate both

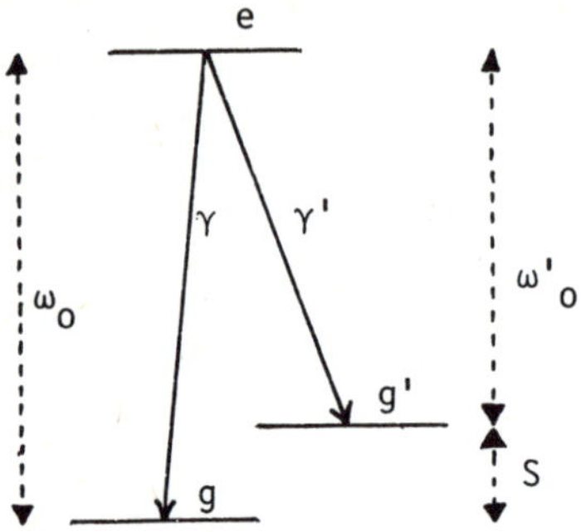

Fig. 1 Energy diagram of the 3-level atomic system

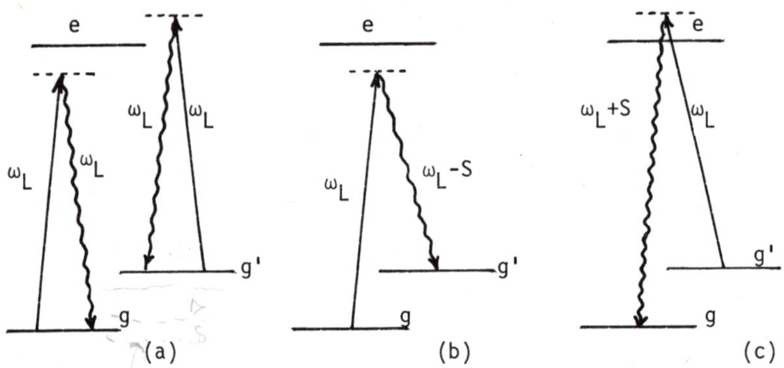

Fig. 2 Lowest order processes corresponding to Rayleigh (a), Raman Stokes (b) and Raman anti Stokes (c) scattering.

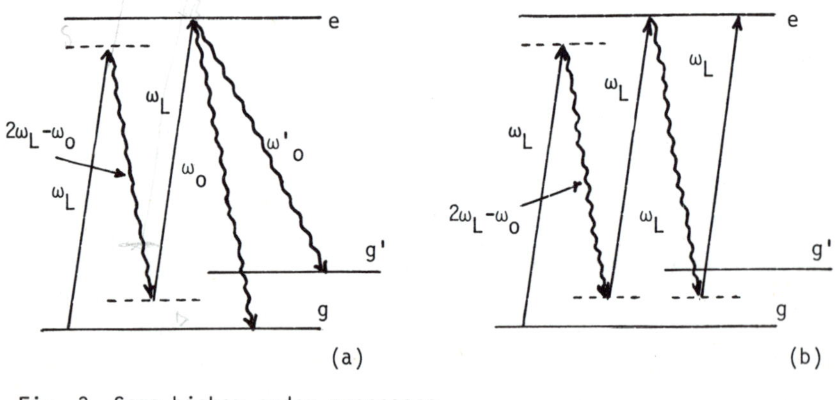

Fig. 3 Some higher order processes

transitions. The emphasis will be put on physical discussions rather than on detailed calculations which can be found in Ref. [4].

2. Lowest order results

Three types of processes can occur at very low intensities (Fig. 2) : Rayleigh type processes (Fig. 2-a) where the atom starts and ends in the same sublevel of the lower state, g or g', absorbing one laser photon ω_L and reemitting one photon which must have exactly the laser frequency because of energy conservation. Raman Stokes processes (Fig. 2-b) where the atom, starting from g, absorbs one ω_L photon and ends in g', emitting a photon with frequency ω_L-S. The symmetric process where the atom starts from g' and ends in g, absorbing ω_L and emitting ω_L+S, is called Raman anti Stokes (Fig. 2-c).

At very low intensities, one expects therefore in the spectrum of the scattered light three delta-functions at ω_L, ω_L-S, ω_L+S, each of them having a weight proportional to the laser intensity I_L. Note also that because of Raman processes, there is an optical pumping of atoms from g to g' or from g' to g. If the interaction time is sufficiently long, a steady state will be reached in which the number of transitions from g to g' balances the number of transitions from g' to g. This means that, in the steady state, the number of photons emitted at ω_L-S and ω_L+S are the same, so that the spectrum is symmetric. Finally, if $\omega'_1 = 0$, i.e. if the laser polarization is such that the transition e-g' cannot be excited, each atom pumped in g' will not be able to leave this state, so that, after a certain time, all atoms will be trapped in g' and the scattered light will vanish.

3. Some higher order processes

If the light intensity is increased, it becomes necessary to consider higher order processes where, instead of interacting with a single laser photon, the atom interacts with two laser photons, three laser photons ... and so on. Some of these higher order processes are sketched on Fig. 3.

For example (Fig. 3-a), the atom can absorb two laser photons (represented by full arrows) and emit two fluorescence photons (represented by wavy arrows). The corresponding amplitude is large when, after the absorption of the second laser photon, the atom reaches the excited state within the natural width Γ of this state. This means that the first fluorescence photon has a frequency $2\omega_L-\omega_0$ (within Γ) and also that the second fluorescence photon has a frequency ω_0 or ω'_0 according as the atom ends in g or in g'. Similar processes occuring from g' lead to fluorescence photons at $2\omega_L-\omega'_0$ and also at ω_0 or ω'_0. We therefore predict four new lines appearing at the four frequencies ω_0, ω'_0, $2\omega_L-\omega_0$, $2\omega_L-\omega'_0$, with an intensity proportional to I_L^2, since two laser photons are involved, and a width of the order of Γ due to the width of the upper level e.

We have also represented on Fig. 3-b an example of a three photon process in which, after having been put in the upper state e as in the two previous examples, the atom emits a fluorescence photon and comes back to e by absorbing a third laser photon ω_L. This can be considered as an "inverse Rayleigh" process from e and gives rise to a new line at ω_L, with an intensity proportional to I_L^3 and a width of the order of Γ.

All these perturbative results which are valid for sufficiently large detunings between the laser and atomic frequencies can be confirmed by a more precise calculation. One finds that there are seven lines in the spectrum of the scattered light. The more intense ones, proportional to I_L, are the Rayleigh (ω_L) and Raman ($\omega_L \pm S$) lines. The Raman lines are not infinitely narrow, as predicted by lowest order theory, but they have a small width of the order of $1/T_p$, where T_p (pumping time between g and g') can be considered as the lifetime of the lower state, much longer however than the lifetime $1/\Gamma$ of the upper state. The Rayleigh line has a structure. One component has a width of the order of $1/T_p$, as the Raman lines. The second component is a $\delta(\omega-\omega_L)$ function corresponding to the coherent scattering by the mean dipole moment driven at ω_L by the laser wave. Then, we have weaker lines, in I_L^2 or I_L^3, at ω_0, ω'_0, $2\omega_L-\omega_0$, $2\omega_L-\omega'_0$, ω_L, with a width of the order Γ, corresponding to the higher order processes discussed above. Finally, one must note that the spectrum is symmetric in steady state. We have already discussed this point for the Raman lines. For the weaker lines, this symmetry is partly due to the fact that, in the second order processes, the fluorescence photons always appear by pairs.

If the laser intensity is still increased, it is clear that the perturbative approach used so far breaks down, especially if the laser is tuned in resonance with one of the 2 transitions.

4. Laser in resonance with one transition, saturating this transition but not the second one

Suppose for example that $\omega_L = \omega_0$, so that the laser is in resonance with e-g, and suppose that $\omega_1 \gg \Gamma$, so that e-g is saturated. The detuning of the laser with the second transition e-g' is equal to the splitting S between g and g'. We will suppose first that $\omega'_1 \ll S$, so that the laser is not sufficiently intense for saturating e-g' : e-g is saturated but not e-g'.

In order to understand what happens in that case, the idea is, in a first step, to forget g' and to apply the well known results of the two-level case to the e-g transition which is saturated. Then, in a second step, one can try to understand the perturbation due to g'.

Let's consider first some energy levels of the combined system atom + laser photons, without any coupling, i.e. with $\omega_1 = \omega'_1 = 0$. Since $\omega_L = \omega_0$, the two states $|g, n\rangle$ (atom in g with n photons) and $|e, n-1\rangle$ (atom in e with n-1 photons) are degenerate . The same result holds for $|g, n+1\rangle$ and $|e, n\rangle$ which are at a distance ω_L above. On the other hand, $|g', n\rangle$ is at a distance S above $|g, n\rangle$, $|g', n+1\rangle$ at a distance S above $|g, n+1\rangle$ (Fig. 4-a).

Now, we introduce the coupling ω_1 between $|g, n\rangle$ and $|e, n-1\rangle$, $|g, n+1\rangle$ and $|e, n\rangle$ ..., but we maintain $\omega'_1 = 0$, i.e. we still ignore any coupling between $|g', n\rangle$ and $|e, n-1\rangle$, $|g', n+1\rangle$ and $|e, n\rangle$ ... The two unperturbed states $|g, n\rangle$ and $|e, n-1\rangle$ transform into two perturbed states $|2, n-1\rangle$ and $|3, n-1\rangle$, separated by ω_1 , and similarly $|g, n+1\rangle$ and $|e, n\rangle$ transform into $|2, n\rangle$ and $|3, n\rangle$. The spontaneous transitions between these two doublets, represented by full arrows on Fig. 4-a, give rise, in the scattered light, to a well known triplet, formed by three lines, one at $\omega_L+\omega_1$ (transition $|2, n\rangle \to |3, n-1\rangle$), one at $\omega_L-\omega_1$ (transition $|3, n\rangle \to |2, n-1\rangle$), one at ω_L (degenerate transitions $|i, n\rangle \to |i, n-1\rangle$ with i = 2, 3). The widths are of the order of Γ, since the width of $|e, n\rangle$ is shared between the two perturbed states $|2, n\rangle$ and $|3, n\rangle$. Since the transition is saturated, there is no mean dipole moment, and consequently, no $\delta(\omega-\omega_L)$ function corresponding to a coherent scattering.

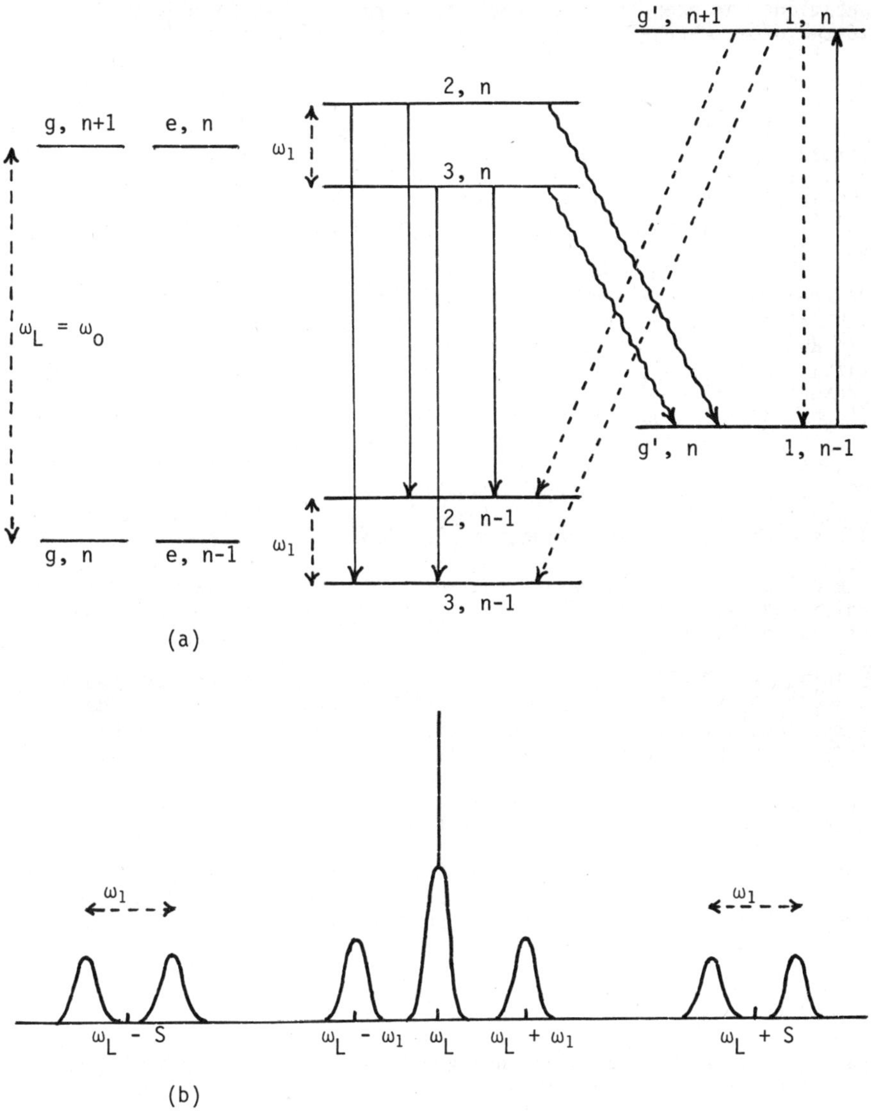

Fig. 4 Raman effect in intense laser fields ($\omega_L = \omega_o$, $\omega_1 \gg \Gamma$, $\omega'_1 \ll S$). Various energy levels of the system atom-laser photons and various transitions between these energy levels (Fig. a) giving rise to the spectrum represented on Fig. b.

Even if $\omega'_1 = 0$, i.e. even if the laser is not coupled to e-g', spontaneous transitions can occur from e to g' with a rate γ'. It follows that there are also two decay channels, represented by wavy arrows on Fig. 4-a, connecting $|2, n>$ and $|3, n>$ to $|g', n>$, and giving rise in the scattered light to two lines at $\omega_L - S \pm \omega_1/2$, with a width of the order of Γ. This doublet structure of the Raman Stokes line may be interpreted as a manifestation of the Autler-Townes splitting on the e-g' transition [5] [6] [7]. The important point is that, once an atom falls back in $|g', n>$, it cannot escape from it when $\omega'_1 = 0$. It follows that optical pumping effects can trap all atoms in g', and now in a very short time, of the order of $1/\gamma'$, since the populations of e and g are very rapidly equalized by the intense resonant laser beam.

We therefore understand why it is important to have a non zero value of ω'_1, which brings in the perturbed state $|1, n>$ associated with $|g', n+1>$ a small admixture of $|e, n>$, of the order of ω'_1/S, allowing new weak transitions $|1, n> \rightarrow |2, n-1>$, $|1, n> \rightarrow |3, n-1>$ and $|1, n> \rightarrow |1, n-1>$ represented in dotted lines on Fig. 4-a, and reintroducing in $|2, n-1>$, $|3, n-1>$ a small population, of the order of $(\omega'_1/S)^2$. These transitions can also be interpreted as anti Stokes processes from g', giving rise to two lines at $\omega_L + S \pm \omega_1/2$. There is also a non-resonant Rayleigh process from g', reintroducing a $\delta(\omega-\omega_L)$ function.

All these results are summarized on Fig. 4-b.

There are always seven lines in the spectrum of the scattered light, one triplet and two doublets. The splitting of the triplet and of the two doublets is now given by the Rabi frequency ω_1.

The intensity of all lines is in $(\omega'_1/S)^2$. The transitions starting from $|2, n>$ or $|3, n>$ have a large transition probability, of the order of 1, but the levels $|2, n>$ and $|3, n>$ have a small population, of the order of $(\omega'_1/S)^2$. On the other hand the population of $|1, n>$ is large, but the transitions starting from this level are weak. This means that optical pumping effects drastically reduce the intensity of the scattered light, in comparison to the 2-level case, by a factor $(\omega'_1/S)^2 \ll 1$.

The $\delta(\omega-\omega_L)$ function does not disappear as in the 2-level case in the high intensity limit. It is mainly due to coherent scattering from g' and represents another consequence of optical pumping effects.

Apart from the $\delta(\omega-\omega_L)$ function, the width of all other lines is of the order of Γ. There are no narrow lines as above. This is due to the finite width of the upper or lower state of the transitions.

Finally, from detailed balance considerations, the spectrum can be shown to be symmetric in the steady state.

5. Very high intensity limit : Laser saturating both transitions

One could think that going to the very high intensity limit, where both ω_1 and ω'_1 are large compared to Γ and S, so that both transitions e-g and e-g' are saturated, can suppress all these optical pumping effects and lead to a saturation of the scattered intensity. We would like now to show that this is not true.

Let's first consider the particular case where S = 0, i.e. where g and g' are degenerate so that $\omega_0 = \omega'_0$ and let's suppose that the laser is in resonance with these 2 transitions, so that $\omega_0 = \omega'_0 = \omega_L$. Since the 2 levels g and g' are degenerate, one can choose any new basis in the lower state by taking any set of 2 orthonormal linear combinations of g and g'. If, for example, one takes

$$|G> = \frac{1}{\Omega_1}\left[\omega_1\ |g> + \omega'_1|g'>\right]$$
$$|G'> = \frac{1}{\Omega_1}\left[\omega'_1|g> - \omega_1\ |g'>\right]$$

where $\Omega_1{}^2 = \omega_1{}^2 + \omega'_1{}^2$,

one can easily show that only the transition e-G is coupled to the laser with a Rabi frequency Ω_1 whereas the second is not. It follows that we are again led to a 2-level system and to optical pumping effects which can trap all atoms in G'.

In fact, S is different from 0, but, in the limit we are considering here, Ω_1 is very large compared to S and Γ, so that we will proceed as follows. First, we treat the effect of the coupling Ω_1, S being neglected. Then, we treat perturbatively the effect of a non-zero value of S.

If both Ω_1 and S are equal to 0, the 3 states $|G, n>$, $|G', n>$ and $|e, n-1>$ are degenerate and similarly $|G, n+1>$, $|G', n+1>$, $|e, n>$ which are at a distance ω_L above (Fig. 5-a).

Let's then consider the effect of the coupling Ω_1 between $|G, n>$ and $|e, n-1>$, $|G, n+1>$ and $|e, n>$. One gets a series of doublets, $|1, n-1>$ and $|3, n-1>$, $|1, n>$ and $|3, n>$ with a splitting Ω_1 . The level $|G', n+1>$ is not perturbed and remains half way between $|1, n>$ and $|3, n>$.

We now treat the effect of S perturbatively. The corresponding operator, $S|g'><g'|$, has in the basis $|1, n>$, $|3, n>$, $|G', n+1>$ both diagonal and off diagonal elements.

The diagonal elements represent first order energy shifts which are small compared to Ω_1 since $S \ll \Omega_1$. One can easily show that $|1, n>$ and $|3, n>$ are shifted by the same amount, different from the shift of $|G', n+1>$. It follows that $|G', n+1>$ is at a distance $S\tau/2$ from the middle of the interval $|1, n>$, $|3, n>$, where the dimensionless parameter τ is easily found to be given by :

$$\tau = \frac{2\omega_1{}^2 - \omega'_1{}^2}{\omega_1{}^2 + \omega'_1{}^2} \quad .$$

If one neglects the off diagonal elements associated with S, the state $|G', n+1>$ remains not coupled to $|1, n-1>$ and $|3, n-1>$ by spontaneous emission, so that one gets only the spontaneous transitions connecting $|1, n>$ and $|3, n>$ to $|1, n-1>$ and $|3, n-1>$, represented by full arrows on Fig. 5-a, and forming a triplet at ω_L, $\omega_L \pm \Omega_1$ and the spontaneous transitions connecting $|1, n>$ and $|3, n>$ to $|G', n>$, represented by wavy arrows on Fig. 5-a and giving rise to two lines at $\omega_L + (\Omega_1/2) - (S\tau/2)$ and $\omega_L - (\Omega_1/2) - (S\tau/2)$.

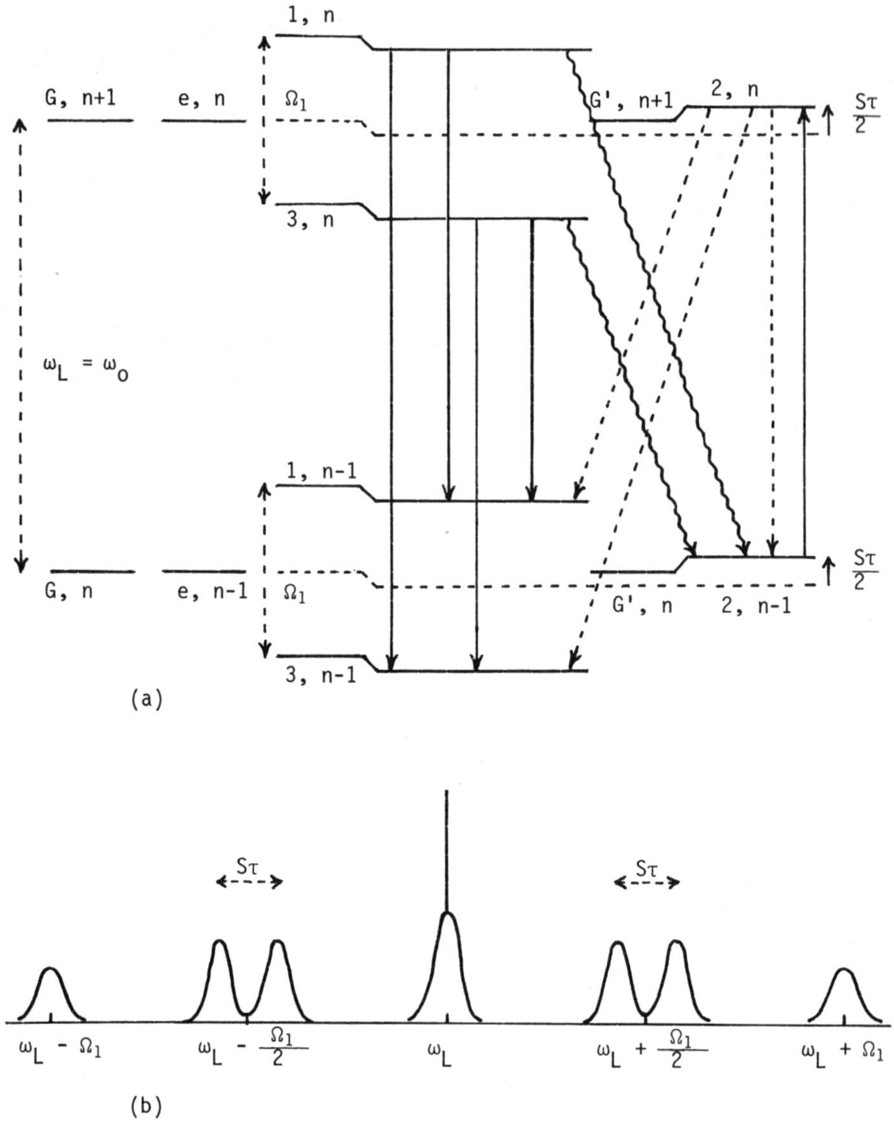

Fig. 5 Raman effect in very intense fields ($\omega\ ,\ \omega'_1 >> S, \Gamma$). Various energy levels of the system atom-laser photons and various transitions between these energy levels (Fig. a) giving rise to the spectrum represented on Fig. b.

Actually, the off diagonal elements associated with S bring in the perturbed state $|2, n\rangle$ corresponding to $|G', n+1\rangle$ a small admixture of $|e, n\rangle$, allowing spontaneous transitions from $|2, n\rangle$ to $|1, n-1\rangle$ and $|3, n-1\rangle$ (represented in dotted lines on Fig. 5-a) and reintroducing in $|1, n-1\rangle$ and $|3, n-1\rangle$ a small population, of the order of $(S/\Omega_1)^2$. These transitions can also be interpreted as non resonant scattering processes from G', giving rise to two lines at $\omega_L - (\Omega_1/2) + (S\tau/2)$ and $\omega_L + (\Omega_1/2) + (S\tau/2)$ and to a $\delta(\omega-\omega_L)$ function corresponding to coherent scattering.

We therefore arrive to the following conclusions concerning the very high intensity limit where both transitions e-g and e-g' are saturated (Fig. 5-b).

We have always seven lines, but now they form three singlets and two doublets.

The main splittings are determined by Ω_1. S only appears in the splittings of the doublets. This means that we have a complete mixing between the initial Rayleigh, Raman Stokes and Raman anti Stokes lines which can be more precisely followed by a computer calculation of the position of the seven lines for an arbitrary intensity. Fig. 11 of Ref. [4] shows the 3 Rayleigh and Raman lines which first split into a triplet and 2 doublets and which progressively transform into 3 singlets and 2 doublets when the laser intensity is increased.

The most important point concerns perhaps the intensity of all lines which is in $(S/\Omega_1)^2$, i.e. in $1/I_L$. The reason is, as above, that strong transition probabilities correspond to weak populations and vice-versa.

This clearly shows the importance of optical pumping effects.

- First, not only the intensity of the lines does not saturate but it decreases as $1/I_L$.

- Second, we have an accumulation of atoms in G', i.e. in a linear superposition of g and g'.

- Third, the $\delta(\omega-\omega_L)$ function, which is due to coherent scattering from G', does not disappear.

Finally, as above, one can show that the width of all other lines is of the order of Γ and that the spectrum is symmetric in the steady state.

In conclusion, we have shown in this paper how it is possible to understand the modification of Raman effect when the laser intensity is progressively increased from very low to very high values.

We have only considered pure radiative effects. Collision processes in 3-level systems can also lead to very interesting effects. By introducing transfers between the various energy levels, they are responsible for important asymmetries in the spectrum of the scattered light. We will not enter into these problems since they are discussed by CARLSTEN and RAYMER [8].

References

1. B.R. Mollow, Phys. Rev., 188, 1969 (1969); see also the list of references in C. Cohen-Tannoudji, Frontiers in Laser Spectroscopy, Les Houches 1975, Session XXVII, ed. R. Balian, S. Haroche and S. Liberman (North Holland, 1977).

2. F. Schuda, C.R. Stroud Jr. and M. Hercher, J. Phys. B7, L 198 (1974). F.Y. Wu, R.E. Grove and S. Ezekiel, Phys. Rev. Lett., 35, 1426 (1975). W. Hartig, W. Rasmussen, R. Schieder and H. Walther, Z. Phys. A278, 205 (1976).
3. W. Heitler, Quantum Theory of Radiation, 3rd edition (Oxford University Press, 1954).
4. C. Cohen-Tannoudji and S. Reynaud, J. Phys.,B10, 365 (1977).
5. S.H. Autler and C.H. Townes, Phys. Rev., 100, 703 (1955).
6. B.R. Mollow, Phys. Rev., A5, 1522 (1972).
7. E.V. Baklanov, Zh. Eksp. Teor. Fiz., 65, 2203 (1973) [Sov. Phys. JETP, 38, 1100 (1973)].
8. J.L. Carlsten and M.G. Raymer, "Laser Spectroscopy," J.L. Hall and J.L. Carlsten, eds., Springer Series in Optical Sciences, Vol. 7, (Springer-Verlag, N.Y., Heidelberg, 1977).

COLLISIONAL AND RADIATIVE EFFECTS IN THREE-LEVEL SYSTEMS [1]

J.L. Carlsten and M.G. Raymer [2]

Joint Institute for Laboratory Astrophysics
University of Colorado and National Bureau of Standards
Boulder, CO 80309, USA

Recently there has been considerable theoretical and experimental interest in the collisional redistribution of near-resonant scattered light [1-5]. When a vapor of three-level atoms is irradiated by a low-intensity laser of frequency ω_L near the resonance frequency ω_{21}, emission occurs near ω_{23} as well as near ω_{21} (see Fig.1). In the absence of collisions, emission (near ω_{23}) occurs only at ω_S, the Stokes frequency, and is known as Raman scattering.

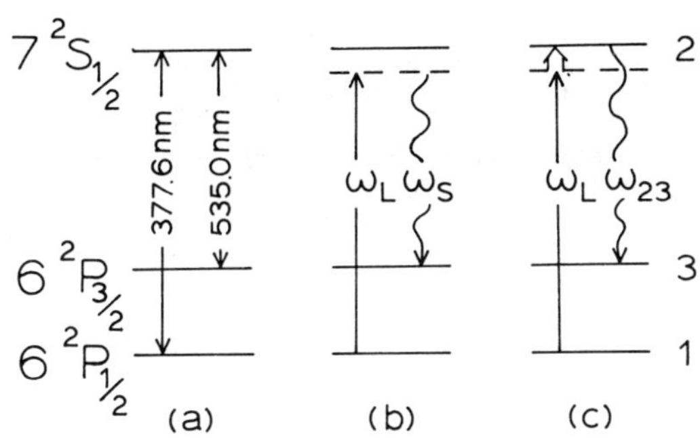

Fig.1 (a) First three energy levels of thallium. With the laser tuned near the 377.6 nm resonance line we observed scattered light near the 535.0 nm line. (b) Schematic representation of electronic Raman scattering at the Stokes frequency ω_S from an incident laser at frequency ω_L. (c) Schematic representation of collision-induced fluorescence. Collisions (which in our case were Tℓ-Ar collisions) transferred Tℓ atoms from the laser-induced virtual level to level 2, resulting in fluorescence at ω_{23}. When level 3 is initially unpopulated, both of these components can become stimulated.

[1] This work was supported by the Office of Naval Research under Contract N00014-76-C-0611 and by the National Science Foundation under grant MPS72-05169, both through the University of Colorado, and by the Lawrence Livermore Laboratory under P. O. 4353803 through the University of California.

[2] Department of Chemistry, University of Colorado, Boulder, Colorado 80309.

When collisions are present, atoms can be transferred from the laser-induced virtual level (broken line) to the real level 2, resulting in emission at ω_{23}, which is known as collision-induced fluorescence. The dependence of the collision-induced fluorescence on such parameters as collision rates, laser intensity and detuning is called the "redistribution function," and is important in the theory of radiative transfer in spectral lines. In an earlier work, we made a detailed study of the redistribution function for scattering from a two-level atom (strontium) and found good agreement with theory [1].

The chief characteristic of light scattering by a *three-level* atom is the possibility of stimulated emission. At low atomic density and low laser intensity the two components in the scattered spectrum arise from spontaneous emission. However at higher intensity (or density), MOLLOW [6] has predicted that both of these emission components can have gain. Therefore the possibility exists for simultaneous generation of stimulated Raman scattering (SRS) and stimulated collision-induced fluorescence (SCF). In an experiment by WYNNE and SOROKIN [7], there were indications that these two processes were occurring but the two components could not be well resolved and the population mechanism for the fluorescence was not determined.

Using a dye laser tuned near the 377.6 nm ($6^2P_{1/2} - 7^2S_{1/2}$) resonance line of Tℓ, we have observed the growth of both of these components (near the 535.0 nm emission line) from the linear regime, where the scattering is spontaneous, to the exponential regime, where the scattering becomes stimulated. In addition, we have studied the collisional dependence of the SCF and SRS. Recently SRS in vapors and gases has been used by a number of researchers as an efficient means of down conversion [8]. We hope to understand the effects of collisions on such stimulated scattering.

The theory of collisional redistribution has been studied both at low and high intensities. For a low intensity monochromatic laser at frequency ω_L, detuned far from resonance, the spontaneous Raman scattering at ω_S has a steady-state intensity (in photons cm^{-3} sec^{-1}) given by [9]

$$I_R = \frac{N}{2}\gamma_N^{23}\frac{\Omega^2}{\Delta^2} , \tag{1}$$

where N is the number density of scattering atoms (in our case Tℓ atoms), γ_N^{23} is the spontaneous decay rate from level 2 to level 3, $\Delta = \omega_{21} - \omega_L$ is the detuning, $\Omega = \mu E/\hbar$ is the Rabi frequency associated with the incident laser field $\vec{E}(t) = \hat{\varepsilon}_L E(t)\cos\omega_L t$, $\hat{\varepsilon}_L$ is the laser's polarization vector, and $\mu = e\langle 2|\hat{\varepsilon}_L\cdot\vec{r}|1\rangle$ is the dipole matrix element between states 1 and 2.

Similarly the collision-induced fluorescence at ω_{23} has a steady-state intensity given by [9]

$$I_F = \frac{N}{2}\gamma_E(\Delta)\frac{\Omega^2}{\Delta^2}\frac{\gamma_N^{23}}{\gamma_N^{23}+\gamma_N^{21}} \tag{2}$$

where $\gamma_E(\Delta)$ is the rate of quasi-elastic collisions (in our case Tℓ-Ar collisions) which make up the energy difference Δ needed to produce an atom in level 2. We have written the collision rate as a function of the detuning to account for the non-Lorentzian dependence outside the impact regime [10].

Effects of spatial degeneracy have not been included in Eqs. (1) and (2).

MOLLOW [6] has predicted that both I_R and I_F will have gain when level 3 is initially unpopulated. For laser intensities where $\Omega^2 < \Delta^2$, we expect the gain of the Raman scattering at ω_S to be

$$g_{SRS} = \frac{\pi}{2} N \frac{c^2}{\omega_{23}^2} \frac{\gamma_N^{23}}{\gamma_R} \frac{\Omega^2}{\Delta^2} \quad , \tag{3}$$

where γ_R is a measure of the Raman line width (to be discussed later). For the collision-induced fluorescence at ω_{23}, the gain is expected to be

$$g_{SCF} = \frac{\pi}{2} N \frac{c^2}{\omega_{23}^2} \frac{\gamma_E(\Delta)}{[\gamma_D/\pi U]} \frac{\Omega^2}{\Delta^2} \quad . \tag{4}$$

Here γ_D is the Doppler line width and U is the peak height of the normalized Voigt profile [11], both for the 2-3 transition. One can think of $\gamma_D/\pi U$ as an effective width for the Doppler-plus-collision-broadened transition, 2-3.

Expressions (1) and (2) for the spontaneous emission and Eqs. (3) and (4) for the gain, can be used with simple photon propagation equations [12] to solve for the single pass stimulated outputs (in photons sec^{-1}) of the SRS and SCF collected by a solid angle α_2:

$$I_{SRS} = \frac{I_R A \alpha_1}{4\pi g_{SRS}} [\exp(g_{SRS}L) - 1] + \frac{I_R}{4\pi} AL(\alpha_2 - \alpha_1) \tag{5}$$

and

$$I_{SCF} = \frac{I_F A \alpha_1}{4\pi g_{SCF}} [\exp(g_{SCF}L) - 1] + \frac{I_F}{4\pi} AL(\alpha_2 - \alpha_1) \quad , \tag{6}$$

where A and L are the area and length of the excitation region, α_1 is the solid angle formed by the excitation region and it is assumed that $\alpha_2 \geq \alpha_1$. In both equations, the second term accounts for the spontaneous emission which exits the sample outside of the gain region, but which still enters our collection angle α_2. We note that Eqs. (5) and (6) predict that both the SRS and SCF will initially have a linear dependence on laser intensity when spontaneous scattering is dominant, but will eventually grow exponentially when the scattering becomes stimulated. Effects of saturation or population depletion have not been included.

The apparatus used in this experiment is similar to that described in detail by CARLSTEN, SZÖKE and RAYMER [1]. A tunable dye laser, pumped by an N_2 laser, was tuned near the $6^2P_{1/2}$-$7^2S_{1/2}$ resonance line of Tℓ at 377.6 nm. The dye laser had a pulse duration of 10 nsec, a spectral width of 0.03 nm and an energy of 50 µJ inside the excitation region. The beam was focused to 330 µm diameter giving a power density of 2 MW/cm^2 inside an oven containing 0.1 torr of Tℓ vapor and 5 to 80 torr of Ar buffer gas. The oven input and output windows were put at an angle to avoid back reflections, which would affect the growth of the stimulated emission. The length of the Tℓ vapor region was ~2.5 cm. The emission region was then imaged onto the slit of a 0.3 m monochromator of 0.06 nm resolution with an f/10 optical system that was capable of viewing the emission at right angles to the laser beam or along the laser beam direction. The output of the photomultiplier, placed at the exit of the monochromator, was fed into a box-car integrator using a 50 nsec integrating time to eliminate dark current. The signals were then recorded on a chart recorder.

When the incident laser is tuned on resonance, only one spectral component, centered at ω_{23}, is observed in the emission. We have studied the dependence of the emission at 535 nm both in the side and forward directions when the laser was tuned to the 377.6 nm resonance line of Tℓ. The results are shown in Fig.2. We see that at low laser intensities, both the side and forward emissions were linear in laser intensity. Above ~10 kW/cm^2, the forward emission became stimulated and eventually saturated, allowing a maximum photon conversion efficiency of 25%. These data were taken at 730°C (10^{15} cm^{-3} of Tℓ atoms) and 20 torr of Ar buffer gas. At 880°C the maximum conversion efficiency was 60%.

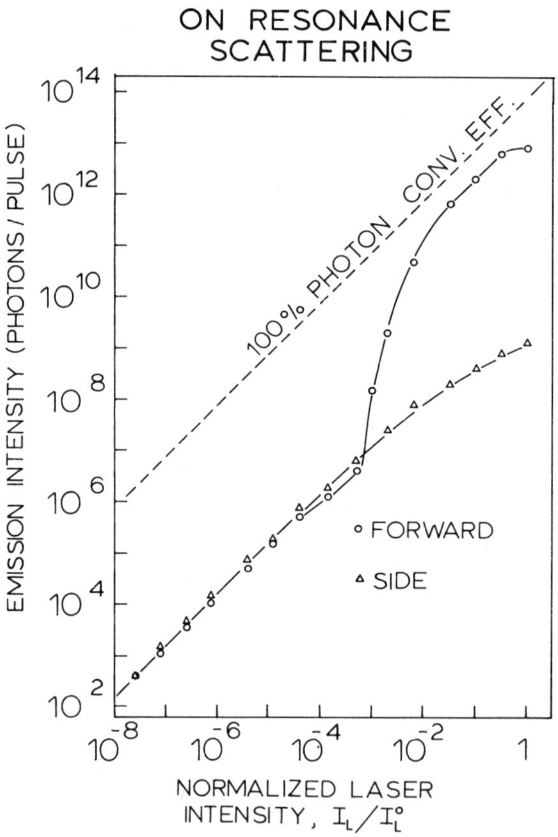

Fig.2 Dependence of scattered light at 535.0 nm with the laser tuned to the 377.6 nm resonance in both the forward direction (along the laser beam) denoted by o and in the side direction denoted by Δ. The laser intensity I_L^0 was 2 MW/cm^2. At low intensities the scattering in both directions is spontaneous and hence linear. Above 10 kW/cm^2, the forward scattering becomes stimulated.

In order to study the collisional effects upon this stimulated scattering, we tuned the laser 0.14 nm to the red side of resonance. We were then able to resolve spectrally the Raman emission at ω_S and the collision-induced fluorescence at ω_{23}. The dependence of these two spectrally resolved components on laser intensity is shown in Fig.3. The Ar pressure for these data was 20 torr and the temperature was 730°C.

In the side direction (Fig.3b) both the Raman scattering I_R and collision-induced fluorescence I_F were linear in laser intensity. By measuring the ratio I_F/I_R, we obtained an absolute measure of the collisional redistribution function. From Eqs. (1) and (2), this ratio is expected to be

$$\frac{I_F}{I_R} = \frac{\gamma_E(\Delta)}{\gamma_N} \tag{7}$$

where $\gamma_N = \gamma_N^{21} + \gamma_N^{23}$ is the radiative decay rate of level 2. Using $\gamma_N = 6.7\times10^7$ rad sec^{-1} [13] and our measured value for I_F/I_R we obtain γ_E(0.14 nm to red) =

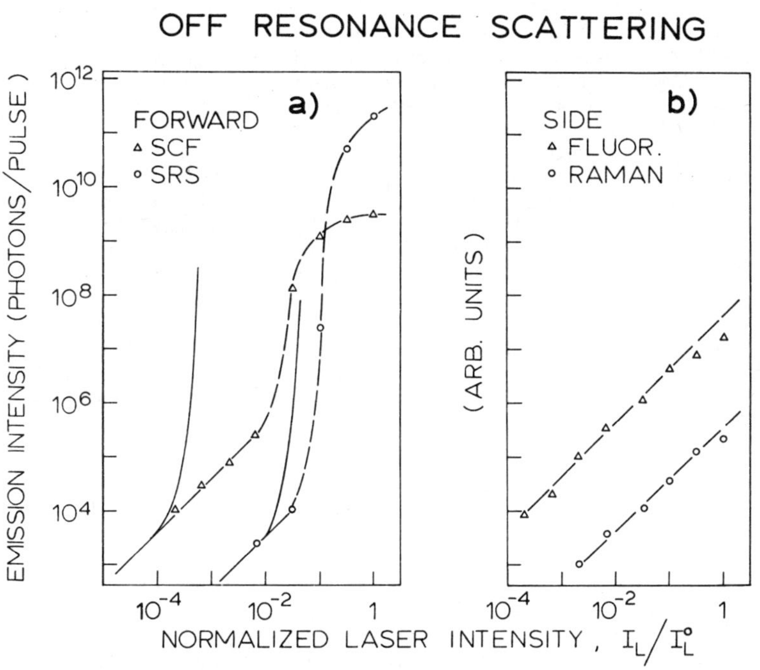

Fig.3 Dependence of Raman scattering and collision-induced fluorescence on laser intensity when the laser was tuned off resonance (0.14 nm to the red side of 377.6 nm resonance line). a) Growth of the Stimulated Raman Scattering (SRS) and Stimulated Collision-induced Fluorescence (SCF) from spontaneous, linear scattering in the forward direction. The solid curves are theoretical curves obtained from Eqs. (5) and (6) for the SRS and SCF, respectively. b) Spontaneous, linear scattering in the side direction for comparison. The laser intensity I_L^0 was 2 MW/cm^2.

7.4×10^9 rad sec^{-1}. This is a factor of 3 larger than the value determined by the emission line shape measurements of CHERON, SCHEPS AND GALLAGHER [14]. According to recent results of NIENHUIS and SCHULLER [10], we expect the two experiments (collisional redistribution and collisional line broadening) to give the same value for $\gamma_E(\Delta)$ over the experimental range that we studied. The existing discrepancy may be due to the transient nature of our experiment as well as neglect of degeneracy in Eq. (7).

In the forward direction as in the side direction, the emission is initially linear at low intensities. However, eventually both the Raman emission and the collision-induced fluorescence grow exponentially when the scattering becomes stimulated. It is interesting to note that while the collision-induced fluorescence is two orders of magnitude larger than the Raman scattering at low laser intensities, eventually at high laser intensities, when saturation occurs, the opposite is true.

Using Eqs. (5) and (6) multiplied by the pulse duration we calculated the expected exponential growth for the Raman scattering and collision-induced fluorescence. The results are shown as the solid curves in Fig.3a. The vertical scale was considered a free parameter, but agreed to within a factor of 4 with an absolute calibration estimate. The laser profile was taken to be uniform in intensity over the 330 μm diameter of the excitation region. It is important to note that in calculating the SRS gain we have used the result of AKHMANOV *et al.* [15] that, for forward SRS in a dispersionless medium, a broad-band laser has the same gain as a monochromatic laser. This is because the intensity variations of the Stokes emission follow those of the pump laser as the two pulses travel with the same velocity through the medium. Thus for γ_R we have used the Raman line width (which is predominantly the Doppler width for the 1-3 separation). This results in an SRS gain which is 240 times larger than that calculated by taking γ_R equal to the laser line width. We see that while the predicted initial exponential growth of the Raman scattering is quite close to our experimental results, the gain for the collision-induced fluorescence is more than an order of magnitude less than predicted by Eq. (6). We do not know the reason for the discrepancy, but possible problems may arise from the use of the steady-state theory as well as from the assumption of a constant spatial profile. Further theoretical analysis will, we hope, include the effects of the laser pulse shape in time and space.

In order to further study the predictions of Eqs. (5) and (6), we have measured the dependence of both the SRS and SCF as a function of Ar pressure. The results are presented in Fig.4. The data were taken at 5, 20 and 80 torr of Ar buffer gas. While the SRS has no pressure dependence, the SCF is highly pressure dependent as expected from Eq. (6). The solid curves in Fig.4b were calculated from Eq. (6) but with the gain decreased by a factor of 30. Except for this overall shift in gain, the fit to the pressure dependence is quite reasonable. Similar agreement was found with the laser tuned to the blue side of resonance but the results were omitted from Fig.4b for clarity.

The authors wish to thank N. Bloembergen, Y.-R. Shen, P. P. Sorokin and A. Szöke for useful discussions on stimulated Raman scattering. In addition the authors acknowledge the helpful comments and suggestions of J. Cooper and W. C. Lineberger.

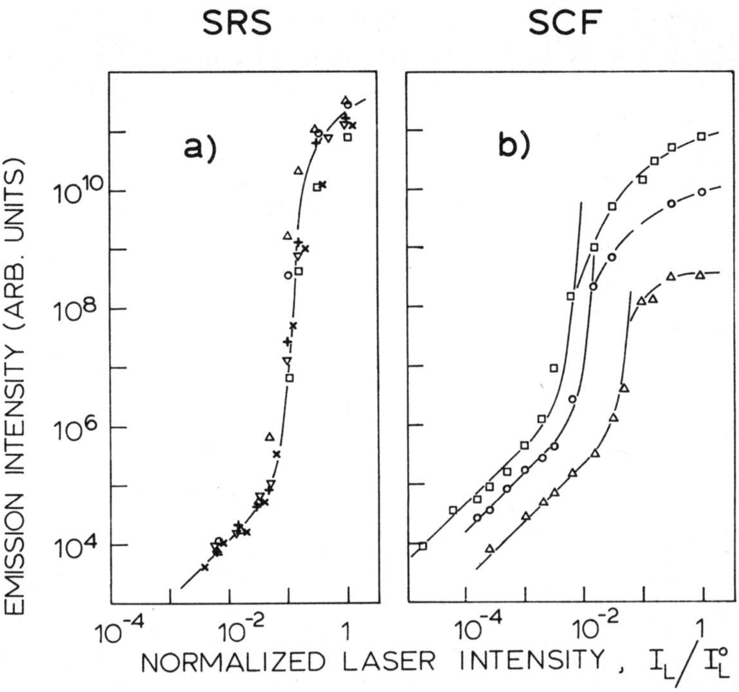

Fig.4 Dependence of stimulated scattering on Ar buffer gas pressure with the laser tuned 0.14 nm to the red side of resonance: □ (80 torr), o (20 torr), Δ (5 torr); with the laser tuned to the blue side of resonance: ∇ (80 torr), x (20 torr), + (5 torr). a) The Stimulated Raman Scattering (SRS) is seen to be independent of Ar pressure, while b) the Simulated Collision-induced Fluorescence (SCF) is highly pressure dependent. Solid lines in b) are theoretical fits to the SCF using Eq. (6) but with the gain decreased by a constant factor of 30. Similar agreement for the SCF was found when the laser was tuned to the blue side of resonance but the results have been omitted for clarity.

References

1. J. L. Carlsten, A. Szöke and M. G. Raymer, Phys. Rev. A 15, 1029 (1977) and references therein.
2. E. Courtens and A. Szöke, Phys. Rev. A 15, 1588 (1977).
3. C. Cohen-Tannoudji and S. Reynaud, J. Phys. B 10, 345 (1977); 10, 365 (1977).
4. J. Liran, L. A. Levin, C. Erez and J. Jortner, Phys. Rev. Lett. 38, 390 (1977).
5. L. Vriens, J. Appl. Phys. 48, 653 (1977).
6. B. M. Mollow, Phys. Rev. A 8, 1949 (1973).
7. J. J. Wynne and P. P. Sorokin, J. Phys. B 8, L37 (1975).
8. See for instance, N. Djeu and R. Burnham, Appl. Phys. Lett. 30, 473 (1977).
9. A. Omont, E. W. Smith and J. Cooper, Astrophys. J. 175, 185 (1972).

10. D. L. Huber, Phys. Rev. 187, 392 (1969); G. Nienhuis and F. Schuller, Physica C (in press).
11. D. Mihalas, Stellar Atmospheres (Freeman, San Francisco, 1970).
12. J. J. Wynne and P. P. Sorokin, Nonlinear Infrared Generation, Topics in Applied Physics, Vol. 16, edited by Y.-R. Shen (Springer-Verlag, Berlin, 1977).
13. M. Norton and A. Gallagher, Phys. Rev. A 3, 915 (1971).
14. B. Cheron, R. Scheps and A. Gallagher, Phys. Rev. A 15, 651 (1977).
15. S. A. Akhmanov, Yu. E. D'yakov and L. I. Pavlov, Sov. Phys. JETP 39, 249 (1974).

V. Optical Transients

SUPERFLUORESCENCE IN CESIUM: COMPARISON WITH THEORY AND APPLICATION TO QUANTUM BEAT SPECTROSCOPY

H.M. Gibbs

Bell Laboratories, Murray Hill, NJ 07974, USA

and

Q.H.F. Vrehen and H.M.J. Hikspoors

Philips Research Laboratories, Eindhoven, The Netherlands

Abstract

The advantages of the $7^2P_{3/2}$ to $7^2S_{1/2}$ transition in Cs for definitive studies of the superradiance of an initially inverted long cylinder are discussed. Quantum beat interferences at both upper and lower state separations are presented. Possible applications of superfluorescence beats to spectroscopy are noted. Observation of a regime of single pulse emission free from significant inhomogeneous dephasing, homogeneous relaxation, or linear diffraction is emphasized. Comparisons of the data are made to the quantum mean-field and semi-classical Maxwell-Schroedinger models of the superfluorescence process. The lack of complete agreement suggests the need for a better treatment of the quantum-to-classical transition and dynamic transverse effects.

1. Introduction

Superfluorescence (SF) is the cooperative emission or superradiance [1] of a sample initially prepared in a completely inverted state [2]. SF is initiated by incoherent quantum spontaneous emission and evolves to highly directional coherent classical emission. Since there is initially no phased array of dipoles, i.e., no macroscopic polarization, it is the initial geometrical asymmetry in gain which leads to the high directionality of emission.

If N inverted atoms are all within a cubic wavelength one might expect a reduction of 1/N in the single-atom spontaneous lifetime τ_o. Near field dipole-dipole dephasing is believed to prevent such small sample SF except in very special geometries [3]. However, Rehler and Eberly [4] have shown that many of the features of small sample SF, i.e., highly directional N^2 emission, persist in extended samples, but the characteristic SF time becomes $\tau_R = \tau_o/\mu N$. The quantity μ is a small number which quantifies the reduction in the effective number of atoms because of the failure of their phases to add constructively over a large sample:

$$\mu \propto \left| \frac{1}{N} \sum_{j=1}^{N} e^{i(\phi_j - \vec{k}\cdot\vec{r}_j)} \right|^2 .$$

Nonetheless, μN may be a large number yielding SF times much shorter than τ_o.

The early small-sample theories predicted sech^2 emission of full width at half maximum of 3.5 τ_R occurring with delay τ_D of several τ_R from the instant of inversion [2]. A quantum mean-field single-mode treatment of SF also led to single pulse sech^2 emission [5]. These treatments neglected spatial variations within the sample. It was then natural for the MIT group [6], which observed ringing in SF emission from hydrogen fluoride, to examine in more detail the propagational ringing discussed earlier by Burnham and Chiao [7] and by McCall [8]. The good agreement between their data and numerical solutions of coupled Maxwell-Schroedinger equations with initial polarization tipping led many to the conclusion that ringing is an inevitable consequence of SF emission in the absence of strong homogeneous relaxation, inhomogeneous dephasing, or diffraction. Bonifacio and Lugiato [5] re-emphasized their prediction of a regime of single pulse SF and clearly stated the conditions for its observation. The present experiment was designed to satisfy their conditions quite closely. A regime of single pulse emission was in fact found in qualitative and semi-quantitative agreement with their predictions.

In addition to the HF and Cs experiments there have been recent experiments in Na [9], Tl and some alkalis [10], and in CH_3F [11]. Although there are interesting aspects to each of these experiments, they do not satisfy the conditions well enough that single-pulse emission could not be explained as ringing averaged or obscured by dephasing or rapid relaxation. Such is not the case here.

2. Advantages of the Cs $7P_{3/2}$ *to* $7S_{1/2}$ *Transition*

The Bonifacio-Lugiato [5] conditions for single-pulse SF are (a) that the Fresnel number $F = A/\lambda L$ be about one to justify a single mode approach, (b) that the transverse excitation pulse be short compared with the emission evolution time τ_D, and (c) that the following time inequalities hold: $\tau_E < \tau_c < \tau_R < \tau_D < T_1, T_2', T_2^*$. The escape time is $\tau_E = L/c$. The Arecchi-Courtens cooperation time is $\tau_c = \sqrt{\tau_E \tau_R}$. The SF time is $\tau_R = \tau_o/\mu N = 8\pi\tau_o/3n\lambda^2 L$ where n is the inversion density over the sample length L. The delay time is of order $\tau_R \ln N$. T_1 and T_2' are the longitudinal and transverse homogeneous relaxation times, and T_2^* is the inhomogeneous dephasing time.

In Cs these conditions have been approximated closely. For each sample length the pump beam diameter is adjusted to make $F = 1$ at th SF wavelength. The 2 ns excitation pulse is much shorter than the single pulse delay times. The excitation is longitudinal but the spatial length of the pulse is much longer than the sample, so the entire sample is excited identically. Using 455.5 nm excitation of an atomic beam, SF at 2931 nm from the $7^2P_{3/2}$ to $7^2S_{1/2}$ state is dominant and the density can be adjusted to yield $\tau_E = 0.067 < \tau_c = 0.18 < \tau_R = 0.5 < \tau_D = 10 << T_1 = 70$, $T_2' = 80$, $T_2^* = 32$ ns; see Fig. 1 and [12] for details of this transition having $\tau_o = 551$ ns and the apparatus.

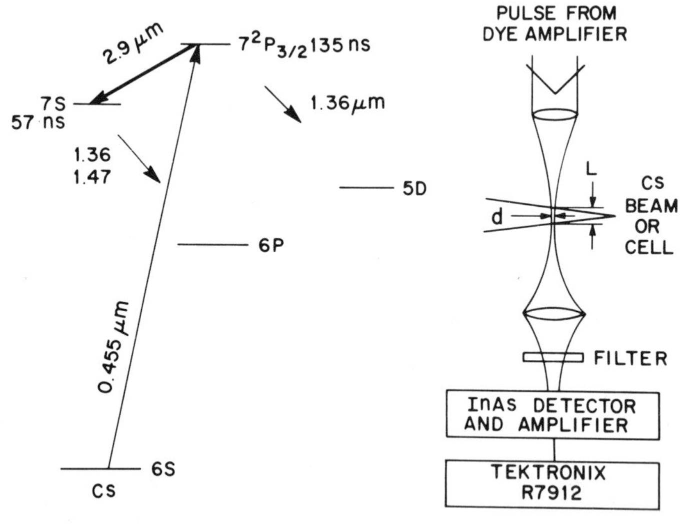

Fig. 1 Schematic diagrams of cesium levels and experimental apparatus.

This transition in Cs has several favorable features. Cesium's heavy mass results in a narrow Doppler width which at 2931 nm yields $T_2^* = 5$ns in a cell. Only weak beam collimation lengthens T_2^* by an order of magnitude. Furthermore, it is relatively easy to produce a dense beam of Cs. The 455.5 nm single-photon excitation wavelength is easily produced by a nitrogen-laser-pumped dye laser. The homogeneous relaxation times from radiative decay are negligibly long, and competition from other transitions is insignificant. Finally Cs has a well measured atomic structure permitting the identification of beat signals and the selection of a simple two-level system for careful single-pulse measurements.

3. Superfluorescence Beats and Spectroscopic Applications

The 2 ns excitation pulse is not transform limited; its width is typically 500 MHz. This is narrow relative to the 9.2 GHz $6^2S_{1/2}$ ground state hyperfine splitting, so that selective excitation from one F state occurs. However, the pulse spectrum overlaps both $7^2P_{1/2}$ hyperfine states separated by 401 MHz. Whenever two states of an atom are simultaneously excited into a coherent superposition, quantum beats occur [13]. If two or more indistinguishable channels exist for absorption and fluorescence to the same final state, the probability for the process is found by first adding the amplitudes and then squaring the sum. This leads to oscillations at excited state energy splittings. An example of quantum beats in SF from the $7^2P_{1/2}$ state is shown in Fig. 2.

Excitation from 6S, F = 3 to 7P, F′ = 2,3,4 results in a superposition of states with frequency differences of 50, 66, and 116 MHz. Figure 3 shows the calculated single-atom quantum beat fluorescence I_F as well as SF data I_{SF}. At delay times for destructive interference in I_F, the magnitude of I_{SF} is greatly reduced also. The macroscopic polarization turns the SF emission on and off as it increases and decreases as a result of the atomic interferences. Since SF is initiated by spontaneous emission, beats are expected in the early emission. Presumably, the cooperative emission which develops later does not stop such interference beating. But it is also possible for two *independent* SF emissions to beat with each other like two lasers since each is coherent. An example is shown in Fig. 4, where both $M_I = -3/2$ and -5/2 substates are excited from 6S to $7P_{3/2}$ and superfluoresce independently to 7S in a 2.8 kOe magnetic field. The resulting beats are close to the calculated transition difference which is dominated by the splitting in the lower state. Single-atom beats are not expected at lower state splittings since in principle one can distinguish the final state of each atom [13]. Lower state beats could appear in many-atom fluorescence if care is taken to prevent motional dephasing [13]. In SF, lower state splittings can be determined by beating two simultaneously superfluorescing transitions or measuring the frequencies of the two alternately.

The theory of quantum beat superfluorescence including frequency pulling effects should be worked out to properly evaluate the spectroscopic potential of SF. At first it may appear that SF beats have no advantages over single-atom beats and that the non-exponential decay of SF increases the complexity of analysis unnecessarily. There are situations where the SF case might have strong advantages, however. SF is capable of converting the isotropic, seconds-long fluorescence of a forbidden transition into highly directional, nanoseconds-long SF [6]. This fantastic enhancement in signal intensity and the capability of observing lower state beats may permit the measurement of splittings by SF which are inaccessible to single-atom techniques. To achieve this, it must be possible to obtain large enough densities to make τ_R no longer than T_1, T_2', and T_2^*.

For details of the Cs quantum beat experiments see [14] and [12].

4. Single-Pulse Superfluorescence

In a magnetic field of 2.8 kOe, the 6S, $M_J = -1/2$, $M_I = -5/2$ state can be selectively excited to $7P_{3/2}$, $M_J = -3/2$, $M_I = -5/2$ which can then decay only to 7S, $M_J = -1/2$, $M_I = -5/2$. In an atomic beam this two level inverted transition is near-ideal for observing SF. The details of the careful two-level preparation and density determination will not be repeated here; see [12] and [15].

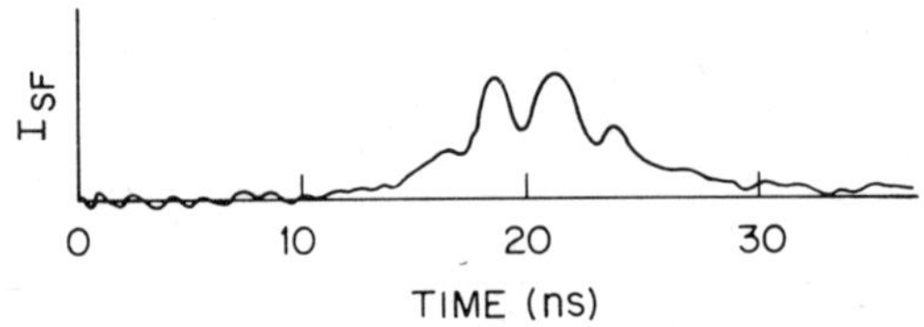

Fig. 2 Quantum beats at 400 MHz in the superfluorescence of a superposition of $7^2P_{1/2}$, F′ = 3 and 4 to $7^2S_{1/2}$. Circularly polarized 459.3 nm excitation from $6^2S_{1/2}$, F = 3 occurred in 2 ns. A 2.6 cm pathlength was traversed across the atomic beam 15 cm from the source.

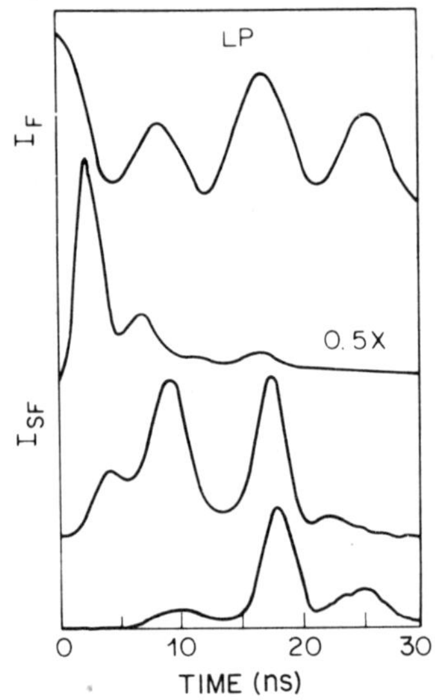

Fig. 3 Quantum beat superfluorescence in a 10 cm cell with 455.5 nm linearly polarized excitation from $6^2S_{1/2}$, F = 3 to $7^2P_{3/2}$, F′ = 2, 3 and 4. The observed SF beats are I_{SF}; I_F is the calculated single-atom fluorescence intensity.

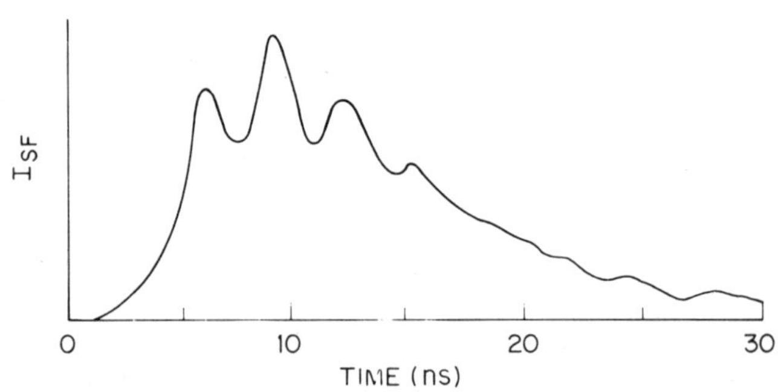

Fig. 4 Beating of the emissions from two incoherently excited M_I substates in a 2.8 kOe field (2.0 cm pathlength atomic beam).

Figure 5 is an example of one of the almost symmetrical pulses often observed. These symmetrical outputs are not quite as typical as those shown in Fig. 6, but their narrower widths suggest that the asymmetric ones are smeared out symmetric ones rather than vice versa. Perhaps the excitation conditions must be optimum for symmetric emission to occur. It is clear from Fig. 5 that some of the pulses are close to sech^2. The density needed for the mean-field [5] delay $\tau_D = \tau_R \ln N$ to agree with that observed is well within experimental uncertainty. The FWHM mean-field width is only 3.5 τ_R, over a factor of two narrower than observed. Propagation and dynamic transverse effects neglected in the mean-field, single-mode approach may be unavoidably significant; Maxwell-Schroedinger simulations indicate such is the case.

Comparison has also been made to numerical simulations of one-way uniform plane-wave coupled Maxwell-Schroedinger equation following the MIT group [6]. The uncertainty in such calculations centers around the value to take for θ_o, the initial polarization tipping angle or electric field area of the small input pulse required to initiate the evolution of these classical equations. MacGillivray and Feld [6] suggest $\theta_o^{MF} = N^{-1/2}(2\pi)^{-1/4}\ (\alpha L)^{-3/4}$; for our conditions $\theta_o^{MF}=3\times10^{-6}$ yielding delays a few times too long and strong ringing (second lobe over 50% of first) where single pulses were observed. Two-way propagation reduces ringing by perhaps a factor of two [16]. Inclusion of dynamic transverse effects would likely reduce the simulated ringing further, but it is unlikely that the delays would be decreased or that the strong asymmetry accompanying strong ringing could be averaged into almost symmetrical pulses. Also the appearance of $\alpha L = T_2^*/\tau_R$ in θ_o^{MF} seems physically unjustified. In the sharp-line limit (long T_2^*), it is surprising for θ_o to depend on the value of T_2^*. A value $\frac{2}{\sqrt{\mu N}}$ for θ_o has been suggested since it is the angle at which classical cooperative emission becomes equal to quantum random emission *in all directions* [4]. This would seem to be unnecessarily large, since the emission along the pencil becomes classical for a much smaller θ_o. The physically more attractive θ_o with some theoretical support [5] is $\theta_o = \frac{2}{\sqrt{N}}$. Speaking loosely, the first photon, which is emitted along the cylinder's solid angle within a time τ_R, produces a tipping of this magnitude. In Fig. 6, θ_o has been treated in two ways. In the left column θ_o is varied to reproduce the observed delay with experimental densities held fixed. In the right column of Fig. 6, $\theta_o = \frac{2}{\sqrt{N}}$ and the densities have been increased by 60% so that simulated delays overlap observed ones; this change in density is about one standard deviation.

The data should be most reliable for delays of 10 to 20 ns. If one fits the 13ns delay curve allowing the density to vary over its uncertainties of +60% and -40%, one finds θ_o varies from 0.05 to 0.0025 to reproduce the observed delay. For comparison $\frac{2}{\sqrt{\mu N}}\approx 0.08$ and $\frac{2}{\sqrt{N}}\approx 0.00024$, so within the quoted density uncertainties the θ_o for a one-way, uniform plane wave simulation lies in between these values and strongly disagrees with θ_o^{MF}. It may appear strange that such simulations agreed so well with the HF data. However, for a given θ_o, the HF density and diffraction loss can be chosen to give the observed delay and ringing. A change from θ_o^{MF} to $\theta_o = \frac{2}{\sqrt{N}}$ can be compensated by a 40% reduction in density. Even $\theta_o = \frac{2}{\sqrt{\mu N}}$ can be used with a factor of 5 change in density. It appears that the density and τ_o, or equivalently αL and T_2^*, are not sufficiently well known for HF to distinguish between these formulae for θ_o.

There is also the possibility that a large θ_o is introduced experimentally. Deliberately introduced feedback does not affect the output until it is a few percent, much larger than introduced in the experiment. No mechanism conceived for producing a large θ_o has proven large enough so far.

Clearly, coupled Maxwell-Schroedinger equations describe correctly the SF after the field at each point in the sample is approximately classical which is the case for most of the evolution time. The lack of agreement between the one-way uniform plane-wave simulations and the data must then arise from two-way or transverse effects or inappropiate initial conditions (θ_o). Uniform plane-wave simulations indicate that propagation effects, obviously neglected in mean-field theory, are impor-

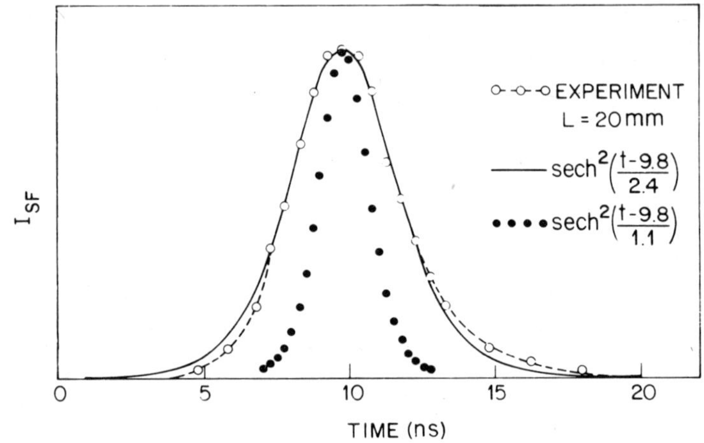

Fig. 5 An example of the almost symmetrical pulses frequently observed (circle points) and comparison with a best-fit sech^2 (solid) and mean-field theory sech^2 (dotted). The $\approx$25% reduction in the observed density required to make the mean-field delay agree with the data is well within experimental uncertainty.

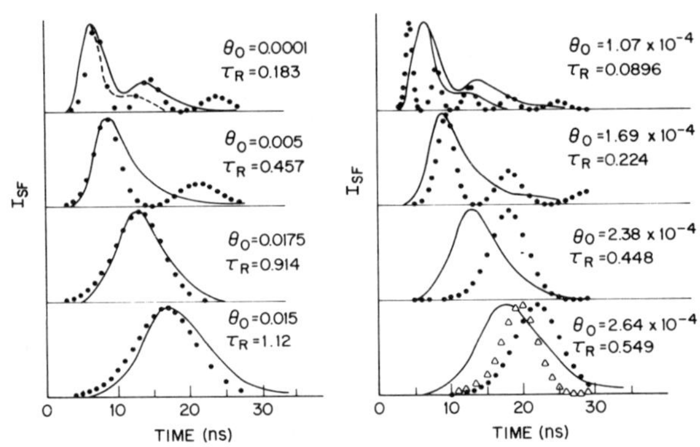

Fig. 6 Comparison of normalized data (solid curves) with simulations of one-way Maxwell equations coupled to Bloch equations. The evolution of the inverted Bloch vector is initiated by a short input pulse of area θ_0. The experimental values of τ_R from the formula $8\pi\tau_0/3n\lambda^2L$ are, from the top down, 0.14, 0.35, 0.71, and 0.87 ns. The values of τ_R in the left part of the diagram are the experimental ones increased by 0.78, the correction to the large Fresnel number μ factor in $\tau_R = \tau_0/\mu N$ found by numerical integration for Fresnel number one; the θ_0 values were chosen to give approximate agreement between the simulated and observed delays. In the right column the τ_R values are about 63% of the experimental values as required so that θ_0 could be given by $2/\sqrt{N}$ and so the delays would not all be much longer than observed. All of these simulations were for the experimental times in ns $T_{ab} = 275.5$, $T_{ac} = 264.7$, $T_{bd} = 57.0$, $T_2^* = 32.0$ except for the lower right hand triangular-points curve for which all the times were 10^5 ns. The upper SF state a can relax to the lower SF state b or to other states c; likewise b can relax to other states d. The long delay simulations still possess ringing; for example, a 26% ring occurs at 40 ns for the $\tau_R = 0.448$ ns curve.

tant as evidenced by the substantial ringing and broader outputs predicted. Perhaps transverse effects succeed in removing the ringing, especially for large θ_0's, but the broadening of the outputs relative to the mean-field value persists. Clearly, when inappropiate approximations are eliminated and correct initial conditions are inserted agreement will be found between coupled Maxwell-Schroedinger equations and the data.

Efforts to reevaluate the limits of validity of such approximations and to include propagation in the quantum approach were reported recently by Haroche [15] and Ressayre and Tallet [15]. Hopefully the increasing availability of good data to be compared with theories will encourage progress toward a much better understanding in the near future.

5. Future Possibilities

Superfluorescence might be used to extend the quantum beat technique to weak transitions and to lower state splittings. Studies of possible chirps in the SF process should be made. Transverse effects should be evaluated to see if they can be as important in SF as in self induced transparency [17]. Already Vrehen has reported some experimental transverse studies for large and unity Fresnel numbers [15]. Understanding of the quantum initiation process and possible observation of associated quantum fluctuations would be of great interest. Perhaps such studies will also be useful in the design of new coherent sources in both the infrared and x-ray regions.

References

1. R. H. Dicke, Phys. Rev. **93,** 99 (1954).
2. For a general introduction see L. Allen and J. H. Eberly, *Optical Resonance and Two-Level Atoms* (John Wiley, N.Y., 1975).
3. R. Friedberg, S. R. Hartmann, and J. T. Manassah, Phys. Lett. **40A,** 365 (1972). R. Friedberg and S. R. Hartmann, Opt. Commun. **10,** 298 (1974).
4. N. E. Rehler and J. H. Eberly, Phyrs. Rev.A **3,** 1735 (1971).
5. R. Bonifacio and L. A. Lugiato, Phys. Rev. A **11,** 1507 (1975) and **12,** 587 (1975). R. Bonifacio, P. Schwendimann, and F. Haake, Phys. Rev. A **4,** 302 and 854 (1971) and references therein.
6. N. Skribanowitz, I. P. Herman, J. C. MacGillivray, and M. S. Feld, Phys. Rev. Lett. **30,** 309 (1973). J. C. MacGillivray and M. S. Feld, Phys. Rev. A **14,** 1169 (1976) and references therein.
7. D. C. Burnham and R. Y. Chiao, Phys. Rev. **188,** 667 (1969).
8. S. L. McCall, Thesis, University of California, 1968, unpublished.
9. M. Gross, C. Fabre, P. Pillet, and S. Haroche, Phys. Rev. Lett. **36,** 1035 (1976).
10. A. Flusberg, T. Mossberg, and S. R. Hartmann, Phys. Lett. **58A,** 373 (1976).
11. A. T. Rosenberger, S. J. Petuchowski, and T. A. DeTemple, in Ref. 12.
12 C. M. Bowden, D. W. Howgate, and H. R. Robl, eds., *Cooperative Effects in Matter and Radiation* (Plenum, N.Y., 1977).
13. Single-atom quantum beats are reviewed by S. Haroche in *High Resolution Laser Spectroscopy* edited by K. Shimoda (Springer, Berlin, 1976).
14. Q. H. F. Vrehen, H. M. J. Hikspoors, and H. M. Gibbs, Phys. Rev. Lett. **38,** 764 (1977).
15. Proceedings of the Fourth Rochester Conference on Coherence and Quantum Optics, 1977.
16. R. Saunders, S. S. Hassan, and R. K. Bullough, J. Phys. A **9,** 1725 (1976) and private communications.
17. H. M. Gibbs, B. Bölger, F. P. Mattar, M. C. Newstein, G. Forster, and P. E. Toschek, Phys. Rev. Lett **37,** 1743 (1976).

COHERENT TRANSIENTS AND PULSE FOURIER TRANSFORM SPECTROSCOPY

R.G. Brewer, A.Z. Genack, and S.B. Grossman*

IBM Research Laboratory
San Jose, CA 95193, USA

Introduction

The early practitioners of pulsed NMR would never have dreamed that their sophisticated coherence techniques would one day be adapted to the optical region. Coherent radiation sources were needed, and the laser had yet to be conceived. We now know that the subject of optical coherent transients has steadily grown, beginning with the photon echo measurements of KURNIT, ABELLA, and HARTMANN [1].

Rather than review the whole field, which would be impossible, we will discuss our experience at IBM with emphasis upon current developments. These studies began about 1971, when SHOEMAKER joined our group, and shortly thereafter the Stark switching technique was born [2].

Since that time, over ten different coherent transients have been monitored by Stark switching. One major objective was to understand the coherence effects themselves which are not identical to NMR transients, and another goal was to use these transient techniques for investigating time-dependent atomic or molecular interactions. Stark switching has not been restricted to molecules. It has also been applied to atoms by P. F. LIAU et al. [3], and to low temperature solids by A. SZABO [4]. In addition, the method is not restricted to one-photon transients, and has been used by M. M. T. LOY [5] of IBM and also by the Bell Laboratories group [3] to observe two-photon transients.

In this talk, we will show how Stark switching can be used to perform "optical pulse Fourier transform spectroscopy" [6]. By this we mean the optical analog of the well-known NMR technique where coherent transients are Fourier transformed to yield an ultra-high resolution spectrum and dynamic process for several lines can be examined simultaneously.

We will also discuss the technique of laser frequency switching which we introduced last year [7]. It is similar to Stark switching but is more universal and will permit a wide variety of dynamic and spectroscopic studies of atoms, molecules and solids in the visible ultraviolet region. It appears possible to measure in a controlled manner dynamic events of atomic and molecular processes on a time scale ranging from milliseconds to about 50 picoseconds.

*Present address: The Aerospace Corporation, Los Angeles, CA 90009

Pulse Fourier Transform Spectroscopy

The method of Fourier transforming transient phenomena from the time to the frequency domain has proven to be an extremely versatile technique in pulsed nuclear magnetic resonance [8]. With it, ultrahigh resolution NMR spectroscopy can be performed quickly and with high sensitivity in a set of densely spaced lines. Since the NMR signals display coherent transient behavior, dynamic information about nuclear spin interactions can be derived in a selective manner for each transition as well.

We discuss here an initial demonstration of the Fourier transform technique in the optical region. By employing suitable coherent optical transient effects in a sample of $^{13}CH_3F$, we are able to resolve Doppler-free spectra in a set of closely spaced lines (Fig.1). The decay characteristics for each transistion are obtained also and reveal the separate contributions of elastic and inelastic molecular collisions. Furthermore, by selectively exciting different molecular velocity groups from the Doppler lineshape, the velocity dependence of these collisional processes can be determined and the long-range force laws deduced.

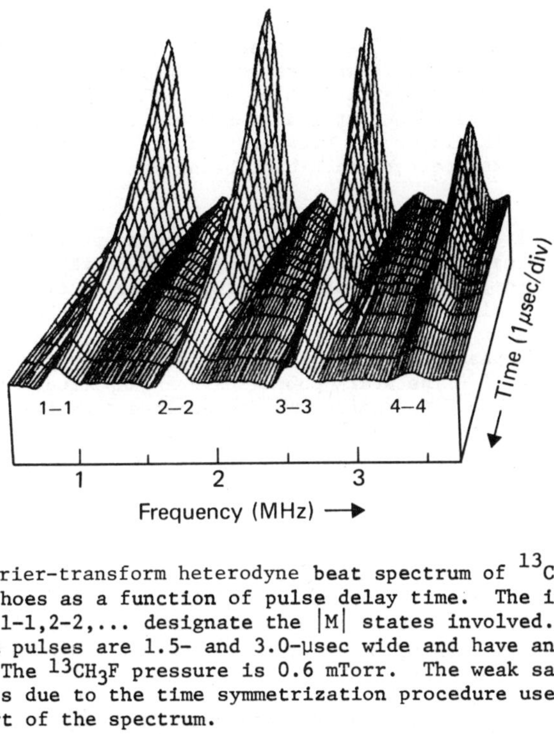

Fig. 1. Fourier-transform heterodyne beat spectrum of $^{13}CH_3F$ derived from two-pulse echoes as a function of pulse delay time. The infrared transitions 1-1,2-2,... designate the |M| states involved. The first and second Stark pulses are 1.5- and 3.0-μsec wide and have an amplitude of 18.8 V/cm. The $^{13}CH_3F$ pressure is 0.6 mTorr. The weak satellite features are artifacts due to the time symmetrization procedure used in obtaining the real part of the spectrum.

Optical transient phenomena are detected by the Stark switching technique [2,6]. Electronic pulses switch the sample into resonance with a cw CO_2 laser beam, thereby placing the molecular transition levels in coherent superposition. For the case of several closely spaced transitions that overlap within their Doppler lineshape, a corresponding number of molecular velocity groups may be prepared simultaneously. After the preparative pulse sequence, these velocity groups radiate at slightly different frequencies due to the Stark shift, producing an interference pattern in time. By interfacing the Stark apparatus with a computer, transient signals can be converted from analog to digital form and Fourier analyzed to yield a many-line spectrum. By varying the pulse delay time, as in an echo experiment, the decay behavior for each line can be mapped in a three dimensional diagram of signal amplitude versus frequency and elapsed time (Fig.1).

Pulse Fourier spectroscopy is performed by monitoring emission signals such as free induction decay (FID) and photon echoes. For this purpose, a two-pulse Stark sequence was chosen, the pulse delay time being τ. Immediately following each pulse, an FID appears while at time $t=2\tau$ an echo is emitted. The $^{13}CH_3F$ vibration-rotation transition excited is the fundamental ν_3 mode where $(J,K)=(4,3)\rightarrow(5,3)$ and the magnetic substate selection rule $\Delta M=0$ applies. This transition is prepared by the CO_2 laser line P(32) at 1035.474cm^{-1} which is frequency-locked to the Lamb-dip of the same $^{13}CH_3F$ transition using a second Stark cell. Hence, eight lines appear symmetrically about the laser frequency in emission, each radiating at a slightly different frequency due to the first order Stark shift $\Delta W=-ME[\mu'K/J'(J'+1)-\mu''K/J''(J''+1)]$ associated with the dc bias field E where $M=-4,-3,\ldots 3,4$. The dipole moments of the upper and lower transition levels are $\mu'=1.9038(6)$D and $\mu''=1.8578(6)$D [9]. Since the sample is prepared with the Stark pulse on but radiates when it is off, the emission is Stark shifted from the laser and with it produces four heterodyne beat frequencies at the photodetector, each beat being due to two transitions $\pm M\rightarrow\pm M$.

The laser beam is expanded with a Galilean telescope from 0.3 to 5cm diameter in order to match the 5.17cm spacing of the Stark plates (length: 45.7cm). By this means, the transverse molecular time of flight is increased to ~100μsec allowing decay time measurements in the range 5 to 20μsec. Since the laser power density is only ~20mW/cm^2, the intensity-dependent dephasing effect [10] noted previously for the echo decay is avoided as well. The Stark pulses have an amplitude of about 18.8V/cm, a duration of 0.5 to 4μsec, and are applied repetitively at a 10kHz rate. Detection is achieved by focusing the transmitted beam onto a Ge-Au photodetector. Methyl fluoride pressures are in the range 0.2 to 1mTorr.

The photodetector output is sampled by a PAR 160 box car integrator, which is gated either for the FID or the echo signal. A small local computer, IBM System 7, generates a digital time base which sweeps the box car in steps and stores its output in digital form. System 7 in turn communicates with larger computers such as the IBM 360 Model 85 or 195 which performs the fast Fourier transform [11], signal averaging, data analysis and long term storage.

An example of the Fourier transform spectrum, the real part, derived from a series of echo signals having different delay times is shown in Fig.1. The four lines are each 170kHz wide (due to the Stark pulse width), are spaced at 0.83MHz intervals (due to the Stark pulse amplitude) and clearly exhibit a Doppler-free behavior as the Doppler width is 66MHz FWHM. Similar spectra are obtained for the FID signals.

As the pulse delay time is advanced, the FID and echo signals exhibit different decay properties. Specifically, the envelope function of the second pulse FID monitors the decay rate Γ_1 for population recovery of the transition levels due to inelastic $^{13}CH_3F$ collisions. On the other hand, the echo envelope function samples the dipole dephasing rates $\Gamma_1+\Gamma$ (in the long-time regime [12]) where Γ is the rate of elastic collisions. These simple exponential decays are independent of Doppler and power broadening or the pulse repetition rate. The dependence of these quantities on molecular velocity can be measured also and is of fundamental interest because it allows an understanding of the force laws that operate, particularly the long range interactions. Experimentally, eight different velocity groups are sampled simultaneously, one for each transition, when the M degeneracy is removed by application of a Stark bias field. Thus, the velocity group selected scales with the magnitude of the Stark bias field E and the M state. The observed dependence of the elastic and inelastic decay rates on v_z, the molecular velocity component along the direction of the laser beam, is shown in Fig.2.

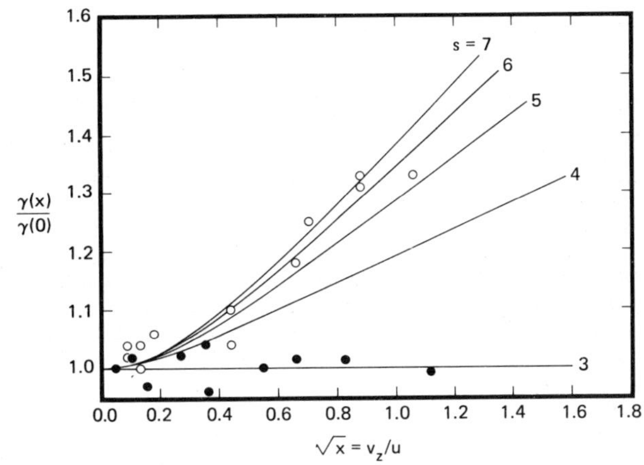

Fig.2. Normalized decay rate $\gamma(x)/\gamma(0)$ as a function of the velocity ratio $\sqrt{x}=v_z/u$. Experiment: elastic scattering (open circles); inelastic scattering (filled circles). Theory: elastic scattering and the s=3 curve for inelastic scattering (curves).

To interpret these results, it is necessary to make contact with scattering theory. For elastic collisions, we follow the approach of BORENSTEIN and LAMB [13] which permits the collision operator of the Boltzmann transport equation to be expressed in terms of a velocity-dependent total cross section σ_T. There results the elastic decay rate

$$\Gamma(\vec{v}) = \quad (|\vec{v}-\vec{w}|)\sigma_T(|\vec{v}-\vec{w}|)F(\vec{w})dw \qquad (1)$$

where in the laboratory frame $\vec{v}$ is the total velocity of the optically prepared molecule and $\vec{w}$ is the velocity of its collision partner. For a central force interaction $V(r)=-C_s/r^s$, we may use the Landau-Lifshitz [14] elastic scattering cross section $\sigma_T=\beta(s)[C_s/h(\vec{v}-\vec{w})]^{2/(s-1)}$. We note that this theory is an approximation as it does not include the complete anisotropic interaction or averaging over the rotational motion. Since an elastic scattering event in CH_3F introduces a characteristic velocity jump in v_z of only ∿85 cm/sec, [10] we can safely assume that a velocity distribution function of the form $f(\vec{v})=F(v_x)F(v_y)f(v_z)$ is maintained with its equilibrium value along the x and y directions while the z component is modified by the optical excitation. This allows us to average (1) over v_x and v_y to obtain an effective elastic decay rate

$$\gamma(x) = e^x \Gamma(a,x) \tag{2}$$

where $x=(v_z/u)^2$, u is the most probable value of v_z in the thermal distribution and $\Gamma(a,x)$ is an incomplete gamma function with $a=(3s-5)/[2(s-1)]$. Eq. (2) is presented in Fig.2 for different force laws. Earlier calculations [15] for inelastic scattering give the same velocity-independent behavior for s=3, the other force laws being less well understood at present.

In principle, the decay rates will depend on both v_z and M. However, from data of the type shown in Fig.2, it is seen that the elastic and inelastic scattering cross sections do not exhibit a significant M dependence. At small v_z, the decay rates for the various M levels are the same, and at high v_z, when the different transitions are tuned to the same velocity group, the decay rates are also the same. Apparently, this is one of the few experimental tests of this point which has been the subject of an earlier controversy [16]. Since CH_3F possesses a permanent electric dipole moment, it is not surprising that the inelastic scattering follows a velocity-independent behavior as expected for a dipole-dipole interaction (s=3). This anisotropic interaction provides the torque required for inducing rotational and reorienting transitions and for symmetric top molecules does not vanish with rotational motion. A similar result has been reported [17] in a velocity-dependent linewidth study of NH_3 using a laser saturation method. What is surprising in this work is that the elastic process behaves differently, the experimental points clustering around the more isotropic second order dipole-dipole interaction (s=6). Additional calculations [18] will be required to explain why the s=3 interaction is not more prominent in elastic collisions and why the observed cross sections for elastic (∿430Å^2) and inelastic (∿500Å^2) scattering (for v_z=0) are roughly the same.

Laser Frequency Switching

We will now discuss the new technique of laser frequency switching which has proved useful in observing coherent optical transient phenomena in atoms, molecules, and solids [7,14,20]. In concept, the method is analogous to pulsed nuclear magnetic resonance techniques [1-3], but in practice, it more closely resembles the Stark switching method. To illustrate its use, we demonstrate photon echoes [1], free induction decay (FID) [21] and nutation effects [1] in numerous lines of the visible electronic transition of I_2 (Fig.3).

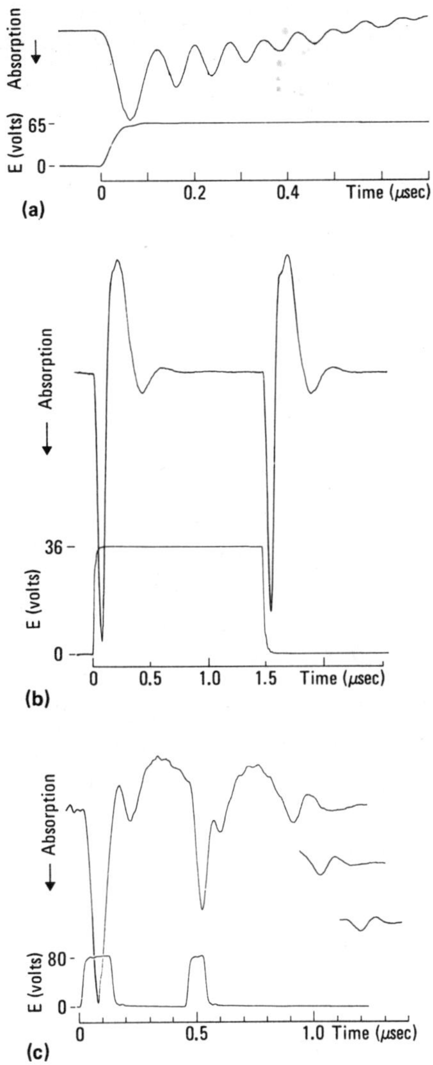

Fig.3. Coherent optical transient phenomena in I_2 vapor: (a) free induction decay where the emission and laser produce a 13-MHz beat; (b) two optical nutation patterns; and (c) photon echoes occurring at ∿1 μsec where the successive echoes decay with increasing pulse delay time. The frequency-switching pulse pattern is displayed in the lower trace of each figure.

In the Stark-switching experiments we have discussed a prescribed sequence of low voltage Stark pulses switched a molecular sample into or out of resonance with an infrared beam from a fixed frequency cw CO_2 laser. Coherent emission or absorption transients were detected in the transmitted beam. A variant of this idea was later realized by HALL [22] who *frequency-switched* the laser instead of the sample. Optical transients arose due to a methane sample located inside the cavity of a frequency modulated 3.39 micron He-Ne laser.

In our configuration, Fig.4, a stable tunable cw dye laser is frequency-switched while the sample's transition frequency remains constant. The sample is now external to the laser system so as to not affect its performance. Coherent transient signals again appear in the forward beam. Frequency-switching is achieved with an electro-optic crystal of ammonium dihydrogen phosphate (ADP) or the deuterated crystal AD*P, which is inside the dye laser cavity and is driven by a sequence of low voltage pulses. The laser frequency follows the refractive index variations induced in the ADP crystal. Hence, the experiment is controlled

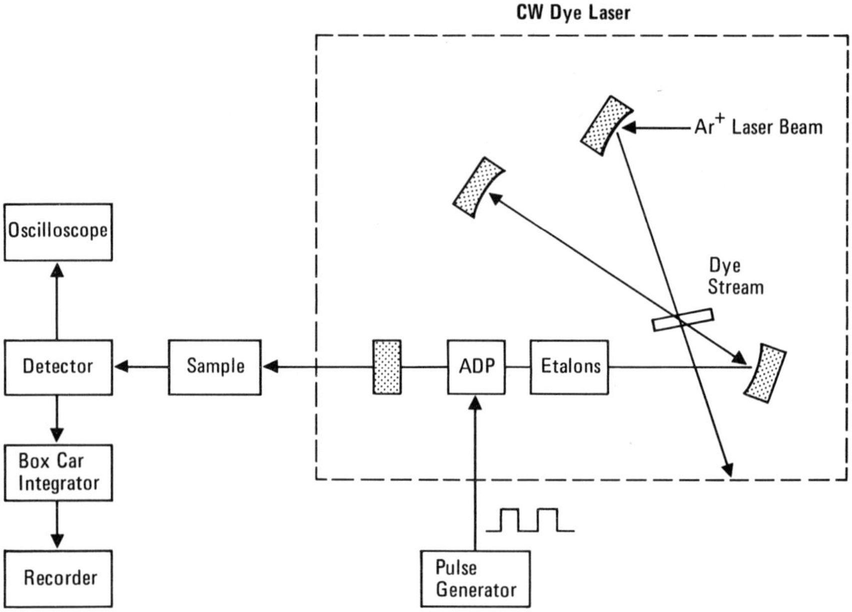

Fig.4. Schematic of the apparatus for observing coherent optical transients using a frequency-switched cw dye laser.

electronically and in such a way that the advantages inherent in the Stark technique are preserved here as well. We find, therefore, that (1) the only transient observed is the desired coherent transient itself; this is not the case with pulsed laser sources as the small coherent transient signal often rides on top of the laser pulse and the two are not easily separated; (2) heterodyne detection is possible because the coherently radiated light propagates with the laser beam in the forward direction and is shifted from it in frequency; this increases the signal amplitude several orders of magnitude and facilitates measuring the decay of emission signals; (3) a further improvement in signal to noise results with signal averaging, which is possible because the pulse sequence is repetitive; and (4) the entire class of coherent optical transient effects can be monitored since the electronic pulse sequence can be tailored to the particular experiment of interest. *Moreover, when these features are combined with the broad tuning range available in a dye laser, it is apparent that coherent transient phenomena can now be observed with ease in a large number of transitions in various atomic, molecular and solid state systems.*

A Spectra-Physics 580A cw dye laser is utilized but modified to include the ADP modulator. The dye is Rhodamine 6G. The output beam is single mode, linearly polarized and has a power up to 100 mW in a beam diameter of 0.5 mm. The collimated beam irradiates in single pass an evacuated and sealed off cell, of 20 cm length, containing I_2 at a vapor pressure (3-150 mTorr) determined by a refrigerated cold finger. Laser tuning by means of an intracavity etalon allows selecting a particular I_2 line where the overlapping Doppler-broadened I_2 hyperfine components span ∿1 GHz (Doppler width : 395 MHz FWHM). Coherent transients in I_2 are seen even at a fraction of 1 mW laser power.

The intracavity ADP crystal is driven by an HP 1900A or 214A pulse generator with a single or double pulse sequence and at a 25 kHz repetition rate. A PIN photo-diode monitors the forward beam, and transients are observed with a Tektronix 7904 sampling oscilloscope or a box car integrator. From the observed FID beat frequency, we find that the ADP electro-optic frequency shift parameter is 0.2 MHz/volt. Hence, ∿30 volt pulses are adequate for nonadiabatically switching the laser frequency outside an I_2 homogeneous linewidth of ∿1 MHz. On the other hand, the laser does not emit a transient signal itself, which otherwise would obscure observations in the sample, because switching occurs inside the dye's homogeneous linewidth of $\gtrsim$200 MHz.

From the multitude of $^{127}I_2$ lines accessible, we selected in these initial studies only one line, $(v,J)=2,59\rightarrow 15,60$ of the electronic transition $X\ ^1\Sigma_g^+\rightarrow B^3\pi_{ou^+}$. It falls at 16,956.43 cm^{-1}, 7.6 GHz to the high frequency side of the sodium D line. The vibration-rotation assignment was verified from the calculated line position and the fluorescence spectrum using a 150,000 resolving power spectrometer.

The three coherent transient effects shown in Fig.3 are (a) FID of an I_2 velocity group that is prepared under steady-state conditions and where the laser frequency is abruptly switched by a step-function voltage pulse; (b) optical nutation patterns arising from an I_2 velocity group that is suddenly excited at the beginning of the switching pulse and another at the end; and (c) the photon echo pulse which follows two short switching pulses. The theory of these processes parallels the molecular infrared case for vibration-rotation transitions where Stark- switching was employed. For electronic transitions, however, we must generalize these density matrix calculations to allow the upper (level a) and lower (level b) transition levels to depopulate at different rates, $\gamma_a\neq\gamma_b$, where γ_a and γ_b are the total decay rates, radiative and nonradiative (elastic and inelastic) of the diagonal density matrix elements [23]. Furthermore, to agree with our experiments, it will be necessary to consider that during *elastic collisions* upper and lower transition states shift by significantly different amounts, so that the off-diagonal element exhibits quantum mechanical phase interruptions rather than classical velocity changes. This subtle point has emerged recently in certain line-broadening theories [24] and is consistent with the results obtained here. It follows that the normalized echo field amplitude will decay with pulse delay time τ as

$$E_c(t=2\tau) = e^{-\gamma t} \quad (3)$$

where $\gamma=(\gamma_a+\gamma_b)/2+\gamma_\phi$ [25] is the rate the optically induced dipole dephases and γ_ϕ is the elastic collision rate for phase interruptions caused by perturber-induced energy level shifts. We note that the infrared echo results [6] represent the other limiting case where elastic collisions are dominated by velocity changes and the echo decay law is not a simple exponential. Our results, therefore, support the BERMAN-LAMB theory [24] for these limiting cases.

The two-pulse sequence of Fig.3(c) also allows a measurement of the rate of population recovery, which results from *inelastic collisions* and radiative decay. The first pulse causes the upper level(a) to gain in population at the expense of the lower level(b) while the second pulse nutation signal is a measure of the extent that the population difference of the transition levels has recovered between the two pulses. The method is described elsewhere [26]. Using the density matrix equations of motion, we find that the normalized second nutation amplitude grows with pulse delay time τ as

$$S_\infty - S(t=\tau) = e^{-\gamma_a t}\left(1+\frac{\gamma_1}{\gamma_a-\gamma_b}\right)+e^{-\gamma_b t}\left(1-\frac{\gamma_1}{\gamma_a-\gamma_b}\right) \quad (4)$$

where S_∞ is the value at $\tau=\infty$, which we identify with the first pulse nutation amplitude. The quantity γ_1 is the decay rate for the single channel (a)$\rightarrow$(b); from the fluorescence intensities of the lines originating in (a), we estimate that $\gamma_1 \sim 0.1\ \gamma_a$. The rate γ_b is, of course, restricted to collisional processes. Eq. (4) suggests that lower and upper state decay rates can be determined independently.

The two pulse nutation measurements are characterized by essentially a single exponential decay in the pressure range 17-130 mTorr. This implies that upper and lower states depopulate by collisions at essentially the same rate so that $\gamma_b \sim \gamma_a$. It follows from (2) that the short time decay rate is $(\gamma_a+\gamma_b+\gamma_1)/2$. The observed value is

$$(\gamma_a+\gamma_b+\gamma_1)/2 = (0.71 + 0.029\ p)\mu\text{sec}^{-1}, \quad (5)$$

where the I_2 pressure p is in mTorr. From the pressure-independent part, we obtain an upper state radiative lifetime of 1.41 μsec. The value is in reasonable agreement with the previous literature [16] and also agrees with our direct fluorescence decay measurements, giving 1.32 μsec at zero pressure. The pressure-dependent part of (3) yields a total *inelastic collision cross section* $\sigma_I=530\text{Å}^2$. This result is about one order of magnitude larger than previous fluorescence measurements [27], which often are insensitive to upper state vibration-rotation quantum jumps of the emitter. In the pressure regime below 17 mTorr, the lower state seems to decay more slowly than the upper state, in accord with (2), but laser jitter and drift prevent quantitative measurements at present. Frequency-locking the dye laser to the I_2 line of interest should remove this difficulty in the future.

The echo measurements reveal a different aspect of the problem, namely, the degree to which coherence is preserved following collisions. We find that the echo decays exponentially as predicted by (1), and no evidence is found for an e^{-Kt^3} decay law at short times (~100 nsec) which would be symptomatic of velocity-changing collisions [26]. The echo decay rate is

$$\gamma = (0.79 + 0.071p)\ \mu\text{sec}^{-1}\ , \tag{6}$$

with p in mTorr of I_2. Utilizing the pressure-dependent parts of (5) and (6) and the relation $\gamma=(\gamma_a+\gamma_b)/2 + \gamma_\phi$, we obtain the *elastic collision cross section* σ_E=780Å^2 associated with the phase interruption rate γ_ϕ. *We believe this to be the first optical coherence measurement of phase interrupting collisions.* While this information is contained in the optical linewidth, it cannot generally be separated from other causes such as power, Doppler and inelastic collision broadening.

Although several papers [27] have dealt with the I_2 relaxation problem in the past, the measurements have been restricted almost exclusively to the upper state and to inelastic collisions, primarily those that terminate spontaneous emission such as predissociation. Some evidence for quasi-elastic I_2 collisions has appeared recently also [28].

These results may obviously be extended in several different directions - to other optically excited systems and to other coherent transient phenomena, in a manner resembling the elegant methods of pulsed nmr.

We express our gratitude to A. Schenzle, P. R. Berman, K. L. Foster, and D. E. Horne for aid and encouragement.

References

1. N. A. Kurnit, I. D. Abella, and A. R. Hartmann, Phys. Rev. Lett. 13, 567 (1964)
2. R. G. Brewer and R. L. Shoemaker, Phys. Rev. Lett. 27, 631 (1971)
3. P. F. Liao, J. E. Bjorkholm, and J. P. Gordon, Phys. Rev. Lett. 39, 15 (1977); *Laser Spectroscopy*, J. L. Hall and J. L. Carlsten, eds. (Springer Series in Optical Sciences, Vol. 7), Springer-Verlag, N.Y., Heidelberg, 1977
4. A. Szabo, to be published
5. M. M. T. Loy; *Laser Spectroscopy*, J. L. Hall and J. L. Carlsten, eds. (Springer Series in Optical Sciences, Vol. 7), Springer-Verlag, N.Y., Heidelberg, 1977
6. S. B. Grossman, A. Schenzle, and R. G. Brewer, Phys. Rev. Lett. 38, 275 (1977)
7. R. G. Brewer and A. Z. Genack, Phys. Rev. Lett. 36, 1959 (1976)
8. I. J. Lowe and R. E. Norberg, Phys. Rev. 107, 46 (1957)
9. R. L. Shoemaker, A. Stenholm, and R. G. Brewer, Phys. Rev. A 10, 2037 (1974)
10. P. R. Berman, J. M. Levy, and R. G. Brewer, Phys. Rev. A 11, 1668 (1975)
11. J. W. Cooley and J. W. Tukey, Math. Comput. 19, 297 (1965)
12. The echo decay function resulting from velocity-changing collisions has the limiting form $\exp(-Kt^3)$ for short times and $e^{-\Gamma t}$ for long times (see Ref.10).
13. M. Borenstein and W. E. Lamb, Jr., Phys. Rev. A 5, 1311 (1972)
14. T. D. Landau and E. M. Lifshitz, *Quantum Mechanics* (Pergamon, Elmsford, N.Y., 1959) p. 416
15. P. W. Anderson, Phys. Rev. 76, 647 (1949); R. J. Cross, Jr., and R. G. Gordon, J. Chem. Phys. 45, 3571 (1966); R. J. Cross, E. A. Gislason, and D. R. Herschbach, J. Chem. Phys. 45, 3582 (1966)
16. H. M. Pickett, J. Chem. Phys. 61, 1923 (1974); W. K. Liu and R. A. Marcus, J. Chem. Phys. 63, 290 (1975)

17. A. T. Mattick, N. A. Kurnit, and A. Javan, Chem. Phys. Lett. 38, 176 (1976)
18. See C. V. Heer, Phys. Rev. A 13, 1908 (1976) for a previous estimate of the elastic cross section.
19. A. Z. Genack, R. M. Macfarlane, and R. G. Brewer, Phys. Rev. Lett. 37, 1078 (1976)
20. A. Zewail, *Laser Spectroscopy*, J. L. Hall, and J. L. Carlsten, eds. (Springer Series in Optical Sciences, Vol. 7), Springer-Verlag, N.Y. Heidelberg, 1977
21. R. G. Brewer and R. L. Shoemaker, Phys. Rev. A 6, 2001 (1972)
22. J. L. Hall in *Atomic Physics 3*, S. J. Smith, G. K. Walters, and L. H. Volsky (Plenum, New York, 1973) p. 615
23. A. Schenzle, S. Grossman, and R. G. Brewer, Phys. Rev. A 13, 1891 (1976)
24. P. R. Berman and W. E. Lamb, Jr., Phys. Rev. A 6, 2435 (1970); P. R. Berman, Phys. Rev. A 5, 927 (1972), and references therein
25. M. Sargent, III, M. O. Scully, W. E. Lamb, Jr., *Laser Physics* (Addison-Wesley, Reading, Mass., 1974) pp. 85-87
26. J. Schmidt, P. R. Berman, and R. G. Brewer, Phys. Rev. Lett. 31, 1103 (1973); P. R. Berman, J. M. Levy, and R. G. Brewer, Phys. Rev. A 11, 1668 (1975), and references therein
27. A. Chutzian, J. K. Link, and L. Brewer, J. Chem. Phys. 46, 2666 (1967); G. A. Capelle and H. P. Broida, J. Chem. Phys. 58, 4212 (1977); J. I. Steinfeld and A. N. Schweid, J. Chem. Phys. 53, 3304 (1970)
28. D. L. Rousseau, G. D. Patterson, and P. F. Williams, Phys. Rev. Lett. 34, 1306 (1975)

HIGH RESOLUTION STUDIES IN FAST ION BEAMS

M. Dufay and M.L. Gaillard

Laboratoire de Spectrometrie Ionique et Moléculaire
(associé au C.N.R.S.), Université de Lyon 1
69621 Villeurbanne, France

1. Introduction

Atomic physics with accelerated ion beams and lasers was initiated by the work of ANDRÄ and his collaborators in 1973 [1]. The primary motivation for those early experiments was to find a solution for the problem of cascade in atomic lifetime measurements by the fast ion beam technique [2]. It was quite clear that, whenever applicable, selective excitation by laser light would be in many ways superior to previous unselective collisional excitation by gaseous or thin solid targets [3]. More recently, the development of high resolution technique applicable on fast beams has given rise to an interesting combination between fast ion beam technology [4] and more traditional laser spectroscopy methods [5]. We will survey the first few results with emphasis on experimental aspects and possible future developments.

2. Time Resolved Spectroscopy

Well before the first apparition of lasers in accelerator rooms, high resolution spectroscopy had been achieved in fast ion beams by time resolved spectroscopy after foil excitation [6]. Being by construction a pulsed, broad band excitation mechanism, beam-foil satisfies the major prerequisites for quantum beat excitation [7] and indeed, quantum beat frequencies between a few ten MHz up to several tens GHz were detected with fast beams several years before the first laser induced quantum beat measurements in a cell [8].

Given the historical success of time resolved spectroscopy with foil excited fast ion beams, experimentalists were naturally inclined to use the same technique again after substitution of selective c.w. laser interaction in lieu and place of foil collisions. The basic set up is displayed in Fig. 1. Time resolved excitation is achieved in a crossed or inclined beam geometry with overlap at the focal point of a slightly converging laser beam. Spontaneous fluorescence photons are detected through a slit parallel to the laser beam. The slit can be moved along the ion beam direction. With a proper choice of slit width, laser and ion beam geometries as well as crossing angle, it has been possible to experimentally resolve quantum beat frequencies in the GHz range [4]. Since the results of such experiments are already well documented in the literature [9-11], we will only summarize here the main advantages and drawbacks of the technique.

Let us first point out a very general advantage of laser excitation of fast beams, namely Doppler tunability. With ions speeding at a few mm/ns (v/c between 1/1000 and 1/100), the Doppler shift is given by the well known formula

$$\lambda_a = \lambda_l \frac{\left(1 - v^2/c^2\right)^{1/2}}{1 - (v/c)\cos\theta}$$

where λ_a is the apparent wavelength of the laser line in the ion rest frame ;
λ_l the wavelength of the laser in the laboratory frame ;
θ the crossing angle between the ion and laser beam (Fig.1)
and can be as high as several angstroms in the visible.

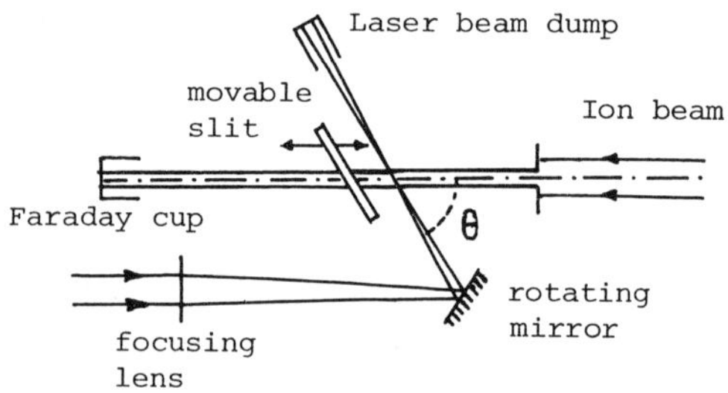

Fig.1 Geometry of time resolved fast beam-laser experiment 1 .

This implies that the tuning of v and eventually of θ is sufficient to give considerable wavelength coverage using fixed wavelength lasers such as commercialized c.w. gas lasers. A further advantage of time resolved spectroscopy, from the point of view of laser technology, is that one can use multimode excitation : the final resolution of the Fourier transform of the quantum beat pattern is independent of the laser linewidth. Both advantages makes quantum beat measurements specially attractive for the spectroscopy of ionized elements since most resonance lines are to be found in the ultraviolet, a spectral range where tunable narrow band laser sources are still scarce whereas a few broadband fixed frequency lasers are already available [12] at least as pulsed sources.

Compared to foil excitation, laser excitation has the further interesting property of providing maximum alignment or orientation of the excited state and thus improves considerably the detectability of quantum beat signals [9]. This is obtained at the cost of a marked decrease in time resolution which limits the applicability of the technique to the measurement of splittings smaller than a few GHz. Other drawbacks are shared by all time differential techniques and have to do with low overall efficiency : low detection efficiency since only a small fraction of the total number of emitted photons is collected through a narrow aperture limiting slit; low excitation efficiency since the time duration of ion-laser interaction in the crossed beam geometry is equal to the transit time of the ions at the laser focus, i.e. well below the excited state lifetime.

The main motivation for recent developments in fast beam laser interaction has been to improve on the sensitivity compared to the quantum beat technique. This has been achieved at the cost of giving up the very high time resolution and relying more specifically on laser properties such as narrow linewidth single mode operation.

3. Fluorescence Spectroscopy

One of the early experiments involving single mode laser excitation on a fast beam was carried out in Berlin by WITTMANN and GAILLARD [13] using the set up of Fig.1. As in the original experiment of ANDRÄ et al. [1], a fast Ba^+ beam was resonantly excited by the 4545 Å line of a 4 Watts argon gas laser. An intra cavity Fabry-Perot etalon was used to reduce the laser linewidth below 1 MHz and Doppler tuning of the resonance was achieved by rotation of the laser beam bending mirror. The fluorescence signal detected at 90° to the ion beam direction during the tuning of the crossing angle θ are given on Fig.2.

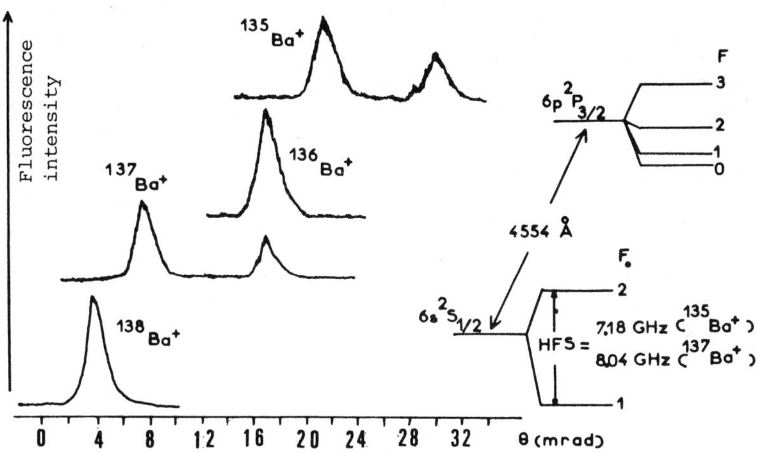

Fig.2 Doppler tuned fluorescence signals obtained with various stable Ba^+ isotopes [13]. The accelerator energy was 320 KeV. The dispersion is of the order of 0.8 GHz/mrad.

The hyperfine structure of the ground state of the odd isotope of Ba^+ is clearly resolved but there is little evidence of the hyperfine structure in the upper state. This is of course due to the Doppler width of the fluorescence which was then measured to be of the order of 1.64 GHz. Such a broad Doppler width is the result of a superposition of several effects :

a/ The laser beam divergence $\Delta\theta$ contributes an effective laser frequency width given by

$$\Delta\nu_1 \sim (v/\lambda)\sin\theta\cdot\Delta\theta$$

i.e. of the order of 1 GHz under the experimental conditions (E = 320 KV, λ = 4545 Å, θ= 30°, $\Delta\theta$ = 1 mrad).

b/ The ion beam divergence $\Delta\phi$ contributes a factor

$$\Delta\nu_2 \sim (v/\lambda)\sin\theta.\Delta\phi$$

Assuming a 1 mrad divergence, this also brings a broadening of the order of 1 GHz.

c/ The distribution of longitudinal beam velocities, which was mainly due in that case to the accelerating high voltage instabilities ($\Delta v/v \sim 10^{-3}$), brings an extra factor $\Delta\nu_3$ given by

$$\Delta\nu_3 \sim (v/\lambda)\cos\theta.\Delta v/v$$

of the order of 1.5 GHz.

From this analysis of the fluorescence line width, it becomes immediately evident that the influence of the first two detrimental effects can be considerably reduced by using a geometry with θ = 0, i.e. a superimposed beam geometry instead of the original crossed or inclined beam geometry. Remains the third contribution which needs further treatment. The high voltage instability which was singled out in the previous experiment as the main contribution to the beam velocity distribution can easily be reduced by use of a better high voltage supply. One must however point out that it is rather difficult to achieve an accelerator energy stability much better than 10^{-4} in the hundred KV range. The only solution is then to use lower energy beams for example in the few KeV range where state of art high voltage supplies have 10^{-6} stability. Under such conditions, the contribution of voltage unstabilities becomes negligible and the residual longitudinal velocity distribution in the beam can become extremely small due to the effect of velocity bunching as recently pointed out by KAUFMAN [14] and independently by WING et al. [15].

This phenomenon is quite specific to accelerated ion beams and has already been used in atomic collision physics. It is most readily explained by considering two identical ions of mass m and charge q in a source at temperature T having velocity components in the z direction of v_i = 0 and $v_i' = (2kT/m)^{1/2}$. After acceleration along the z direction by an electrostatic potential drop V, their final velocities are respectively

$$v_f = (2qV/m)^{1/2} \qquad \text{and} \qquad v_f' = (v_f^2 + v_i'^2)^{1/2}$$

The kinematic compression of the velocity distribution which occurs upon acceleration is then measured by the ratio

$$\Delta v_i/\Delta v_f = 2(qV/kT)^{1/2}$$

which can easily be as big as 10^3 for a source temperature of several thousands degrees and accelerating voltage of several kilovolts. Of course, this compression occurs only for the longitudinal component.

In conclusion, narrow fluorescence line width should be observable in a superimposed beam geometry, when both ion and laser beams travel along the same z-axis. The discussion by KAUFMAN [14] shows that residual second order effects limit the resolution in the few MHz range assuming a perfectly stable high voltage supply and laser line. Several experiments have been set up in order to measure this effect [16] and convincing evidence has been reported recently [15], including the observation of nicely resolved hyperfine structure patterns in the fluorescence of odd isotope Xe^+ beams [17]. Those first few results clearly indicate that the direct fluorescence technique is quite suitable for high sensitivity, high resolution spectroscopy in the superimposed ion-laser beam geometry. Some of the practical limitations were however pointed out in the discussion, the main concern here being that at high energies, accelerating voltage instabilities will wash out the velocity bunching effect and thus considerably broaden the fluorescence line profile.

4. "In-Flight" Lamb-Dip Spectroscopy

Since high resolution is difficult to achieve on a high energy beam by direct fluorescence, it became quite attractive in that case to search for laser induced non linear effects such as saturated absorption or three level resonances which are by now well established tools for sub-Doppler spectroscopy [18]. Although it is quite clear after the successful experiments of HALL and collaborators, which have been reported independently during this conference [19], that saturated absorption signals can be observed in a crossed beam geometry, we tend to favour again the superimposed beam geometry for obvious reasons of sensitivity and Doppler tunability.

Whereas the fluorescence technique makes little use of the ion motion, the "in flight" Lamb-dip technique [20] takes full advantage of the time resolution along the beam in order to single out two successive resonant interaction zones between the moving ions and the forward propagating laser field as indicated on Fig.3. In the first zone A, the single mode laser interaction is

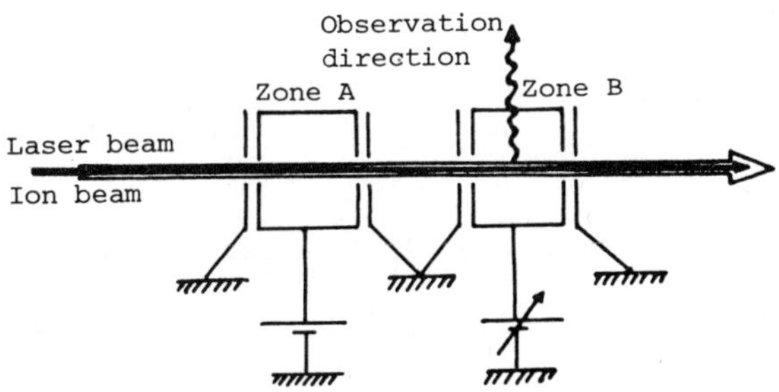

Fig.3 Schematic "In flight" Lamb-dip set up

used in order to "label" one velocity class in the otherwise broad longitudinal beam velocity distribution. In the second zone B, this labelling is probed by a second laser interaction. Critical for the success of the experiment is the fact that the mean beam velocity can be altered quite abruptly along the ion beam pass by simple application of accelerating or decelerating fields over very short distances [21]: it is possible to modify the Doppler tuning of the laser during the time of flight of a given ion.

Up to now, the experiment has been carried out with three level systems of the folded type shematized on Fig.4. The mean velocity of the beam is adjusted so that in zone A the laser resonates at frequency ω_{12} with a given velocity class. During the interaction, optical pumping occurs out of level 1 into level 3 and the "labelled" velocity class appears as a "hole" in the velocity distribution of ions in level 1 and a corresponding "spike" in the velocity distribution of level 3. The probing can be done by tuning the mean velocity of the beam in zone B and monitoring the fluorescence which then displays on top of the broad Doppler profile a dip (respectively a spike) as the laser is Doppler tuned across the ω_{12} transition (respectively ω_{23} transition). Quite clearly, the complete experiment is just the "in flight" version of such well known saturation spectroscopy techniques as "Lamb dip" [22] or "inverted Lamb dip" [23], detected on the fluorescence side light as first demonstrated by FREED et al.[24].

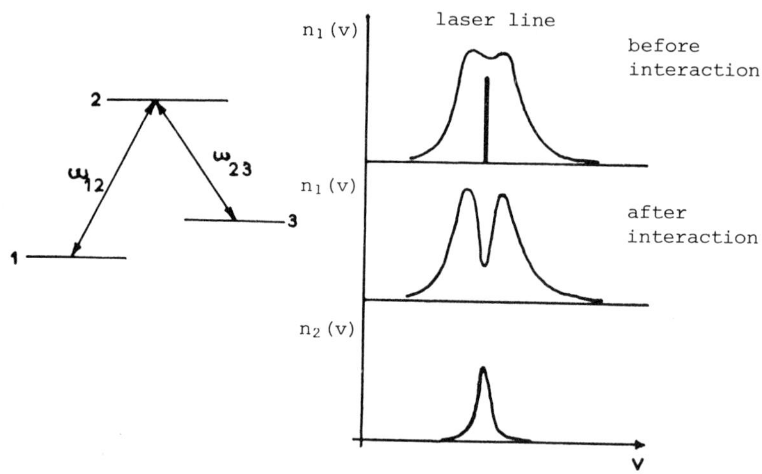

Fig.4 Three level system and hole burning model. $n_1(v)$: number of ions in level 1 with velocity v. $n_2(v)$: number of ions in level 2 with velocity v.

As a test for the "in flight" experiment, we used again a single mode argon laser for resonant excitation of a $^{138}Ba^+$ beam [20]. The direct fluorescence signal obtained by Doppler tuning in a single zone (Fig.5a) clearly displays the broad velocity distribution of the beam with the characteristic signature of the high frequency ripple of the accelator high voltage supply. With two zones (Fig.5b), optical pumping from the 6s $^2S_{1/2}$ ground state into the metastable 5d ^2D state manifests itself as expected by a narrow dip on the fluorescence of the beam in the second zone. A complete theoretical analysis of the experimental situation has been carried out [25]. Of main relevance for high resolution spectroscopy are the results summarized on Fig.5c which shows the variation with laser power of the HWHM of the dip, basically limited by the natural lifetime of the upper level of the resonant transition.

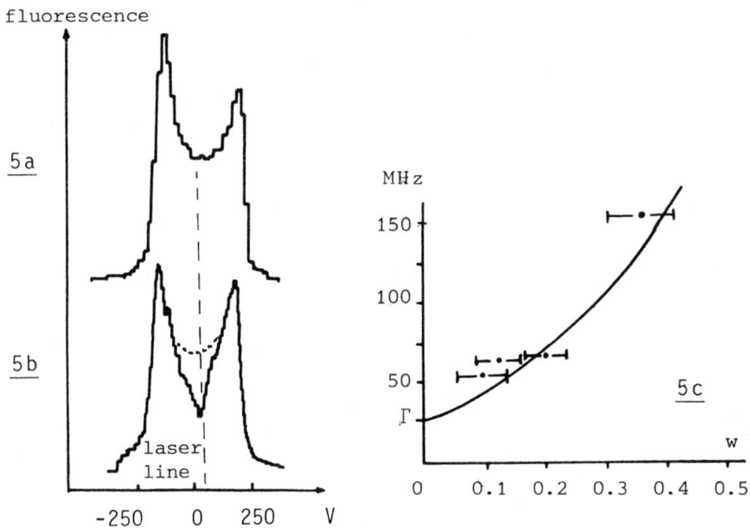

Fig.5 Experimental observation of the hole drilled in the velocity distribution of the ground level.
5a : Fluorescence in zone B with zero applied voltage in zone A.
5b : Fluorescence in zone B with hole burning in zone A (dispersion 2.6MHz/V).
5c : Width of the Lamb dip as a function of laser power ($w = 5.43\sqrt{P}$, P watts); theoretical curve from [25]; dots are experimental points.

The first spectroscopic application of this technique to the analysis of the hyperfine structure of the 6p $^2P_{3/2}$ level of the odd isotopes of Ba^+ has already been published [26] and will not be discussed here. As an example of application we will rather concentrate on the results of a further experiment using fast Rb^+ ions. Figure 7 gives a schematic description of the experimental set up and Fig.6, the relevant part of the energy level diagram of Rb^+. As demonstrated by previous measurements on fast beams [27], it is possible to populate in a Rb^+ beam the metastable level 5s 3/2 $[3/2]^0_2$ by collision in a gas cell. The Ar^+ laser line at λ4764.862 Å can then be Doppler tuned into resonance with the ionic transition at λ4775.949 Å [28] in the superimposed beam geometry when the ion velocity reaches 0.697 mm/ns (i.e. at a total accelerating energy of 218.64 KeV for $^{87}Rb^+$ or 213.62 KeV

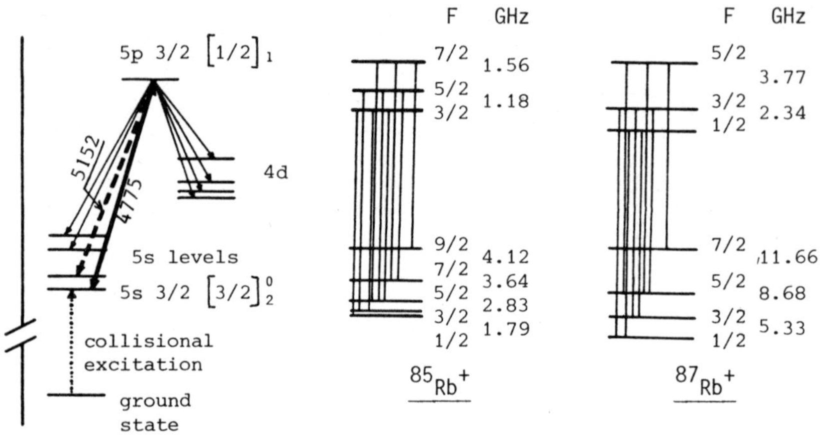

Fig.6 Part of the level diagram of Rb II (from [28]) and hyperfine structure of the 5s 3/2 $[3/2]^0_2$ - 5p 3/2 $[1/2]_1$ transition in both stable isotopes.

for $^{85}Rb^+$). Decay of the ions excited to the 5p 3/2 $[1/2]_1$ level occurs either back to the metastable 5s and 4d levels or via radiative cascades down to the ground state. The beam fluorescence is monitored through an interference filter centered on the transition 5s 3/2 $[3/2]^0_1$ - 5p 3/2 $[1/2]_1$ at λ5152.115 Å.

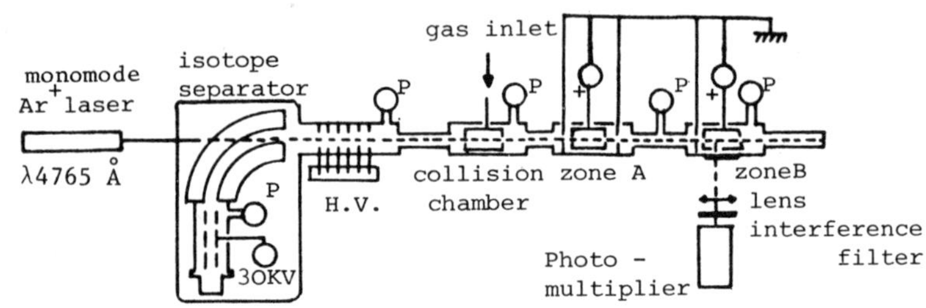

Fig.7 Experimental set up used for the Rb^+ experiment. P : vacuum pumps ; H.V. : accelerator high voltage.

Typical results of direct fluorescence experiments obtained for the two Rb^+ stable isotopes by beam velocity tuning are displayed on the top of Fig.8. In flight Lamb dip measurements were then carried out by Doppler tuning the laser in zone B to the center of one of the fluorescence hyperfine components and then scanning the beam velocity in zone A. This procedure has the built-in advantage of displaying the dips and peaks of the Lamb dip experiment on a flat background. It appears quite clearly on the panoramic scans of Fig.8 that the observed pattern depends strongly upon the selection of the probing transition in zone B. High precision scans were carried out for each isotope and for

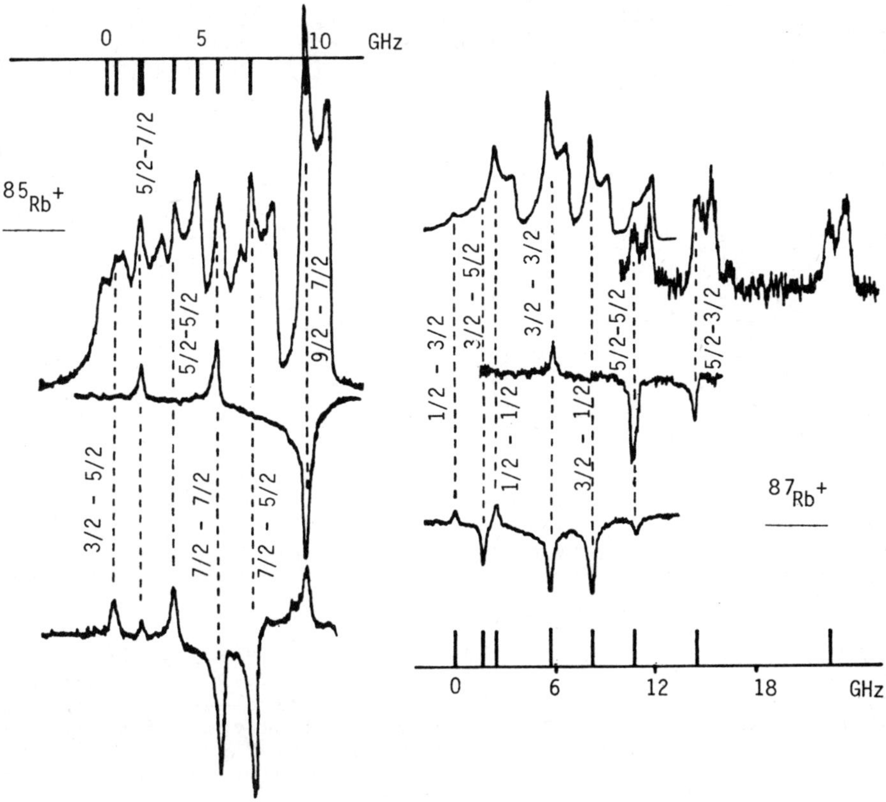

Fig.8 Experimental results on the hyperfine structure of the 5s 3/2 $[3/2]^0_2$ - 5p 3/2 $[1/2]_1$ transition in $^{85}Rb^+$ and $^{87}Rb^+$. Upper trace : direct fluorescence. Lower two traces : Lamb dip signals with various choice of probing transition. Also indicated is the position of each hyperfine transition on a frequency scale.

each possible choice of the probing transition at low laser power in order to minimize the power broadening. Due to a compromize between line narrowing and fluorescence count rate, most scans were recorded with a FWMH of 150 MHz although the ultimate achievable line width is probably much smaller. The dip (and peak) positions were measured on a voltage scale which corresponds to the difference of electrical potential between zone A and zone B, i.e. to the difference in energy of the ion in the two zones. In order to convert the dip (or peak) intervals into absolute energy splittings, we need an accurate dispersion factor to be deduced from the Doppler formula, namely

$$\Delta\nu = - \frac{\lambda_1}{\lambda_a^2} \frac{c + v}{m_o c v} \Delta E$$

The final results on the hyperfine splittings in both the 5s 3/2 $[3/2]^0_2$ and 5p 3/2 $[1/2]_1$ levels are summarized in Table 1 and compared with earlier measurements [29] obtained by classical optical spectroscopy.

Table 1 Hyperfine splittings in Rb^+ 5s and 5p levels (GHz)

Level	$^{85}Rb^+$	$^{87}Rb^+$	Previous measurements
5s 3/2 $[3/2]^0_2$			
F 1/2 → 3/2	1.790 ± 0.011	5.333 ± 0.016	
3/2 → 5/2	2.826 ± 0.009	8.682 ± 0.025	
5/2 → 7/2	3.640 ± 0.011	11.665 ± 0.033	
7/2 → 9/2	4.118 ± 0.012		
5p 3/2 $[1/2]_1$			
F 1/2 → 3/2		2.335 ± 0.007	2.36 [a]
3/2 → 5/2	1.176 ± 0.005	3.778 ± 0.012	3.81 [a]
5/2 → 7/2	1.558 ± 0.005		

a : Ref.29

One obvious advantage of the technique is that it does not rely on laser wavelength tuning and is in fact rather insensitive to laser frequency shifts. Furthermore it is clear that fluctuations in the absolute accelerating voltage do not affect the final accuracy on level intervals since the relative positions of the observed dips or peaks depends only on the difference in voltage between the labelling zone and the probing zone. On the negative side, the fast beam version of Lamb dip spectroscopy is only suitable for the measurement of relative energy level positions and not for absolute wavelength determination. Isotopic shift measurements for example should rather be carried out by direct fluorescence on a velocity bunched beam.

5. Final Remarks

Although high resolution spectroscopy with laser excited fast ion beams may appear a relatively new outgrowth of laser spectroscopy, it has not been possible to cover, within this introductory survey, all aspects of the field. At least, a brief mention should be made of infrared-X-ray double resonance study of the Lamb shift in high Z ions [30] or of Rydberg state studies in Hydrogen and Helium [31] which pioneered the use of the superimposed beam geometry.

Further methodological developments are to be expected in the near future : it has been suggested that fast beams could be used in order to achieve resonance enhancement in Doppler free two-photon spectroscopy [32] or in order to observe spatially resolved transient effects such as optical nutation and free precession

[21,33] . Fast beam techniques should permit widespread use of non optical detection techniques as demonstrated in Ref. 15,25,31.

The applicability of the various methods is of course limited by the available laser wavelength range, but as demonstrated here in the Rb^+ experiment, as well as in many previous laser experiments [2,17,27,30,31] , it is possible to extend considerably the range of application by use of a combination of collisional and laser excitation in cells. Application to fast beam of neutral atoms, singly and multiply ionized elements [34] as well as singly ionized molecules [15] have already been published. Feasibility of pulsed laser excitation has also been checked [35].

As suggested by OTTEN [16], "in flight" spectroscopy technique should be quite suitable for the study of fast unstable isotope beams obtained with "on line" isotope separators near reactors or high energy accelerators. The experience already gathered with stable ion beams indicates that the high density of laser excitation techniques could be used to study flux of unstable isotopes as low as 10^7 or 10^6 ion/s. In order to underline the difficulty of the measurement, let us point out that, in such a beam of singly charged ions (mass around 140) accelerated up to 25 KeV, there will be no more than a handful of ions for every meter of the beam path (to be compared against the residual gas density of 10^{10} molecules/cm^3 for a beam line vacuum of about 10^{-6} Tor.).

6. References

1 H.J.Andrä, A.Gaupp, K.Tillman, W.Wittmann, Nucl.Inst.Methods 110, 453 (1973).

2 H.Harde, G.Güthohrlein, Phys.Rev.A 10, 1488 (1974).

3 I.Martinson, A.Gaupp, Phys.Reports 15 C, 114 (1974).

4 H.J.Andrä, Proceedings of the 4th Int. Conf. on Atomic Physics in "Atomic Physics 4", Plenum Press, p.635 (1975).

5 J.J.Snyder,J.L.Hall, Proceedings of the 2nd Int. Conf. on Laser Spectroscopy, Springer Verlag, p.6 (1975).

6 H.J.Andrä, Phys.Rev.Let. 25, 325 (1970).

7 H.J.Andrä, Physica Scripta 9, 257 (1974).

8 S.Haroche, in "High Resolution Laser Spectroscopy, edited by K.Shimoda, Springer Verlag (1976).

9 L.Henke, Diplomarbeit (unpublished), Freie Univ.Berlin (1975).

10 M.Kraus, Diplomarbeit (unpublished), Freie Univ.Berlin (1975).

11 H.J.Andrä, in "Beam Foil Spectroscopy", edited by I.A.Sellin and D.J.Pegg, Plenum Press, p.835 (1975).

12 J.J.Ewing, C.A.Brau, in "Tunable Lasers and Applications", edited by A.Mooradian, T.Jaeger and P.Stokseth, Springer Verlag, p.21 (1976).

13 W.Wittmann, M.L.Gaillard, unpublished material (1974), see Ref.4.

14 S.L.Kaufman, Optics Comm. 17, 309 (1976).

15 W.H.Wing, G.A.Ruff, W.E.Lamb, J.J.Spezeski, Phys.Rev.Let. 36, 1488 (1976).

16 E.W.Otten, Proceedings of the 5th Int. Conf. on Atomic Physics in "Atomic Physics 5", Plenum Press, p.239 (1977).

17 T.Meier, H.Hünnermann, H.Wagner, Optics Comm. 20, 397 (1977).

18 See Ref. 8.
19 J.C.Bergquist, S.A.Lee, J.L.Hall, these proceedings.
20 M.Dufay, M.Carré, M.L.Gaillard, G.Meunier, H.Winter, A. Zgainski, Phys.Rev.Let. 37, 1678 (1976).
21 H.Winter, M.L.Gaillard, Zeit.Phys. A 281, 311 (1977).
22 R.A.Macfarlane, W.R.Bennett Jr, W.E.Lamb, Appl.Phys.Let. 2, 189 (1963).
23 P.H.Lee, M.L.Skolnick, Appl.Phys.Let. 10, 3641 (1967).
24 C.Freed, A.Javan, Appl.Phys.Let. 17, 53 (1970).
25 F.Beguin, M.L.Gaillard, H.Winter, G.Meunier, J.Phys.Paris (in press).
26 H.Winter, M.L.Gaillard, J.Phys. B London (in press).
27 M.L.Gaillard, H.J.Andrä, A.Gaupp, W.Wittmann, H.J.Plohn, J.O. Stoner, Phys.Rev. A 12, 987 (1975).
28 J.Reader, G.Epstein, J.Opt.Soc.Am. 63, 1153 (1973).
29 H.Kopfermann, A.Steudel, J.O.Trier, Zeit.Phys. 144, 9 (1956).
30 H.W.Kugel, M.Leventhal, D.E.Murnick, E.K.N.Patel, O.R.Wood, in "Proceedings of the 2nd Int. Conf. on Laser Spectroscopy", Springer Verlag, p.465 (1975).
31 P.M.Koch, L.D.Gardner, J.E.Bayfield, in "Beam Foil Spectroscopie" edited by I.A.Sellin and D.J.Pegg, Plenum Press, p.829 (1976).
32 R.Salomaa, S.Stenholm, Opt.Comm. 16, 292 (1976).
33 L.Allen, B.Allen, P.L.Knight, Opt.Comm. 20, 150 (1977).
34 J.D.Silver, N.A.Jelley, L.C.McIntire in abstracts of the 4th Int. Conf. of Atomic Physics (Berkeley), p.205 (1976).
35 M.L.Gaillard, H.J.Plöhn, H.J.Andrä, D.Kaiser, H.H.Shultz in "Beam Foil Spectroscopy" edited by I.A.Sellin and D.J.Pegg, Plenum Press, p.853 (1976).

LASER SATURATION SPECTROSCOPY IN THE TIME-DELAYED MODE [1]

M. Ducloy [2]

Laboratoire de Physique des Lasers, Université Paris-Nord
93430 Villetaneuse, France

and

M.S. Feld [3]

Department of Physics and Spectroscopy Laboratory
Massachusetts Institute of Technology, Cambridge, MA 02139, USA

ABSTRACT

This paper describes a new class of techniques, called time-delayed laser saturation spectroscopy, which combine frequency and time-domain methods of laser spectroscopy to provide a way of studying a molecular system as it evolves from an initially-prepared stationary state to a second, final state. The specific example analyzed here is three-level free induction decay, in which the time dependent gain of a Doppler-broadened molecular transition is probed after the sudden termination of an intense field resonating with a coupled transition. The calculated lineshape features expected under different experimental conditions are described, and some of these are demonstrated in NH_3. The experiments clearly separate the contributions of population saturation and Raman-type processes in the time evolution of the lineshapes, and yield the first measurement of the alignment relaxation rate in the ground electronic state of a molecule.

1. Introduction

Over the past decade powerful laser saturation techniques for inducing narrow resonances in Doppler-broadened systems have become a major tool for studying atomic and molecular structure in the frequency domain [1]. On the other hand, the recently developed coherent transient techniques have been very useful in obtaining new information about laser interactions and relaxation processes in the time domain [2]. The purpose of this paper is to show that these two approaches can be merged to form a new class of techniques which yields information in both fre-

[1] Supported in part by NSF and ARO (Durham).

[2] This work was performed while the author was a Visiting Scientist at MIT.

[3] Alfred P. Sloan Research Fellow.

quency and time domains and thus extends the range of available information.

As an example, consider a conventional saturated absorption experiment in which the saturation of a Doppler-broadened transition by an intense monochromatic field is observed by studying the narrow resonance induced in the transmission of a tunable probe beam (Fig. 1.).

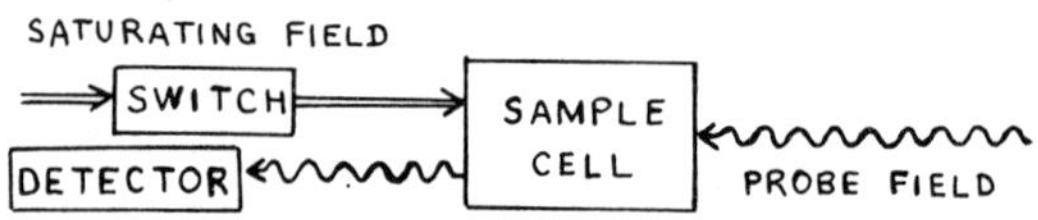

Fig.1 Example of the time-delayed technique. The configuration shown is that of time-delayed Lamb dip.

Now suppose that the saturating field is suddenly switched off, and the lineshape of the narrow resonance is probed a fixed interval of time later. As the time delay is increased the change signal will become smaller, corresponding to the decay of the saturated molecules and their return to equilibrium. Furthermore, as shown below the *shape* of the resonance may evolve from its steady state form. This information can be combined to form a surface in a coordinate system having axes: probe intensity (z-axis), frequency detuning (x-axis), time delay (y-axis) (Fig. 2).

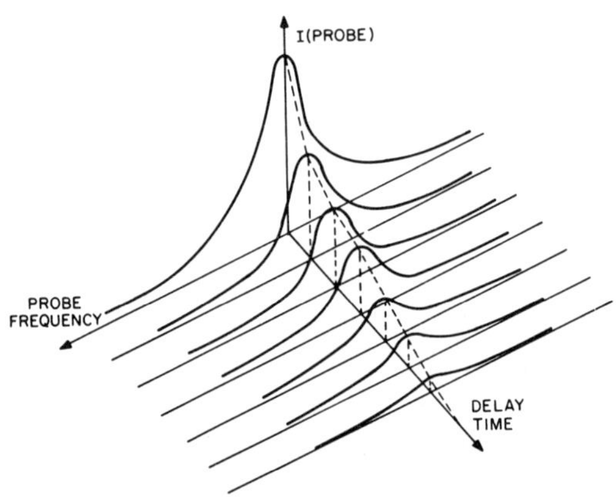

Fig.2 Three dimensional representation of the time-delayed change signals.

Sections parallel to the x-axis give the resonance lineshape at delayed times. Similarly, sections parallel to the y-axis give the free decay of the system at various values of frequency detuning.

Experiments of this type are inherently different from conventional free induction decay experiments [2], where the probe field does not interact with the molecules and serves only as a local oscillator for heterodyne detection. Thus, in such experiments there is no resonant behavior as the probe field is tuned. For the same reason the new technique is not an analog of pulse Fourier transform spectroscopy [2], and the time-delayed lineshapes may bear no relationship to the corresponding decay times. In fact, different portions of the lineshape may decay at different rates, resulting in a deformation of the overall lineshape. Lineshape evolution of this type becomes particularly important when the physical processes contributing to the change signal decay at different characteristic rates (*e.g.* T_1 processes vs. T_2 processes), or the lineshape exhibits Ramsey-type fringes due to decay of a phase-coherent contribution. Other features of the time-delayed change signals which occur at high saturation field intensities include power broadening and dephasing, and oscillatory behavior related to the dynamic Stark effect.

The wide range of techniques to which time-delayed saturation spectroscopy can be applied includes free decay, optical nutation and photon echoes in two and three level systems (both cascade and folded), for probe and saturating waves either co- or counter-propagating [3]. In the following we shall concentrate on free decay in Doppler-broadened three level systems.

2. Theory of Free Decay in Three Level Systems [4,5]

When an intense, monochromatic laser field, E_2, frequency Ω_2, resonates with one of the transitions (0-2, see Fig. 3) of a Doppler-broadened gas, the populations of levels 0 and 2 are altered over a narrow range of axial velocities centered about v_2, satisfying the resonance condition $\Omega_2 - k_2 v_2 = \omega_2$ (ω_2 molecular frequency, $k_2 = \Omega_2/c$).

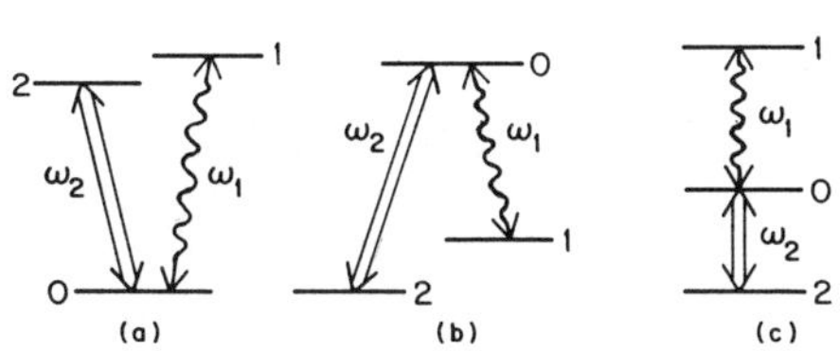

Fig.3 Energy level diagrams.

This resonant change in the velocity distribution can manifest itself in the spectral profile of a Doppler-broadened transition, 0-1, sharing a common level with 0-2. If a weak monochromatic field, $E_1(\Omega_1)$, colinear with the intense field, probes the 0-1 transition, center frequency ω_1, a sharp change in transmission will occur when ω_1 is tuned into resonance with the molecules of velocity v_2:

$$\Omega_1 = \Omega_1(\varepsilon) \text{ where } \Omega_1(\varepsilon) = \omega_1+\varepsilon k_1(\Omega_2-\omega_2)/k_2 \tag{1}$$

(ε = +1 and -1 for co-propagating and counter-propagating waves, respectively). This effect, called laser induced line narrowing, has been the subject of numerous theoretical and the experimental investigations devoted to studying the steady state change signals at the probe transition [1]. It is now well known that the change signal lineshape cannot be accounted for in terms of population considerations alone, and that coherent Raman-type processes such as two-photon transitions play an important role [6]. Thus, for example, the widths of the change signals in the forward (ε = +1) and backward (ε = -1) directions can differ considerably. This and other lineshape asymmetries have been useful in extracting detailed information about collisional and radiative decay processes.

Our purose is to analyze the transient behavior of the change signal lineshape when the saturating field is suddenly terminated at time t = 0. In the thin sample approximation the probe change signal is proportional to the velocity-integrated value of Im(P), with $P(\Omega_1)$ the amplitude of the optical polarization induced by the probe field. In the slowly-varying envelope approximation P obeys an oscillator equation of the type [4,5]

$$\dot{P}+LP = i\mu_1^2E_1(n_1-\sigma_{00})+i\mu_1\mu_2E_2\sigma_{21}, \tag{2}$$

where $L = \gamma_{01}+i(\Omega_1-\omega_1-k_1v)$ and $\hbar = 1$; n_j is the thermal (or background) population of level j; μ_j is the 0-j dipole moment matrix element; σ_{ij} is the envelope of the i-j density matrix element (σ_{00}, population of level 0 as influenced by E_2; σ_{21}, coherence induced by Raman-type transitions) and γ_{ij} is its relaxation rate. As can be seen, P can be excited by both population and Raman-type driving terms.

The solution of (2) in the transient regime (E_2 = 0, t>0) is given by

$$P(t) = P_\ell+P(0)e^{-Lt}-i\mu_1^2E_1\int_0^t\Delta n(t')e^{-L(t-t')}dt', \tag{3}$$

where P_ℓ is the linear (unsaturated)polarization, P(0) the initial value (t = 0) of the saturated polarization, and

$$\Delta n(t) = [\sigma_{00}(0)-n_0]e^{-\gamma_0 t}, \tag{4}$$

with γ_0 the population decay rate of level 0. The last term in (3) describes the coupling of P with σ_{00}. Since $E_2 = 0$ for t>0, transient Raman processes do not occur (as they would in three level optical nutation). But their influence is contained in the initial polarization, along with that of the saturated population of level 0:

$$P(0) = i[-\mu_1^2 E_1 \Delta n(0) + \mu_1 \mu_2 E_2 \sigma_{21}(0)]/L. \quad (5)$$

The resulting probe lineshapes, obtained by integrating (3) over velocity, depend on the relative direction of the two waves.

In the following the lineshape features will be described for a folded three level system (Figs. 3a,b). In the cascade case (Fig. 3c) the lineshapes are the same, except that the roles of forward and backward signals are interchanged.

2.1 Counter-Propagating Waves ($\varepsilon = -1$)

For weakly saturating pump field the lineshape is given by the real part of g ($\varepsilon = \pm 1$):

$$g(-) = \frac{e^{-[\Gamma + i\delta(-)]t}}{\Gamma + i\delta(-)} + \frac{e^{-\gamma_0 t} - e^{-[\Gamma + i\delta(-)]t}}{\Gamma - \gamma_0 + i\delta(-)}, \quad (6)$$

where $\delta(-)$ is the probe field detuning $[\delta(\varepsilon) = \Omega_1 - \Omega_1(\varepsilon)]$, and $\Gamma = \gamma_{01} + k_1\gamma_{02}/k_2$. The first term in g(-) describes the decay of the initial polarization. Its decay rate, Γ, consists of two terms, the ordinary relaxation rate, γ_{01}, and a Doppler dephasing contribution, $k_1\gamma_{02}/k_2$. This latter term is due to the velocity spread $\Delta v = \gamma_{02}/k_2$ of the excited molecules, which gives rise to a corresponding spread in the reradiated frequencies.

The second term in (6) comes from the coupling of P with the saturated level population, and decays at characteristic rate γ_0.

The first term exhibits Ramsey-type fringes and the associated line narrowing, as is characteristic of the time-delayed lineshape resulting from the decay of a phase-coherent polarization. This type of behavior can be observed in g(-) for $\Gamma \lesssim \gamma_0$. As seen in Fig. 4, for $t \gtrsim 1/\Gamma$ the line narrows below its natural width.

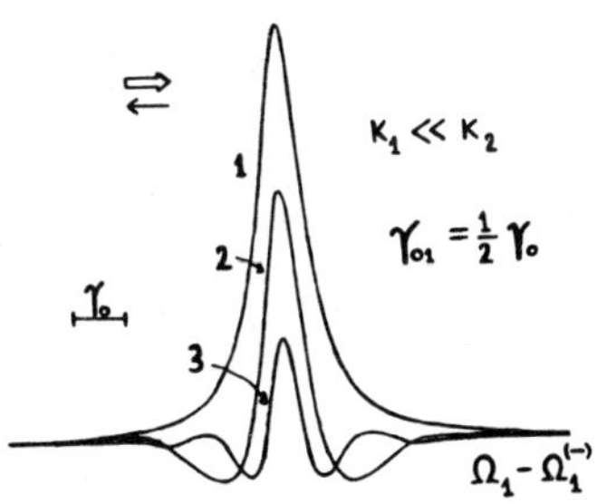

Fig.4 g(-) change signal for $\Gamma = \gamma_0/2$. Time delays: $t_1 = 0$, $t_2 = 2/\gamma_0$, $t_3 = 4/\gamma_0$.

This extreme narrowing is due to the selection of long-lived molecules by the time-delayed measurement process.

In the case of $\Gamma>\gamma_0$ the polarization contribution is dominated by the population term for $t\gtrsim 1/\Gamma$, leading to a different type of behavior. There are no fringes and the line remains Lorentzian, narrowing from an initial width Γ to an eventual width $\Gamma-\gamma_0$ at $t>1/\Gamma$.

In contrast, when there are strong phase-changing collisions $[\gamma_{ij}>>(\gamma_i+\gamma_j)/2]$ the lineshape is not deformed during the transient evolution (Fig. 5b).

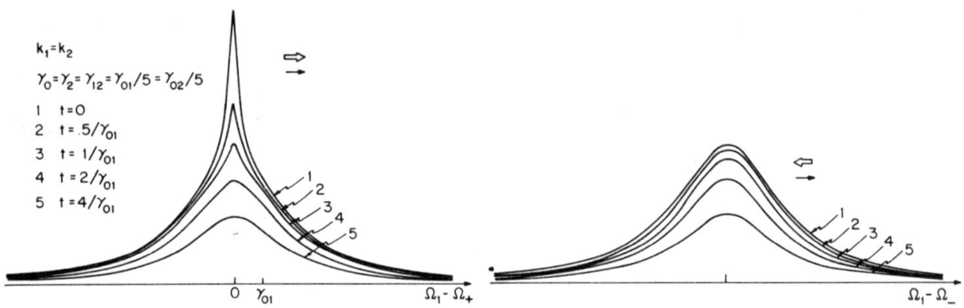

Fig.5 Forward (a) and backward (b) change signals when phase-changing collisions predominate.

The resonance, a Lorentzian of width Γ, then decays with the population time constant:

$$g(-) \simeq e^{-\gamma_0 t}/[\Gamma+i\delta(-)]. \tag{7}$$

Finally, note that in all cases Raman-type contributions are absent in g(-), as can be seen by the absence in (6) of terms containing γ_{12}, the Raman-coherence decay rate. These contributions cancel in the velocity integration because of the destructive interference arising from their strong velocity dependence.

2.2 Co-Propagating Waves (ε = +1)

In contrast to 2.1, in the forward direction the change signal is influenced by Raman-type processes, which are not averaged out in the velocity integration. It also depends on whether $k_2 \gtrless k_1$.

For $k_2 \leq k_1$ and no phase-changing collisions $[\gamma_{ij} = (\gamma_i+\gamma_j)/2]$ the lineshape [4] is a single Lorentzian of width characterized by Raman-type processes (and therefore narrower than the $\varepsilon = -1$ case), which is undistorted during the decay. However, this is not the case when phase changing collisions dominate ($\gamma_{01},\gamma_{02} >> \gamma_0,\gamma_{12}$). For example, in the important special case of $k_1 \simeq k_2$ the change signal lineshape is given by

$$g(+) \simeq \frac{e^{-\gamma_0 t}}{\Gamma + i\delta(+)} + \frac{\gamma_0}{\Gamma}\frac{e^{-\Gamma t}}{\gamma_{12} + i\delta(+)} \quad . \tag{8}$$

The first term describes a broad resonance, induced by population saturation, which decays at characteristic rate γ_0. It is identical to the corresponding g(-) lineshape, (7). The second term is a narrow resonance of width γ_{12}, induced by Raman-type processes, which decays at the much faster rate $\Gamma = \gamma_{01}+\gamma_{02}$, characteristic of the decay of the initial polarization. This leads to the remarkable conclusion that after a time $\sim 1/\Gamma$ the narrow contribution decays away and the forward change signal evolves to a broad Lorentzian identical to that of the backward signal, (6). An example is shown in Fig. 5a.

When $k_2 > k_1$ there is an additional contribution to the change signal, due to the decay of a new velocity group prepared in the steady state ($t \leq 0$). This velocity group, associated with the resonance condition for Raman-type processes, is centered at v_{12}, defined by $\Omega_2 - \Omega_1 - (k_2 - k_1)v_{12} = \omega_2 - \omega_1$. (This is the energy conservation condition for $2 \to 1$ two quantum transitions.) This extra term, which exists even when the medium is transparent to the saturating field ($n_2 = n_0 \neq n_1$), is proportional to the real part of

$$g'(+) = \kappa \frac{e^{-[\gamma + i\delta(+)]t/\kappa}}{[\gamma + i\delta(+)]^2} \quad , \tag{9}$$

with $\kappa = (k_2 - k_1)/k_2$ and $\gamma = \kappa\gamma_{01} + k_1\gamma_{02}/k_2$. As can be seen in Fig. 6, the corresponding time-delayed lineshape exhibits Ramsey-type fringes, as is characteristic of a pure phase-coherent contribution.

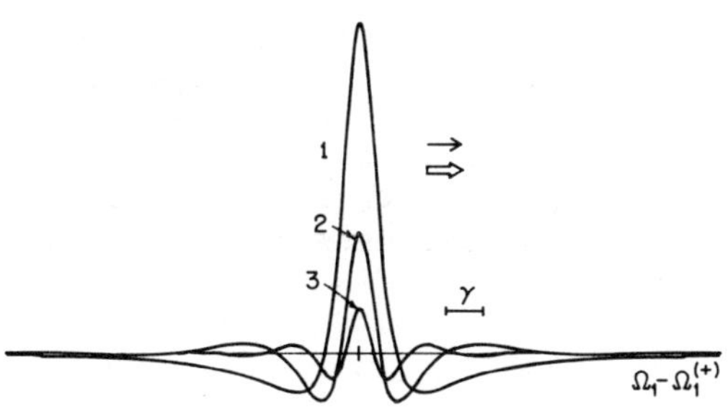

Fig. 6 g(+) change signal for $n_2 = n_0 \neq n_1$, $k_2 = 2k_1$, $\gamma_{ij} = \gamma$. Time delays: $t_1 = 0$, $t_2 = 0.5/\gamma$, $t_3 = 1/\gamma$.

2.3 Saturation Effects

The above results describe the case in which the saturation induced by the intense laser field is small. The additional lineshape effects occurring in the case of strong saturation include power broadening and dephasing, and dynamic Stark splittings exhibiting a novel type of oscillatory decay. A complete discussion of these effects in given in Ref. 5.

3. Experiments in NH_3 [ν_2 asQ(8,7) transition]

3.1 Experimental set-up (Fig. 7)

In the experiments the ν_2 asQ(8,7) transition of NH_3 was saturated and probed using two c.w. N_2O lasers oscillating on the P(13) line (λ = 10.78 μm), which falls within the NH_3 Doppler profile.

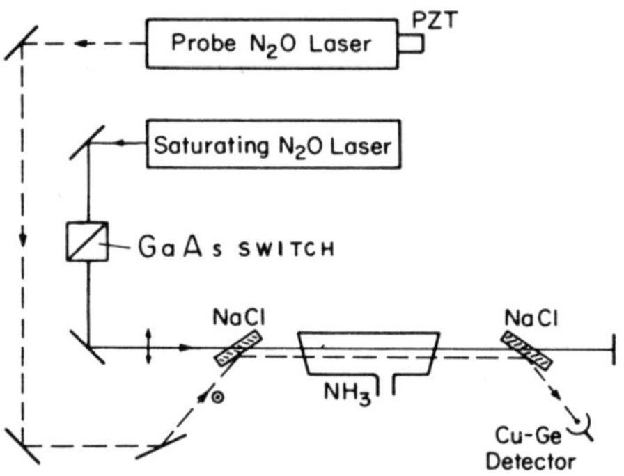

Fig.7 Experimental set-up. The GaAs switch was absent in the steady-state experiments.

A 10 cm glass sample cell with NaCl end windows was used. Holding the frequency of the saturating N_2O laser fixed, the probe was tuned (tuning range, ±35 MHz) by means of calibrated PZT. A flip mirror was used to reverse the direction of the saturating beam and thereby select forward or backward configurations. The two laser beams were linearly polarized at right angles, and Brewster angle NaCl beam splitters were used to overlap the beams before the cell and separate them afterwards. The probe beam was monitored using a He-cooled Cu-Ge detector. Steady-state experiments were performed by chopping the saturating beam at a low frequency (~1 kHz) and using phase sensitive detection. The transient change signals were observed by turning on and off the saturating beam with an external electro-optic modulator, a GaAs crystal to which high voltage square pulses were applied (risetime 30 ns, duration 10 μs, repetition rate 1 kHz), thus inducing a fast rotation of the polarization of the c.w. saturating beam. A subsequent analyze , a Brewster angle silicon plate, yielded the square pulses of the saturating beam. Probe signals were analyzed with a Boxcar integrator operated in two different modes: either monitoring the decaying signal at a fixed probe frequency, or scanning the probe frequency at fixed time delays.

3.2 Steady-State Lineshapes [7]

Experimental lineshapes for co- and counter-propagating waves are shown in Figs. 8a,b.

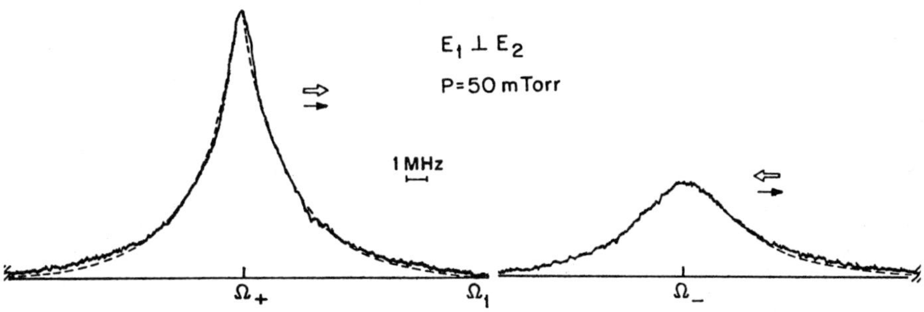

Fig.8 Steady state forward (a) and backward (b) observed change signals. The dashed line shows the theoretical fit.

Since the NH_3 transition is degenerate ($J = 8 \rightarrow 8$) and the two beams have perpendicular linear polarizations, the saturating field can be considered to induce $\Delta M = 0$ transitions, and the weak field then probes $\Delta M = \pm 1$ transitions. In the absence of M-changing collisions the system decomposes into two groups of coupled three-level systems having the common level in the ground (g) and excited (e) states, respectively. The Raman coherence responsible for the forward-backward asymmetry of the signals of Fig. 8 is thus the coherence between adjacent M-sublevels.

As is characteristic of systems with strong phase-changing collisions [(7) and (8)], the backward change signal of Fig. 8 is a single Lorentzian, whereas the forward signal contains an additional narrow feature. However, a quantitative analysis of the experimental results requires further consideration, since in the degenerate NH_3 transition strong M-changing collisions couple together the independent three level systems. As is well known from the symmetry properties of collisional relaxation processes in gases [8], allowance must then be made for different relaxation rates of the various multipole moments of each level.

Using the tensorial formalism [8] to account for these features, the following expression for the change signals can be derived [9]:

$$G(\varepsilon) = L(2\gamma_{eg}^{1}) \sum_{\alpha=e,g} \left\{ \frac{96}{\gamma_\alpha^{o}} - \frac{38}{\gamma_\alpha^{2}} + \frac{\varepsilon+1}{2}[57L(\gamma_\alpha^{2})+L(\gamma_\alpha^{1})] \right\}, \quad (10)$$

where $L(x) = [x+i\delta(\varepsilon)]$. γ_{eg}^{1} is the relaxation rate of the optical polarization and γ_α^{k} the decay rate of the kth order multipole in level α: $k = 0$, population; 1, orientation; 2, alignment. Thus, in the presence of M-changing collisions the backward signal is still a single Lorentzian of width $2\gamma_{eg}^{1}$, but the forward one now consists of several Lorentzian components associated with the various tensorial moments. In the present experiments the width of the narrow Raman-type contribution of the forward signal is primarily determined by alignment relaxation processes (γ_α^{2}).

The observed width of the backward signal gives γ_{eg}^{1} = 24±1 MHz/torr, consistent with earlier Lamb dip measurements. Previous microwave experiments in the NH_3 ground state have shown that due to the small inversion splitting of 0.8 cm^{-1}, inelastic collisions are the predominant relaxation mechanism (which implies $\gamma_g^{k} = \gamma_g^{0}$) and that $\gamma_g^{0} = \gamma_{eg}^{1}$ [10]. This is not true for the excited state, where the inversion splitting is much larger, ∿36 cm^{-1}. In the next section we show that transient experiments yield γ_e^{0} = 3.5±0.6 MHz/torr. The best fit of the experimental forward signal to (10) (dashed curve of Fig. 8) then gives γ_e^{2} = 6±1 MHz/torr [7].

To demonstrate the inherent polarization dependence of the change signal lineshapes, similar experiments were done in which the probe field polarization was oriented *parallel* to that of the saturating field by inserting a half-wave plate in the path of the saturating beam. A small misalignment was introduced to separate the beams. As seen in Fig. 9, in this case the forward change signal is narrower than that observed for crossed polarizations.

(The change signal linewidths observed in the backward direction are the same, as predicted by (11).) This can be understood as follows: When the polarizations are parallel both fields obey $\Delta M = 0$ selection rules, and so no Raman coherence can be induced. However, since two e.m. fields of the same polarization and propagation vector are equivalent to an amplitude modulated excitation, a new type of contribution now arises, originating in the coherent modulation of the level populations at the frequency difference between pump and probe fields. This effect causes additional narrowing of the forward signal if $\gamma_e^{0}<\gamma_e^{2}$. Using the tensorial formalism, the change signal lineshape is now given by [9]

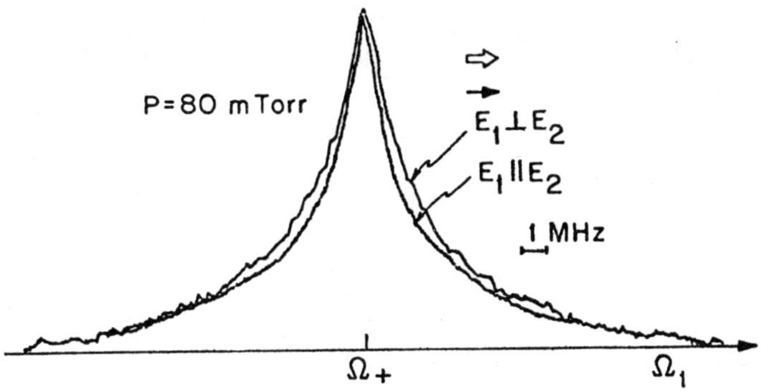

Fig.9 Forward change signal for parallel polarizations of probe (E_1) and saturating (E_2) fields. The corresponding signal for $E_1 \perp E_2$ is shown for comparison, normalized such that the peak heights are equal.

$$G_{11}(\varepsilon) = \mathcal{L}(2\gamma_{eg}^{1}) \sum_{\alpha=e,g} \left\{ \frac{96}{\gamma_\alpha^0} + \frac{76}{\gamma_\alpha^2} + \frac{\varepsilon+1}{2}[96\mathcal{L}(\gamma_\alpha^0) + 76\,\mathcal{L}(\gamma_\alpha^2)] \right\}. \quad (11)$$

The observed narrowing of the lineshape for parallel polarization is another clear evidence of M-changing collisions, since for $\gamma_\alpha^\kappa = \gamma_\alpha^2$ (10) and (11) predict the same lineshape, contrary to the experimental results.

From the above results it follows that in the asQ(8,7) excited state the cross-section for elastic M-changing collisions is about two-thirds as large as that for inelastic collisions. To our knowledge, this is the first measurement of the alignment relaxation rate in the ground electronic state of a molecule. The present experimental technique does not require Stark or Zeeman tuning nor fluorescence detection, and thus compliments the well-established Hanle effect and double resonance techniques of optical pumping [8].

3.3 Time-Delayed Lineshapes

As mentioned above, the experiments have been carried out in two different ways: Either the probe frequency is fixed and the transient signal is observed as a function of time (Fig. 10), or the transient lineshape is studied by tuning the probe frequency at a given delay time after the saturating field has been interrupted (Fig. 11).

Figure 10 shows the decay signals at line center, and Figs. 11a,b show the transient lineshapes observed at different time delays. The simple three-level model of Section 2 predicts that forward

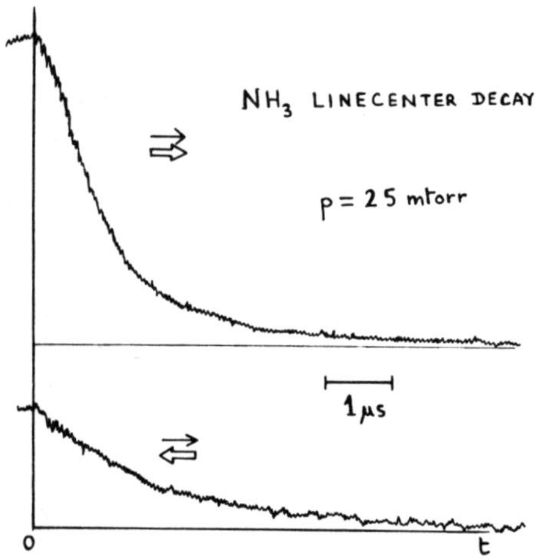

Fig.10 Time domain study of forward (upper trace) and backward (lower trace) change signals. In these traces the probe field is tuned to line center (δ = 0). As can be seen, the backward decay is a single exponential, whereas the forward decay contains fast and slow exponential components.

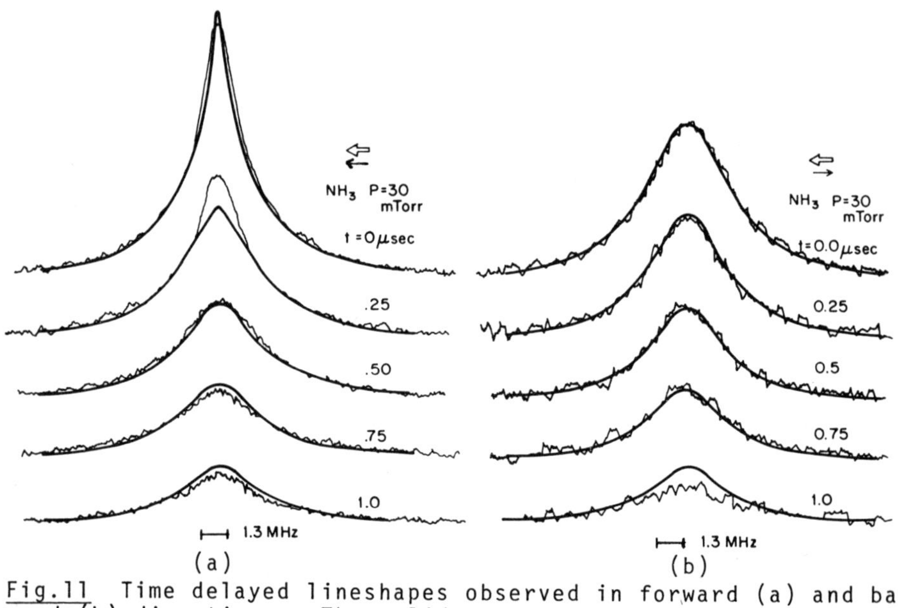

Fig.11 Time delayed lineshapes observed in forward (a) and backward (b) directions. The solid curves give the theoretical fit.

and backward signals differ by the narrow Raman-type cortribution, and this contribution decays away rapidly, leaving a slowly decaying population-induced component which is the same in both propagation directions [(7) and (8) and Fig. 5]. These predictions are confirmed by Figs. 10 and 11. However, quantitative comparison of experiment with theory must take into account the level degeneracy, the existence of M-changing collisions, and also the power-broadening induced by the pump field. As explained in the previous section, level degeneracy and re-orienting collisions can be included by means of the tensorial formalism [9]:

(i) In the same way as in the simple theory, Raman-type contributions have a characteristic decay rate $2\gamma_{eg}^{1}$ (= 48 MHz/torr). A theoretical study valid for arbitrary intensities of the saturating field [5] shows that this decay rate is shortened by power dephasing. For instance, in Fig. 11 power dephasing (for I = 70 mW/cm^2 and p = 30 mtorr) contributes 0.5 MHz to the Γ decay rate.

(ii) Population saturation contributions are the same in both propagation directions and have a Lorentzian lineshape of width $2\gamma_{eg}^{1}$. A noteworthy feature is that their decay rate is not sensitive to power dephasing [5]. Eq. (10) shows that various multipole moments in excited and ground states contribute to the corresponding components of the change signal amplitudes and thus, there should be several time constants associated with the decay of the backward signal. However, due to the short lifetime of the ground state, the amplitude of its steady-state contribution is very small compared to that of the excited state population. (The ratio is about $58\gamma_e^0/96\gamma_g \simeq 9\%$.) In addition, its decay is very fast. On the other hand, in the excited state, (10) shows that the ratio of alignment to population cortributions is $38\gamma_e^0/96\gamma_e^2 \simeq 23\%$. The excited state population therefore provides the dominant contribution to the decaying change signals, and lasts for the longest time. γ_e^0 can thus be extracted from the slowly-varying component of the curves of Fig. 10. The pressure dependence of this decay rate is shown in Fig. 12.

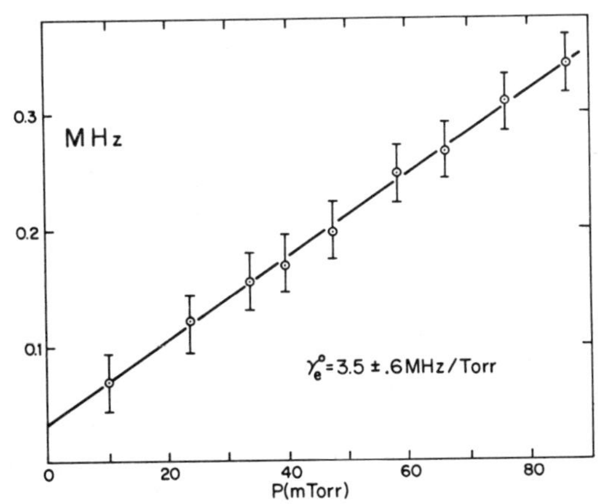

Fig.12 $(2\pi T_e^0)^{-1}$ vs. pressure. The data was taken from the slowly-decaying component of the forward change signal (upper trace of Fig. 10).

The zero-pressure value (∿0.04 MHz) is due to molecular transit time effects.

Since all the parameters are known, the theoretical predictions can be compared with the experimental lineshapes (Fig. 11). The agreement is quite good, except for the short term (t = 0.25 μs) behavior of the forward signal, where the time resolution of the electronic detection system was not sufficiently high to follow the fast decay of the narrow component. The success of the model adopted here, which does not include velocity changing collisions, indicates that in the range of pressures observed ($p \gtrsim 30$ mtorr), contributions from such collisions are negligible in NH_3.

4. Conclusion

These first results demonstrate the power of time-delayed saturation spectroscopy. By combining frequency and time-domain measurements in NH_3 we have obtained a complete set of information which improves our knowledge of the dynamics of the excited vibrational state, providing the first measurement of a molecular alignment relaxation rate in an infrared transition.

We have already pointed out the applicability of time-delayed saturation spectroscopy to optical nutation and photon echoes in multilevel systems. To underscore the generality of the technique, notice that it provides a precise analysis of the dynamics of atomic and molecular systems: Scattering and diffusion mechanisms, thermalization by velocity-changing collisions and radiative transfer, velocity-dependence of such processes can all be studied Extension to four-level systems, where pump and probe transitions have no common level, should lead to similar information about inelastic collisions. These kinds of experiments in excited neon are now underway at Université Paris-Nord (France).

Acknowledgments

We are grateful to our colleagues José Leite, Antonio Sanchez and Dan Seligson for their contributions to various parts of this work.

References

1 See *e.g.* V.S. Letokhov and V.P. Chebotayev, Nonlinear Laser Spectroscopy, Springer Series in Optical Sciences, Vol. 4, (Springer-Verlag, 1977).

2 See the contribution by R.G. Brewer in this volume.

3 A time-delayed Lamb dip experiment has been performed using a pulsed dye laser to produce counter-propagating beams: T.W. Hänsch, I.S. Shahin and A.L. Shcawlow, Phys. Rev. Lett. 27, 707 (1971). It should be noted that pulsed lasers are not

ideal for such studies. The present approach has the advantage that the system can be prepared in a well defined steady state, and the weak c.w. field can precisely probe its decay. The importance of studying both co- and counter-propagating change signals is also emphasized.

4 M. Ducloy and M.S. Feld, J. de Phys.-Lettres (Paris) 37, L-173 (1976); M. Ducloy, J. Leite, R. Sheffield and M.S. Feld, Bull. Am. Phys. Soc. 21, 599 (1976).

5 M. Ducloy, J.R.R. Leite and M.S. Feld, to be published.

6 Both single and double-quantum Raman-type processes can occur in coupled systems of this type having a resonant intermediate state. See, for example, M.S. Feld in Fundamental and Applied Laser Physics, edited by M.S. Feld, N.A. Kurnit and A. Javan (Wiley, New York, 1973), pp. 369-420.

7 J.R.R. Leite, M. Ducloy, A. Sanchez, D. Seligson and M.S. Feld, to be published.

8 For a general review of tensorial formalism and multipole moments see A. Omont in Progress in Quantum Electronics, Vol. 5, p. 69 (Pergamon Press, Oxford, 1977). For their application in laser spectroscopy see B. Decomps, M. Dumont and M. Ducloy in Laser Spectroscopy of Atoms and Molecules p. 283 (Topics in Applied Physics, Vol. 2, Springer-Verlag, 1976).

9 M. Gorlicki and M. Ducloy, to be published.

10 G.M. Dobbs, R.H. Micheels, J. Steinfeld, J.H.S. Wang and J.M. Levy, J. Chem. Phys. 63, 1904 (1975).

OPTICAL TRANSIENTS IN DOPPLER-FREE TWO-PHOTON EXCITATION

B. Cagnac, M. Bassini, F. Biraben, G. Grynberg

Laboratoire de Spectroscopie Hertzienne de l'E.N.S.
Université Pierre et Marie Curie, 4 place Jussieu
Paris 5ème, France

Since the first experiment of TORREY [1] the transient phenomena have been widely investigated in the radiofrequency range [2] and they furnish many methods for the measurement of relaxation times. The coherent field of the Laser permits now to generalize in the optical range many experiments which were possible earlier in the radiofrequency range only. The first example of transient phenomena in optics has been the generalization of the spin-echoes method of HAHN [3] to the photon-echoes method by ABELLA, KURNIT and HARTMANN [4].

BREWER and SHOEMAKER have observed other kinds of transients such as optical nutation [5] and free induction decay [6]. LOY has also observed a population inversion by optical adiabatic rapid passage [7] and a two-photon transient [8].

1. Interest of Two-Photon Excitation

Most of these experiments of optical transients have been performed on single photon transitions ; and they select among the atoms one particular velocity group, the Doppler shift of which gives the frequency tuning. These resonant atoms produce the predominant effect, and it is impossible to observe the atoms belonging to other velocity groups, for which the frequency detuning is not zero. But the theoretical shape of the transients depends on this frequency detuning and the comparison with experiments could be interesting.

The cancellation of the Doppler shifts in two photon excitation with two counterpropagating beams [9,10,11] permits that problem to be overcome. In fact the two photon method appears to be of particular interest in the investigation of *off-resonance transients*, because it permits to excite all the atoms with the same frequency detuning. Figure 1 shows our notations : the Laser at circular frequency ω_L can induce the two photon transition from the ground state g to the excited state e ; $\hbar\omega_{ge}$ represents the energy gap of the transition and we define the energy detuning of the transition $\delta'\omega = 2\omega_L - \omega_{ge}$. We note $\delta'\omega$ with the dot in order to avoid the confusion with the Laser detuning, which has the half value. The two photon excitation permits to excite all the atoms with the same detuning $\delta'\omega$.

An other interest of the two photon excitation is the facility to detect the excited atoms by the observation of the spontaneously reemitted light toward an intermediate level called r on Fig. 1, because the reemitted wavelength is quite different from the Laser wavelength and it can be easily separated from the stray light of the Laser with a monochromator. This method of detection constitutes also a difference by comparison with the preceeding

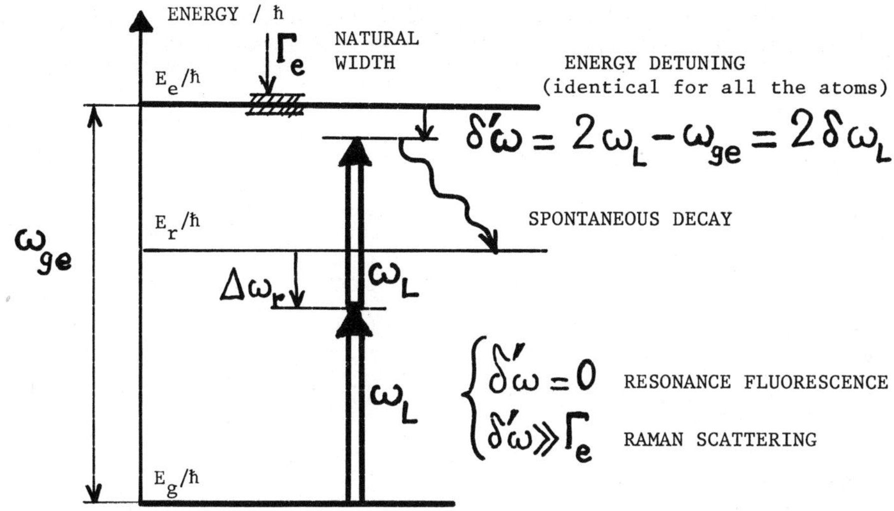

Fig. 1 Energy diagram and notations

transient-experiments either in the radiofrequency range, or in the Infra-Red range.

Finally we are able to observe the transient behaviour of the population of the excited state for a definite energy detuning $\delta'\omega$, common for all the atoms, and to study the continuous variation of these transients
- from the case $\delta'\omega = 0$ corresponding to resonance fluorescence
- to the case $\delta'\omega >> \Gamma_e$ (natural width of the excited level) this case corresponding to the Raman scattering.

The theory of these two-photon transients is quite identical to the theory of the one-photon transients provided that the energy defects $\Delta\omega_r$ in the relaying levels of the transition are sufficiently important. This energy defect is defined as $\hbar\Delta\omega_r = \hbar\omega_L - (E_r - E_g)$; it is represented on Fig. 1 in the case of a particular r level (but in real cases we can find several relaying levels). Because of the uncertainty principle the atoms stay in the r level during a very short time Δt of the order of the inverse of the energy defect $\Delta\omega_r$. If this time $\Delta t \simeq 1/\Delta\omega_r$ is much shorter than all other times characteristic of the problem (lifetimes, collision-times, and so one) we can forget the r level and express the direct relation between the ground level g and the excited level e using the two-photon operator Q, which has been extensively studied by GRYNBERG [10,12] :

$$Q = \vec{\varepsilon}_1.\vec{D}\,\frac{1}{\hbar\omega_L - \mathcal{H}_0}\,\vec{\varepsilon}_2.\vec{D} + \vec{\varepsilon}_2.\vec{D}\,\frac{1}{\hbar\omega_L - \mathcal{H}_0}\,\vec{\varepsilon}_1.\vec{D}$$

where $\mathcal{H}_0$ is the hamiltonian of the free atom
$\vec{D}$ is the electric dipolar operator
$\vec{\varepsilon}_1$ and $\vec{\varepsilon}_2$ are the polarizations of the direct and reflected light beams.

By comparison with the theory of one photon transients, we must operate only two modifications : 1°) we must replace the matrix elements of the electric dipolar operator by the matrix elements of this two-photon operator Q - 2°) we must replace the Laser detuning $\delta\omega_L$ by the energy detuning $\delta'\omega$, which is twice as great. The sudden irradiation of the atoms or the sudden cut-off of the light raise quite different problems, and they will be discussed separately. But before, we will explain the experimental technique, which is common for the two processes.

2. Experimental Set-up

The experimental set-up is very similar to the one which has been previously used [13]. For these experiments we need a better stability and a easy control of the dye-Laser frequency. We did not obtain performances as good as in other laboratories [14], but the frequency jitter of our Laser is now reduced to one MHz and the frequency remains stable within 1 MHz during one hour. This value is smaller than the current value of the natural width in the optical range, and it is sufficient for our experiment. We determine the Laser frequency by pressure tuning of the external Fabry-Pérot étalon, which is used for its control, or by changing slightly the working point on the side of the response curve of this etalon.

Figure 2 shows the experimental set-up. As in the current Doppler-free two photon experiments, the transitions are produced in the focus of a lens, where the reflected beam is also focused by a concave mirror ; and the photons spontaneously reemitted from the excited level are collected by a large lens and received by the Photo-Multiplier after wavelength selection in the Monochromator. But for the observation of transients we must add a fast optical shutter : an acousto-optical modulator is placed between the experimental cell and the concave mirror. The Brillouin scattering by an ultrasonic wave permits to deflect sharply the return beam in or out the experimental cell.

For this transient experiment we have chosen one of the easiest two photon transitions : in sodium from the 3S ground state to the 4D excited level [15]. The lifetime of this excited level, which is about 50 ns [16], is fairly longer than the risetime of the light-modulator (about 14 ns). We detect the photons reemitted from the 4D level to the 3P levels with the wavelengths 5682 and 5688 Å ; the number of detected photons in such an experiment is of the order of 10^6 per second in the center of the two-photon line. But we wish to study off-resonance transients, and, when the energy detuning $\delta'\omega$ is equal to ten or more natural widths, the number of detected photons decreases to 10^4 or 10^3 per second. It follows that during the observation time (four lifetimes, or 200 ns) there are rather few chances to detect one photon.

It is the reason why we produce the light pulses with a rate of the order of 10^6 per second. A Time-to-Amplitude Converter permits the measurement of the arrival time of each photon in relation to the switching on or switching off of the light pulse. A photo-cell receives a part of the return beam only, and gives the zero time to the Time-Amplitude Converter. A Multi-Channel-Analyser records the numbers of detected photons versus the arrival time ; each channel corresponds to 3.3 ns. So are produced the experimental curves that we present in the following.

3. Sudden Irradiation of the Atoms

The transients produced by the sudden irradiation are quite similar to the transients observed in radiofrequency experiments. As you know, in Nuclear Magnetic Resonance with a spin $^1/_2$, the evolution of the atomic spins between the two orientations is a sinusoïdal oscillation with the Rabi frequency

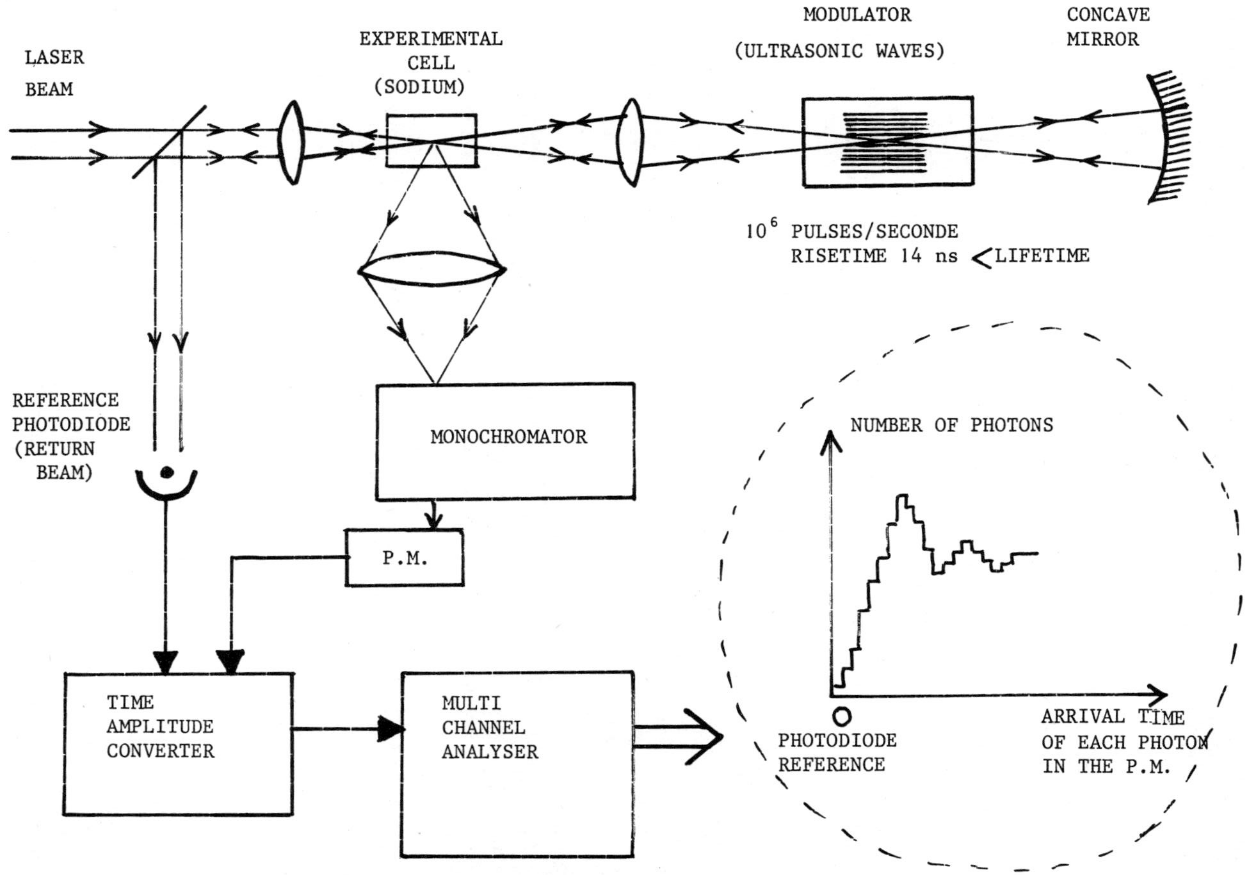

Fig. 2 Experimental set-up

$\omega_1 = \sqrt{(\gamma H_1)^2 + \delta\omega^2}$, where $\delta\omega$ is the radiofrequency detuning, H_1 is the magnetic field of the radiofrequency wave and γ is the gyromagnetic ratio of the spins. If we apply suddendly the radiofrequency field, all the spins oscillate together with the same phase ; and one observes this macroscopic evolution of all the spins together, which decreases with a time constant determined by the relaxation times of the spins, because the relaxation processes introduce the disorder between the spins.

It is well known that, for any problem of transition involving only two levels, the equations are the same as in the problem of the spin $\frac{1}{2}$. For an electric dipolar transition in the optical range it is easy to calculate an optical nutation with the circular frequency $\omega_1 = \sqrt{(\mu E_L/\hbar)^2 + \delta\omega_L^2}$, where $\delta\omega_L$ is the Laser detuning, E_L is the electric field of the Laser wave and μ is the dipolar momentum of the atom in this transition (strictly speaking, that is to say the matrix element of the electric dipolar operator between the two concerned levels). In the experiments of BREWER and SHOEMAKER [5] one selects the velocity group for which $\delta\omega_L = 0$, and one observes the nutation frequency which is proportional to the Laser field E_L.

In the two-photon case, we must operate the two modifications noticed at the end of section 1 and we obtain the nutation frequency

$$\omega_1 = \sqrt{\left(\frac{\langle e|Q|g\rangle}{\hbar} E_1 E_2\right)^2 + \delta'\omega^2}$$

where E_1 and E_2 are the electrical fields of the two counterpropagating light beams. In our present experiments with the c.w. dye Laser, the two photon transition is far below the saturation ; that is to say the two photon transition probability $\mathcal{P}^{(2)}$ is much smaller than the natural width Γ_e :

$$\mathcal{P}^{(2)} = \frac{4}{\Gamma_e} \left| \frac{\langle e|Q|g\rangle}{\hbar} E_1 E_2 \right|^2 \ll \Gamma_e$$

We deduce that the first term inside the square root of ω_1 is also much smaller than the natural width Γ_e. If we wish to observe these damped oscillations, the circular frequency ω_1 must be much greater than the damping time constant, that is to say much greater than the natural width Γ_e. In these conditions we must choose the energy detuning $\delta'\omega \gg \Gamma_e$ and the circular frequency ω_1 is practically equal to the detuning $\delta'\omega$.

This simple result in the case of transients far below the saturation is very easy to understand ; the frequency ω_L of the c.w. dye Laser is precisely defined, but this electromagnetic field is sharply introduced on the atoms, which see the Fourier transform of this sharp pulse. In the calculation two components only of this Fourier spectrum have predominating contributions : 1°/ the component at the doubled Laser frequency $2\omega_L$ because it has the biggest intensity - 2°/ the component at the frequency ω_{eg} corresponding to the atomic energy gap, because it produces a resonant process. We observe the beats between these two Fourier components.

Figure 3 shows the experimental results obtained in our laboratory. The different curves of this figure have been obtained for different frequencies of the Laser. The corresponding energy detuning $\delta'\nu = \delta'\omega/2\pi$ is indicated beside each experimental curve (it is twice the Laser frequency detuning). The curve a corresponds to the resonance case $\delta'\nu = 0$; it is a linear combination of exponential functions. The other curves have been obtained for increasing values of the energy detuning $\delta'\nu$ from 6 MHz, that is to say twice the natural width, to 30 MHz that is to say ten times the natural width. In these Raman cases, we observe the beating oscillations with the designed frequencies equal

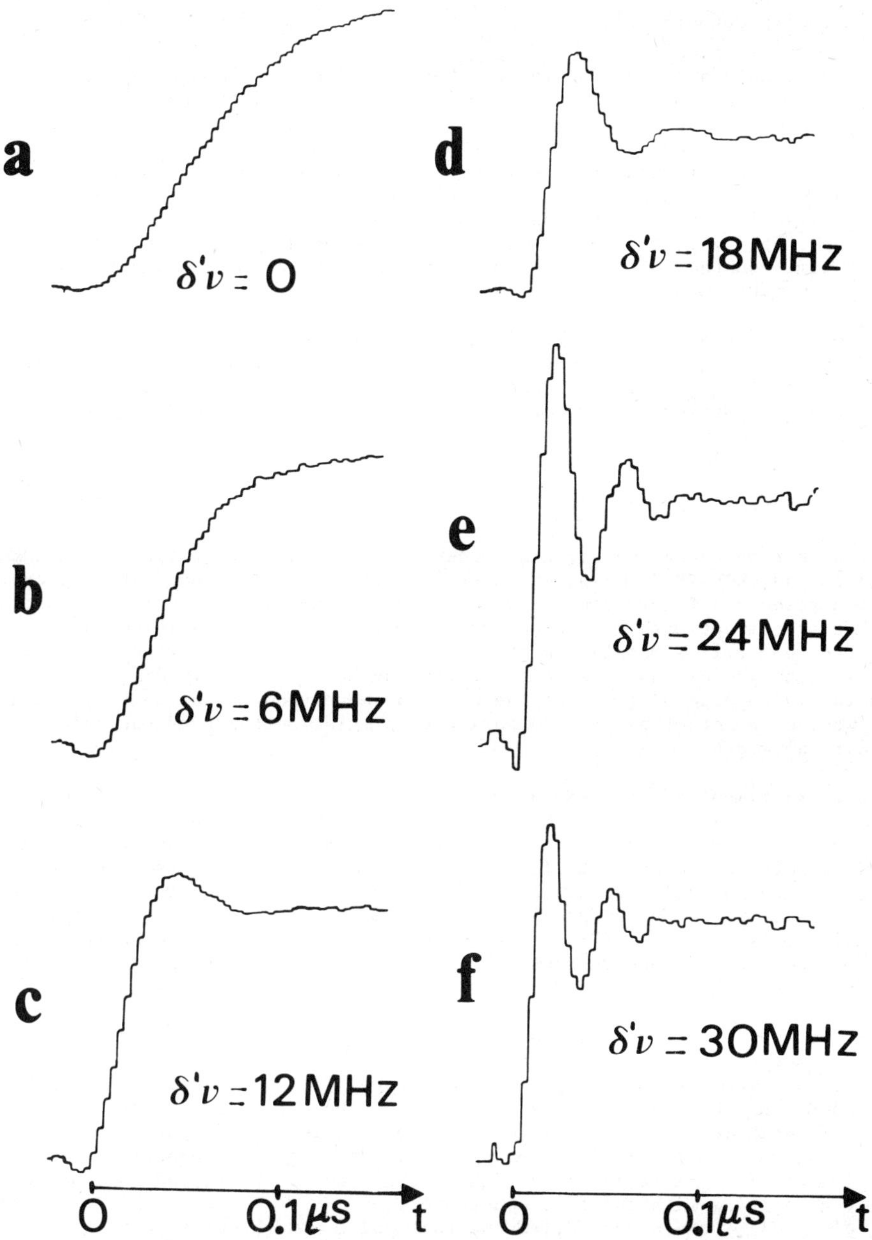

Fig. 3 Experimental curves obtained by sudden irradiation for different values of the energy detuning $h\delta'\nu$ (number of spontaneously scattered photons versus the time : 0 start of the switching on ; rise time $\sim$ 14 ns)

to the energy detuning $\delta'\nu$.

In the ideal case, the damping of these oscillations should be produced only by the spontaneous radiative decay of the atoms from the excited level, and the amplitude of the oscillations should decrease following an exponential curve of time constant $T_2 = 2T_1$ ($T_1 = 1/\Gamma_e$ being the radiative lifetime, 50 ns in our case). But in practice we were limited by the transit time of the atoms through the focus of the light beam. We have two other independant estimations of this transit time from the dimension of the waist of the Laser beam, and from the line shape [17,18]. So we can explain the damping of the oscillations with a time constant of the order of 40 ns. But strictly speaking the damping curve is not exponential [19].

An experiment done by LOY [8] is directly related to ours. He uses also a two photon technique near resonance, but with two light sources of different frequencies and intensities. These conditions leave a residual Doppler effect and lead to important light-shifts, which do not permit to define precisely the energy detuning. It is also worth comparing our experiment with another one done with gamma rays using Mössbauer effect [20].

4. Sudden Cut-off of the Light

The interest on these phenomena has been raised by the experiment of WILLIAMS, ROUSSEAU and DWORETSKY [21], who have observed the time dependence difference between resonance fluorescence and Raman scattering, after a pulse excitation. In this experiment with single photon absorption, the iodine molecules were submitted to various Doppler shifts : the observed phenomena were averaged on the velocity groups, and the Raman scattering was only observed for values of detuning which were at least of the order of magnitude of the Doppler width. The two-photon method permits to work with precise detuning of the order of the natural width.

This experiment of WILLIAMS and coworkers gave rise to numerous theoretical works on time resolved Raman scattering [22]. In order to understand these transient phenomena, one must be conscious that they depend essentially on the imperfection of the optical shutter : if the light pulse is a perfect square pulse with an instantaneous cut-off, the decay of the scattered light is always described by an exponential curve with a time constant equal to the radiative lifetime $T_1 = 1/\Gamma_e$ of the excited level, whatever the energy detuning may be. In practice the cut-off is never rigorously instantaneous but the incident light decreases during a finite time θ, and we must distinguish two stages in the decaying process : 1) the first one during the time interval θ (we suppose $\theta < T_1$). 2) the second one after the time θ, and during which the decay follows always an exponential curve with the time constant T_1 . New problems arise during the first stage. The detailed decay curve calculated during this cut-off time θ depends on the exact shape of the light pulse. Nevertheless for a smooth decrease of the incident light from a constant level to zero level, it is possible to note a general behaviour :
- if $\delta'\omega \ll 1/\theta$, the atoms cannot follow the too fast pulse and they decay from the begining with the time constante T_1
- if $\delta'\omega \gg 1/\theta$, the atoms follow the fast pulse and the most part of them fall down to the ground level during the time θ (the comparison can be done with the adiabatic passage in N.M.R.)
- if $\delta'\omega \sim 1/\theta$, the atom-decay is faster than T_1 but slower than θ ; and one must observe two curves of similar amplitudes during the two successive stages, their relative amplitudes depending on the product $\delta'\omega.\theta$.

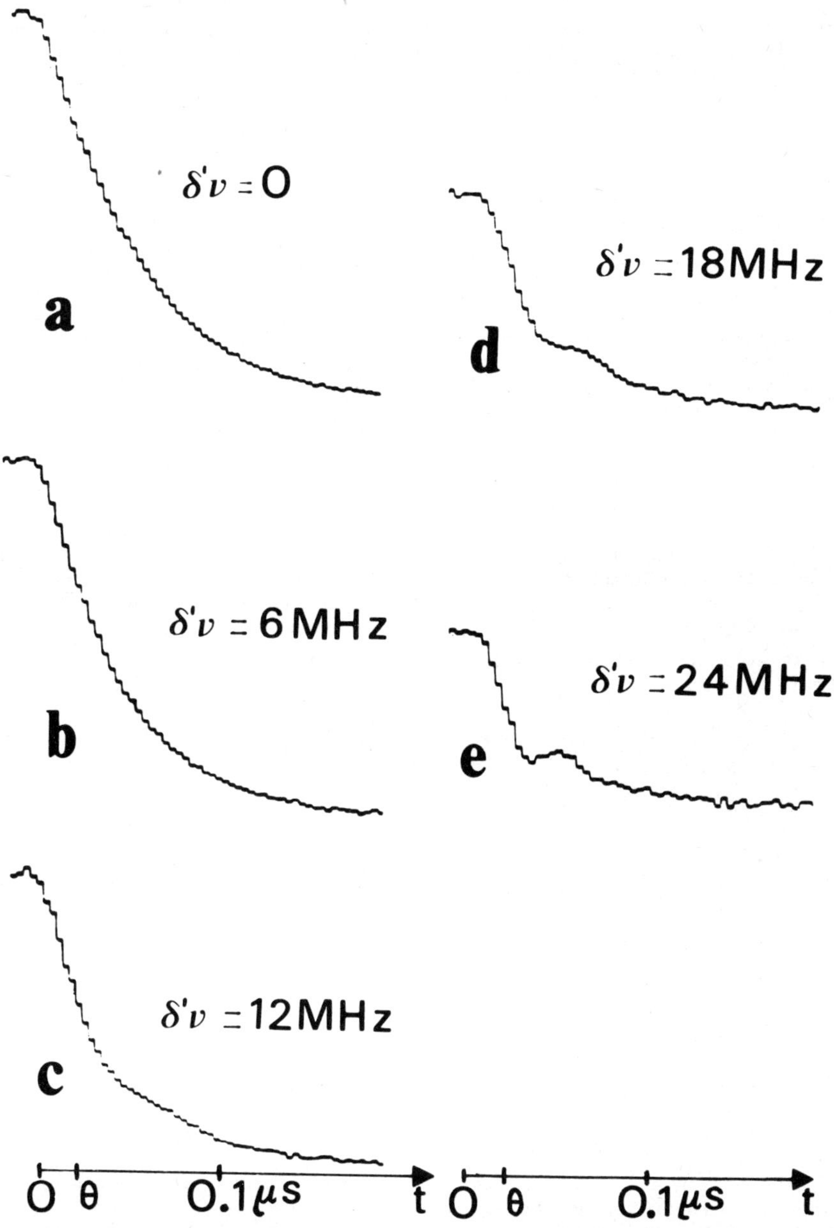

Fig. 4 Experimental curves obtained by the cut-off of the light for different values of the energy detuning $h\delta'\nu$ (number of spontaneously scattered photons versus the time : 0 start of the cut-off ; θ = 20 ns end of the cut-off time).

Figure 4 shows the experimental curves obtained in our laboratory. Each curve is obtained for a different definite value of the energy detuning $\hbar\delta'\omega = h\delta'\nu$, as on Fig. 3. In this experiment the cut-off time is of the order of θ = 20 ns ; it corresponds to six channels of the Multichannel Analyser. We see the regular decay following the lifetime on the curve a corresponding to the resonant case. But we see a faster decay on the curve b corresponding to $\delta'\nu$ = 6 MHz. The two parts of the decay curve with similar amplitudes are very well demonstrated on the curve c corresponding to the detuning $\delta'\nu$ = 12 MHz, that is to say the product $\delta'\omega.\theta = 2\pi\delta'\nu.\theta = 1.5$.

We see also on the curves d and e the decreasing amplitude of the slow part of the curve, corresponding to increasing energy detuning. Unfortunately we see also on these curves a parasitic effect : the ringings which appear on these curves are due to the imperfectness of our experimental set-up. Our modulator does not stop entirely the Laser light and the minimum value of the return beam intensity is not zero. The sudden passage from a stationary Raman scattering of big intensity to another stationary Raman scattering with small intensity produces an oscillating transient very similar to the ones presented in § 3. If you substract these parasitic ringings, the experimental curves of Fig. 4 give a good understanding of the transformation of the transients from the resonance fluorescence case to the Raman scattering case.

I believe that this experiment demonstrates the interest of the Doppler-free two-photon excitation for the investigation of transients. In the future these transients can be used for measurements of relaxation times as in the radiofrequency range. They will permit to measure the longitudinal relaxation time T_1 as well as the transversal time T_2, and that can be interesting in collisions experiments.

References

1 H.C. Torrey, Phys. Rev., 76, 1059 (1949)

2 A. Abragam, The principle of nuclear magnetism (Clarendon press, Oxford)

3 E.L. Hahn, Phys. Rev. 80, 580 (1950)

4 I.D. Abella, N.A. Kurnii and S.R. Hartmann, Phys. Rev., 141, 391 (1966)

5 R.G. Brewer and R.L. Shoemaker, Phys. Rev. Lett., 27, 631 (1971)

6 R.G. Brewer and R.L. Shoemaker, Phys. Rev., A6, 2001 (1972)

7 M.T. Loy, Phys. Rev. Lett., 32, 814 (1974)

8 M.T. Loy, Phys. Rev. Lett., 36, 1454 (1976)

9 L.S. Vasilenko, V.P. Chebotaev and A.V. Shishaev, JETP Lett. 12, 161 (1970)

10 B. Cagnac, G. Grynberg and F. Biraben, J. Physique, 34, 845 (1973)

11 Doppler-free two-photon review papers :
B. Cagnac, Proceedings of SICOLS (Mégève 1975) Lectures Notes in Physics (Springer) 43, 165
N. Bloembergen and M.D. Levenson in High Resolution Laser Spectroscopy, p. 315, Springer-Verlag (Berlin New York 1976)
B. Cagnac, Proceedings of 5th Int. Conf. on Atomic Physics (Berkeley 1976) (Plenum) page 147
G. Grynberg and B. Cagnac, Reports on Progress in Physics, 40, 791 (1977)

12 G. Grynberg, Thèse, Paris (1976) CNRS AO 12497 ;
G. Grynberg, F. Biraben, E. Giacobino and B. Cagnac, J. Physique, 38, 629 (1977)

13 F. Biraben, E. Giacobino and G. Grynberg, Phys. Rev. A12, 2444 (1975)
E. Giacobino, F. Biraben, G. Grynberg and B. Cagnac, J. Physique, 38 623 (1977)
14 R.E. Grove, F.Y. Wu and S. Ezekiel, Opt. Engineering, 13, 531 (1974)
J.J. Snyder and J.L. Hall, Proceedings of SICOLS (Mégève 1975) Lectures Notes in Physics (Springer) 43, 6
R.L. Barger, J.B. West and T.C. English Appl. Phys. Lett., 27, 31 (1975)
M. Steiner, H. Walther and K. Zygan, Opt. Comm. 18, 2 (1976)
15 T.W. Hänsch, K. Harvey, G. Meisel and A.L. Shawlow, Opt. Comm., 11, 50 (1974)
F. Biraben, B. Cagnac and G. Grynberg, C.R. Acad. Sc. 269B, 51 (1974)
16 F. Karsten and J. Schramm, Z. Physik, 195, 360 (1970)
D. Kaiser, Phys. Lett. 53A, 61 (1975)
17 Bordé C. , C.R. Acad. Sc., (1976)
18 F. Biraben, Thesis Paris (1977)
19 M. Bassini, Thèse de 3ème cycle Paris (1977)
20 F.J. Lynch, R.E. Holland and M. Hamermesh, Phys. Rev. 120, 513 (1960)
21 P.F. Williams, D.L. Rousseau and S.H. Dworetsky, Phys. Rev. Lett., 32 196 (1974)
22 J.O. Berg, C.A. Langhoff and G.W. Robinson, Chem. Phys. Lett., 29, 305 (1974)
S. Mukamel and J. Jortner, J. Chem. Phys. 62, 3609 (1975)
R.C. Hilborn, Chem. Phys. Lett., 32, 76 (1975)
A. Szöke and E. Courtens, Phys. Rev. Lett., 34, 1053 (1975)
H. Metiu, J. Ross and A. Nitzan, J. Chem. Phys. 63, 1289 (1975)
H.J. Kimble and L. Mandel, Opt. Comm. 14, 167 (1975)
D.L. Rousseau and P.F. Williams, J. Chem. Phys. 64, 3519 (1976)
J.M. Friedman and R.M. Hochstrasser, Chem. Phys. 6, 155 (1974)

COHERENT OPTICAL SPECTROSCOPY OF MOLECULES UNDERGOING RESONANCE SCATTERING AND RADIATIONLESS TRANSITIONS: THE RIGHT-ANGLE PHOTON ECHO

A.H. Zewail

Arthur Amos Noyes Laboratory of Chemical Physics [1]
California Institute of Technology
Pasadena, CA 91125, USA

Before lasers were brought to the laboratories of chemists and physicists, molecular relaxations were identified using broad band chaotic light excitation. Knowledge of the radiative decay from the photon counting rate and the quantum yield gives information about one of the most fundamental processes in chemical dynamics; radiationless transitions. From the last two decades much is known about these processes, which simply result in a change of molecular electronic state without absorption or emission of photons. and encompass a wide class of phenomena such as autoionization in atoms, predissociation, molecular electronic relaxation and energy transport in condensed media. What is not known, at least experimentally, can be outlined with the following questions:

(1) Do radiationless transitions depend on the nature of the exciting photon field?

(2) What is the exact nature of the state we excite?

(3) What is the influence of the ensemble optical coherence on the evolution of nonradiative processes?

(4) Is there a threshold for an "ergodic" behavior in large molecules?

In this paper I will (a) present new laser techniques for probing the optical dephasing of molecules undergoing radiationless transitions. The method, which utilizes three optical pulses, detects the photon echo on the spontaneous emission at right-angles to the exciting beam. This way the echo appears as a temporally burned hole in the emission, which can be detected for as few as 10^4 molecules; (b) present our recent experiments [1] on the probing of molecular states of large molecules by narrow (10^{-4} cm^{-1}) and wide-band laser excitation; (c) examine the nature of resonance emission and resonance scattering of molecules by photons within the framework of a single quantum mechanical process, and finally (d) discuss what we can learn about the fundamental nature of radiationless relaxations using coherent optical spectroscopy.

---

[1] Contribution No. 5634.

1. The Right-Angle Photon Echo [2]

Inhomogeneously broadened electronic transitions contain information about both intra- and intermolecular processes. The question, of course, is how to find and untangle these resonances by simple optical methods that give both the spontaneous decay and the collision induced dephasing caused by the elastic and inelastic scatterings in gases or solids. In contrast to conventional techniques, the field of nonlinear optical spectroscopy is indeed capable of locating very weak transitions and time resolving these radiative and nonradiative processes in selectively excited levels. The photon echo [3] decay provides the total dephasing time (T_2) while the incoherent resonance decay gives optical T_1 directly.

The method [2] described here utilizes three optical pulses ($\pi/2$, π, $\pi/2$) rather than two ($\pi/2$, π), and exploits the spontaneous emission (not absorption) which can usually be obtained from electronically excited systems in a straightforward manner. The pulses were generated using the technique of frequency switching [4], and the emission was detected at right-angles to the exciting beam [5]. The third pulse simply tilts the formed echo from the XY plane to the Z-axis of the rotating frame, i.e., converting the optical polarization induced by the laser into a change in the population difference between the ground and the electronic state. The photon echo of the optically pumped two levels can therefore be monitored by detecting the excited state population (diagonal element of the density matrix). This technique offers several advantages. First, the echo can be obtained on any vibronic line of the spectra since the spontaneous emission at right angles to the exciting beam can be resolved. Because of this high emission sensitivity, as few as 10^4 molecules can be detected. Second, the technique provides an easy way for detecting the optical coherence (i.e., dephasing and spontaneous processes) in excited ensembles without high demands on the quality of the material (e.g., solids that do not have good optical properties at low temperatures). Third, this three-pulse photon echo method together with the incoherent resonance decay (IRD) method [5], which utilizes only the first $\pi/2$ pulse, give both the optical T_1 and T_2.

In what follows, we shall discuss the theoretical development necessary to describe the detection of photon echoes with a probe pulse. Although the third pulse makes the physics different, the mathematical approach is that of photon echo theory [3]. For a selective laser excitation between the ground $|g\rangle$ and excited vibronic level $|v\rangle$ only the two-level polarization and the population difference need to be considered in describing the coherent coupling. The total wavefunction and dipole matrix element of the system can be written as follows:

$$|\psi(t)\rangle = a(t)\, e^{-i\omega_g t} |\psi_g\rangle + b(t)\, e^{-i\omega_v t} |\psi_v\rangle \tag{1}$$

$$\langle \mu \rangle = \mu_{ab}\, a(t)\, b^*(t)\, e^{+i\Delta\omega_t t} + \text{c.c.} \tag{2}$$

where $\omega_g = E_g/\hbar$ and $\Delta\omega_t$ is the transition frequency. The molecular density matrix takes the following form

$$\underset{\approx}{\rho} = \frac{1}{2}\begin{bmatrix} 1 + R_3 & R_1 - iR_2 \\ R_1 + iR_2 & 1 - R_3 \end{bmatrix} \tag{3}$$

where R_1 and R_2 (polarization) are given in terms of the cross-terms of (1), and R_3 is the population difference, i.e., $\rho_{gg} - \rho_{vv}$ (aa* - bb*) [6]. In the optical region where spatial effects could be important, the radiation field which propagates say along the z-direction must contain both the phase and amplitude time dependence:

$$E(z, t) = \epsilon(z, t) \cos(\omega t - kz + \phi(z, t)) \ . \tag{4}$$

If the first two pulses entering the sample have the same frequency but different propagation directions (i.e., $\vec{k}_{\pi/2} \neq \vec{k}_{\pi}$ and $|\vec{k}_{\pi/2}| = |\vec{k}_{\pi}|$), and for samples with no spatial dispersion, one can show that a pure density matrix has a diagonal element of one-half and an off diagonal element at time τ_2 after the second pulse of the form

$$\rho_{gv}(z, t) = i \exp[i\{\Delta\omega(\tau_1 - \tau_2) + (2\vec{k}_{\pi} - \vec{k}_{\pi/2}) \cdot \vec{r}_z\}] \tag{5}$$

where $\vec{r}_z$ specifies the position of the optically absorbing molecules in the sample and $\Delta\omega = \Delta\omega_t - \omega$. Note that ρ_{gv} determines the magnitude of the in-plane polarization which in turn determines the magnitude of the echo. Even though the spontaneous emission is radiated in all directions, the detected incoherent signal depends on the spatial phase matching that determines the formation of echo. Therefore, the effect of the third probe pulse is to induce the following population density

$$\langle\rho_{vv}^{(z)}(t)\rangle = \tfrac{1}{2}(1 - \exp\{-|\tau_1 - \tau_2|/T_2^*\}) \ . \tag{6}$$

This on-resonance (for the exciting pulses) equation which was obtained for an averaging over a Lorentzian line shape demonstrates the principle of right-angle photon echo: independent of the magnitude of the inhomogeneous decay time (T_2^*), all the molecules will be in the ground state $\rho_{vv} = 0$, when the echo is formed. On the other hand, at longer or short enough times for τ_2 the spontaneous emission will be at the saturation level. Therefore, <u>the echo induced in the forward direction "burns a hole" in the emission at right angle</u> to the exciting beam. The decay of the echo by optical T_2 and T_1 can easily be incorporated into (6).

Theoretically one must consider the propagation of the polarization as the third probe pulse scans the echo. Defining the duration of the first $\pi/2$ pulse as 0 to t_1 and the π pulse as t_2 and t_3, the population difference and the laser induced polarization (R_1, R_2 and R_3 in FVH picture [6]) can be written as:

$$R(t) = \underset{\approx}{G}_{t3}(E)\, \underset{\approx}{G}_{32}(\pi)\, \underset{\approx}{G}_{21}(D)\, \underset{\approx}{G}_{10}(\pi/2)\, R(0) \tag{7}$$

where the operation starts with the $\pi/2$ propagator from time 0 to time t_1. All these operations are 3×3 matrices which describe the pulses, the decay (D) and the echo formation (E) as the polarization evolves in time. Notice the probe laser pulse will operate on the net component of the polarization in the XY plane to produce R_3 which lies along the Z axis of the FVH frame. This $\pi/2$ tilting is similar to that used in observing spin echoes in excited states [7].

Because we are switching the laser between two sub-ensembles, the echo shape as a function of the separation τ_2 between the π and probe pulses can be obtained from (7), since every propagator has different boundary conditions. As usual we assume that t_{01}, t_{23}, $t_{probe} < T_2$, T_1, and that the molecules that contribute to the formation of the echo "see" only $\pi/2$ and π pulses. Since the laser pumps a fraction of the inhomogeneous resonance the echo shape signal is simply given by

$$S(\Delta, \tau_2, \tau_1) = A\, e^{-|\tau_2 - \tau_1|/\sigma}\, e^{-(\tau_2+\tau_1)/T_2} \cos \Delta (\tau_2 - \tau_1) \tag{8}$$

where Δ is the off-resonance frequency and σ is the echo width in time for a Lorentzian distribution of molecules that are excited by the laser. The constant A is conditioned by the inhomogeneous resonance shape. This result resembles that obtained for the field created by a polarization in free induction decay theory [8]. One notes that: (a) beating at frequency Δ is expected, (b) the maximum of the echo is at $\tau_1 = \tau_2$, (c) the echo shape is essentially a modulated envelope since $T_2 > \sigma$, and (d) the decay ($e^{-2\tau/T_2}$) of the normalized photon echo signal is, of course, independent of the more rapid inhomogeneous dephasing encountered in free induction decay.

Figure 1 depicts the right-angle photon echo observed in iodine as the probe pulse is swept through the echo position at $\tau_1 = \tau_2 = 180$ nsec. As predicted by (8), the beat pattern is clearly seen. At longer time τ and large Δ, up to eight oscillations have been observed. The decay of the echo at 10 mtorr gas pressure gives $T_2 = 500 \pm 50$ nsec. At this pressure we have also measured the decay of the photon echo in the forward direction using only two pulses. The decay of the "forward" echo gives T_2 which is in agreement with previous work [4] and with the probe pulse results [2]. Furthermore, as expected, the decay is sensitive to the gas pressure and gives information about the radiative and radiationless processes [2] of

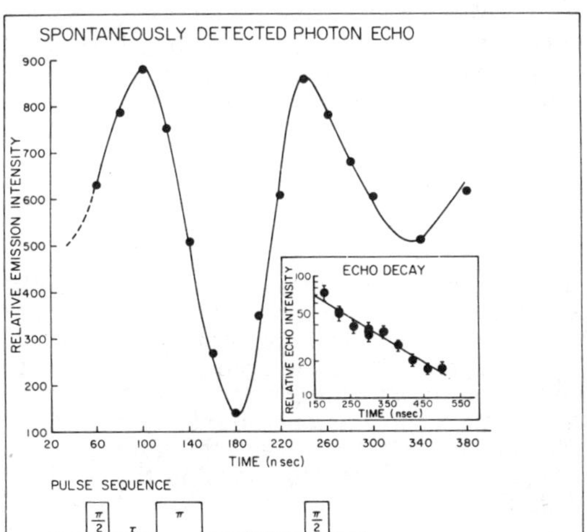

Fig. 1 The right-angle photon echo of iodine gas. The spontaneous emission was detected while exciting the gas with a single-mode of a tunable dye laser. For details of the experimental setup, see references 2 and 5. The insert depicts the echo decay maximum as a function of time ($\tau_1 = \tau_2$)

the optically pumped two levels. The echo position in all our measurements is in good agreement (within ~ 3%) with the predicted value. The duration of the echo σ is also consistent with the theory. Two additional features of the echo are noteworthy. First, the echo disappears when the probe pulse is not in the train. This is because the detector at right-angles to the laser beam cannot "see" the in-plane polarization. Second, the echo amplitude decreases when the pulse width and/or the optical field ϵ is changed from optimum values. This is due to the change in the areas of the $\pi/2$ and π pulses which is given by $(2\mu/\hbar)\int_{-\infty}^{\infty} \epsilon\, dt$.

In the solid phase, the echo is formed because of the ensemble inhomogeneity due to the crystal environment. For a mixed crystal of pentacene in p-terphenyl at 1.7 K, a signal was seen on the emission spectrum of the first electronic origin. These preliminary results [9] showed a beat pattern which, as we shall discuss later, might reflect the intramolecular level structure in such large molecules.

2. Radiationless Transitions in Molecules

In large molecules (where the density of vibrational modes at the optical excitation energy is very large) one usually finds that the lifetime of the excited state is shorter than the radiative lifetime (obtained from oscillator strength measurements) and the quantum yield is less than one. On the other hand, in small molecules (e.g., diatomics) the measured lifetime is longer than the radiative one while the quantum yield is close to unity. This paradox originates from radiationless processes [10] within the molecule, and was resolved by incorporating different coupling mechanisms among the so-called Born-Oppenheimer singlet and triplet states. In the ROBINSON-FROSCH and the BIXON-JORTNER models [11], the molecule has a "primary" state (e.g., the singlet state which carries most of the oscillator strength) ϕ_p which couples with many isoenergetic vibronic levels $\{\phi_\ell\}$ from the ground singlet state or a nearby triplet state. The coupling matrix element $v_{p\ell}$ and the density of states $\rho(E_p)$ will determine the routes of the nonradiative decay. In many ways this scheme resembles FANO's description [12] for the autoionization of helium. However, in molecules, irreversible electron relaxation (T_1 mechanism) deactivates excited levels with no ionization or bond breakage. This nonradiative irreversible process occurs only when the vibrational levels form an effective continuum around ϕ_p. The coupling between ϕ_p and $\{\phi_\ell\}$ will then have an energy distribution that is Lorentzian [11, 12] in energy. In cases where the triplet state is very close to the singlet, ϕ_p couples discretely to a sparse triplet levels to form the molecular eigenstates:

$$\psi_m = \alpha_{\beta m}\phi_p + \sum_\ell \beta_{\ell m}\phi_\ell \quad . \tag{9}$$

These states will have different radiative decay constants when they couple to the field states. In this limit, the large molecule act as being small and one should be able to "pick-up" these true molecule eigenstates or true eigenpackets [13] (collection of ψ_m's) depending on the frequency spectrum of the exciting radiation [14]. Therefore, narrow and wide band laser excitation of large molecules, together with the measurements of optical dephasing by the photon echo method should help to answer the questions raised earlier in this paper. Our approach [1] was to examine the radiationless transitions and the optical dephasing in small (e.g., I_2) intermediate (e.g., NO_2), and large (e.g., pentacene) molecules.

2.1 Optical Dephasing of Small Molecules in Molecular Beams

In iodine the energy spacing between the molecular eigenstates is large enough that the observed decay in a narrow or wide band laser excitation gives the same emission rate. Recently, we have studied the decay from the transition at 5897.5 Å at different gas pressures and at zero pressure in a molecular beam [15]. The zero-pressure decay time extrapolated from the Stern-Volmer plot gives $T_1 = 1.29 \pm 0.05$ μsec when the molecule is excited by the single mode (resolution of Ca. one part in 10^8) of a tunable dye laser. The molecular beam results gives $T_1 = 1.24 \pm 0.02$ μsec which is in excellent agreement with the extrapolated "bulb" results. This indicates that the beam is collisionless as far as T_1 measurements are concerned, and that the quenching cross section for the pressure induced non-radiative broadening is $\sigma = 70 \pm 2$ Å^2.

To obtain the optical dephasing in the beam, the forward coherent signal which rides on the top of the laser was detected. This way we observed the optical free induction decay in the molecular beam of iodine following the single-mode laser excitation. Such measurements allows one to obtain T_2, which when compared with T_1 measurements [16] enables us to learn about the radiationless transitions in molecules. The heterodyne signal together with the IRD signal is depicted in Fig. 2. The results demonstrate that coherent optical transients in molecular beams can be observed even if the molecular density is low. The beat frequency follows the frequency of the electro-optic switching as in bulb experiments (see Fig. 2). The decay, however, exhibits the power saturating term which we corrected for by measuring the Rabi frequency in the bulb and scaling it by the square root of the power density ratio (beam-to-bulb). Considering the time of ground state feeding and the excited state decay when the molecules flies across the laser beam, we concluded that $2T_1 \approx T_2$ at zero pressure. These findings indicate that there is no intramolecular dephasing that destroys the optical coherence of the prepared state in such small molecules; a situation that is not necessarily the case for large molecules like pentacene or even NO_2. The latter case was reported by W. Demtröder in the proceeding of this conference and by ZARE et al. [17].

2.2 Molecular Eigenstates vs Born-Oppenheimer States in Large Molecules

The vibronic structure in a large molecule like pentacene with 102 optical modes belongs to the statistical limit where the density of vibrationally hot ground states in the neighborhood of ϕ_p is very large. However, pentacene is known to have nearby BO triplet states. Moreover, not all the modes of the $\{\ell\}$ states are <u>active</u> in the coupling because of the symmetry requirement and/or the Franck-Condon overlap. Therefore, the coupling between ϕ_p and the discrete $\{\ell\}$ levels will give rise to molecular eigenstates that have a resonance width determined by the strength of the coupling to the molecular continuum formed on the ground state and to the continuum of the radiation field. Narrow band excitation should therefore select these states while wide band excitation should prepare some eigenpacket that evolves in time differently. The T_1 decay in the narrow band excitation will reflect the relative amplitude of the singlet state in the molecular eigenstates. Excitation of a large number of ψ_m-s yields a short decay (like a dephasing decay process) and a long decay of essentially the state ψ_m, in addition to a complicated beat pattern which depends on the level structure [10]. The amplitude ratio of the fast decay to long decay depends on the number of states involved in the coupling. This explains why it is sometimes not feasible to see the long decay in such experiments.

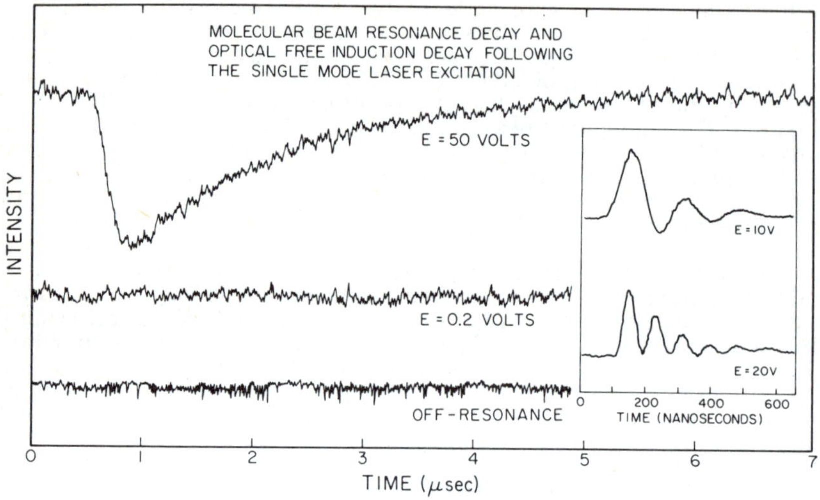

Fig. 2 The resonance decay and the optical free induction decay of iodine molecular beam. The decay observed at the electro-optic voltage of 50 V gives optical T_1 in the beam. Notice that the heterodyne signal of the free induction decay follows the switching frequency

Figure 3 depicts the T_1 decay observed in pentacene in p-terphenyl (1.7 K) as a function of the laser bandwidth. Several observations [1, 9] were made: (1) the decay changes from 15 μsec to 25 nsec when the laser effective width changes from 6 MHz to 18 GHz; (2) at high temperatures, the signal disappears (see Fig. 3); (3) there is a magnetic field effect on the decay in the narrow band laser excitation; (4) the decay signal is absent when the molecule is excited above the 0, 0 electronic origin, and the emission was detected at the same d.c. level as that of the origin; (5) the build up time of the signal gets longer when the laser power is reduced; (6) the decay time does not change when the power level of the laser is reduced in the range we studied; (7) the T_1 decay is exponential and changes as we scan the narrow-banded laser in the manifold of the first electronic origin; (8) the narrow band coherent transients give a T_2 of 45 nsec when the laser is on-resonance with the 0, 0 transition, but less than a few nanoseconds when the laser is in resonance with the vibronic line at 267 cm^{-1} above the 0, 0 transition; (9) the signal gets larger at higher switching frequencies; and finally (10) the measurement of $1/T_2$ vs temperature shows a transition temperature for the loss of coherence; the system response is almost flat at low temperatures and undergo the transition at ≈ 3.7 K. These observations are all consistent with the above mentioned theory which indicates that the preparation of molecular eigenstates can be done if the laser width is narrow enough and the resonances are not smeared to form a continuum.

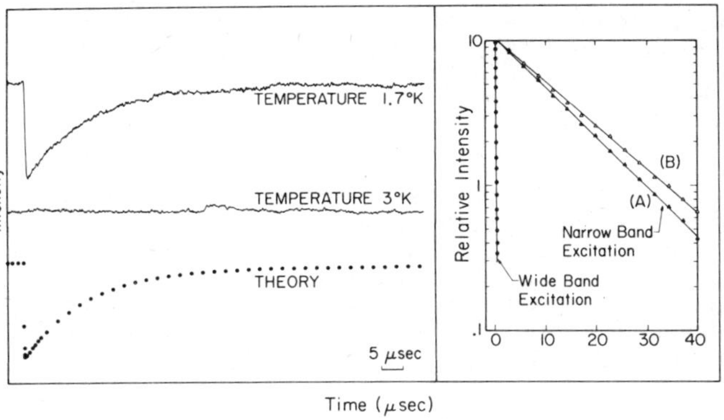

Fig. 3 The resonance decay of pentacene in p-terphenyl crystals at 1.7 K and at higher temperature (> 3 K). The insert shows the observed decay as a function of the bandwidth of the laser and the magnetic field effect. (A: zero field, B: 0.4 KG). For details see reference 1.

On the other hand, the broad band excitation prepares eigenpackets which as mentioned before decay by a fast dephasing component ($\sim v_{p\ell}^2 \rho$) and a low amplitude slow-decay component. More recently, LIM has found a similar but less dramatic effect in tetracene in mixed crystals at low temperatures [18]. It was concluded that the second triplet is very close to the first singlet state such that the level structure in the neighborhood of ϕ_p is very sparse and the vibrational relaxation is slow. It should be mentioned that in pentacene low-pressure gas and at zero pressure, a transient spectrum [19] of hot excited levels in the neighborhood of ϕ_p was found when the system was excited by broad band excitation. The decay time (100 μsec) becomes shorter at high pressures, again consistent with our conclusion [20].

2.3 Effect of the Radiation Field: From Heitler's Limit to the Molecular Decay Limit

The above discussion does not explicitly include the radiation field power spectrum. When the field is included in the formalism, both resonance emission and scattering can be handled. The paper of C. COHEN-TANNOUDJI in this book discusses the strong signal limit. In this limit, in which the induced transition rate is compared to, or greater than the decay rate, new additional Rabi peaks are expected. The width of the resonance transition is the intrinsic one while those of Rabi satellites are slightly broader (on resonance). The weak field limit, on the other hand, gives a decay constant that depends on the relative frequency width of the radiation field to the molecular resonance. The on- and off-resonance characteristics of the decay in isolated molecules have been treated by many investigators [21]. More recently, the decay and scattering properties of systems having collisional broadenings have also been investigated [22].

The total response of the molecule-light system can be obtained since the homogeneous resonance of the molecule is Lorentzian [22, 23]. Taking the distribution in the frequency of the photon packet to be Gaussian, the convolution theorem yields the following response function:

$$F(t) = C \int_{-\infty}^{t} \bar{\epsilon}(t') \, \bar{G}(t-t') \, dt'$$

$$= e^{(\Gamma_m/\Gamma_L)^2} \, e^{-\pi \Gamma_m t} \left\{ 1 + \mathrm{erf}\left(\frac{\pi}{2} \Gamma_L t - \Gamma_m/\Gamma_L \right) \right\} \tag{10}$$

where Γ_m is the linewidth (FWHM) of the Lorentzian and Γ_L is the linewidth parameter of the Gaussian light source. The functions $\bar{\epsilon}$ and $\bar{G}$ are the Fourier transform of $\epsilon(\omega)$ and the molecular Green operator $G(\omega)$, and C is a constant. Note that the imaginary part of $G(\omega)$ is a Lorentzian resonance. Figure 4 gives the response function as a function of Γ_L/Γ_m. It is clear that as the frequency width of the light spans a wider range of energy compared to the molecular resonance energy width, the decay becomes molecular

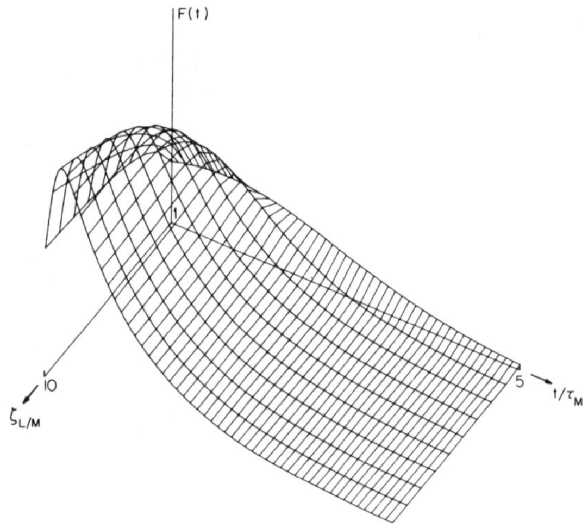

Fig. 4 Plots of the response function F(t) for the system (light-molecule) as a function of time and the ratio ζ, of Γ_L to Γ_m.

in nature. On the other hand, if the light frequency width is smaller than the resonance width, Heitler's limit is recovered and the decay is dominated by the light source. A close examination of (10) indicates that after one molecular lifetime from t = 0 (i.e., after the build-up zone and the nonexponential region) the decay is molecular when the ratio of the light frequency width to Γ_m is 3.2 within a tolerance of 5%. One can therefore utilize these findings to obtain the correlation time for the laser involved, and to distinguish between the scattering by and the emission from molecules excited by weak fields [24].

In iodine we now know that the observed T_1 decay is a molecular one. This is because (a) Stern-Volmer plot gives a cross section and a zero pressure decay time that is in excellent agreement with the wide band laser excitation results and the molecular beam results; and (b) the homogeneous resonance width measured by the photon echo and the FID is consistent with the T_1 decay results. This means that the "effective" laser width is larger than the molecular width by a factor of 3.2 (i.e., 10 MHz or more) in the pressure range studied [15]. Utilizing this effective width of the laser, which was used in the pentacene experiment, and knowing T_2 of the transition, one concludes that the observed decay time (15 μsec) can at most be influenced by a factor of 2-3, considering the finite width of the laser source [25]. As pointed out before, in the strong coupling limit the decay is totally molecular. These conclusions about the effect of the radiation field on the decay characteristics are, therefore, consistent with our earlier findings and explanations of the preparation of molecular eigenstates in large molecules. More detailed analysis of this treatment in multilevel structure is under study.

In conclusion, coherent optical spectroscopy is a useful tool for probing radiationless transitions, resonance scattering and decay, and the ergodic behavior in molecules and in molecular beams. Experiments dealing with the nature of molecular eigenstates in molecules like NO_2 and in beams (pentacene and NO_2) are now in progress. Because the optical phase coherence of these states can be monitored in a selective way, we hope to learn more about the origin of intra- and intermolecular dephasing processes. These processes are very important to understand and also play a central role in describing different phenomena such as laser-induced chemistry and multiphoton dissociation of molecules, as discussed by Professor Bloembergen in this meeting.

Acknowledgments

Professor Willis Lamb has discussed the contents of this paper with me in great detail. I am very grateful to him and to Professor N. Bloembergen for their interest in this work and for very fruitful and stimulating discussions. I also wish to thank Professor J. Jortner for the careful reading of the manuscript and for valuable suggestions.

References

1. A.H. Zewail, T.E. Orlowski, K.E. Jones: Proc. Natl. Acad. Sci. (USA) 74, 1310 (1977)
2. A.H. Zewail, T.E. Orlowski, K.E. Jones, D.E. Godar: Chem. Phys. Lett. 48, 256 (1977)
3. N.A. Kurnit, I.D. Abella, S.R. Hartmann: Phys. Rev. Lett. 13, 567 (1964); and Phys. Rev. 141, 393 (1966)
4. R.G. Brewer and A. Genack: Phys. Rev. Lett. 36, 959 (1976)
5. A.H. Zewail, T.E. Orlowski, D.R. Dawson: Chem. Phys. Lett. 44, 379 (1976)
6. R.P. Feynman, F.L. Vernon, Jr., R.W. Hellworth: J. Appl. Phys. 28, 49 (1957)
7. W.G. Breiland, C.B. Harris, A. Pines: Phys. Rev. Lett. 30, 158 (1974)
8. R.G. Brewer, R.L. Shoemaker: Phys. Rev. A6, 2001 (1972); F.A. Hopf, R.F. Shea, M.O. Scully: Phys. Rev. A7, 2105 (1973)
9. T.E. Orlowski, A.H. Zewail (to be published)
10. For a review see: S.A. Rice: In: Excited States, Vol. 2, ed. by E. Lim (Academic Press, New York 1975) p. 111; J. Jortner, S. Mukamel: In: The World of Quantum Chemistry, ed. by R. Dandel, B. Pullman (Reidel Publishing Company 1974)
11. G.W. Robinson, R.P. Frosch: J. Chem. Phys. 37, 1962 (1962); M. Bixon, J. Jortner: J. Chem. Phys. 50, 4061 (1969)
12. U. Fano: Phys. Rev. 124, 1866 (1961)
13. W. Lamb, Jr.: (private communication)
14. W.H. Rhodes: J. Chem. Phys. 50, 2885 (1969)
15. A.H. Zewail, T.E. Orlowski, R.R. Shah, K.E. Jones: Chem. Phys. Lett., in press

16. S.Ezekiel, R.Weiss: Phys. Rev. Lett. 20, 91 (1968); and Appl. Phys. Lett. 21, 320 (1972)
17. R.Zare: (private communication)
18. E.Lim: (to be published)
19. B.Soep: Chem. Phys. Lett. 33, 108 (1975); R.K.Sander, B.Soep, R.N.Zare: J. Chem. Phys. 64, 1242 (1976)
20. The relationships between the measurements of T_1, T_2 and the density of level structure of the molecular eigenstates will be discussed elsewhere. H.deVries, T.E.Orlowski, D.A.Wiersma, A.H.Zewail: (to be published)
21. J.M.Friedman, R.M.Hochstrasser: Chem. Phys. 6, 155 (1974); F.A.Novak, J.M.Friedman, R.M.Hochstrasser: (to be published); S.Mukamel, J.Jortner: J. Chem. Phys. 62, 3609 (1975); J.O.Berg, C.A.Langhoff, G.W.Robinson: Chem. Phys. Lett. 29, 305 (1974)
22. S.Mukamel, A.Ben-Reuven, J.Jortner: Phys. Rev. A12, 947 (1975); R.M.Hochstrasser, F.A.Novak: Chem. Phys. Lett. 48, 1 (1977); J.Carlsten, M.G.Raymer: In: Laser Spectroscopy, Vol. 7, ed. by J.L.Hall and J.Carlsten (Springer Series in Optic Sciences, Springer-Verlag, New York, Heidelberg 1977)
23. C.A.Langhoff, G.W.Robinson: Mol. Phys. 28, 249 (1973)
24. A.Nichols, D.Godar, A.H.Zewail: (unpublished work)
25. For long time pulses, the Fourier transformed width is ~ 10 KHz, which is much less than the (jitter included) effective width of the laser. In the pentacene system, one sees clearly the decay when the long time pulse is on or off. For short pulses, the pulse-on decay is similar to the long pulse case but the pulse-off decay exhibits a fast decay component that we ascribe due to population inversion in the originally pumped group of molecules [9].

VI. High Resolution and Double Resonance

INFRARED-MICROWAVE DOUBLE RESONANCE

K. Shimoda

Department of Physics, University of Tokyo, Bunkyo-ku, Tokyo 113, Japan
and Institute of Physical and Chemical Research, Wako-shi, Saitama 351, Japan

1. Introduction

Double resonance may be defined as an effect with two resonances of different frequencies in a three-level or a four-level system. We find that the off-resonance behavior of double resonance merges inherently into the stimulated Raman effect, two-photon transition, light modulation effects, optically induced anisotropy and rotation of polarization etc.

The same subject of infrared-microwave (IR-MW) double resonance was discussed in the first conference on laser spectroscopy at Vail [1]. Subsequently a number of theoretical and experimental studies on IR-MW double resonance have been reported [2,3]. The effects of double resonance may most easily be interpreted in terms of rate-equation approximation. Earlier experiments of double resonance were therefore performed by using one weak radiation to probe the effect induced by another strong radiation. We showed experimental results in the previous conference [1] that the higher-order and quantum interference effects are dominant when the relaxation rates are comparable or slower than the rates of induced transitions in the case of a three-level system. In the case of double resonance in a four-level system, on the other hand, the double resonance effect may be interpreted by considering only population changes of each levels, because molecular coherence is not usually transferred with the collision-induced transitions. Then the velocity-dependent analysis of rate equations gives a satisfactory approximation.

IR-MW double resonance is particularly useful for investigation of vibrational and rotational transitions in molecules. Assignments of infrared and microwave transitions, observation of weak transitions, detection of microwave transitions in a vibrationally excited state, and studies of relaxation processes or collision-induced transitions by the method of IR-MW double resonance have so far been demonstrated in many systems of molecules. It is true that experiments on IR-MW double resonance in the initial stage were performed in the system in which the involved energy levels had been fairly well known, although they were thereby confirmed and more accurately measured. More specifically, double resonance spectroscopy has so far been studied mostly on linear molecules and symmetric top molecules in which the rotational levels are simple. But now the technique of IR-MW double resonance has been so developed that really new informations on molecules can be derived from its observation.

It is not the purpose of this paper to review a wide variety of recent works, but two typical works of my colleagues are reported.

As mentioned above, spectroscopic investigations by the method of IR-MW double resonance in other laboratories were so far limited in symmetric top and linear molecules. We have studied asymmetric top molecules H_2CO [4], HDCO [5], HCOOH [6] and HCCCHO [7] in sequence of complexity. In the following section, double resonance study of propynal by TAKAMI shows how the rotation-vibration transitions and microwave transitions are assigned, resulting in the determination of molecular constants.

Vibrational and rotational relaxations of molecules in gas can be studied in some detail by observing transient double resonance effects. In particular, ammonia has been most extensively studied because of its good coincidence with a few laser lines of CO_2 and N_2O. Besides, strong microwave lines between inversion doublets corresponding to different rotational states provide a large number of four-level systems for IR-MW double resonance experiments. In section 3, double resonance studies of relaxation processes among vibration, rotation, and inversion levels of ammonia by SHIMIZU and his collaborators are given.

2. Double Resonance Spectroscopy of Propynal

Microwave spectroscopy of vibrationally excited molecules can not be carried out practically with a conventional microwave spectrometer, when the vibrational energy is higher than $\sim 1000\ cm^{-1}$. Use of a high-temperature absorption cell is restricted by technical difficulties and thermal dissociation of the sample. The method of IR-MW double resonance allows sensitive detection of microwave transitions in such a vibrationally excited state with an accuracy comparable to the conventional method of microwave spectroscopy for the ground state [1].

We have recently observed many microwave transitions of propynal, HCCCHO,

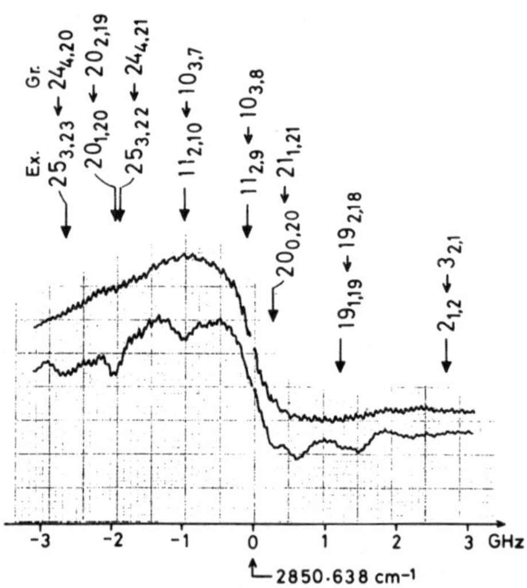

Fig. 1 Infrared absorption of DCCCHO in the Zeeman-tuned range of the 3.51 μm He-Xe laser. The upper curve shows the laser power, and the lower curve is the transmitted power through a 1-m long absorption cell with DCCCHO at 0.7 torr.

and propynal-d_1 DCCCHO, in the vibrational $v_2 = 1$ state in order to determine their structures in the excited state.

Prior to our work the infrared absorption bands of propynal were observed by a grating spectrometer without giving any assignments of their rotational structures [8]. The ν_2 band (CH stretching of the aldehyde group) is close to the frequency of the He-Xe laser of 2850.638 cm^{-1}. Infrared absorption of HCCCHO within a Zeeman tuning range ±3 GHz of the He-Xe laser is shown in [7], and that of DCCCHO is shown in Fig. 1. Infrared absorption as shown by the difference between the two curves in Fig. 1 has some peaks corresponding to the rotational structure of the band which are not resolved with this resolution of Doppler linewidth. Laser Stark spectroscopy reveals only Stark-sensitive lines which are somewhat better resolved. Stark spectrum with a low Stark field shows up lines associated with a K-type doubling of small separation. At a higher Stark field, it reveals lines of the second-order Stark effect corresponding to the transitions between levels with large K-type splittings. The levels of small K-type splittings are ascribed to $K_{-1} > 2$, if J is not very large, and to $K_{-1} = 2$ when J is small. The levels of large K-type splittings have $K_{-1} < 2$, or $K_{-1} = 2$ when J is large.

In the case of HCCCHO, one of the observed lines was resolved into three Stark components which was tentatively assigned as the $3_{1,3}(v_2 = 1) \leftarrow 4_{2,2}(gr)$ transition. In the case of DCCCHO, two of the Q-branch lines were tentatively assigned as $20_{1,20}(v_2 = 1) \leftarrow 20_{2,19}(gr)$ and $19_{1,19}(v_2 = 1) \leftarrow 19_{2,18}(gr)$. These assignments were then confirmed by observation of IR-MW double resonance with the expected microwave transitions in the ground state which had been determined by microwave absorption. The centrifugal distortion constants of DCCCHO in the ground state, however, were not known. Then we have measured about 150 lines of DCCCHO in the millimeter range up to 180 GHz, and determined its rotational constants and centrifugal constants.

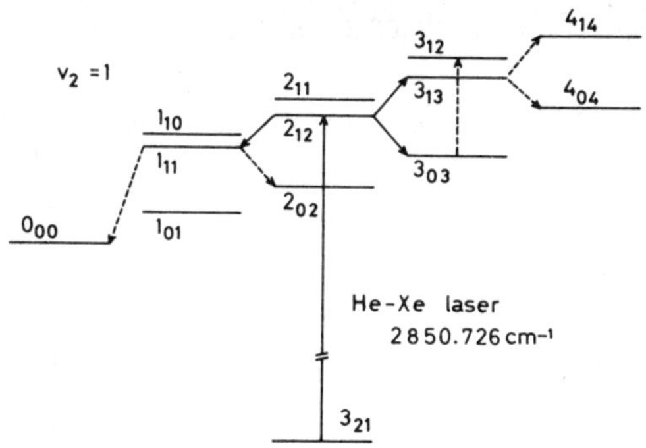

Fig. 2 Energy levels of DCCCHO involved in the IR-MW double resonance and triple resonance, in which the infrared transition is $2_{1,2}(v_2 = 1) \leftarrow 3_{2,1}$ (ground) and the microwave transitions in the $v_2 = 1$ state. Broken lines indicate the microwave transitions which can be observed by triple resonance.

Since one of the observed infrared lines was definitely assigned, it was rather easy to assign other lines. The assignments of the infrared transitions and the connecting microwave transitions in the ground state were subsequently confirmed experimentally by the observation of three-level double resonance signals. The upper levels of the above-mentioned infrared absorption are connected to different rotational levels in the $v_2 = 1$ state with corresponding microwave transitions as shown with solid arrows in Fig. 2 for DCCCHO. Using a frequency swept microwave radiation in the IR-MW double resonance experiment, we have observed microwave transitions in the $v_2 = 1$ state of DCCCHO as shown in Table 1.

Table 1 IR-MW double resonance of DCCCHO for observation of rotational transitions in the $v_2 = 1$ state

IR transition $v_2 = 1 \leftarrow 0$		MW transition $v_2 = 1$	
Frequency[GHz]*	Assignment	Rotational assignment	Frequency[MHz]
-1.88	$20_{1,20} \leftarrow 20_{2,19}$	$19_{2,17} \leftarrow 20_{1,20}$	50 883.30
-0.91	$11_{2,10} \leftarrow 10_{3,7}$	$11_{2,10} \rightarrow 12_{1,11}$	71 841.22
1.15	$19_{1,19} \leftarrow 19_{2,18}$	$19_{1,18} \leftarrow 19_{1,19}$	55 394.45
2.61	$2_{1,2} \leftarrow 3_{2,1}$	$3_{1,3} \leftarrow 2_{1,2}$	25 835.37
		$2_{1,2} \rightarrow 1_{1,1}$	17 230.48
		$2_{1,2} \rightarrow 3_{0,3}$	35 065.98

*Add 85 459.997 GHz (2850.638 cm^{-1}). ν_2 = 2857.86 cm^{-1}

The double resonance signals in HCCCHO are much stronger than those in DCCCHO. Once a strong double resonance signal is observed in a three-level system, it is possible to observe triple resonance by using one of the tran-

Table 2 Observed microwave transitions of HCCCHO in the $v_2 = 1$ state

a-type transition	Frequency[MHz]	b-type transition	Frequency[MHz]
$2_{1,1} - 1_{1,0}$	18 973.17 tr	$2_{1,1} - 2_{0,2}$	63 772.65 tr
$2_{1,2} - 1_{1,1}$	18 316.54 tr	$3_{1,2} - 3_{0,3}$	64 266.344 dr
$3_{1,2} - 2_{1,1}$	28 458.788 dr	$2_{1,2} - 3_{0,3}$	34 822.63 tr
$3_{1,3} - 2_{1,2}$	27 473.741 dr	$3_{1,3} - 4_{0,4}$	25 016.026 dr
$4_{1,3} - 3_{1,2}$	37 943.206 dr	$4_{1,4} - 5_{0,5}$	15 054.16 tr
$4_{1,4} - 3_{1,3}$	36 630.197 dr	$19_{2,17} - 20_{1,20}$	46 279.679 dr
$5_{1,4} - 4_{1,3}$	47 426.06 tr	$20_{1,18} - 21_{1,21}$	42 219.612 dr
$5_{1,5} - 4_{1,4}$	45 784.82 tr	$20_{1,19} - 19_{2,18}$	29 674.68 tr
$20_{1,19} - 20_{1,20}$	68 517.847 dr	$25_{2,23} - 26_{1,26}$	28 046.777 tr
$21_{1,20} - 21_{1,21}$	75 335.399 dr	$24_{3,22} - 25_{2,23}$	68 367.97 tr
$20_{2,18} - 20_{2,19}$	9 030.22 tr		
$25_{2,23} - 25_{2,24}$	20 794.48 tr		

sitions shown with broken arrows in Fig. 2. In Table 2 are listed values of microwave transition frequencies in the $v_2 = 1$ state of HCCCHO observed by the methods of double resonance (dr) and triple resonance (tr). Rotational analysis by using these values has yielded a set of provisional values of molecular constants of HCCCHO in the $v_2 = 1$ state. Those values in MHz are $A = 67\ 938.6 \pm 6.4$, $B = 4\ 823.87 \pm 0.69$, $C = 4\ 496.66 \pm 0.69$, $\Delta_J = -\ 0.015 \pm 0.007$, $\Delta_{JK} = -\ 0.75 \pm 0.26$, $\Delta_K = 15.8 \pm 5.3$, $\delta_J = 0.0000 \pm 0.0002$, and $\delta_K = -\ 0.16 \pm 0.05$. The centrifugal constants were not determined previously [7], but the above-mentioned values of centrifugal constants must have larger errors than those shown, because observed microwave transitions in the vibrationally excited state are subject to rather large perturbations from other vibrational states.

A smaller number of observed rotational transitions in DCCCHO makes it difficult at present to calculate precise values of rotational constants of this molecule. The reasons for weaker signals in DCCCHO are that (1) vibrational frequencies of the ν_9 and ν_{12} modes are lower and reduce the ground-state population, and (2) the electric quadrupole hyperfine structure due to D gives rise to unresolved splittings in IR and MW transitions. Triple resonance experiments with a low-temperature cell at 200° K in order to improve the signal-to-noise ratio are now in progress.

3. IR-MW Double Resonance Studies of Collisional Relaxation in Ammonia

Observation of double resonance signals in a four-level system can be applied to study relaxation processes in the system. The technique of IR-MW double resonance provides powerful methods of studying collision-induced transitions particularly among vibration-rotation levels of molecules in gas. Collisional excitation transfer among rotation-inversion levels of NH_3 was extensively studied by OKA by using four-level MW-MW double resonance [9].

The progress in laser Stark spectroscopy has revealed several pairs of good coincidence between laser lines and the absorption lines of NH_3 [10]. A few useful pairs in the 10 μm band are listed in Table 3. The first pair in the table was used by SHIMIZU and OKA in their first IR-MW double resonance ex-

Table 3 Coincidence between laser and ammonia lines

Laser Transition	Laser Frequency[cm^{-1}]	Ammonia Transition	Ammonia Frequency[cm^{-1}]
N_2O P(13)	927.7417	$^{14}NH_3$ ν_2 asQ(8,7)	927.7420
CO_2 R(42)	988.6466	$^{15}NH_3$ ν_2 asR(2,0)	988.6472
$^{13}CO_2$ R(18)	927.3004	$^{15}NH_3$ ν_2 asQ(5,4)	927.2992*

*Indirectly measured with a larger uncertainty of ±0.0006 cm^{-1} compared to others.

periment [11]. Collisional transfer between the J = 8, K = 7 inversion doublets and the J = 8 ± 1, K = 7 levels were observed. The preference rule for changes in rotational quantum numbers at collision, $\Delta J = 0, \pm 1$, $\Delta K = 0$, and a↔s (antisymmetric↔symmetric), was confirmed. This investigation has been extended to a large number of rotational states by improving the sensitivity of double resonance so as to detect a change of 0.01% in the absorption constant of the microwave transitions between the inversion doublets [12].

When the upper inversion level of the J = 8, K = 7 state is pumped by the N_2O P(13) laser line, many microwave inversion lines are found to change their intensities because of collision-induced transfer of the population. The non-thermal distribution of the pumped level is transferred to other levels (1) by collision-induced transitions and their cascading transitions, (2) by intermolecular energy transfer between the colliding molecules, and (3) by the temperature rise due to absorption of the laser power. This thermal effect in double resonance was investigated by KREINER et al. [13]. Population transfer between inversion doublets was discussed by LEMAIRE et al. [14], disregarding the thermal effect. Careful measurements of the pressure dependence by KANO et al.[15] have discriminated the thermal effect from more or less direct processes of energy transfer between the pumped and the observed levels.

The double resonance experiment by using the second coincidence in Table 3 is particularly important for the study of relaxation processes because the K = 0 level is pumped where the partner of inversion doublet is missing. The relaxation effect in this case does not therefore involve transfer of energy between inversion doublets.

In order to further discriminate the effects of intermolecular and intramolecular energy transfer, behavior of double resonance in a mixture of $^{14}NH_3$ and $^{15}NH_3$, when either $^{14}NH_3$ or $^{15}NH_3$ was pumped, was studied by MORITA et al. [16]. Investigations of pressure dependence and J-, K-dependence of these double resonance signals have revealed vibration-rotation and vibration-vibration processes as well as rotation-rotation and inversion processes of collisional relaxation.

The rotational states of ammonia are classified into ortho states (K = 3n) and para states (K = 3n ± 1) according to the spin statistics of three protons. Double resonance signals observed in microwave transitions of ortho (para) ammonia, when the para (ortho) ammonia is pumped, have been found to behave like the double resonance signals observed in the mixture of $^{14}NH_3$ and $^{15}NH_3$. Since exchange of ^{14}N and ^{15}N is extremely slow in ammonia, those observed effects are ascribed to intermolecular energy transfer which is mostly the vibration-vibration process of energy transfer. Thus the apparent transfer between ortho and para states is not due to the intramolecular conversion between the ortho and para states.

Selective energy transfer between inversion doublets has been found to be less dominant. This is probably because the inversion doubling in ammonia (0.8 cm^{-1}) is much smaller than the kinetic energy (210 cm^{-1}).

These findings have also been confirmed by time-resolved observation of microwave absorption following the resonant infrared laser pulse.

4. Future Trends

Vibrational and rotational states of molecules in gas will be best studied in detail by the methods of IR-MW double resonance. Development of tunable lasers will make their wider applications possible.

High-resolution spectroscopy of molecules in the excited state will reveal finer structures of higher-order effects, and show up dynamic pictures of the molecule. Relaxation processes in other gases will be studied, and quantitative measurements of the rate of collision-induced transition between any specified vibration-rotation levels will become possible. Not only the state-dependence but also the velocity-dependence of the collision-induced transition are being investigated.

Although the double resonance technique promises such investigations because of its high sensitivity, further improvements are desired before its practical application. The method of polarization spectroscopy will not be very useful in the infrared. Instead, stable lasers are available in the infrared and they ensure comparable sensitivity in spectroscopic detection, as long as the Kramers-Kronig relation holds at least in approximation.

Higher-order resonance, and modifications or combinations of double resonance with other nonlinear effects are the interesting subjects which may not be called double resonance. Coherent transient and other quantum interference effects in double resonance are the subjects of further research.

References

1. K. Shimoda; in *Laser Spectroscopy*, ed. by R. G. Brewer and A. Mooradian (Plenum, New York, 1974) pp.29-44
2. K. Shimoda: in *Laser Spectroscopy of Atoms and Molecules*, ed. by H. Walther (Springer, Berlin, 1976) pp.197-252
   V. P. Chebotayev: in *High Resolution Laser Spectroscopy*, ed. by K. Shimoda (Springer, Berlin, 1976) pp.201-251
   J. I. Steinfeld and P. L. Houston: in *Laser and Coherence Methods in Spectroscopy*, ed. by J. I. Steinfeld (Plenum, New York, to be published) and references therein
3. Some other recent papers are:
   S. M. Freund, T. Oka: Phys. Rev. A, 13, 2178 (1976)
   F. Herlemont, J. Thibault, J. Lemaire: J. Mol. Spectrosc. 61, 138 (1976)
   F. Herlemont, J. Thibault, J. Lemaire: Chem. Phys. Letters, 41, 466 (1976)
   F. Herlemont, J. Lemaire: CR Acad. Sci. Paris, B282, 511 (1976)
   H. Jones, F. Kohler, H. D. Rudolph: J. Mol. Spectrosc. 63, 205 (1976)
   H. Jones: Z. Naturforsch. 31a, 1614 (1976)
   W. A. Kreiner, T. Oka: Can. J. Phys. 53, 2000 (1975)
   B. Macke: Appl. Phys. 13, 271 (1977)
   M. Takami: Jpn. J. Appl. Phys. 15, 1063 and 1889 (1976)
   T. Tanaka, C. Yamada, E. Hirota: J. Mol. Spectrosc. 63, 142 (1976)
4. M. Takami, K. Shimoda: Jpn. J. Appl. Phys. 10, 658 (1971)
   K. Shimoda, M. Takami: Opt. Commun. 4, 388 (1972)
   M. Takami, K. Shimoda: Jpn. J. Appl. Phys. 11, 1648 (1972)
5. M. Takami, K. Shimoda: Jpn. J. Appl. Phys. 12, 603 (1973)
6. M. Takami, K. Shimoda: Jpn. J. Appl. Phys. 13, 1699 (1974)
7. M. Takami, K. Shimoda: J. Mol. Spectrosc. 59, 35 (1976)

8. J. C. D. Brand, J. K. G. Watson: Trans. Faraday Soc. 56, 1582 (1960)
9. T. Oka: J. Chem. Phys. 49, 13 (1967); 48, 4919 (1968); 49, 3135 (1968)
10. Y. Ueda, K. Shimoda: in *Laser Spectroscopy*, ed. by S. Haroche et al. Lecture Notes in Physics, vol.43 (Springer, Berlin, 1975) pp.186-197
11. T. Shimizu, T. Oka: Phys. Rev. A2, 1171 (1970)
12. S. Kano, T. Amano, T. Shimizu: Chem. Phys. Letters, 25, 119 (1974)
13. W. A. Kreiner, A. Eyer, H. Jones: J. Mol. Spectrosc. 52, 420 (1974)
14. J. Lemaire, J. Houtiez, F. Herlemont, J. Thibault: Chem. Phys. Letters, 19, 373 (1973)
J. Lemaier, J. Thibault, F. Herlemont, J. Houtiez: Mol. Phys. 27, 611 (1974)
15. S. Kano, T. Amano, T. Shimizu: J. Chem. Phys. 64, 4711 (1976)
16. N. Morita, S. Kano, Y. Ueda, T. Shimizu: J. Chem. Phys. 66, 2226 (1977)

RF SPECTROSCOPY IN A LASER CAVITY: "PURE" NUCLEAR QUADRUPOLE SPECTRA

E. Arimondo, P. Glorieux, and T. Oka

Herzberg Institute of Astrophysics, National Research Council of Canada
Ottawa, Ontario, Canada

1. RF Spectroscopy

The intensity of radiofrequency absorption spectrum is very small for two reasons;

(a) the low energy of photon $h\nu$,

(b) the small population difference $h\nu/kT$ in LTE.

In molecular beam resonance methods one successfully avoids these points by (a) detecting molecules rather than radiation and (b) using state selectors to produce non-thermal distribution. The infrared-radiofrequency double resonance in a laser cavity, used in this paper, avoids these points by (a) detecting IR laser radiation rather than the RF radiation, and (b) producing a non-thermal population by infrared pumping.

The basic setup is shown in Fig.1. A sample cell containing molecules at a low pressure ($\sim$ 10m Torr) is placed in the cavity of an infrared laser (CO_2, isotopic CO_2 and N_2O) and radiofrequency radiation ($\sim$ 3 Watts) is applied to the molecules while the laser power is monitored by a detector. At a RF resonance, the load characteristics of the gas vary and a sharp variation of the laser power is detected.

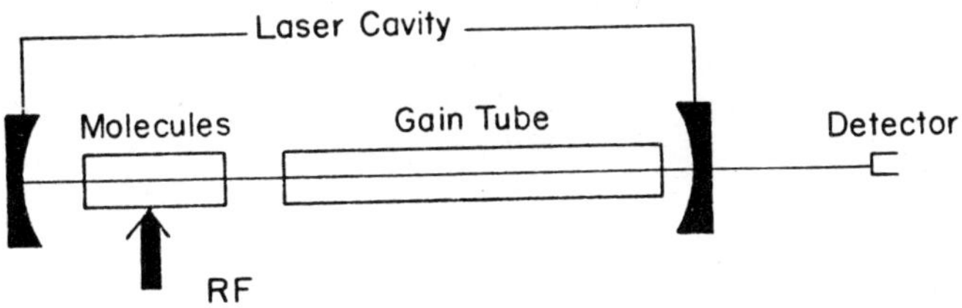

Fig.1 Basic experimental setup for RF spectroscopy inside a laser cavity

2. How does it work?

The response of molecules to the IR and RF radiations can be seen by using the energy-population diagram shown in Fig.2. The laser radiation interacts with a group of molecules with certain velocity components $\pm v$ and bleaches two holes in the Maxwellian velocity profiles in the ground state and produces two spikes in the excited state, thus creating a very non-thermal population distribution. When the RF resonance is tuned to resonance in the ground state (ω_{12}) or excited state (ω_{34})

these holes or spikes, respectively, are transferred to other levels thus changing the load characteristics of the gas. Both resonances increase molecular absorption and thus decrease the laser power. It is difficult to discriminate between resonances ω_{12} and ω_{34} experimentally. The increase in sensitivity over the conventional radiofrequency absorption method is the product of the following three factors; (a) $\Omega(IR)/\omega(RF) \gtrsim 10^5$, (b) $kT/2h\nu \gtrsim 3\times10^4$ and (c) gain due to laser nonlinearity $\gtrsim 10$.

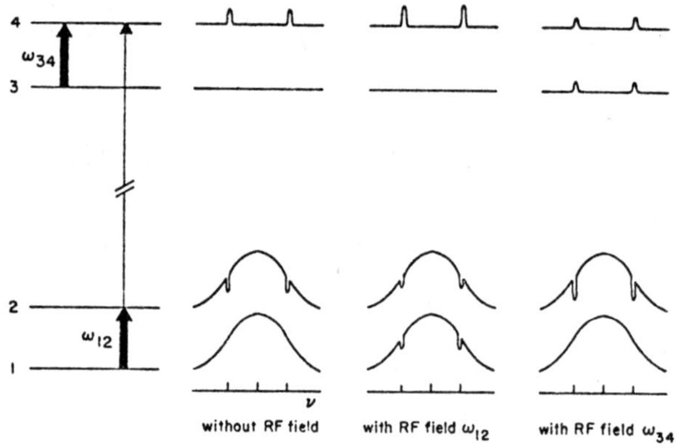

Fig.2 Operation of infrared-radiofrequency double resonance in saturated regime

3. "Pure" Quadrupole Resonance

A nucleus with $I \geq 1$ possesses an electric quadrupole moment and its energy level is split into $2I+1$ levels with different M_I through the interaction with the field gradient of electron cloud (Fig.3). The RF transitions between these split levels are normally observed in solid[1]. The high sensitivity of laser technique enables us to observe them in gases.

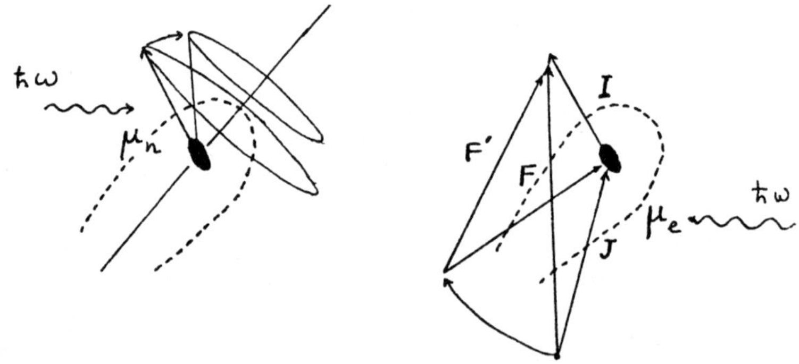

Fig.3 Pure quadrupole resonance in solid (left) and in gas (right)

In gas phase the nuclear spin angular momentum $\vec{I}$ is coupled with the rotational angular momentum $\vec{J}$ to form the total angular momentum

$$\vec{F} = \vec{J} + \vec{I} \qquad (1)$$

This coupling makes the quadrupole coupling energy different for each rotational level of a molecule. Therefore quadrupole resonance in gas is reduced from that of solid not only by the lower density of nuclei ($\sim 10^{-10}$) but also by the dilution into many rotational levels ($\sim 10^{-3}$). However, much of this is compensated by the transition moment; while in solid the radiation reorients nuclei by using the nuclear magnetic moment μ_n, in gas the radiation can reorient the molecular frame by using the electric dipole moment μ_e. This increases absorption by a factor of $(\mu_e/\mu_n)^2 \sim 10^{10}$. An observation of the straightforward radiofrequency spectrum was reported by Sterzer and Beers[2] but has not been pursued since, because of the weakness of the signal and the complicated spectrum pattern.

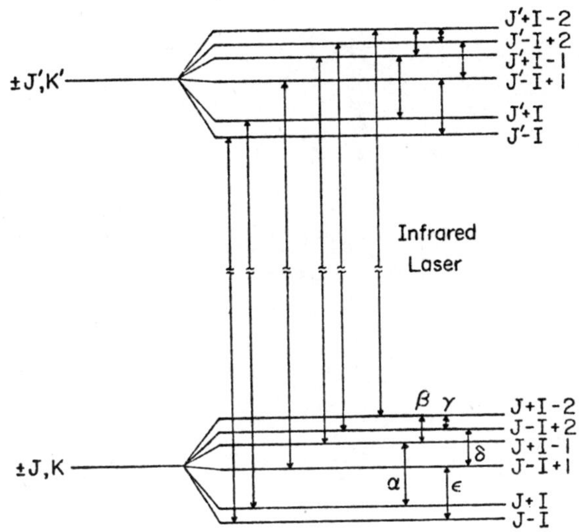

Fig.4 Splitting of a set of vibration-rotation levels of CH_3I due to eqQ of the iodine nucleus

As indicated in Fig.4 the quadrupole coupling splits each level into a sextet. Normally the laser radiation pumps all the infrared transitions and we observe a set of five radiofrequency resonances (α, β, γ, δ, ε) in the excited state and in the ground state. Fewer resonances are observed when the laser does not pump all of the transitions. The splitting of the energy level is given to the first order by

$$\Delta E = eqQ\left[\frac{3K^2}{J(J+1)} - 1\right] f(I,J,F) \qquad (2)$$

Since the Casimir factor f(I,J,F) gives relative spacing and the direction cosine factor changes the magnitude, we can uniquely assign quantum numbers J, F, and K from the observed spectrum. This in turn gives definite assignments of the infrared transitions and related far infrared laser transitions observed by Dyubko et al.[3] and by Chang and McGee[4].

In order to observe the "pure" quadrupole spectrum it is essential that each level has double parity so that $\Delta J=0$ transitions are allowed.

4. Observed Spectrum

Approximately 400 resonances have been observed using 45 infrared transitions. A few examples are shown below.

A. General pattern

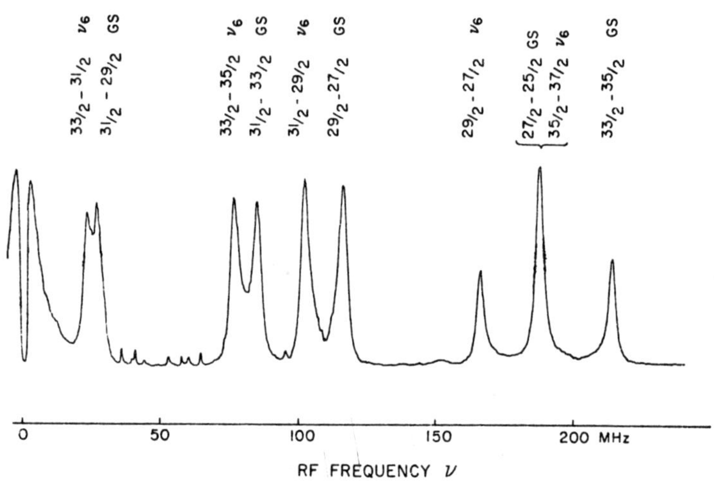

Fig.5 An example of the observed resonances for CH_3I. A coincidence of the CO_2 P(32) 10.72 μm line with the $\nu_6 \leftarrow 0$ $^rR(15,5)$ transition is used. Sample pressure $\sim$ 10 m Torr. Note the pairs of quintets for the ground state and for the excited state, very similar in shape.

B. Resolution

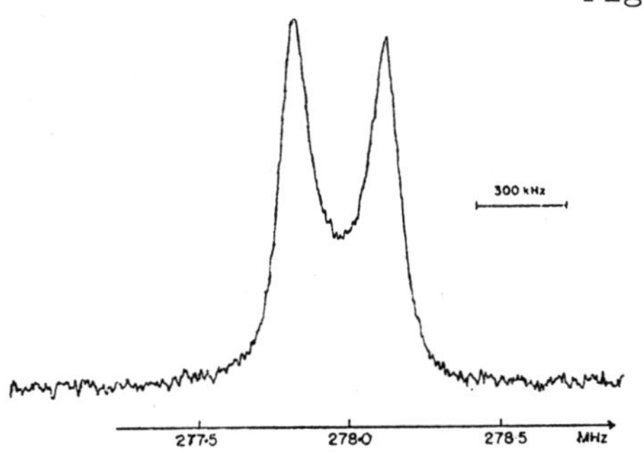

Fig.6 An example of high J resonances for CH_3I where the excited state resonance and the ground state resonance appear close as a doublet. A coincidence between the CO_2 P(34) 9.68μm line and the $2\nu_3 \leftarrow 0\ {}^{q}P_3(41)$ transition is used. Pressure 2m Torr. Time constant of detection 30 msec.

C. Sensitivity

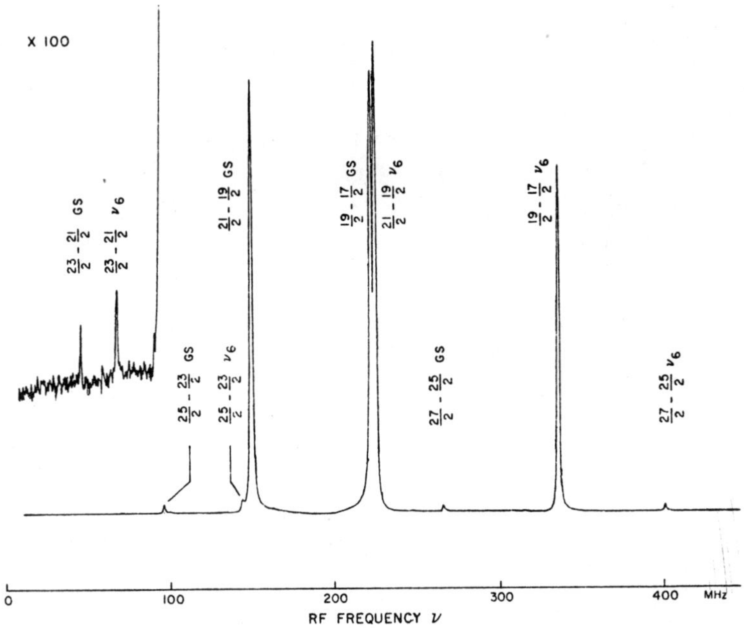

Fig.7 An example of patterns where strong signals and weak signals are observed by the same laser line. Note that the ×100 magnification of the record reveals weaker lines still with good signal-to-noise ratios. A coincidence between the CO_2 P(4) 10.44 μm line and the $\nu_6 \leftarrow 0\ {}^{r}Q(11,9)$ transition of CH_3I was used. Pressure ∿ 8m Torr. Time constant of detection ∿ 30 msec.

5. Collision-Induced Resonances

The sensitivity of the method is so high that we can observe RF resonances associated with the levels which are not directly pumped by the laser but are connected to the pumped levels through collision-induced transitions. Fig.8 gives an example. The appearance of these collision-induced signals not only helps the analysis of the spectrum by increasing the number of observable resonances but also gives interesting information on the nature of collision-induced transitions between rotational levels.

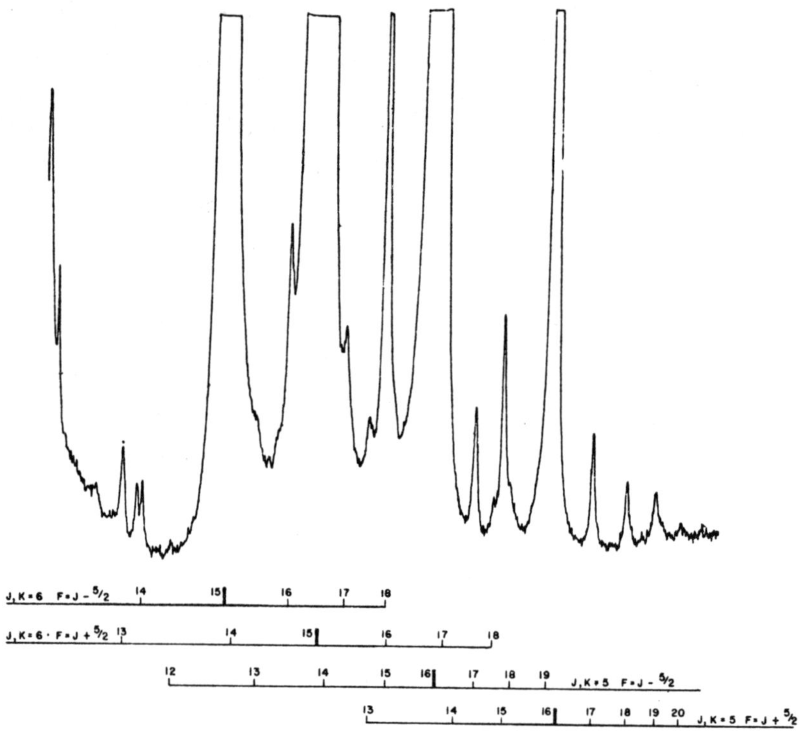

Fig.8 An example of traces showing collision-induced quadrupole resonances. The normal resonances associated with directly pumped levels are blown to off-scale. The many smaller signals which appear as satellites to the main lines are collision-induced signals. A coincidence between the N_2O P(25) line and the $\nu_6 \leftarrow 0$ ^rP (16,5) line of CH_3I is used. Sample pressure is 20m Torr

Our analysis of the collision-induced resonances using rate equations indicates that in order for such resonances to be observable, the following two conditions have to be satisfied:

(a) Collision-induced transitions between hyperfine components of different rotational levels obey some "selection" rules.[5]

(b) The collision-induced transitions occur without large velocity changes.

The laser pumps molecules and bleaches holes in the ground state and creates spikes in the excited state as shown in Fig.2. These holes and spikes are then transferred by collisions to other levels. Condition (a) is necessary in order that such transfer creates non-thermal population distribution between the levels connected by the RF radiation. Our result can be explained by assuming that the rotational transitions with $\Delta F=\Delta J$ have higher probabilities than the corresponding $\Delta F\neq\Delta J$ transitions, that is, collision-induced transitions occur without changing the angle between the $\vec{I}$ and $\vec{J}$ vectors. Condition (b) is necessary in order that the result of collision-induced transitions affects the laser operation. There has been already experimental evidence for (b)[6,7,8]. The same quadrupole resonances as shown in Fig.5 are shown in Fig.9 on the next page with their collision-induced satellites.

6. Hot Bands

The high sensitivity of the method enables us to observe quadrupole resonances using hot band infrared transitions. From such observation, vibrational dependence of quadrupole coupling constants is determined. The vibrational states for which quadrupole resonances are observed are summarized in Fig.10.

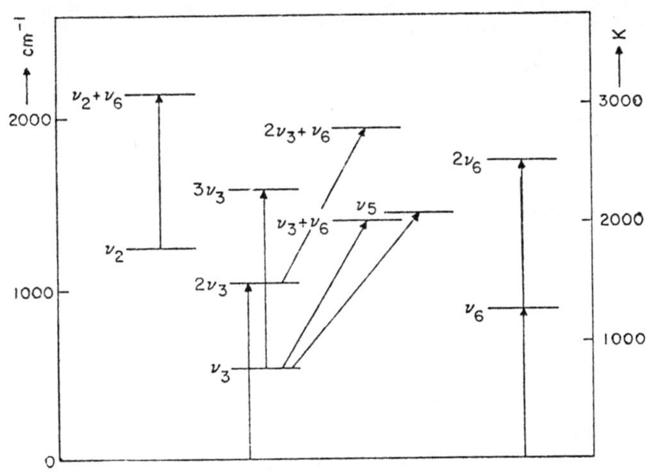

Fig.10 Infrared transitions of CH_3I used for observation of quadrupole resonances. The $2\nu_3\leftarrow 0$ and $3\nu_3\leftarrow\nu_3$ transitions are parallel bands ($\Delta K=0$) which appear in 9 μm region whereas all the other transitions are perpendicular bands ($\Delta K=\pm 1$) appearing in 10 μm region.

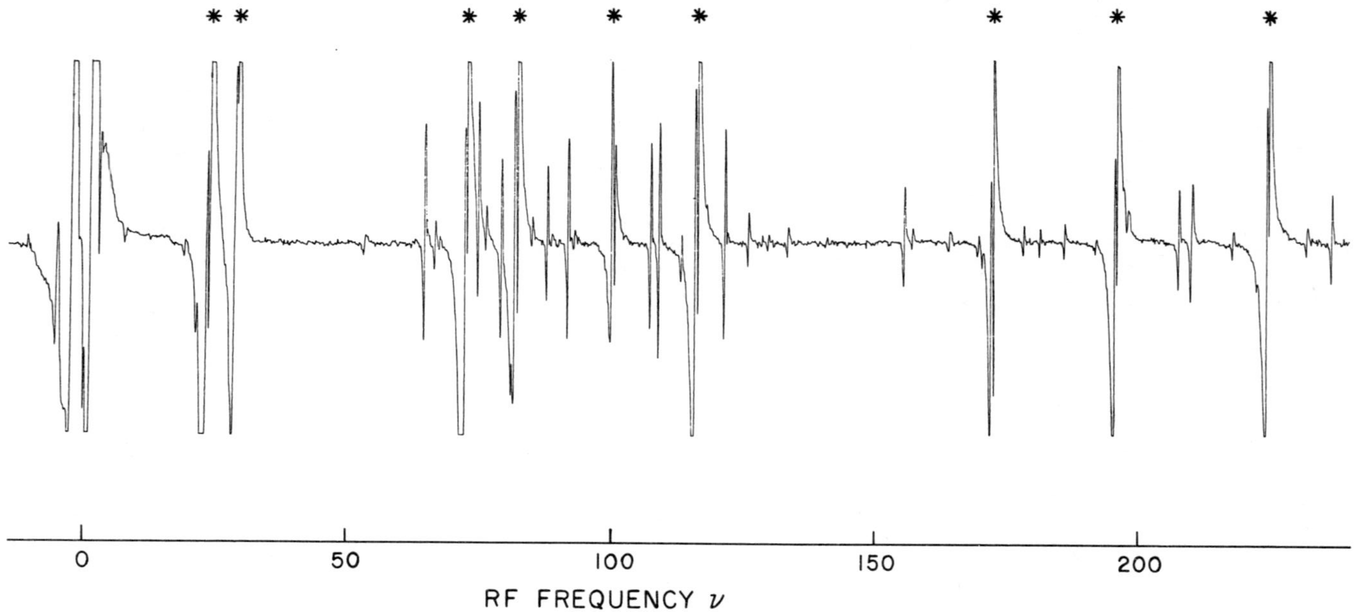

Fig.9 Quadrupole resonances of CH_3I using the CO_2 P(32) 10.72 μm laser line. This is the same spectrum as shown in Fig.5 except amplification of the normal signals (marked with *) now reveals spectrum of weaker collision-induced signals. Sample pressure ∿ 10 m Torr. Frequency modulation at 15 kHz and the time constant of detection is 100 msec.

7. Analysis

The observed quadrupole resonances have been analyzed by using the first order electric nuclear quadrupole interaction Hamiltonian given in Eq.(2) together with its higher order corrections and the magnetic spin-rotation Hamiltonian

$$E_{mag} = \frac{1}{2}\left[C_N+(C_K-C_N)\frac{K^2}{J(J+1)}\right][F(F+1)-I(I+1)-J(J+1)] \qquad (3)$$

It was necessary to include the vibrational and rotational dependence of eqQ. Some examples of vibrational and rotational dependence are shown in Figs.11 and 12. The vibrational dependence is linear as expected. The rotational dependence includes Hougen's term[10], the effect of which is rather clearly shown as the observed singularity in ΔeqQ shown in Fig. 11. Our results agree with previous microwave[11] and molecular[12] beam results but cover much higher rotational and vibrational states.

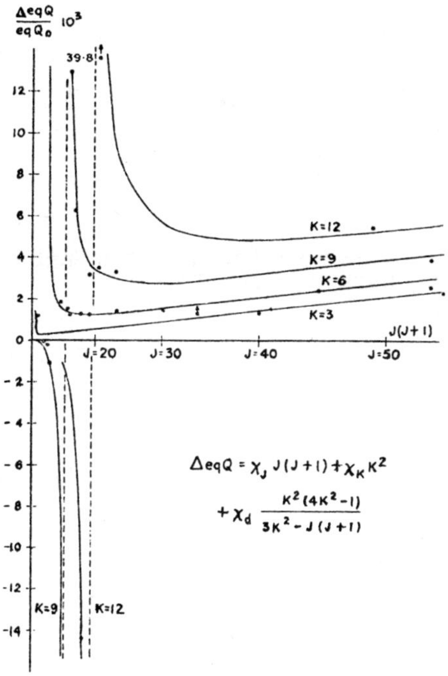

Fig.11 Rotational dependence of eqQ. The black circles indicate observed values and the curves show theoretical fits. K=3n resonances were taken for convenience.

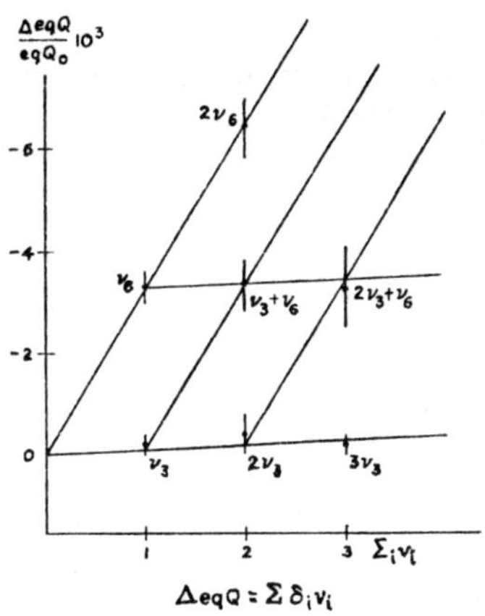

Fig.12 Vibrational dependence of eqQ.

8. Multiphoton Processes etc.

In addition to the RF resonances described above, we observe many signals caused by combinations of IR-, RF-, and IR-RF multiphoton processes shown in Fig. 13[13,14,15]. These multiphoton signals were used to measure infrared transitions accurately.

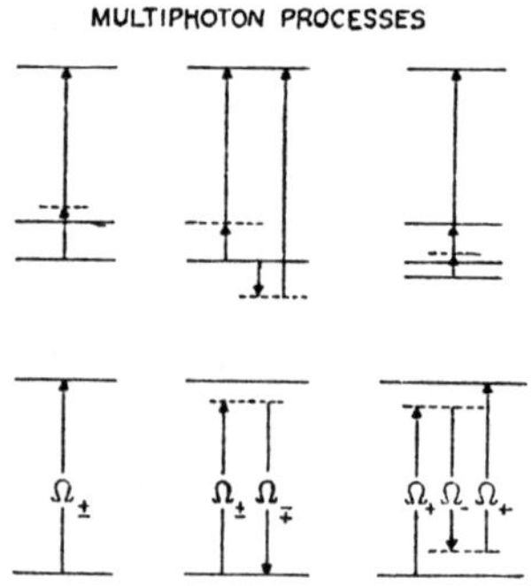

Fig.13 Various IR-RF multiphoton processes (upper) and velocity-tuned IR multiphoton processes (lower). Combinations of these processes give many sharp signals.

Often a peculiar line shape is observed for RF resonances. An example is given in Fig.14. We have not completely understood this shape.

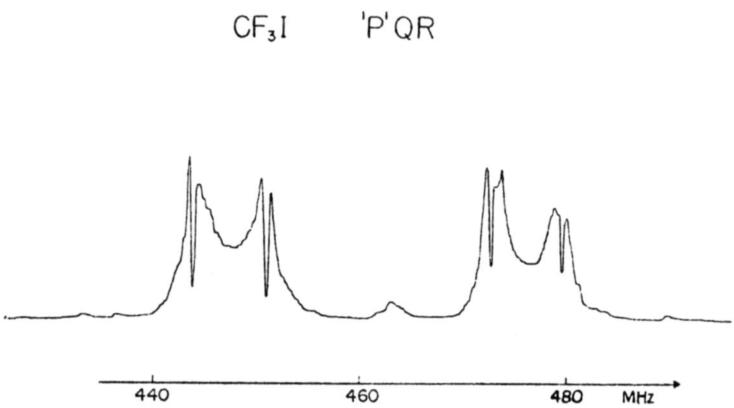

Fig.14 An example of dispersive structures at centres of RF resonances. Laser line CO_2 R(24) 9.4 μm. Pressure of CF_3I ∿ 6 m Torr.

9. Other RF Spectra

Some other examples of RF spectra are shown below.

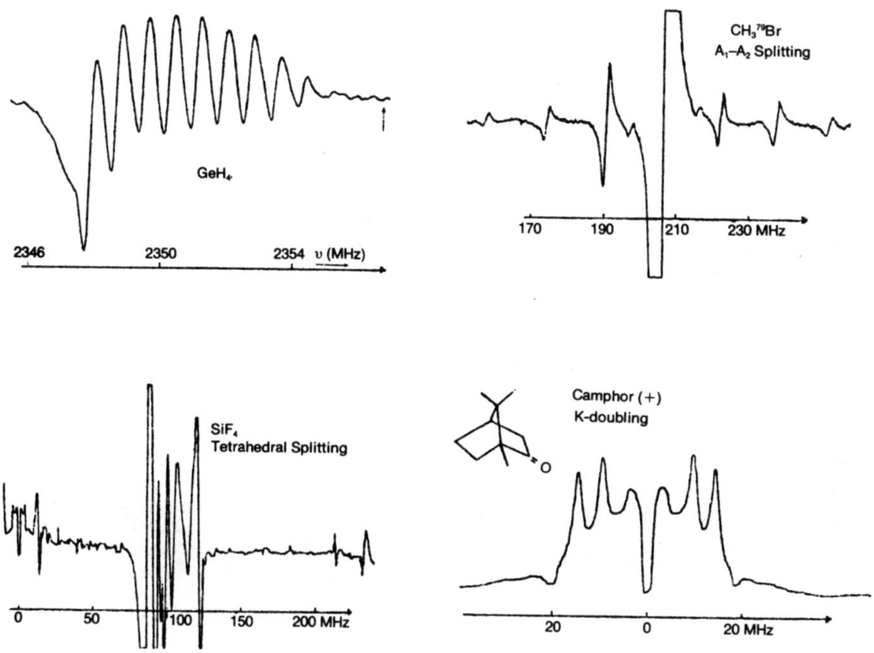

Fig.15

The GeH_4 and SiF_4 spectra were recorded by W. Kreiner and by G.W. Hills, respectively.

10. Conclusion

We have developed a very sensitive technique for radiofrequency spectroscopy. Several obvious applications are the study of the following subjects:

(1) "Forbidden" transitions

(2) Unstable molecular species such as free radicals, charge transfer complexes and molecular ions

(3) Collision-induced rotational transitions

(4) Multiphoton and other molecular processes inside the laser cavity.

(5) Higher order effect such as hexadecapole interaction.

References

1. H.G. Dehmelt and H. Krüger, Z. Physik 129, 401 (1949);
   R.V. Pound, Phys. Rev. 79, 685 (1950).

2. F. Sterzer and Y. Beers, Phys. Rev. 100, 1174 (1955).

3. S.F. Dyubko, L.D. Fesenko, O.I. Baskakov and V.A. Svich, Zh. Prikl. Spektrosk. 23, 317 (1975).

4. T.Y. Chang and J.D. McGee, IEEE J. Quant. Electr. QE1, 62 (1976).

5. T. Oka, Adv. Atom. Mol. Phys. 9, 127 (1973).

6a. M. Ouhayoun and C. Bordé, C.R. Acad. Sci. 274B, 411 (1972);
   M. Ouhayoun, Méthode de Spectroscopie sans langeur Doppler de niveaux excités de systèms moléculaires simples, 155 (CNRS) Paris (1973).

6b. S.M. Freund, J.W.C. Johns, A.R.W. McKellar, and T. Oka, J. Chem. Phys. 59, 3445 (1973);
   J.W.C. Johns, A.R.W. McKellar, T. Oka, and M. Römheld, J. Chem. Phys. 62, 1488 (1975).

7. T.W. Meyer and C.K. Rhodes, Phys. Rev. Lett. 32, 637 (1974);
   W.K. Bischel and C.K. Rhodes, Phys. Rev. A 14, 176 (1976).

8. R.G. Brewer, R.L. Schoemaker and S. Stenholm, Phys. Rev. Lett. 33, 63 (1974);
   R.L. Schoemaker, S. Stenholm and R.G. Brewer, Phys. Rev. A10, 2037 (1974).

9. J.L. Hall and J.A. Magyar, "Topics in Applied Physics" 13, 174, Springer-Verlag (1976).

10. J.T. Hougen, J. Chem. Phys. 57, 4207 (1972).

11. J. Burie, D. Boucher, J. Demaison and A. Dubrulle, Mol. Phys. 32, 289 (1976).

12. Y. Morino and C. Hirose, J. Mol. Spectrosc. 22, 99 (1967).

13. S.M. Freund, M. Römheld, and T. Oka, Phys. Rev. Lett. 35, 1487 (1975).

14. T. Oka, "Frontiers in Laser Spectroscopy", Vol. 2, p.531, North-Holland Publishing Co. (1977).

15. J. Reid and T. Oka, Phys. Rev. Lett. 38, 67 (1977).

INVESTIGATION OF THE FINE STRUCTURE SPLITTING OF RYDBERG STATES

G. Leuchs

Sektion Physik, Universität München

H. Walther

Sektion Physik, Universität München and Projektgruppe für Laserforschung
Max-Planck-Gesellschaft zur Förderung der Wissenschaften e.V.
8046 Garching, FRG

Introduction

The study of highly excited Rydberg states has recently obtained large attention partially since an efficient isotope selective ionization is possible via those states [1] . In addition, their investigation provides useful information on the atomic structure. An electron in a shell with a high principal quantum number is a sensitive probe for the interaction with the ionic core of the atom. Measurements of these Rydberg states give valuable data on quantum defects [2] , anomalies in fine structure splitting [3, 4, 5] , polarizabilities [6, 7] , configuration interactions [8, 9] , ionization potentials [10] etc. Since the lifetime of Rydberg states increases with n^3 high lying levels cannot be observed using fluorescence techniques but only by detecting electrons which are either ejected by multi-photon ionization, collisions, or field ionization. Normally, the number of emitted electrons has been measured, however, further information on the Rydberg atom can be obtained when other properties of the electrons are observed, e.g. angular distribution. In principle, such electrons are able to provide the same information as emitted photons, however, the experimental set-up required is somewhat more complicated. In the following quantum beat and double resonance techniques will be discussed. Detection is performed via the electrons generated either by field or photo ionization. In the case of the double resonance experiment it is proposed to observe the change of the angular distribution of the emitted electrons.

Quantum Beat Method

In the standard quantum beat experiment light pulses are used to excite a coherent superposition of two closely spaced levels. Detection is performed by observing the temporal behaviour of the fluorescence. The modulated part of the signal is given by

$$I_{mod} \propto \langle g|\underset{\sim}{D}|1\rangle \, \rho_{12} \, \langle 2|\underset{\sim}{D}|g\rangle \, , \qquad (1)$$

where $\underset{\sim}{D}$ is the dipole operator, $|g\rangle$, $|1\rangle$, $|2\rangle$ represent the ground and the two excited states respectively and ρ_{12} is the nondiagonal element of the density matrix. The time dependence of ρ_{12}, given by $\rho_{12} \propto \exp\ i\omega_{12}t$, provides direct information on the energy splitting, $\hbar\,\omega_{12}$ of the two levels.

Quantum beats can also be observed by means of absorption. In this case, the detection is performed by a second light pulse which measures the absorption of the system starting from the two intermediate levels $|1\rangle$ and $|2\rangle$ to a third level $|3\rangle$.

The quantum beats are obtained by measuring the absorption as a function of the time delay between the two light pulses. Similarly to (1) the resulting modulation of the absorption probability, A_{mod}, is

$$A_{mod} \propto \langle 3 | \underset{\sim}{D} | 1 \rangle \rho_{12} \langle 2 | \underset{\sim}{D} | 3 \rangle . \tag{2}$$

However, ρ_{12} has now the time behaviour $\rho_{12} \propto \exp [i \omega_{12} t_o]$ where t_o is the time interval between the two light pulses. As for the exciting light pulse, it is necessary that the duration of the second light pulse, Δt, fulfils the condition $\Delta t < \omega_{12}^{-1}$. A quantum beat experiment using this absorption method has been performed by DUCAS et al. [11] .

In the case of high Rydberg states the quantum beats can neither be observed in fluorescence nor in absorption to higher bound states. However, when the second step is a bound-free transition to the continuum the quantum beats remain observable. It can be shown [12] that the electron yield has a modulated part, which is given by

$$Y^{\vec{k}_o}_{mod} = ne \frac{d\Omega_{\vec{k}_o}}{4\pi R} \int_0^\infty \langle \psi_{\vec{k}} | \underset{\sim}{D} | 1 \rangle \rho_{12} \langle 2 | \underset{\sim}{D} | \psi_{\vec{k}} \rangle k^2 dk \tag{3}$$

where $\vec{k}$ is the wave vector of the free electron, $\Psi_{\vec{k}}$ represents the continuum state of the electron and the integration has to be carried out for constant $\vec{k}_0 = \vec{k}/k$. R is the distance from the excitation volume to the detector and $d\Omega_{\vec{k}_0}$ the corresponding solid angle. ρ_{12} has still the same form as in (2). The order of magnitude of the expected modulation may be as high as 30 % of the background signal [12] .

In order to determine fine structure splittings using this method a variable delay of the second pulse of up to 1 µs may be necessary. This is not easy to be accomplished. Thus, it is of considerable advantage when the second laser pulse is replaced by an electric field pulse so that the electrons are emitted due to field ionization. If the number of electrons is plotted as a function of the delay between the exciting laser pulse and the ionizing electric field pulse, a modulation is observed which determines the splitting between levels $|1\rangle$ and $|2\rangle$. This may represent the fine structure splitting of Rydberg states. For this method, it is essential that the amplitude of the field pulse has to be kept sufficiently small so that a finite value for the ionization probability is obtained otherwise the ionization is saturated and the time dependence of ρ_{12} is no longer detectable.

Double Resonance Method

Several investigations on Rydberg states have been performed using optical microwave double resonance. In most cases electric dipole transitions between levels with different orbital angular momentum have been investigated. The detection of the microwave transitions has either been performed by observing the change of the fluorescence intensity [5, 13, 14, 15] or the electron emission induced by field ionization [7, 16] . The observation of fluorescence is limited to lower lying levels. The use of field ionization for detection has the disadvantage that the ionizing electric field is applied when the microwave field is still present. Therefore, the measurements may be influenced by the Stark effect. To overcome this we propose to use the angular distribution of photoionization for the detection of the microwave resonance. When photoelectrons are produced by resonantly enhanced two photon transitions into the continuum their angular distribution depends sensitively on the intermediate state. In our experiments, described below, we demonstrate that this angular distribution can easily be detected.

When a microwave transition to another state is performed these intermediate states will be mixed resulting in a change of the angular distribution of the electrons.

Experimental Set-Up and Experiments

Rydberg states of Sr and Na have been investigated. The energy level scheme of Na is shown in Fig.1.

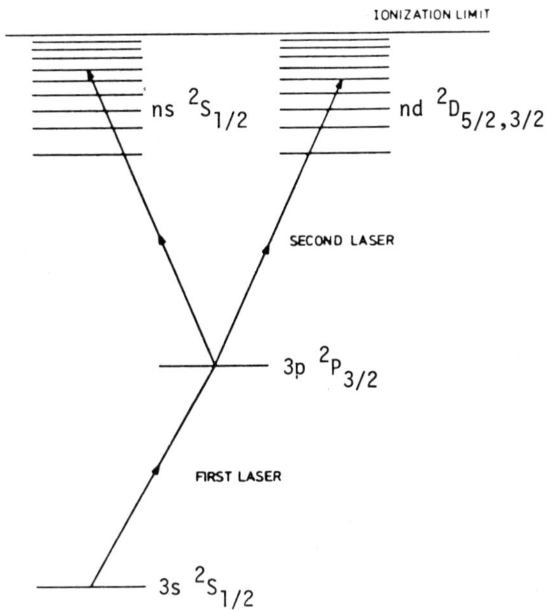

Fig.1 Energy level scheme of Na showing the transitions used in the experiment

Using a two step excitation with two N_2-laser pumped dye lasers ns-and nd-states were populated via the $3^2P_{1/2}$ or $3^2P_{3/2}$ level. A schematic of the experimental set-up is shown in Fig.2. The two dye lasers were simultaneously pumped by the same N_2-laser. Dye laser 1 was tuned to the NaD_2-line (λ = 5889 Å) and dye laser 2 was scanned over the ns and nd Rydberg series. The spectral width of the dye lasers was 0.2 Å, the output power 10 kW and the pulse duration 4 ns. The radiation of both lasers was directed at right angle to the atomic beam. The atomic density of the beam was of the order of 10" atoms/cm^3. The initial excitation of the $3^2P_{3/2}$ state of Na was checked by observing its

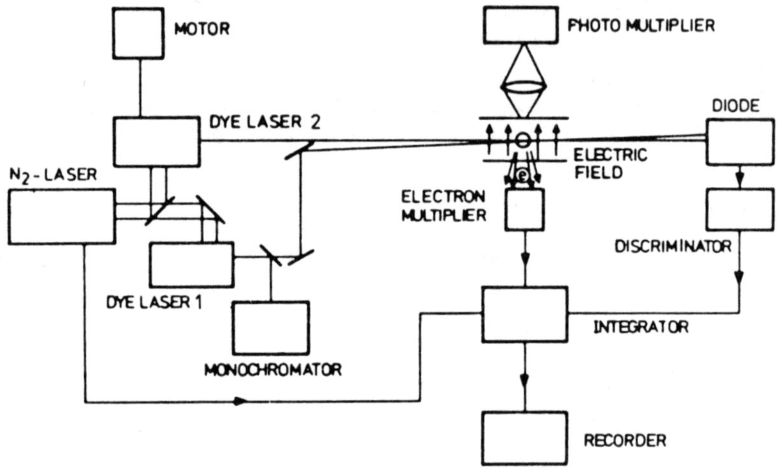

Fig.2 Experimental Set-Up

fluorescence by a photomultiplier. Field ionization was utilized to detect the excitation of Rydberg states. A schematic of the excitation region of the atomic beam is shown in more detail in Fig.3.

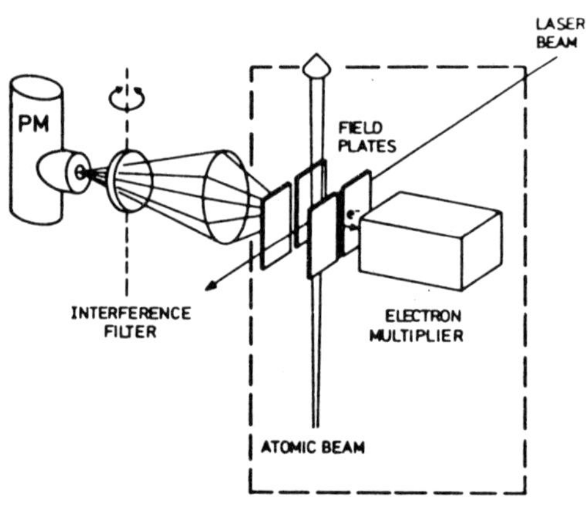

Fig.3 Schematic of the excitation region

The illuminated section of the atomic beam was situated inside a parallel plate capacitor providing a homogeneous electric field. Electrons escaping through the mash in the capacitor plate were detected by an electron multiplier. The electron signal was then sampled in a gated integrator. The gate pulse was generated by a photodiode observing the laser light. The integrated electron signal was recorded as a function of the wavelength of dye laser 2.

The absorption spectrum of the unperturbed atom is obtained by applying a pulsed electric field with a time delay of some hundred nsec in respect to the laser pulse. Using this technique the spectra of the Na Rydberg states shown in Fig. 4 have been taken for three different values of the electric field strength. It is clearly seen that the ionization limit is shifted by a variation of the field amplitude. Up to n $\simeq$ 30 s- and d-states could be selectively excited. The resolution of the measurements is determined by the laser linewidth.

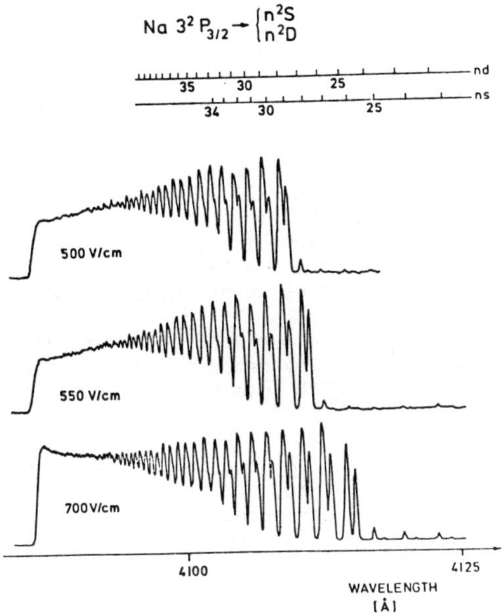

Fig. 4 Field ionization signal as a function of wavelength of the exciting laser for three different electric field strengths

Using short laser pulses a coherent superposition of Rydberg fine structure states is excited. With the electric field strength adjusted so that the fine structure levels of the Rydberg state have a finite transition probability to the continuum, the quantum beats between the fine structure levels have been measured in the way discussed above. For the measurement the delay between excitation and field ionization was controlled by the sweep of a multichannel analyzer working in the multiscaling mode. An example of the results obtained is shown in Fig. 5. The upper trace is a quantum beat signal of the $22\,^2D_{3/2,\,5/2}$ levels of Na. The lower trace shows a similar measurement on the $25\,^2S$ level. This result demonstrates that no perturbing effects are present

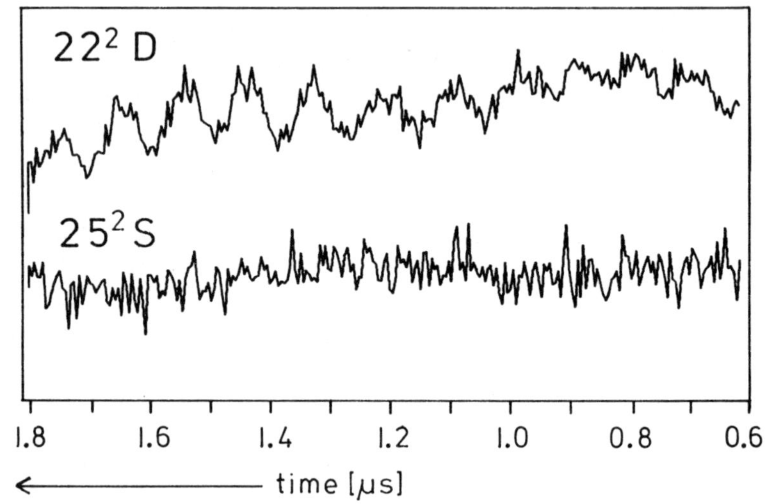

Fig. 5 Quantum beat signal between the Na 22 ^{2}D levels observed by means of time delayed field ionization. The polarization of the laser beam was perpendicular to the direction of the applied electric field. Rotating the direction of polarization by 90 degrees changes the phase of the beat signal by 180 degrees as expected by theory.

simulating a quantum beat signal. To our knowledge our measurements are the first observations of quantum beats by means of field ionization. The fine structure splitting between the 22 ^{2}D levels following from the beat signal in Fig. 5 is (9.1 ± 0.2) MHz.

In addition, photoionization can be used to detect the excitation of Rydberg states. The upper spectrum in Fig. 6 was taken with the technique described above. The electric field strength, however, was 25 V/cm so that field ionization takes place only for $n > 60$. The lines which are observed for $n < 60$ are due to electrons produced by photoionization. The purpose of the electric field is to collect all emitted electrons and accellerate them towards the detector. The lower spectrum in Fig. 6 shows the recording obtained without any electric field applied. Therefore, only those electrons are observed which are ejected towards the electron multiplier. The polarization vector of the exciting laser light was parallel to the direction of observation. It should be noted that as a function of the wavelength the signal never vanishes completely which is a result of double quantum transitions starting from the $3^2P_{3/2}$ state to the continuum. In the lower recording the s-state lines are very weak compared to the corresponding d-state lines. This is due to the inhomogeneous angular distribution of the photoelectrons which varies for different photoionization channels [17] . For resonant intermediate ns-states the angular distribution of the electrons emitted by the two photon transition is proportional to $\cos^2\vartheta$ while for resonant nd-states the distribution is given by a sum of $\cos^n\vartheta$ terms containing even power up to $n = 6$. ϑ is the angle between the direction of the polarization of the incident light and the direction into which the electron is ejected. Hence, the electrons emitted from d-states are more directional and a greater number of them reaches the detec-

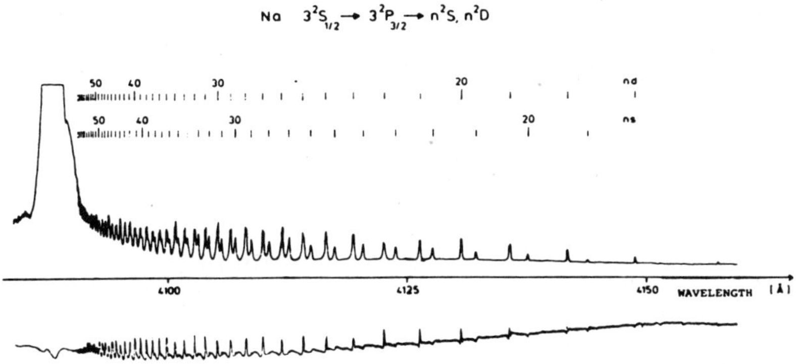

Fig. 6 Photoionization signal as a function of wavelength of the exciting laser with and without electric field (upper and lower recording respectively). The lower recording demonstrates that for nd intermediate states the photoelectrons are emitted into a smaller solid angle than for ns states.

tor as compared to electrons emitted from s-states. In the upper spectrum, however, the intensity ratio of s- and d-states is only about two. Here the effect of the angular distribution vanishes because the small electric field, applied shortly after the laser pulse, collects all electrons, no matter in which direction they were ejected.

Our experiments show that the angular distribution of the photoelectrons can easily be detected and therefore utilized to perform optical double resonance experiments. Preliminary measurements in this direction are under way.

References

1. V.S. Letokhov, Les Houches XXVII, Frontiers in Laser Spectroscopy Vol. 2, p. 771 North Holland Publishing Company, Amsterdam 1977
2. R.R. Freeman, D. Kleppner, Phys. Rev. A 14, 1614 (1976)
3. K.C. Harvey, B.P. Stoicheff, Phys. Rev. Lett. 38, 537 (1977)
4. C. Fabre, M. Gross, S. Haroche, Opt. Comm. 13, 393 (1975)
5. S. Svanberg, P. Tsekeris, W. Happer, Phys. Rev. Lett. 30, 817 (1973)
6. K. Fredriksson, S. Svanberg, Z. Physik A 281, 189 (1977)
7. T.F. Gallagher, L.M. Humphrey, R.M. Hill, W.E. Cooke, S.A. Edelstein, Phys. Rev. A 15, 1937 (1977)
8. P. Esherick, J.J. Wynne, J.A. Armstrong, Opt. Lett. 1, 19 (1977)
9. P. Esherick, J.J. Wynne, Comments Atomic Mol. Phys. 7, 43 (1977)
10. R.W. Solarz, C.A. May, L.R. Carlson, E.F. Worden, S.A. Johnson, J.A. Paisner, Phys. Rev. A 14, 1129 (1976)
11. T.W. Ducas, M.G. Littman, M.L. Zimmerman, Phys. Rev. Lett. 35, 1752 (1975)
12. R. Zygan-Maus, Diplomarbeit, München (1977)
13. J. Farley, R. Gupta, Phys. Rev. A 15, 1952 (1977)
14. J. Farley, P. Tsekeris, R. Gupta, Phys. Rev. A 15, 1530 (1977)
15. T.F. Gallagher, R.M. Hill, S.A. Edelstein, Phys. Rev. A 13, 1448 (1976)
16. C. Fabre, P. Goy, S. Haroche, J. Phys. B 10, L 183 (1977)
17. P. Lambropoulos, private communication

LASER-INDUCED LINE SHIFTS AND DOUBLE-QUANTUM LAMB DIPS

R. Keil and P.E. Toschek

Institut für Angewandte Physik, Universität Heidelberg
6900 Heidelberg, FRG

1. Introduction

There has been recently considerable interest in light-induced resonance frequency shifts of atoms and molecules, since they can give rise to systematic errors in high-resolution spectroscopy. Light shifts occur when atomic or molecular oscillators interact with off-resonant light fields. It is useful to think of these shifts as being generated by the exchange of *virtual quanta* between the light field and the oscillator [1]. From this point of view one may acknowledge that they constitute a class of fundamental phenomena of intrinsic importance rather than a mere experimental nuisance.

Light-induced frequency shifts are detected by studying the fluorescence of the driven oscillators, or by the absorption of a second light beam, the "probe" light. In the latter case, the line shape of what is in fact a "double-resonance" signal is shown in Fig. 1 for the simplest conditions (i.e. homogeneous line broadening, population in the lowest level only, weak probe field).

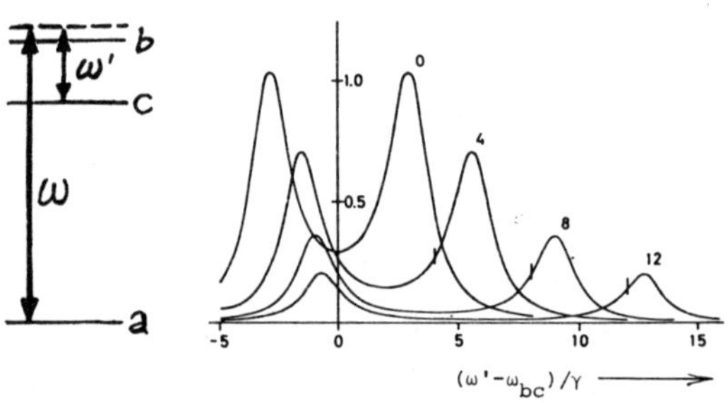

Fig. 1: Line shapes of double-resonance signals $I(\omega'-\omega_{bc})$ for various values of $\omega - \omega_{ab}$. Vertical bars: $\omega'-\omega_{bc} = \omega-\omega_{ab}$. After [2].

The vertical bars give the tuning of the intense light field. For a large detuning, the deviation of the line centres of the doublet from the oscillator resonance and from the light frequency represents the light shift, where the two resonances correspond to fluorescence and Raman scattering, respectively. With the intense light close to resonance, line splitting due to the dynamic Stark effect will eventually become dominant at a high enough light flux.

Light shifts have been observed in the past in RF double-resonance experiments by ARDITI and CARVER [3], and in particular by COHEN-TANNOUDJI [4]. Subsequently, observations were reported with the use of pulsed high-power lasers e.g. by BONCH-BRUYEVICH and collaborators [5], and by PLATZ [6]. The latter investigations were hampered by the spectral quality of the light sources, which was poor compared with present sources.

At the Second International Conference on Laser Spectroscopy in Megeve, 1975, BJORKHOLM and LIAO have reported modifications of two-photon spectra by virtual single photon transitions to a close intermediate level [7]. With detunings of the light fields by 5 to 25 GHz from the single-photon resonance, the observed shifts were some hundred MHz. An experimental problem in this type of study is caused by spatial inhomogeneities due to the strong focusing of the beams.

2. Experiment

We have chosen a different approach, which combines a highly homogeneous intracavity beam of stable frequency and a bandwidth of the order of 100 kHz with frequency controls of high stability and spectral reproducibility. With this system, we have investigated by a *cross-saturation technique* [8] optical light shifts which occur *inside* a Doppler-broadened line. The interacting medium is neon in a helium-neon discharge (s. Fig. 2). The relevant transitions,quasi-resonant with the shift-generating red light and the IR probe

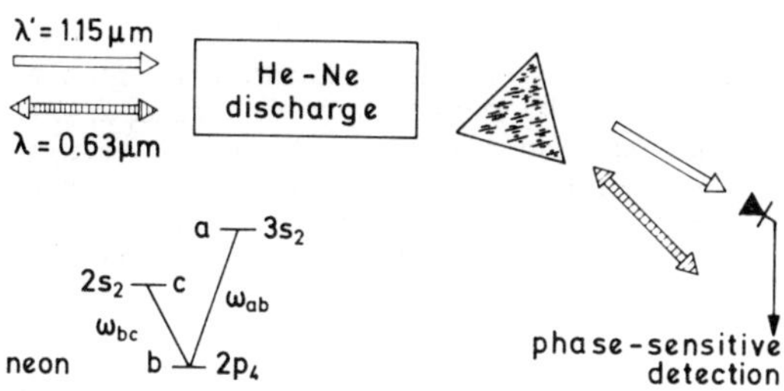

Fig. 2: Scheme of cross-saturation experiment.

light, are $3s_2$ - $2p_4$ and $2s_2$ - $2p_4$, respectively. The experimental setup (s. Fig. 3) is analogous to the one employed in the past for the observation of the dynamic Stark splitting [9] ; it is, however, supplemented by two heterodyne frequency control systems:

The light of the red high-intensity laser is mixed, on the surface of a fast photo diode, with the output of an iodine-stabilized He-Ne laser. The difference frequency is divided by a preselected number, and the resulting ratio is compared with a quartz-generated frequency. Using this method, the light frequency may be reproducibly set to any value in the range of ± 600 MHz about line centre.

The heterodyne frequency of the IR laser is counted in.1-sec intervals. The counting rate is converted into an analogue signal which is fed to the x-input of an x-y recorder.

The IR amplification (or absorption) in the probe cell is observed via a double phase-lock detection scheme which enables us to suppress (1) the linear, power-independent part of the susceptibility, and (2) the tuning dependence of the IR power. This is achieved by switching (1) the red laser on and off, and (2) the probe light to alternating paths inside or ouside the probe cell.

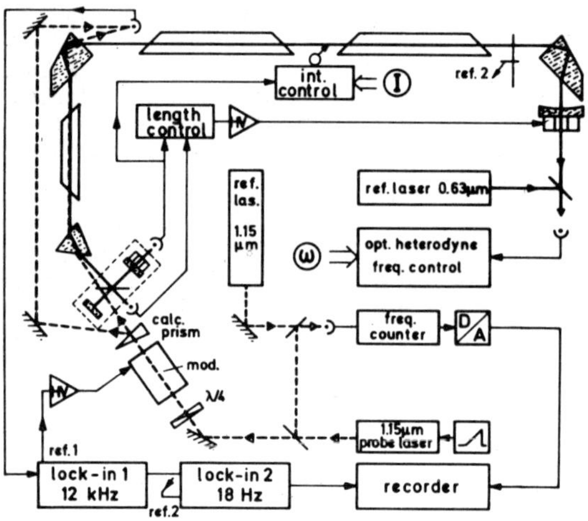

Fig. 3: Experimental setup.

To more easily interpret the spectra, we apply a well-known trick: at a certain partial pressure in the probe cell, the medium becomes transparent for the red light, i.e., the populations of the a = $3s_2$ and b = $2p_4$ levels become equal. This condition eliminates all contributions due to *real* transitions among these levels and renders a line shape which is determined only by *virtual* transitions a-b which impose the light shift upon the neighbor resonance b-c.

3. Light Shift Results

Typical recordings are shown in Fig. 4. The line shape is qualitatively understood as the difference between a split line and a first-order Lorentzian, where the frequency distance of the negative lobes reflects the dynamic Stark splitting of resonant atoms. With the power flux of the red light up to $90 W/cm^2$, we observe light shifts of several MHz. With a constant detuning of the red laser frequency, we notice an approximately linear dependence of the light shift on the light flux (s. Fig. 5).

One expects no net light shifts for a flat velocity distribution. However, there is a contribution which results from the slopes of the Gaussian distribution whose tuning spectrum, about line centre, is the derivative of a Gaussian. In our experiment, there is also the counterpropagating component of the red standing wave present, which gives rise to an additional contribution. The lowest perturbation order in which this contribution appears is the fifth order (corresponding to the fourth power in the red field) which refers to a three-photon process or, more specifically, to a two photon light shift, where one shifting photon is taken from the saturator wave E_+ traveling to the right, and the second photon is taken from the saturator wave E_- traveling to the left.

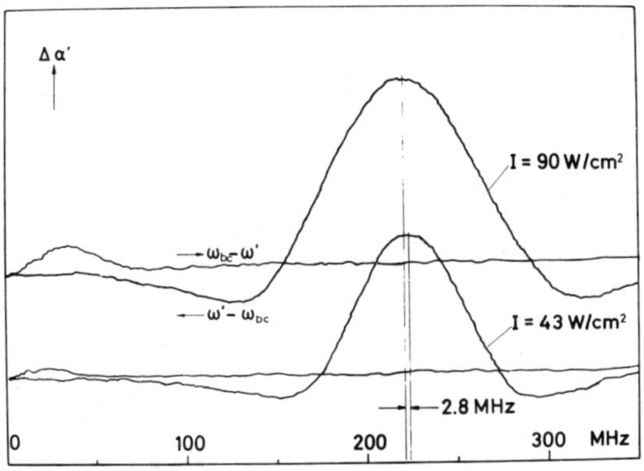

Fig. 4: Cross-saturation spectra $I'(\omega')$, detected by probe-beam absorption, for $n_a=n_b$ at two power levels of red laser (ω).

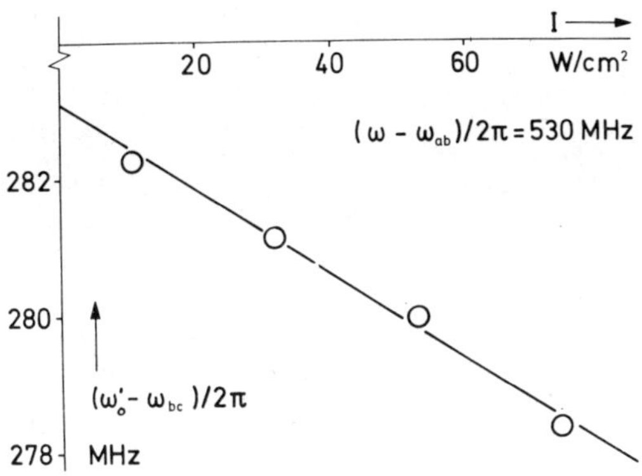

Fig. 5: Peak frequency detuning of cross-saturation spectra vs. light flux of red laser, $I(\omega)$.

We consider only the fifth-order result of both contributions and allow the width γ of the two-photon Lorentzian to be power-broadened. This is appropriate as long as the power broadening is small compared with the Doppler width of the line. Then, the tuning spectrum of the light shift is |10|:

$$\delta\Omega=\delta\Omega_0\left(\Omega - \frac{k - k'}{8k} \frac{\Omega}{(\gamma/ku)^2 + (\Omega/ku)^2}\right)e^{-(\Omega/ku)^2} \qquad (1)$$

where $\Omega = \omega - \omega_{ab} = (k/k')(\omega'-\omega_{bc})$, and ω, k, and ω', k' are the light frequency and wave number of saturator and probe wave, respectively, ω_{ij} is the frequency of transition i-j, and u is the most probable molecular speed. A spectrum which was calculated according to this expression - and which consequently includes one scaling factor - is shown in Fig. 6, in addition to experimental data. The width γ was derived from the observed two-photon Lamb dip (see below).

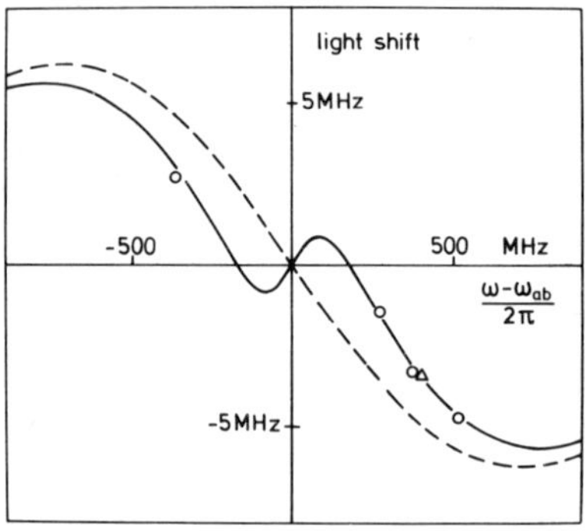

Fig.6: Light shift at 75 W/cm^2 vs. tuning of red laser across Doppler-broadened line (Δ: control measurement). Full line: calculated spectrum (1), broken line: contribution taking into account only the red light co-propagating with the IR probe beam.

The two-photon light shift which, to our knowledge, has not been observed before, is not a small correction but rather a qualitatively different contribution. This results from the three physically discernible waves which participate in the interaction.

Recently, STENHOLM |11| has hinted to the usefulness of a resonant intermediate level in two-photon spectroscopy: an increase of up to 10 orders of magnitude in the signal may result. The data show, that for central tuning, no prohibitive light shifts occur, and the two-photon light shift caused by a standing wave even may efficiently cancel the remaining shift within a range of some 100 MHz.

4. Higher-Order Lamb Dips

In contrast with light shifts, which are caused by the *virtual* exchange of photons, it is the phenomenon of the LAMB dip |12| , which is generated by a change of the emission rate of *real* quanta due to the simultaneous interaction of some atoms with the counterpropagating parts of the standing wave. It would be redundant to stress the importance of this resonance for the purpose of laser stabilization and high-resolution spectroscopy. In this respect, little attention has been devoted to analogous higher-order phenomena, some of which have been predicted by Fudjio SHIMIZU |13|. FREUND, RÖMHELD, and OKA have observed resonances of this type in the interaction of methylfluoride (CH_3F) with a microwave and with the standing IR wave (9.4 μm) of a CO_2 laser |14|. The interpretation of their results involves (a) single-photon transition amplitudes, (b) IR-microwave two-photon amplitudes, and (c) IR three-photon amplitudes, which can be resonant with some atoms according to their Doppler tuning. IR-microwave two-photon Lamb dips from the interaction b - b were observed, and also cross-resonances ot the types a - b and b - c.

Let us turn to experiments which include only optical radiation fields. Recently, WOERDMAN and SCHUURMANS were able, in a spectroscopic investigation of the sodium dimer, Na_2 [15], to excite single-photon transitions and two-photon transitions within a single dye laser tuning spectrum making use of an intermediate level placed almost half-way between the levels of the two-photon transition. The population of the uppermost level was detected by recording its fluorescent decay. The observed signals include the Doppler-free two-photon line (which is, of course, the equal-frequency limit of a two-photon Lamb dip), a single-photon Lamb dip, and a crossover resonance from the interaction of a two-photon amplitude with a single-photon amplitude.

BAKLANOV, BETEROV, CHEBOTAYEV, and DUBETSKY have reported the observation of a resonance in the output of a weak probe He-Ne laser at 1.15 μm which was irradiated by an intense beam of 1.5 μm light [16]. The main contribution to this phenomenon is from an interference term in the susceptibility closely related to that one which generates a two-photon Lamb dip.

5. Results on Two-Photon Lamb Dips

In our experiments, we have studied the amplitudes of the light-shifted signals which were discussed above as a function of the detuning of the red laser (s. Fig. 7). It turns out experimentally and also from a fifth-order calculation that the susceptibility contribution which is singled out by the experimental procedure has a Gaussian tuning dependence of its magnitude, with a

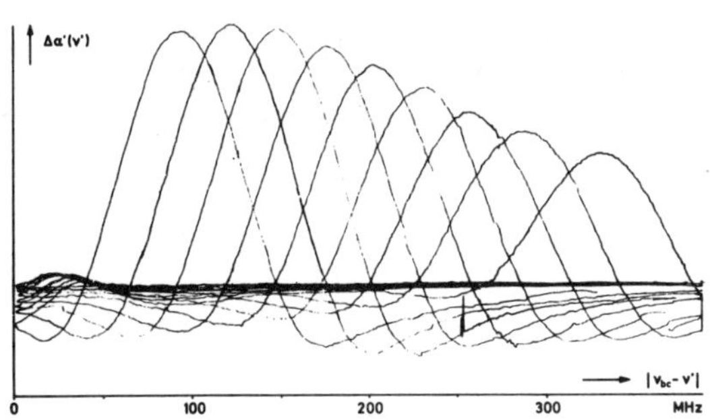

Fig. 7: Cross-saturation spectra $I'(\omega')$ with stepwise tuned frequency $\omega-\omega_{ab}$.

Lorentzian dip superimposed (s. Fig. 8). The latter is a genuine two-photon Lamb dip, as can be visualized by a pertubation chain (s. Fig. 9). Analogous two-photon amplitudes, each of which includes the square of the amplitude of one of the two counter-propagating waves, mutually saturate a dipole density. This corresponds to the conventional Lamb dip, where analogous single-photon amplitudes of the counterpropagating waves saturate a population difference (an inversion density). In contrast, in the dip of [16], the mutual saturation of the dipoles occurs by *differently* ordered two-photon amplitudes, or by a two-photon amplitude and a sequence of single-photon amplitudes.

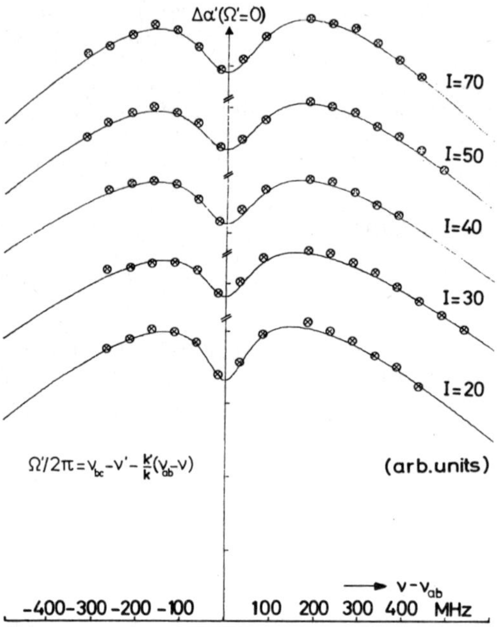

Fig. 8: Peak signal $I'(\omega_0')$ vs. $\omega-\omega_{ab}$ for various light flux levels of red laser $I(\omega)$. Lines: Fitted Gaussians with superimposed Lorentzians.

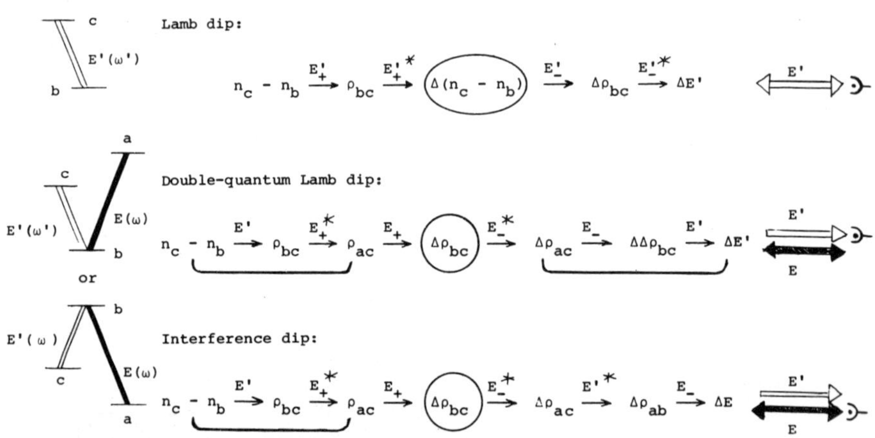

Fig. 9: Lowest-order interactions which contribute to Lamb dips and two-photon Lamb dips. Third line: interference of non-identical amplitudes, which contributes to resonance of [16].

The width of the observed dip in the limit of small saturator light flux is

$$\gamma_0 = \frac{k}{k-k'}\gamma_{ac} = \frac{k}{k'}\frac{k}{k-k'}\Gamma_N - \frac{k}{k'}\gamma_{bc}\ , \qquad (2)$$

where γ_{ij} is the relaxation rate of the transition i-j. In the present experiment, the slightly pressure-broadened γ_{bc} is approximately 37 MHz, and Γ_N is known from previous observations |9,17|. We have studied the width of the dip as a function of the saturator power. The observed power broadening corresponds to seventh-order and higher-order effects. When the usual power-broadening factor |18,9| is tentatively applied,

$$\gamma = \gamma_0 \sqrt{1 + (\kappa\Omega_R/\Gamma_N)^2}, \qquad (3)$$

where $\kappa^2 = (k'/k)(k-k')/k$, and $\Omega_R = p_{ab}E/\hbar$, reasonable agreement with the observed widths is achieved without fit of a free parameter (s. Fig. 10). It may be worth noting that at low powers the resonance - taking the particular homogeneous line width of the given transition into account - is one of the narrowest spectral resonances ever observed.

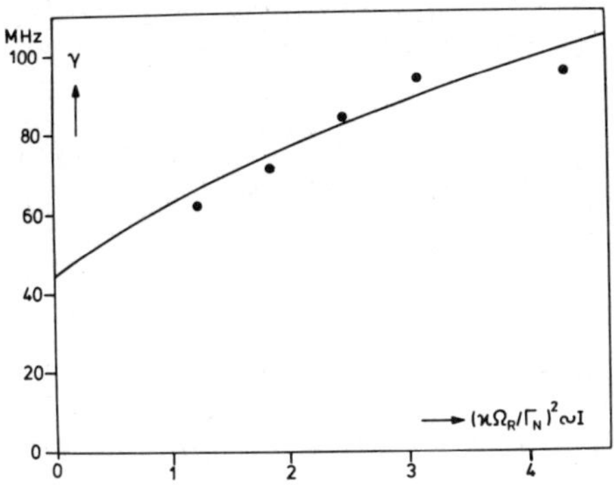

Fig. 10: Width (HWHM) of two-photon Lamb dip vs. red light flux I(ω).

In conclusion, one may anticipate that higher-order Lamb dips and related resonances will, in the future, play a role in high-resolution spectroscopy and applications which seems not less important than that of LAMB's original discovery.

Acknowledgements

We very much appreciate stimulating discussion with Claude Cohen-Tannoudji, Paris, with Stig Stenholm and Rainer Salomaa, Helsinki, and with Paul Berman, New York.

The construction of the iodine-stabilized He-Ne reference laser was performed in collaboration with Reinhard Mundt.

This work was supported by the Deutsche Forschungsgemeinschaft.

References:

|1| e.g., C. Cohen-Tannoudji, Cargèse Lectures in Physics, Vol. 2, New York, Gordon and Breach, p. 347, 1968

|2| K. Shimoda, "Laser Spectroscopy of Atoms and Molecules", Topics in Appl. Physics, Vol. 2, p. 197, Springer, Berlin 1976

|3| M. Arditi and T.R. Carver, Phys.Rev. 124, 800 (1961)

|4| C. Cohen-Tannoudji, Annales de Physique 7, 423 and 469 (1962)

|5| E.B. Aleksandrov, A.M. Bonch-Bruyevich, N.N. Kostin, V.A. Khodovoi, JETP Letters 3, 53 (1966); A.M. Bonch-Bruyevich, N.N. Kostin, V.A. Khodovoi, JETP Letters 3, 279 (1966).

|6| P. Platz, Appl.Phys.Letters 14, 168 (1969) and 16, 70 (1970)

|7| P.F. Liao and J.E. Bjorkholm, Phys.Rev.Letters 34, 1 (1975)

|8| Th. Hänsch and P. Toschek, IEEE J.Quantum Electronics QE-4, 467 (1968)

|9| A. Schabert, R. Keil, and P.E. Toschek, Appl.Phys. 6, 181 (1975)

|10| P.E. Toschek, Laser Report 9-1977, Institut für Angewandte Physik, Univ. Heidelberg. - R. Salomaa, M. Sargent, III, and P.E. Toschek, to be published.

|11| S. Stenholm, private communication

|12| W.E. Lamb, Jr., Phys.Rev. 134, 1429 (1964)

|13| F. Shimizu, Phys.Rev. A, 10, 950 (1974)

|14| S.M. Freund, M. Römheld, and T. Oka, Phys.Rev.Letters 35, 1497 (1975)

|15| J.P. Woerdman and M.F.H. Schuurmans, Opt.Communications, 21, 243 (1977)

|16| Ye.V. Baklanov, I.M. Beterov, V.P. Chebotayev, and B.Ya. Dubetsky, Appl. Phys. 11, 75 (1976)

|17| A. Schabert, R. Keil, and P.E. Toschek, Opt.Communications 13, 265 (1975)

|18| B.J. Feldman, and M.S. Feld, Phys.Rev. A5, 899 (1972)

VII. Laser Spectroscopic Applications

COHERENT ANTI-STOKES RAMAN SPECTROSCOPY

J.-P. Taran

Office National d'Etudes et de Recherches Aérospatiales (ONERA)
92320 Châtillon, France

1. Introduction

The use of coherent anti-Stokes Raman scattering (CARS) has revolutionized the field of Raman scattering. Its major achievements are in the areas of chemical analysis and temperature measurements in reactive gaseous media [1-3], fluorescence-free spectroscopy of biological samples [4], and, recently, ultra-high resolution Doppler-free spectroscopy [5]. CARS offers many advantages over spontaneous Raman scattering, among which intense collimated signal beams and virtual elimination of fluorescence and stray light ; its most severe drawback is interference between species, which reduces the detectivity.

Indeed, in spite of the huge signal strength, detectivity in CARS remains limited by the presence of the diluent nonresonant background. Two methods have been proposed and used to reduce this background. In the first one, an additional resonance is brought into destructive interference with the background by use of a third laser [6] ; in the second one (RIKES) [7], one takes advantage of the spatial properties of the susceptibility tensor. Limited gains are anticipated from these modifications, especially in the gas phase. Another solution, which involves using electronic enhancement of the Raman resonances, holds more promise. One anticipates gains of 10^2 or over, which should bring the detectivity from 10^3 ppm down to 10 ppm in gas mixtures at STP. This point has just been verified in I_2 vapor in our laboratory. Similar improvements had been observed in solutions [8].

Finally, a generalization of the CARS concept to the probing of other higher order nonlinear processes should greatly facilitate the acquisition of data important to physical chemists. An outstanding example is the possible study of hyper-Raman active modes of vibration.

2. Recent Progress in Conventional CARS

In the field of chemical analysis and temperature measurements, measurement accuracy is a determining factor and instrumental reliability is vital. However, CARS has been known since the beginning as giving very poor signal reproducibility. In general the fluctuations result from frequency drifts in the lasers and instabilities in the spatial quality of their beams. One year ago, peak to peak fluctuations of the signal to reference ratio were about 40 % [1]. They have now been reduced under 10 % by use of a stable and compact design, and thanks to a specially designed ratioing electronics that gates the detectors and rejects anomalous reference readings ; the latter are often traced back to mode-hopping in the ruby laser, which causes anti-Stokes variations over 30 %. In stable mixtures, each spectral element of χ is now collected with an accuracy of about ± 3 %, and the readings are stable over several days ; this makes CARS a reliable instrument for chemical analysis. Figure 1 gives a typical CO spectrum recently obtained in our diffusion flame.

Several new problems have recently appeared [9]. Of particular importance is the dependence of Raman linewidth Γ upon temperature. Most CARS spectra in flames are taken with a laser linewidth sufficient to resolve the Q-branch envelope, but not the individual lines within it, so that Γ cannot be determined. However, the amplitude, and to some extent the shape, of the Q-branch envelope depend on Γ [10]. Therefore, it is essential, either to measure Γ prior to performing data analysis, or to assume a realistic functional dependence vs temperature for it. Failure to do so will lead to an appreciable error ($\simeq$ 30 %) in the density of the gas as evaluated from the line contour, and to a minor error (5 - 10 %) on the temperature [9].

3. Resonance CARS

3.1. Theory

Resonance enhancement of CARS has been observed recently in liquids [8]. This effect is somewhat more difficult to see in gases and its interpretation is delicate. Using a density matrix approach, DRUET has recently derived an expression of the nonlinear CARS susceptibility [11, 12] which is correct only if elastic collision broadening is negligible. A modification of her calculation which takes elastic collisions into account (following Shen's derivation of Raman cross sections [13]) leads to [14] :

$$\chi = \chi_R + \chi_{NR} \quad ; \quad \chi_R = \frac{N}{3!\hbar^3} \times \frac{1}{(\omega_{ba} - \omega_L + \omega_s - i\Gamma_{ba})} \tag{1}$$

$$\times \Bigg[\rho^{o}_{aa} \sum_{n'} \Big(\overset{A}{\frac{P_{an'} P_{n'b}}{\omega_{n'a} - \omega_a - i\Gamma_{n'a}}} + \frac{P_{an'} P_{n'b}}{\omega_{n'b} + \omega_a + i\Gamma_{n'b}} \Big) \times \sum_{n} \Big(\frac{P_{bn} P_{na}}{\omega_{na} + \omega_s - i\Gamma_{na}} + \overset{A}{\frac{P_{bn} P_{na}}{\omega_{na} - \omega_L - i\Gamma_{na}}} \Big)$$

$$- \rho^{o}_{bb} \sum_{n'} \Big(\overset{B}{\frac{P_{an'} P_{n'b}}{\omega_{n'a} - \omega_a - i\Gamma_{n'a}}} + \frac{P_{an'} P_{n'b}}{\omega_{n'b} + \omega_a + i\Gamma_{n'b}} \Big) \times \sum_{n} \Big(\overset{B}{\frac{P_{bn} P_{na}}{\omega_{nb} - \omega_s + i\Gamma_{nb}}} + \frac{P_{bn} P_{na}}{\omega_{nb} + \omega_L + i\Gamma_{nb}} \Big) \Bigg]$$

This expression is given for steady state, assuming that there only exists one Raman resonance (we ignore splitting of rotational origin), that the interaction is weak so that higher order corrections to the initial population fractions ρ^{o}_{aa} and ρ^{o}_{bb} in states $|a\rangle$ and $|b\rangle$ can be ignored ; ω_L , ω_s and ω_a are the usual CARS frequencies ; $\omega_{n'a}$ is the absorption frequency from $|a\rangle$ to $|n'\rangle$. In the usual case where the number density N of the Raman resonant molecules is small compared to that of the diluent nonresonant species, χ_{NR} is mainly contributed by the latter and is a constant ; note, however, that the two-photon nonresonant terms contributed by the resonant molecules, which are lumped together with χ_{NR} , actually bring small corrections to the expression of χ_R in (1).

The representation of the molecular perturbation by the field is done traditionally by the diagrams of Fig. 2a and b. Far from the electronic resonance, the diagram of Fig. 2a, which represents a parametric process, is often incorrectly used as the only representation. The diagram of Fig. 2b is equally important and is, as a matter of fact, the only process taking place exactly at the vibrational resonance. The major drawback of these diagrams in CARS is that the arrows are sometimes wrongly interpreted as one photon transition events, and that the diagrams themselves no longer make sense in the vicinity of electronic resonances.

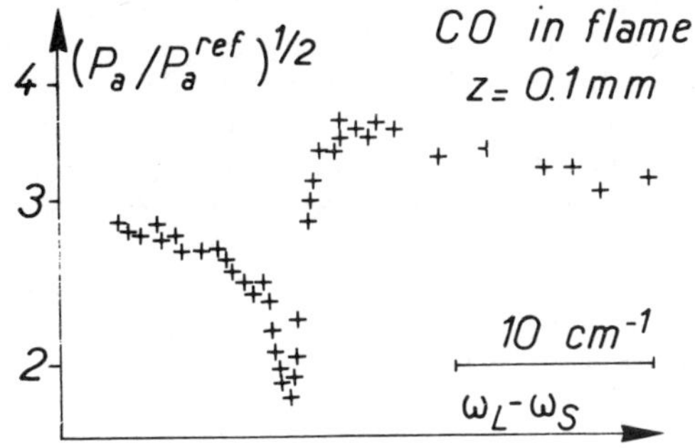

Fig. 1 — Spectrum of CO in our diffusion flame, 100 μm from the spherical burner surface ; fuel is ethylene glycol ; 4 shots are averaged per data point ; temperature is approximately 500 K and concentration 5 %.

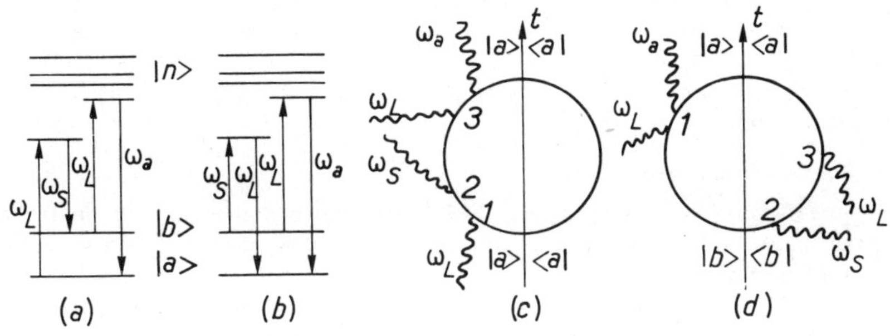

Fig. 2 — Representations of CARS ; a, b : classical, associated to the Raman resonant susceptibility ; c, d: diagrammatic perturbation representations taken from [15, 16]. *Photon absorption and emission are operations linear in the fields. Propagation theory has been used to directly obtain the polarization components at* ω_a .

A less misleading model is suggested by YEE et al. [15, 16]. In this approach, one uses diagrams to depict the time-ordered representations of the perturbations to the wave function and its complex conjugate. The simple rules by which these diagrams are interpreted and used to evaluate susceptibilities are a natural extension of those used by OMONT et al. [17] to calculate transition rates in spontaneous scattering. The time evolution of the wave function is plotted on the left hand side and that of its complex conjugate on the right hand side (Figs. 2c, d). Absorption and emission are represented in the standard way [16, 17].

The diagram then depicts the evolution of the density matrix as a function of time. By carrying out the time-ordered calculations, an accurate correspondence of each term of Eq. (1) to a specific time-ordered diagram is obtained. For example the diagrams for the fully resonant CARS (electronic as well as vibrational) consist of the two shown in figures 2c, d. The first involves transitions which take the molecule from the ground state back to the ground state (terms A in Eq. (1)). This involves interactions solely on one side of the diagram in the specific time-ordering indicated, i.e. the absorption operation at ω_L followed by emission at ω_S and then a second absorption at ω_L with the final emission at the anti-Stokes frequency. During the interaction, the complex conjugate wave function remains unperturbed. The other resonant diagram takes the molecule from the excited vibrational state to the ground state and involves two interactions on the wave function and two on the complex conjugate ; on one side absorption at ω_S followed by emission at ω_L and on the other side absorption at ω_L followed by emission at ω_a (terms B in Eq. (1)). Since these diagrams provide in general complex polarizations, they both provide storage and energy exchange. It is also clear that these diagrams are not simple photon emission or absorption representations, but representations of a time sequence of photon creation and annihilation operations, which are linear in the field and result in a polarization. To discuss energy interchange, it is necessary to use Maxwell's Eqs., in which both the phases of the fields and propagation vector matching are determining factors.

There are a total of 24 possible configurations for ω_L and ω_S on diagrams of this type in order to obtain a polarization at ω_a and starting from $|a\rangle\langle a|$; six are of the parametric type and are obtained from different time sequences of vertices 1, 2, 3 on Fig. 2c, three on the ket and three on the bra ; the remaining 18 are derived from Fig. 2d (but with $|a\rangle\langle a|$ as the initial state). Initial state $|b\rangle\langle b|$ also contributes 24 analogous terms. If three pump frequencies were present $(\omega_L, \omega_L', \omega_S)$, there would be twice as many terms for $\overset{o}{\rho}_{aa}$ and $\overset{o}{\rho}_{bb}$.

3.2. Spectral Properties

Predicting the features to be found in a CARS spectrum and their contour in the absence of electronic resonance enhancement is straightforward. The electronic terms involved in the factors proportional to $\overset{o}{\rho}_{aa}$ and $\overset{o}{\rho}_{bb}$ in (1) are easily shown to be nearly equal, so that χ_R is thus actually proportional to $\overset{o}{\rho}_{aa} - \overset{o}{\rho}_{bb}$. If thermodynamic equilibrium exists within each molecular degree of freedom, then the populations are easily calculated using appropriate Boltzmann factors, and the spectral contour of $|\chi|$ — which is the quantity accessible to the measurements — can be calculated readily for the case of an isolated Raman line depicted in (1). If several resonances are present, an algebraic sum over them must be taken to obtain the total χ, but the problem remains identical in essence. In gases, off-electronic resonance contours have been calculated routinely for several years [1].

These standard rules break down whenever any of the three light frequencies involved is tuned into resonance with a discrete absorption line. Consider, for simplicity, the case of a diatomic molecule such as N_2 at low temperatures (hence $\rho_{bb} \sim 0$), and assume that ω_L is fixed and in resonance with a transition connecting a particular vibro-rotational level (v'', J) of the $X^1\Sigma_g^+$ state to the upper state $B^3\Pi_{ou}^+$. Only P(J) or R(J) transitions are allowed for that particular resonance. It is then clear that :

1. Selection rules for the other three transitions involved, ie nb, bn' and n'a in the notation of (1), impose a rotational quantum number of J or J − 2 for level b if ω_L is resonant with a P transition, J or J + 2 if ω_L is resonant with an R transition.
2. As $\omega_L - \omega_S$ is scanned through one of the ω_{bq} Raman resonances, ω_S becomes automatically one photon resonant with the ω_{nb} transition ; this resonance, however, only enters into the second B term in (1), which has negligible weight in χ because we

have assumed ρ^{o}_{bb} to be negligible.

3. Unless a fortuitous coincidence takes place, ω_a will not be in exact resonance with any of the allowed $\omega_{n'a}$ transitions, because of anharmonicity and rotation vibration coupling ; this implies that a second pump laser at ω'_L which could be tuned independently of ω_L would be advantageous.

4. As ω_S is scanned, both ω_S and ω_a are swept across many electronic resonances pertaining to levels distinct from the upper and lower Raman levels under study ; the contribution of these resonances to χ will often be negligible, since the possible Raman transitions associated with these other levels will not in general be excited resonantly by $\omega_L - \omega_S$.

We thus arrive at the following qualitative description for the spectral features to be found in the spectrum :

1. the spectrum is composed of doublets , O + Q or Q + S if ω_L is in resonance with a P or an R line respectively ;
2. there is one doublet for the fundamental band and one for each of its overtones ;
3. except for these few lines, all the other lines composing the Raman spectrum, which are still present, form a set of weakly resonance - enhanced O, Q and S branches ;
4. if the temperature is raised, the spectrum can become extremely complicated ; furthermore, the possibility of fortuitous resonances from ω_a should not be disregarded.

In addition to predicting the features in the CARS spectrum, one can also describe their spectral contour and lineshifts. The first experimental evidence of resonance CARS has been reported recently for the liquid phase [8], leading to the observation of unusual dispersive properties. Liquids differ markedly from gases because the molecules cannot rotate freely, and also because the absorption lines usually merge into a continuum. Therefore, the Raman resonance pertaining to a particular vibrational mode is unique (no rotational splitting) and is electronically enhanced as a whole. However, (1) is adequate for the liquids as well as gases, and has been used [18, 19] to carry out a numerical analysis of the spectral properties reported in [8]. A much simpler discussion can in fact be given by means of the circle model introduced in [1, 20] for nonresonant CARS. This discussion will be conducted here in the context of chemical analysis, where one generally tries to detect the resonance of the species of interest in the presence of a nonresonant background which is contributed by the diluent species for the most part.

The circle model is based upon the vector representation of χ in the complex plane, recognizing the fact that the associated vector describes a circle when $\omega_L - \omega_S$ is varied about ω_{ba}. This resonance is here assumed to be isolated, with Lorentzian broadening Γ_{ba}. Then χ_R in (1) is written $\chi_R = N\alpha_R \Gamma_{ba} / (\omega_{ba} - \omega_L + \omega_S - i\Gamma_{ba})$ where α_R stands for the electronic terms.

In the nonresonant case, α_R depends little on the frequencies, and is proportional to the spontaneous Raman cross section $d\sigma / d\Omega$. Two cases are interesting. If the concentration is large, the situation of Fig. 3a applies. Apart from a slight asymmetry caused by the interference with χ_{NR} , the line closely fits the square root of a Lorentzian. If the concentration is small, the line contour draws close to $\mathcal{R}e(\chi)$ (Fig. 3b). In either case, the difference between the maximum and the minimum is strictly equal to the magnitude of the vibrational resonance, i.e. one circle diameter $N\alpha_R$.

When enhancement of χ_R through electronic resonances is produced, a complex situation appears since several of the terms in α_R can become large simultaneously.

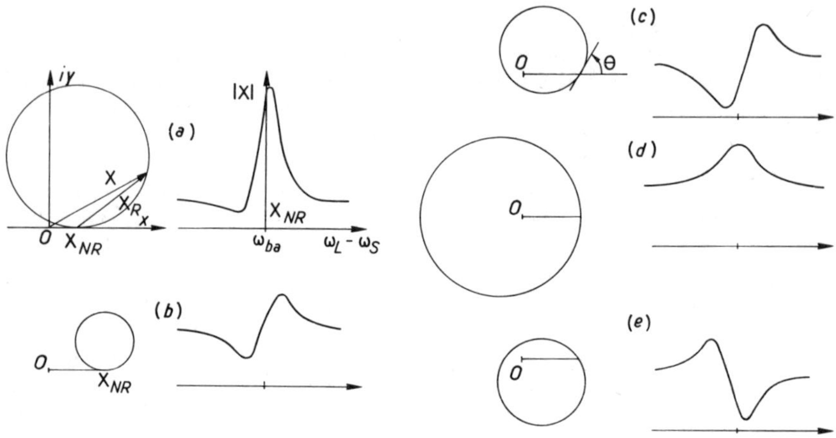

Fig. 3 — Circle diagram and dispersive profile of $|\chi|$ *; a, b : nonresonant case ; c, d, e : about an electronic resonance.*

The discussion, however remains simple for the conditions assumed in the presentation of the spectral content of χ (ie ω_L near resonance, ω_a off resonance). Then α_R is independant of ω_S , ω_a and we can write $\alpha_R \propto 1/(\omega_{na}-\omega_L-i\Gamma_{na}) = \rho \exp(i\theta)$. It can be shown that a similar expansion is still legitimate in liquids if the Raman line is much narrower than the absorption profile.

For ω_L fixed, it is apparent that χ still describes a circle :

1. χ_{NR} being contributed mostly by the diluent species, we may assume it to be constant and ignore the variations of the contribution of the species to be detected as ω_L , ω_S and ω_a are varied across its electronic resonances ;
2. the circle is rotated by argument θ about the extremity of χ_{NR} ;
3. its diameter scales as ρ .

The rotation of the circle proceeds counter-clockwise as ω_L increases past ω_{na} and θ is precisely 90° on resonance if we assume the product of transition moments to be real. Figures 3c, d, e illustrate the cases $\omega_L - \omega_{na} < 0$, $= 0$, > 0 . Cases c and e produce lineshapes which are mirror images of one another with respect to ω_{ba} . Case d is interesting : depending on the magnitude of the Raman resonance with respect to χ_{NR} , the line, which is symetric with respect to ω_{ba}, appears as a maximum or a minimum. The situation represented here corresponds to the maximum. The situation where a minimum is present was found experimentally in [8], but was incorrectly presented as inverse Raman scattering (this effect then appearing as a saturation mechanism for CARS due to excessive pump strength). Note finally that in case d, if the circle radius is precisely equal to χ_{NR} , the line vanishes completely.

3.3. Experimental results in I_2 vapor

Resonance enhanced CARS has been observed recently at ONERA in I_2 vapor at 1.5 Torr [21] using flash-pumped dye lasers. These lasers are capable of near-diffraction-limited output with a linewidth of 0.1 cm^{-1} and peak powers of 10 - 100 kW. Their beams are combined by means of a dichroic filter mounted on a stable support with 10 μrad. alignment sensitivity. The I_2 is contained in a cell filled with air and temperature-controlled to vary the partial pressure. The beams are focussed into the cell by means of a 10 cm focal length lens. The cell length is 10 cm ; this value is a compromise between the need to avoid generation of a nonresonant signal off cell windows and the need to minimize optical density at all wavelengths.

Excitation was done in a spectral region easily covered by Rhodamines 6 G and B and for which the I_2 absorption lines are identified. The vicinity of the Na D line at 597 nm was chosen for ω_L. The latter was tuned as close as possible to the Na line by comparison though a monochromator and two solid Fabry-Perot filters of 0.1 and 1 mm, with a precision estimated at ± 0.5 cm^{-1}. The spectrum obtained for the 5th overtone is plotted on Fig. 4 (top). The inset shows a fragment of the I_2 high-resolution absorption spectrum [22] with the rotational assignment of [23]. The CARS lines were assigned using the constants of [24]. The ambiguity concerning the exact value of ω_L is further complicated by the presence of weak, unidentified features, especially between the R119 and P114 lines, and by the collisional broadening by air. However, features 2 and 3 on the spectrum can be assigned to $Q_{1\text{-}7}$ (67) and $Q_{1\text{-}7}$ (62) respectively with reasonable degree of confidence. Lines 1,4 and 5 cannot be assigned at the present time (they may be associated with one of the unidentified absorption lines or an anti-Stokes resonance). Feature 6 is probably a fragment of the preresonance enhanced $Q_{0\text{-}6}$ branch,

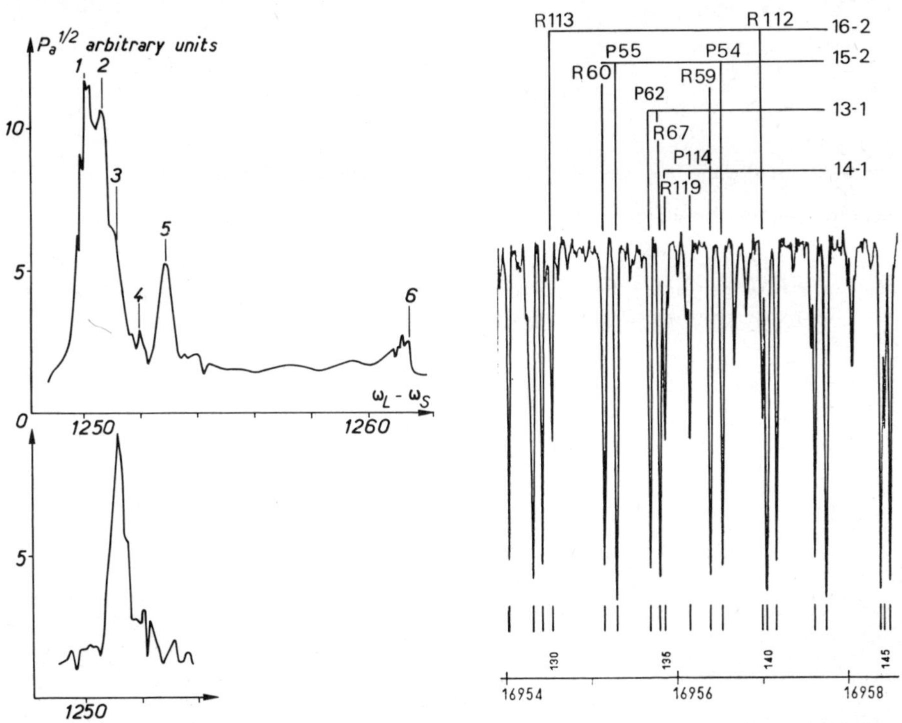

Fig. 4 — CARS spectrum of overtone 5 in I_2 (left) and absorption spectrum about the NaD line at 16956.1 cm^{-1}.

the Q(O) line lying at 1261.3 cm^{-1}. If ω_L is downshifted by 0.3 cm^{-1}, the bottom spectrum is obtained, leaving only the Q_{1-7} (62) line and feature 6 (not drawn here) providing further support for the above interpretations and giving 16 955 . 9 cm^{-1} as a probable value for the initial ω_L.

We verified that no appreciable change in the spectral contour would occur when the I_2 pressure was varied and when the laser powers were varied, indicating that linear absorption was negligible and that the ac Stark effect and saturation were unimportant, in spite of the large power densities in the focal region (~ IGW/cm^2). Finally, the lines seen here are as strong as the Q-branch of the O_2 in the cell, indicating that enhancements of about 100 can be expected from electronic resonances in I_2 and other gases. However, we were not able to detect any CARS signal from the 2 ν_2 band of NO_2 with ω_L tuned in the same spectral region (the ruby laser set-up used for conventional CARS work in the flame also failed to produce a detectable spectrum with NO_2). A thorough discussion of the problems of absorption, dispersion, population perturbation was done by DRUET [11].

4. New Developments : Hyper-Raman Spectroscopy or CAHRS

The introduction of coherent scattering to the field of hyper–Raman spectroscopy holds considerable promise : faster data acquisition, superior resolution, and possibility of studying gases [24]. Using a classical approach [2], we arrive at :

$$P_a^{CAHRS} \approx \left(\frac{\pi}{2\lambda^3F^3}\right)^2 \left(\frac{3\pi^3\omega_a^3}{c^3}\right)^2 \left(\frac{N}{m|D|}\left(\frac{\partial\beta}{\partial q}\right)^2\right)^2 P_o^4\, P_s$$

where the pump ω_o, ω_S and the signal ω_a are arranged according to Fig. 5, m is the reduced mass of the vibrator, $D = \omega_{ba}^2 - (2\omega_o-\omega_s)^2 - i\,\Gamma_{ba}(2\omega_o-\omega_s)$, $\partial\beta/\partial q$ is the derivative of the hyperpolarizability with respect to normal coordinate, λ the mean value of wavelength and F the f-number of the focused beams. Phase matching was assumed. Note that there exists a strong dependence on f-number.

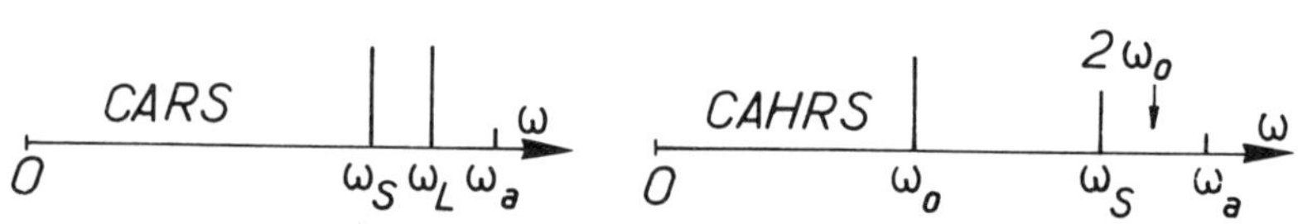

Fig. 5 – Spectra of radiation fields in CARS and CAHRS.

In conventional CARS, one had [2].

$$P_a^{CARS} \approx \left(\frac{2}{\lambda}\right)^2 \left(\frac{4\pi^2\omega_a}{c^2}\right)^2 \left(\frac{N}{m|D|}\left(\frac{\partial\alpha}{\partial q}\right)^2\right)^2 P_L^2\, P_s$$

using polarizability α, assuming phase matching and with $D = \omega_{ba}^2 - (\omega_L - \omega_S)^2 - i\Gamma_{ba}(\omega_L - \omega_S)$. It is interesting to compare the two situations in terms of anti-Stokes yield. Assuming that a Yag laser set-up is used for both experiments, with frequency doubling for generation of ω_L in CARS and without for ω_0 in CAHRS, we have typically P_0 = 10 MW, P_L = 1 MW, F = 100, $\lambda \sim 0.5\ \mu m$. Assuming $(\partial\beta/\partial q)/(\partial\alpha/\partial q) \simeq \beta/\alpha \simeq 10^{-6}$ [25], we arrive at $P_a^{CAHRS}/P_a^{CARS} \sim 10^{-5}$. This means that CAHRS in gases is very much in the realm of possibilities. An attempt at seeing it was therefore made on CO with the CARS set-up in operation at Quantel Corporation. The pressure was 15 atm., the other important parameters having the values given above. The coherence length being only 1 mm, phase matching was obtained by allowing a small angle between the ω_0 and ω_S beams ; F could not be made smaller, because of breakdown. We failed to observe any of the N, P, R or T branches expected for CO, although, according to our estimation, they should have been detected. Possible explanations are that we have overestimated $\partial\beta/\partial q$. Another likely possibility is strong perturbations to molecular energy levels by the intense fields as a result of the ac Stark effect [26] leading to a large broadening of the resonances and a reduction in their intensities. We also tried ethylene, which is another interesting candidate possessing a strong hyper-Raman resonance [27]. However, experiments in that gas proved to be unfeasible because of the low threshold for stimulated Brillouin scattering. This drawback is common to many organic gases, and it will be important for future experiments to provide for a protection of the laser oscillators against the reflected energy.

5. Conclusion

It is felt that conventional CARS is close to being used routinely for the study of reactive media. The extension into the field of resonance Raman scattering also holds considerable promise, both for analytical chemistry and for a spectroscopy free of redistribution emission. Finally, in spite of severe experimental difficulties, hyper–Raman spectroscopy by CARS is an exciting area for future research.

References

1. F. Moya, S. Druet, M. Péalat and J.-P. E. Taran, "Flame Investigation by Coherent Anti-Stokes Raman Scattering", AIAA 14th Aerospace Sciences Meeting, Washington D.C., (1976), AIAA Paper 76-29.

2. F. Moya, "Application de la diffusion Raman anti-Stokes cohérente aux mesures de concentrations gazeuses dans les écoulements", Thesis, Orsay, 5 March 1976 ; available as ONERA Technical Note no. 1975-13.

3. J.W. Nibler, J.R. McDonald and A.B. Harvey, "CARS Measurements of Vibrational Temperatures in Electric Discharges", Optics Com., **18**, 371 (1976).

4. R.F. Begley, A.B. Harvey and R.L. Byer, Appl. Phys. Letters **25**, 387 (1974).

5. M.A. Henessian, L. Kulevskii, R.L. Byer, J. of Chem. Phys. **65**, 5530 (1976) ; V.I. Fabelinsky, B.B. Krynetsky, L.A. Kulevsky, V.A. Mishin, A.M. Prokhorov, A.D. Savel'ev, V.V. Smirnov, Opt. Comm. **20**, 389 (1977).

6. R.T. Lynch, Jr., S.D. Kramer, H. Lotem and N. Bloembergen, Optics Com. **16**, 372 (1976).

7. D. Heiman, R.W. Hellwarth, M.D. Levenson, and G. Martin, Phys. Rev. Letters **36**, 189 (1976).

8. J. Nestor, T.G. Spiro, and G. Klauminzer, Proc. Natl. Acad. Sci. **73**, 3329 (1976).

9. O. Schnepp, F. Slabodsky, and M. Péalat, to be published.

10. "Handbook of Infrared Radiation from Combustion Gases", C.B. Ludwig, W. Malkmus, J.E. Reardon and J.A.L. Thomson, Edited by R. Goulard and J.A.L. Thomson NASA SP 3080, Washington, D.C., 1973.

11. S. Druet, "Diffusion Raman anti-Stokes cohérente au voisinage des résonances électroniques et réalisation d'un montage expérimental", Thesis, Orsay, 17 september 1976, available as ONERA Technical Note no. 1976-6.

12. S. Druet, "Resonant CARS in Gases", presented at the 5th International Raman Spectroscopy Conference, Freiburg, 2-8 September 1976, paper 8A 1145.

13. Y.R. Shen, Phys. Rev. **B9**, 622 (1974).

14. S.A.J. Druet, B. Attal, T.K. Gustafson, and J.-P. E. Taran, to be published.

15. S.Y. Yee and T.K. Gustafson (submitted to Phys. Rev. A).

16. S.Y. Yee, T.K. Gustafson, S.A.J. Druet and J.-P. E. Taran (submitted to Optics Comm.).

17. O. Omont, E.W. Smith, and J. Cooper, The Astrophysical Journal **175**, 185-199 (1972).

18. R.T. Lynch, Jr., H. Lotem, and N. Bloembergen, in "Lasers in Chemistry", The Royal Institution, London, 31 May- 2 June 1967, p. 1.

19. J. Hertz, A. Lau, M. Pfeiffer, and W. Werncke, in "Lasers in Chemistry", The Royal Institution, London, 31 May- 2 June 1967, p. 399.

20. J.-P. E. Taran, in "Tunable Lasers and Applications", Loen, Norway, 1976, Editors A. Mooradian, T. Jaeger and P. Stokseth, Springer Verlag, Berlin, Heidelberg, New York 1976.

21. B. Attal, O. Schnepp and J.-P. E. Taran, to be published.

22. S. Gerstenkorn and P. Luc, "Atlas du spectre d'absorption de la molécule d'iode", Laboratoire Aimé Cotton CNRS II, 91405 Orsay (France).

23. R.L. Brown and W. Klemperer, J. Chem. Phys. **41**, 3072 (1964).

24. F. Moya, O. Schnepp, and J.-P. E. Taran, unpublished.

25. D.A. Long and L. Stanton, Proc. Roy. Soc. London **A 318**, 441 (1970).

26. Q.H.F. Vrehen and H.M.J. Hickspoors, Optics Com. **21**, 127 (1977).

27. J.F. Verdieck, S.H. Peterson, C.M. Savage, and P.D. Maker, Chem. Phys. Letters **7**, 219 (1970).

LASER SPECTROSCOPY RELEVANT TO STRATOSPHERIC PHOTOCHEMISTRY

R.T. Menzies

Jet Propulsion Laboratory, California Institute of Technology
Pasadena, CA 91103, USA

1. Introduction

Due to the recent advances in high resolution laser spectroscopy which have been discussed elsewhere in this conference, it seems almost old-fashioned to state that we have been doing Doppler-limited spectroscopy in studies of certain stratospheric molecules. However, there is a good reason to study Doppler-limited spectra in this case: that is, these molecules as they exist in the stratosphere have linewidths which are Doppler-broadened at the least, and the application of these spectral studies to the measurements of stratospheric species by remote sensing infrared techniques can be made almost directly.

To begin, let us consider a few of the reaction families which are of importance in their effect on the ozone layer. Then we will discuss spectral studies of certain constituents which appear in these reactions. The catalytic destruction of ozone by the NOX and HOX families of free radicals:

$$HO + O_3 \rightarrow HOO + O_2 \qquad (1)$$

$$HOO + O \rightarrow HO + O_2 \qquad (2)$$

$$\text{net: } O_3 + O \rightarrow O_2 + O_2$$

$$NO + O_3 \rightarrow NO_2 + O_2 \qquad (3)$$

$$NO_2 + O \rightarrow NO + O_2 \qquad (4)$$

$$\text{net: } O_3 + O \rightarrow O_2 + O_2$$

has been considered by many as a source of fundamental concern over a buildup in high altitude aircraft traffic [1]. The stratospheric nitric acid plays an important part in these catalytic cycles because it acts as a reservoir for both OH and NO_2 [2]:

$$HO + NO_2 \overset{M}{\rightarrow} HNO_3 \qquad (5)$$

$$HNO_3 + h\nu\ (\lambda < 330\ \text{nm}) \rightarrow HO + NO_2 \qquad (6)$$

In this manner, nitric acid can tie up two reactive radicals for relatively long periods of time (3 days) before being photodissociated. The catalytic destruction of ozone by the CℓX free radicals [3]:

$$C\ell + O_3 \rightarrow C\ell O + O_2 \tag{7}$$

$$C\ell O + O \rightarrow C\ell + O_2 \tag{8}$$

net: $O_3 + O \rightarrow O_2 + O_2$

has also been an important concern due to the fact that man-made aerosol propellants are a probable source of chlorine atoms in the stratosphere [4]. The molecule $C\ell ONO_2$, or chlorine nitrate, has recently been suggested as an important reservoir for both CℓO and NO_2 radicals [2]:

$$C\ell O + NO_2 \rightarrow C\ell ONO_2 \tag{9}$$

$$C\ell ONO_2 + h\nu\ (\lambda < 350\ \text{nm}) \rightarrow C\ell O + NO_2 \tag{10}$$

To date, stratospheric measurements of several of these important species are rather sparse or nonexistent. One valuable measurement technique involves taking absorption or emission spectra in the infrared, where all of these species except oxygen atoms have strong absorption bands. The spectra, in order to eliminate confusion, would optimally be taken with instruments whose spectral resolution is equivalent to the Doppler limited linewidths. However, in order to make sense out of data taken with such instruments, laboratory spectra must be generated for these species with matching spectral resolution and extremely good frequency accuracy.

We have been studying the spectra of ozone, CℓO, and chlorine nitrate at JPL, using both gas and semiconductor diode lasers. I will discuss this work, along with some excellent diode laser spectroscopy of ozone and nitric acid which is taking place at NASA Langley Research Center. Related work is also underway at NASA Goddard Space Flight Center and at other institutions.

2. Infrared CℓO Spectroscopy

At the time the study of the CℓO fundamental vibration-rotation band was started, in the fall of 1975, no definitive measurement of this stratospheric trace species had been made. Studies of its microwave and ultraviolet spectra existed, but there had been no observation of its infrared band. The work of COXON, et. al., in the ultraviolet had predicted an infrared band center near 850 cm^{-1} [5], and a diode laser spectroscopy experiment was initiated to study this band, with the objectives of more accurate frequency determinations, and a determination of the band intensity. The apparatus is shown in schematic form in Fig. 1.

During the initial search for CℓO lines, the monochromator was usually bypassed by inserting appropriately aligned mirrors in front of it. Then the diode laser current was ramp-scanned, with a small sinusoidal dither superimposed on the ramp. The dither produced a frequency oscillation whose peak-to-peak amplitude was about 30 MHz (roughly equivalent to the Doppler broadened half-width of the CℓO lines), and a phase synchronous detector was used to isolate a derivative signal caused by an absorption line. This

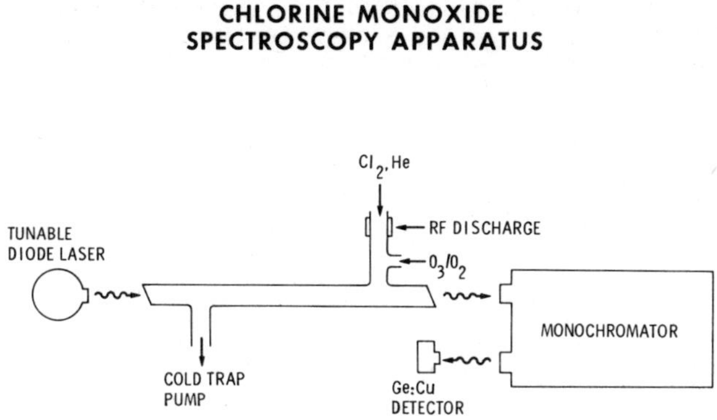

Fig.1 Chlorine monoxide spectroscopy apparatus. The total pressure in the flow tube was 7 Torr, with the O_3/O_2 mix being 1% O_3. The flow tube length was 50 cm.

technique was used because of its sensitivity in past diode laser spectroscopy experiments. The amount of generated CℓO was small (about 10^{-3} atm-cm), due to its short lifetime. Thus the medium strength lines were expected to produce a peak absorption of only a few percent. Once a derivative signal due to CℓO was detected, the laser radiation was then passed through the monochromator to isolate the wavelength. The system response was calibrated by measuring the direct absorption spectrum of NH_3 in the same spectral range, if a suitable NH_3 absorption line could be found, and then comparing the NH_3 and CℓO derivative signals. The frequencies of a few CℓO lines were determined to within ± 0.005 cm^{-1} by measuring their frequency displacement from nearby NH_3 lines, whose wavelengths have been measured by CURTIS [6]. Fig. 2 indicates an example of such a comparison.

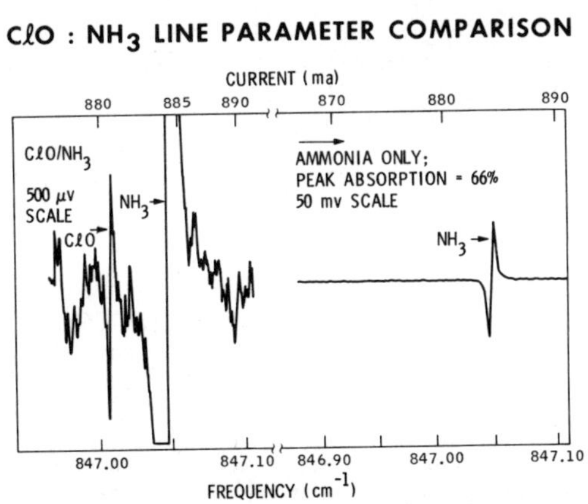

Fig.2 CℓO:NH_3 line parameter comparison. The scale at the top is laser tuning current. This ammonia line, from [6], is the sP(6,5) line. The CℓO line is the Cℓ-35 isotope Ω = 3/2, R(3/2) transition.

The parameter values for the $C\ell O$ fundamental band which were deduced from this work, and from microwave observations of two rotational transitions [7], are listed below in Table 1. A further description of this work can be found in [8].

Table 1 Band center and rotational constants for the $C\ell O$ fundamental band

	$^{35}C\ell^{16}O$	$^{37}C\ell^{16}O$	
ν_0 (Ω = 3/2)	843.961	836.786	cm^{-1}
B' + B"	1.23362	1.21286	cm^{-1}
B' - B"	-0.00593	-0.00578	cm^{-1}
ν_0 (Ω = 1/2)	841.839	834.59	cm^{-1}
B' + B"	1.23653	1.21560	cm^{-1}
B' - B"	-0.00593	-0.00578	cm^{-1}

3. Ozone Spectroscopy

High resolution infrared instruments, such as those using heterodyne detection, have the ability to measure ozone altitude profiles with a minimum of interference from other species. The ozone lines in the region of the ν_3 band near 9.5 µm are collision broadened up to about 35 km altitude, and by scanning a heterodyne radiometer through a selected absorption line with sufficient spectral resolution, the relative contributions of ozone molecules at various altitudes to the total absorption line shape can be determined [9]. The radiometric determination of altitude profiles can be accomplished by observing solar absorption [10, 11] or thermal radiation from the ozone itself, from either ground-based or airborne platforms. Another technique which can produce the same results with greater flexibility (global coverage, day/night operation) involves the use of an active, nadir looking tunable laser system which measures differential absorption as the laser beam propagates to the ground and is scattered back [9]. The 35 km altitude limitation on profiling is not severe, since the ozone number density falls off rapidly with increasing altitude in this region, and the profile is relatively well-behaved and determined by chemical reactions. Other techniques, such as backscattered ultraviolet (BUV) sensing and microwave radiometry, can be used to obtain altitude profile information above 35 km.

Figure 3 is a portion of the ozone solar absorption spectrum near the CO_2 P(24) laser line, indicating the presence of an ozone line at 1043.183 cm^{-1}, or 0.02 cm^{-1} above the center of the P(24) line. The upper trace indicates zenith solar absorption due to the ozone in the 33-35 km altitude region, for a mid-latitude summer model [12] ozone profile with a (number density) peak near 22 km. The lower trace indicates solar absorption due to the total ozone column through the atmosphere. Also shown are solar absorption measurements, made on January 15, 1976, over Pasadena, using a heterodyne radiometer with 30 MHz IF bandwidth. It is clear that the ozone at higher altitudes contributes to absorption only in small spectral regions, such that independent measurements in wing regions (e.g., 1043.16 cm^{-1}) and in line center regions (e.g., 1043.183 cm^{-1}) can be used to discriminate between low and high altitude ozone. However, very accurate _a priori_ knowledge of the pertinent ozone

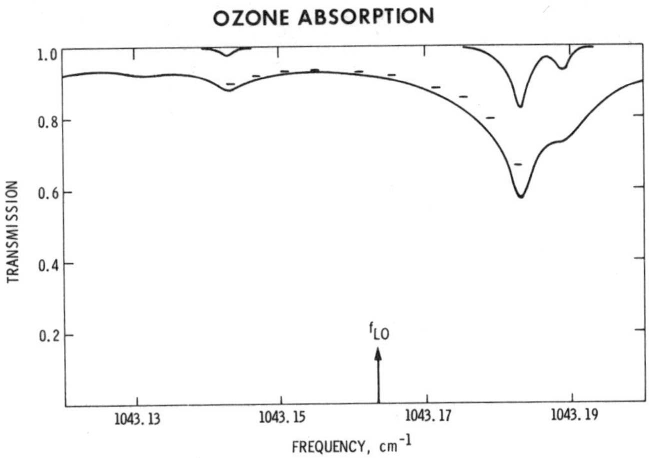

Fig.3 Zenith solar absorption due to ozone, in the frequency region near the $^{12}C^{16}O_2$ laser line at 1043.163 cm^{-1}. The solid lines are calculations based on the mid-latitude summer model (see text). The dashed lines are measurements using a heterodyne radiometer, adjusted to remove slant path and carbon dioxide absorption effects.

line parameters is required before an accurate altitude profile can be constructed. For example, if the ozone line at 1043.183 cm^{-1} were thought to be at 1043.187 cm^{-1} instead, the measurements in Fig. 3 would indicate a much larger amount of ozone in the 20 - 30 km region than actually existed. The lack of complete tunability of the local oscillator makes the situation more severe, since absorption lines which lie just beyond the tuning range and affect the absorption within the tuning range are lines whose center frequencies and pressure broadening parameters must be known very accurately.

Ozone spectroscopy in the 9.5 μm region of the ν_3 band is being conducted with tunable lasers both at JPL and at NASA Langley Research Center, with the objective of using the data to make accurate atmospheric measurements with high resolution instruments. The JPL work has involved the use of grating tunable CO_2 waveguide lasers in order to determine with CO_2 laser frequency standard accuracy the parameters of ozone lines near CO_2 transition frequencies [13]. At Langley, semiconductor diode lasers have been used to generate spectra such as that shown in Fig. 4 [14]. The diode laser spectra were generated using nearby CO_2 laser frequencies as markers, and etalons to calibrate the tuning rates of the diodes as they are tuned away from the CO_2 transition frequencies. The ozone spectrum from 1040 to 1050 cm^{-1}, a good spectral region for stratospheric measurements, has been generated in this manner. The center frequencies of several of the strong lines shown in Fig. 4 differ significantly from those listed in the AFCRL ozone line compilation [15]. For example, the $[(J', K_a', K_c') \rightarrow (J'', K_a'', K_c'')]$ (10, 5, 6) → (9, 5, 5) transition is listed in [15] at 1048.678 cm^{-1}, but is found to be 1048.6752 cm^{-1} according to the CO_2 waveguide laser spectroscopy [13], and 1048.676 cm^{-1} according to the diode laser spectroscopy [14]. As another example, the (9, 3, 6) → (8, 3, 5) transition is listed at 1048.730 cm^{-1} in [15], but is found to be at 1048.723 cm^{-1} from the diode laser measurements. These discrepancies, although of minor importance when using standard infrared instruments, are significant when heterodyne detection, with its high spectral resolution, is employed. Continued work in this area and in laboratory studies of other

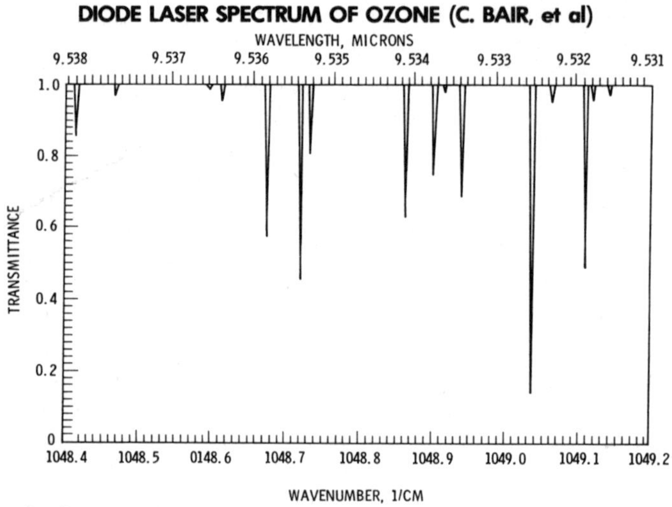

Fig.4 A composite diode laser spectrum of a portion of the ozone ν_3 band (courtesy of C. Bair, NASA Langley Research Center). The absorption cell pressure and length were 0.25 Torr and 10 cm, respectively.

molecular spectra are very important in supporting the atmospheric measurements.

4. Spectra of Other Stratospheric Molecules

The nitric acid molecules have a medium strength band in the 11.3 micrometer region, which has been used to identify nitric acid in the stratosphere [16]. A portion of this band, from 891 to 899 cm^{-1}, has been studied in detail using diode laser spectroscopy [17]. The resulting spectra show a multitude of lines, some of which are well resolved in the Doppler limit. This work can be useful in selecting a spectral region for nitric acid measurements with, e.g., a stratospheric heterodyne radiometer.

Recently published infrared spectra of chlorine nitrate, taken with a laboratory scanning Michelson interferometer [18], indicate sharp features in the ν_3 and ν_4 bands at 809.3 cm^{-1} and 780.2 cm^{-1}, which become more outstanding with increased spectral resolution. The highest resolution measurements in this work were taken at 0.0625 cm^{-1}, which is still quite low compared with the resolution achieved with the laser techniques which we have described. A laser spectrum of chlorine nitrate in this frequency region would be very valuable in assessing the ultimate sensitivity of a laser measurement instrument to stratospheric chlorine nitrate.

5. Planned Stratospheric Measurements

A balloon-borne heterodyne radiometer, shown in Fig. 5, is to be launched during the late summer or early autumn of 1977, with the goal of measuring chlorine monoxide in the stratosphere. The instrument will observe the solar spectrum in a narrow frequency interval around 853.15 cm^{-1}, where an isolated $C\ell O$ absorption line exists ($^{37}C\ell^{16}O$, $\Omega = 3/2$, R(13-1/2). The local oscillator is a $^{14}C^{16}O_2$ laser, tuned to the 00°1 - I band, P(16) line at 853.181 cm^{-1}. The frequencies of these laser lines have been recently measured with very high accuracy [19]. The photomixer used in this instrument is a high speed mercury-cadmium telluride photodiode, fabricated at the MIT Lincoln

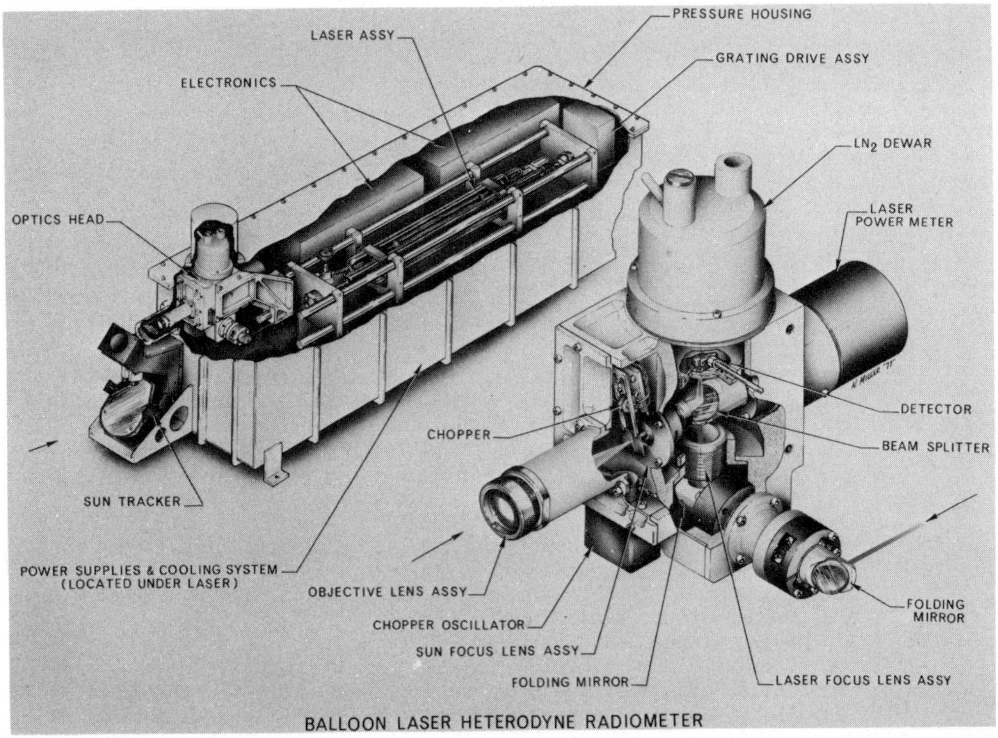

Fig.5 A balloon-borne laser heterodyne radiometer, constructed at JPL to be used for measurements of trace stratospheric constituents.

Laboratories by D. SPEARS. When observing the sun near sunrise or sunset, when a long ($\sim$ 500 km) stratospheric path can be observed, the expected sensitivity of this instrument to chlorine monoxide in the 35 - 40 km altitude region is 10^7 cm^{-3} minimum detectable number density. This limit is determined by the frequency roll-off of the photomixer and the IF preamplifier noise. The ultimate sensitivity limit of the type of instrument should be much less. It is planned to use this instrument for several other measurements in the future, each time making use of the high resolution laboratory spectroscopy to select optimum spectral regions in which to operate.

6. Acknowledgments

The author wishes to acknowledge the following contributors to the spectroscopic work which is discussed in this paper: J. S. Margolis, E. D. Hinkley, and M. S. Shumate from the Jet Propulsion Laboratory; and C. Bair, P. Brockman, R. K. Seals, Jr., and F. Allario from NASA Langley Research Center.

A portion of the work described in this paper was supported by the National Aeronautics and Space Administration, under Contract NAS 7-100.

References

1. *Proceedings of the Fourth Conference on the Climatic Impact Assessment Program*, T. M. Hard and A. J. Broderick, Editors (U.S. Department of Transportation, Washington, D.C., 1975).
2. H. Johnston, "Photochemistry in the Stratosphere," in the Proceedings of the Conference on *Tunable Lasers and Applications*, Loen, Norway, 1976, edited by A. Mooradian, T. Jaeger, and P. Stokseth (Springer-Verlag, Berlin) 1976.
3. R. J. Cicerone, R. S. Stolarski, and S. Walters, Science 185, 1165 (1974.
4. M. J. Molina and F. S. Rowland, Nature 249, 810 (1974).
5. J. A. Coxon, W. E. Jones, and E. G. Skolnik, Can. J. Phys. 54, 1043 (1976).
6. J. Curtis, "Vibration-Rotation Bands of NH_3 in the Region 670 cm^{-1} to 1860 cm^{-1}," Ph.D. dissertation, Ohio State University, Columbus, Ohio, 1974.
7. R. Kakar, E. Cohen, and M. Geller, "The Millimeter Spectrum and Molecular Parameters of C O in the V = 0 and V = 1 Vibration States," unpublished.
8. R. T. Menzies, J. S. Margolis, E. D. Hinkley, and R. A. Toth, Appl. Opt. 16, 523 (1977).
9. R. T. Menzies and M. T. Chahine, Appl. Opt. 13, 2840 (1974).
10. R. T. Menzies and R. K. Seals, Jr., "Ozone Monitoring with an Infrared Heterodyne Radiometer," to be published in Science.
11. M. A. Frerking and D. J. Muehlner, Appl. Opt. 16, 526 (1977).
12. R. A. McClatchey, R. W. Fenn, J. E. A. Selby, F. E. Volz, and J. S. Garing, *Optical Properties of the Atmosphere (Revised)* (Air Force Cambridge Research Laboratories, Bedford, MA, AFCRL-TR-71-0279, 1971).
13. R. T. Menzies, Appl. Opt. 15, 2597 (1976).
14. C. Bair, private communication.
15. R. A. McClatchey, W. S. Benedict, S. A. Clough, D. E. Burch, R. F. Calfee, K. Fox, L. S. Rothman, and J. S. Garing, *AFCRL Atmospheric Absorption Line Parameters Compilation* (AFCRL-TR-73-0096, Jan., 1973).
16. D. G. Murcray, A. Goldman, A. Csoeke-Poeckh, F. H. Murcray, W. J. Williams, and R. N. Stocker, J. Geophys. Res. 78, 7033 (1973).
17. P. Brockman, C. H. Bair, and F. Allario, "High Resolution Spectral Measurement of the HNO_3 11.3 Micron Band Using Tunable Diode Lasers," unpublished.
18. R. A. Graham, E. C. Tuazon, A. M. Winer, J. N. Pitts, Jr., L. T. Molina, L. Beamon, and M. J. Molina, Geophys. Res. Lett. 4, 3 (1977).
19. C. Freed, private communication.

LONG RANGE INTERACTION BETWEEN CW LASER BEAMS IN AN ATOMIC VAPOR[1]

W. Happer and A.C. Tam

Department of Physics, Columbia University
New York, NY 10025, USA

Can photons in a vacuum exert forces on each other? This is an old question and the standard modern answer is yes, but since the forces are due to the exchange of virtual electrons and positions[1], the forces are extremely weak and have never been observed experimentally. We would like to describe some recent experiments which demonstrate that photons in laser beams can exert very strong forces on each other in an atomic vapor [2]. These forces are due to the exchange of real atoms rather than virtual electrons and positrons.

The phenomenon which first alerted us to the existence of these forces is illustrated in Fig.1.

A beam of linearly polarized light from a CW dye laser is focused into a glass cell containing sodium vapor at atomic number density of about 10^{13} atoms/cm^3. The laser frequency is detuned by about 1.5 GHz to the high frequency side of the D_1 resonance line of sodium at 5896Å, and the laser power is about 100 mW. The beam initially narrows down into a self-trapped filament of the type first described by Bjorkholm and Ashkin in 1974 [3]. However, after a few centimeters of propagation in the sodium vapor the beam splits up into two daughter beams, which continue to propagate as self-trapped filaments through the vapor. As one can see from Fig.1, the breakup angle increases and the point of breakup moves toward the entrance window of the cell as the cell temperature and the sodium number density increase.

Beam breakup is not a new phenomenon and it has been reported many times in the literature, especially for intense pulsed lasers [4]. However, this "normal" beam breakup can result in two, three, or many daughter filaments and the number of filaments is roughly proportional to the intensity of the parent laser beam. For sufficient detuning $\Delta\nu$ of the laser frequency to the high frequency side of the D_1 or D_2 resonance lines of sodium (typically, for $\Delta\nu \gtrsim$ 3GHz), we were able to observe this normal multiple filamentation. However, when the laser was only slightly detuned above the D_1 resonance line (typically 1GHz $\leq \Delta\nu \leq$ 2GHz) the beam always broke up into two daughter filaments, and no breakup at all could be observed for D_2 light in the same tuning

[1] This work was supported by the U. S. Air Force Office of Scientific Research under Grant AFOSR-74-2685

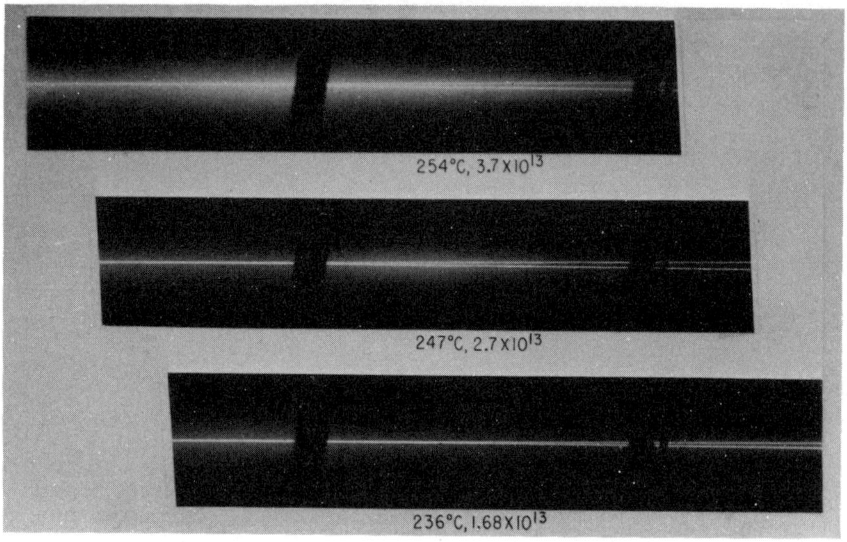

Fig.1 Observed binary-breakup of a linearly polarized parent beam, incident on the Sodium vapor from the left, into two coherent daughter beams of opposite circular polarizations, equal intensities and symmetrical deviations from the parent direction. Cell temperatures and Sodium densities are given.

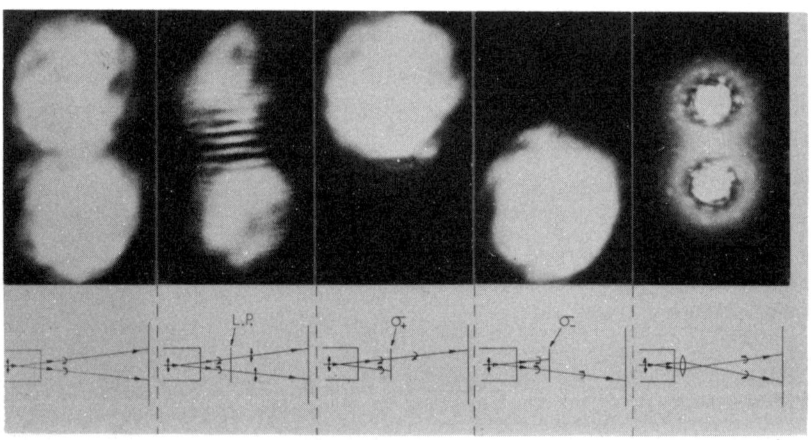

Fig.2 Observed patterns when the daughter beams hit the screen. From left to right: no polarizer, linear polarizer, σ_+- polarizer, σ_-- polarizer, and a positive lens (of f = 2" and placed about 2" from the cell exit window) is inserted between the sodium cell and the screen.

range. A further puzzle was that the binary beam breakup only occured with linearly polarized incident light or for slight elliptical polarization. No binary beam breakup occurred for circularly polarized light.

An even more surprising characteristic of the binary beam breakup is illustrated in Fig.2.

The daughter beams turn out to be 100% circularly polarized in opposite senses. Because of the strong diffraction spreading of the self-trapped daughter beams after they leave the cell, there is some overlap of the beams when they strike a screen located about two meters beyond the cell. If a linear polarizer is inserted in the beams, pronounced two-slit interference fringes are observed on the screen, as shown in Fig.2. This proves that the beams are coherent with each other and are not some kind of Raman sideband. By examining the beams with a spectrum analyzer we were able to confirm that the beams had the same frequency as the parent beam from the laser. Because the laser beam simultaneously breaks up into two circularly polarized daughter beams we have proposed that this phenomenon be called "self circular birefringence."

After some thought about the experiments shown in Fig.1 and Fig.2 we began to suspect that the two beams might actually be exerting force on each other. We tested this naive idea in the most direct way by aiming two beams at each other in a cell as shown in Fig. 3.

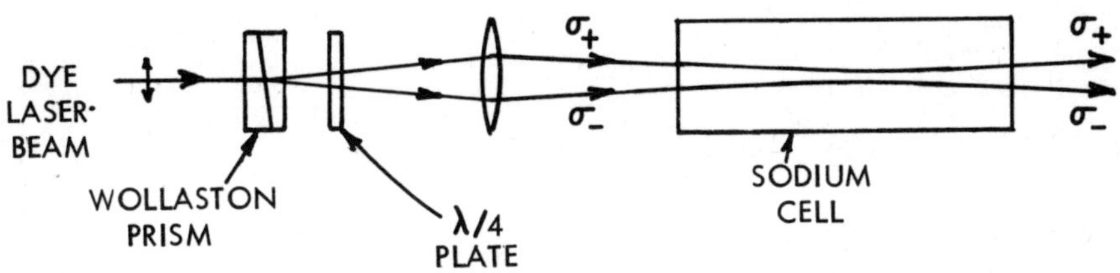

Fig.3 Schematics of the experimental arrangement to observe the repulsion between σ_+ and σ_- laser beams; laser frequency is 1.5 GHz above the center of D_1 resonance. The angle between beams before entering cell is about 0.002 radian.

The results of our colliding beam experiments were unambiguous. We found that beams of opposite circular polarization repel each other, while beams of the same circular polarization attract each other.

How is one to understand this curious force between self-focused beams in sodium vapor? We believe that the force is transmitted between the beams by sodium atoms in much the same way that virtual photons are believed to transmit the force between charged particles, or phonons are believed to transmit the force between Cooper pairs in a superconductor.

The simplest way to understand the force on a light beam is to first consider the force exerted by a light beam on an atom.

Once we have gained a clear understanding of the forces exerted by a light beam on an atom we may assert that the atoms exert an equal and opposite force on the light beam, and we may inquire how forces between light beams can be mediated by the exchange of atoms.

The force exerted by the electric field $\vec{E}$ and the magnetic field $\vec{B}$ of a light beam on an atom is just the well-known Lorentz force

$$\vec{F} = \int\left(\rho\vec{E} + \frac{\vec{j} \times \vec{B}}{c}\right) dV \tag{1}$$

where ρ is the charge density within an atom and $\vec{j}$ is the corresponding current density. We may expand (1) in a multipole expansion by writing

$$\vec{E}(\vec{r}) = \vec{E} + \vec{r} \cdot \vec{\nabla}\vec{E} + \ldots \tag{2}$$

where $\vec{E}$ and its spatial derivatives are evaluated at the center of the atom, $\vec{r} = 0$, in the right-hand side of (2). We may write an expansion similar to (2) for $\vec{B}(\vec{r})$. Then (1) becomes

$$\vec{F} = \int\left(\rho\vec{r} \cdot \vec{\nabla}\vec{E} + \frac{\vec{j}}{c} \times \vec{B} + \ldots\right)dV \tag{3}$$

In (3) we have retained only the lowest-order, non-vanishing terms in $\vec{E}$ and $\vec{B}$, which are now independent of $\vec{r}$. We note that the atomic dipole moment $\vec{D}$ of the atom is

$$\vec{D} = \int\rho\vec{r}\, dV \tag{4}$$

Also, if we multiply the equation of continuity

$$\frac{\partial\rho}{\partial t} + \vec{\nabla} \cdot \vec{j} = 0 \tag{5}$$

by $\vec{r}$ and integrate over the volume of the atom we have

$$\frac{\partial\vec{D}}{\partial t} = \int\vec{r}\,\frac{\partial\rho}{\partial t}\, dV = -\int \vec{r}\vec{\nabla} \cdot \vec{j}dV = \int \vec{j}dV \tag{6}$$

where we have integrated by parts to obtain the right-hand side of (6). Using (6) and (4) we find the (3) becomes

$$\vec{F} = \vec{D} \cdot \vec{\nabla}\vec{E} + \frac{\vec{D} \times \vec{B}}{c} + \cdot \cdot \cdot \tag{7}$$

One can show that the additional terms in (7) involve the electric quadrupole moment, the magnetic dipole moment and all of the other higher-order atomic moments. We shall therefore neglect these terms and limit our discussion to the forces associated with the electric dipole moment $\vec{D}$ of the atom.

We may transform (7) into a more convenient form by using Faraday's law

$$\frac{\partial E_j}{\partial x_i} - \frac{\partial E_i}{\partial x_j} = -\frac{1}{c}\,\varepsilon_{ijk}\,\frac{\partial B_k}{\partial t} \tag{8}$$

Here ε_{ijk} is the unit antisymmetric tensor ($\varepsilon_{123} = 1$), and in (8) and in subsequent formulas we shall use the summation convention that a sum over repeated dummy indices is understood. Then

$$\vec{D} \cdot \vec{\nabla}\vec{E} = D_i \frac{\partial E_j}{\partial x_i}\,\hat{i}_j = D_i \frac{\partial E_i}{\partial x_j}\,\hat{i}_j - \frac{1}{c} D_i \varepsilon_{ijk} \frac{\partial B_k}{\partial t}\,\hat{i}_j \tag{9}$$

Substituting (9) into (7) we find

$$\vec{F} = D_i \frac{\partial E_i}{\partial x_j}\,\hat{i}_j + \frac{1}{c}\frac{\partial}{\partial t}(\vec{D} \times \vec{B}) \tag{10}$$

We may drop the second term on the right of (10) since its time average over a few optical cycles is negligible. Thus the average force acting on an atom in the dipole approximation is

$$\vec{F} = D_i \frac{\partial E_i}{\partial x_j}\,\hat{i}_j \tag{11}$$

We now write the electric field of the light wave as

$$\vec{E}(\vec{r},t) = \vec{\varepsilon}(\vec{r},t)e^{i(\vec{k}\cdot\vec{r}-\omega t)} + \text{c.c.} \tag{12}$$

where $\vec{\varepsilon}$ is a slowly varying amplitude in space and time, $\vec{k}$ and $\vec{\omega}$ are the mean propagation vector and frequency of the light, and c.c. denotes the complex conjugate of the preceding expression. We assume that the electric field (12) induces a dipole moment

$$\vec{D}(t) = \vec{\mathcal{D}}(t)e^{-i\omega t} + \text{c.c.} \tag{13}$$

in an atom located at $\vec{r} = 0$. The amplitude of the dipole moment is related to the amplitude of the electric field by the polarizability tensor α of the atom

$$\vec{\mathcal{D}}(t) = \alpha\,\vec{\mathcal{E}}(0,t) \tag{14}$$

The polarizability tensor depends strongly on frequency in the neighborhood of an atomic resonance line, as is illustrated in Fig.4. If we substitute (13) and (12) into (11) we find that the force on an atom, averaged over a few optical cycles, is

$$\vec{F} = \left[\mathcal{D}_i \frac{\partial \mathcal{E}_i^*}{\partial x_j}\,\hat{i}_j - i\vec{k}\vec{\mathcal{D}} \cdot \vec{\mathcal{E}}^*\right] + \text{c.c.} \tag{15}$$

We note that the polarizability tensor α, which is defined by (14), can be written as the sum of a Hermitian and an antiherma-tian part

$$\alpha_{ij} = \alpha_{ij'} + i\,\alpha_{ij''} \tag{16}$$

where

$$\alpha_{ij'} = \frac{\alpha_{ij} + \alpha_{ji^*}}{2} \tag{17}$$

$$\alpha''_{ij} = \frac{\alpha_{ij} - \alpha_{ji^*}}{2i} \tag{18}$$

Substituting (16) into (15) we find that

$$\vec{F} = \vec{F}_c + \vec{F}_d \tag{19}$$

where

$$\vec{F}_c = -\vec{\nabla}u \tag{20}$$

$$u = -\mathcal{E}^*_i\,\alpha'_{ij}\,\mathcal{E}_j \tag{21}$$

$$\vec{F}_d = 2\vec{k}\,\mathcal{E}^*_i\,\alpha''_{ij}\,\mathcal{E}_j + \left[i\,\hat{i}_j\,\frac{\partial\mathcal{E}^*_i}{\partial x_j}\,\alpha''_{ik}\mathcal{E}_k + \text{c.c.}\right] \tag{22}$$

We shall call these the conservative and dissipative forces exerted by a light beam on an atom. These forces have a simple physical interpretation. The effective Hamiltonian of an atom in a light beam is known to be[5]

$$\delta\mathcal{H} = \delta\mathcal{E} - i\hbar\frac{\delta\Gamma}{2} = -\mathcal{E}^*.\alpha.\vec{\mathcal{E}} \tag{23}$$

We see that the potential u of (21) is the light shift $\delta\mathcal{E}$ (actually, the expectation value of the light shift operator). Furthermore, the dissipative force (22) is (for $\vec{\nabla}\vec{\mathcal{E}} = 0$)

$$\vec{F}_d \simeq \hbar\vec{k}\,\delta\Gamma \tag{24}$$

that is, it is the momentum per photon $\hbar\vec{k}$ times the photon absorption rate $\delta\Gamma$ of the atom. Thus, the dissipative force is the momentum transfer to the atom from the absorbed photons. We add in passing that the term involving α'' and $\vec{\nabla}\vec{\mathcal{E}}$ in (22) is a correction to account for the fact that the local direction of the photon momentum is not exactly parallel to $\vec{k}$ near the edges of the light beam where $\mathcal{E}$ changes rapidly with position.

We note that the conservative force is the same as the force exerted by an electric field on a piece of dielectric. In our everyday experience we are used to thinking of these forces as attractive (for example the attraction of small bits of lint or paper to a plastic comb which has been charged by friction on a dry day). This is because most objects of everyday experience have positive dielectric constants (or positive polarizabilities α') and the potential u of (21) is an attractive potential well. However the polarizability of an atom can be positive or negative (or complex) as is shown in Fig.4. Thus, an atom can be either attracted or repelled by the electric fields of a laser beam.

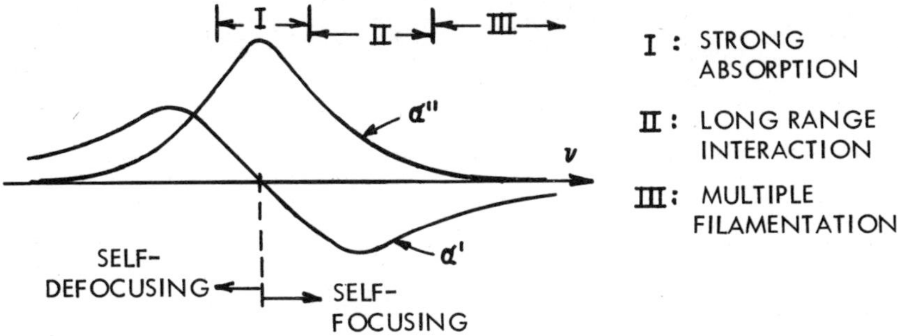

Fig.4 Characteristic interaction regimes for laser beam (of frequency ν) in sodium vapor near the D_1 absorption line. Polarizability $\alpha = \alpha' + i\alpha''$.

Since the gradient of $\mathcal{E}$ is large and transverse to the direction of propagation near the edge of a light beam, we expect the conservative force to be transverse to the direction of propagation and to act only near the surface of the laser beam. Of course the atoms will exert an equal and opposite force on the light beam and the "reaction" of the conservative force on the light beam will be transverse and will rotate the total beam momentum without changing the magnitude of the momentum. Thus, the conservative force conserves the number of photons in the laser beam.

In contrast, the dissipative force is mainly parallel to the propagation vector as indicated by (24). Therefore, the reaction of the dissipative force on the light beam is antiparallel to the photon momentum, and it will tend to diminish the magnitude of the total photon momentum without changing the direction of the momentum. Of course this comes about because the dissipative force is due to the absorption of photons from the light beam. This leads to an attenuation of the light beam. Another characteristic feature of the dissipative force is the fact that it is a bulk phenomenon which can act inside the laser beam where $\nabla\vec{\mathcal{E}} = 0$.

The preceding arguments are quite general and they apply to all vapors. However, the self-circular birefringence of sodium vapor is due to the special structure of the sodium atom, whose lowest energy levels are illustrated in Fig.5.
Note that a spin down atom in the ground state can only absorb spin up photons, while a spin up photon can only absorb spin down photons. This situation is particularly favorable for optically pumping [5] with circularly polarized light, since circularly polarized D_1 light beams will tend to spin-polarize sodium atoms along the direction of the photon spin. If we denote the mean photon spin by $\langle\vec{S}\rangle$ and the mean atomic spin by $\langle\vec{J}\rangle$, then one can show [5] that the effective Hamiltonian (23) for a sodium atom is

$$\delta\mathcal{H} = -\vec{\mathcal{E}}^{*}.\alpha.\vec{\mathcal{E}} = -\alpha_o\ |\mathcal{E}|^2(1 - 2\langle\vec{J}\rangle \cdot \langle\vec{S}\rangle) \qquad (25)$$

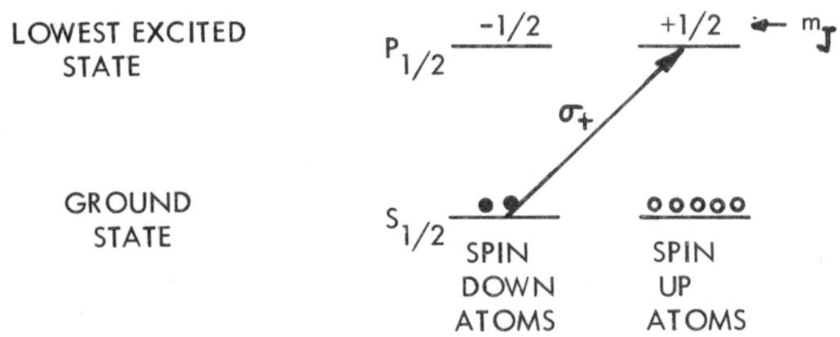

Fig.5 Energy levels of an alkali atom (neglecting hyperfine structure)

Thus, the forces (20) and (22) depend very strongly on the relative orientation of the photon and atomic spins. Maximum forces occur when the atomic and photon spins have their maximum possible values of 1/2 and 1 respectively and when the spins are antiparallel. When the spins have their maximum possible values but are parallel the forces are zero.

We may now construct a qualitative picture of how laser beams exert forces on each other by exchanging atoms in an atomic vapor. As sketched in Fig.6, atoms emerging from a σ_+ light beam will have their spins polarized parallel to the beam momentum. Those atoms which happen to fly over to the σ_- light beam will exert a maximum repulsive force on the σ_- light beam while they are entering the beam because the atomic spin is antiparallel to the photon spins of the σ_- beam.

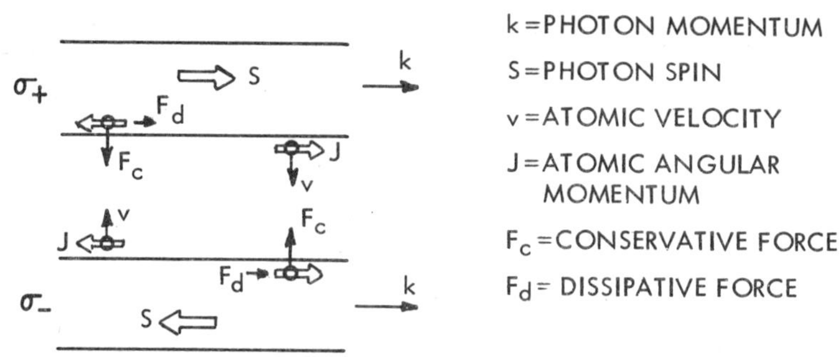

Fig.6 Schematic explanation of forces between σ_+ and σ_- laser beams due to exchange of atoms.

Note that the force is repulsive provided that the optical frequency is above resonance, where $\alpha' < 0$ (see Fig.4). If the optical frequency were below resonance, the atoms would be attracted into the beam and the σ_+ and σ_- beams would exert attractive forces on each other, since $\alpha' > 0$. However, frequencies below resonance are not suited for colliding beam experiments like those sketched in Fig.3 since no self-trapped filaments are formed, and, on the contrary, the self-defocusing rapidly leads to a loss of beam definition.

A quantitative analysis of the long range forces between laser beams in atomic sodium vapor is a complicated problem because one must account for optical pumping, the spatial profile of a self-trapped laser beam, the atomic hyperfine structure and other effects which we have ignored in our discussion so far. However, one can obtain a useful semiquantitative estimate of the splitting angle of the daughter beams by the following reasoning.

We have shown that the conservative force on an atom can be represented as the negative gradient of a potential, the light shift. Thus the total force on the atom in some segment of one of the laser beams is

$$\vec{F}_a = \int N(-\vec{\nabla}u)dV = \int u\, \vec{\nabla}N\, dV \tag{26}$$

where N is the number density of atoms. For circularly polarized atoms we must think of N as the number of atoms of the appropriate spin polarization. The last equality in (26) can be obtained by integration by part. Since the force on the atom must be equal and opposite to the force on the light, the force on the light is

$$\vec{F}_\ell = - F_a = \int -u\, \vec{\nabla}N = \int \alpha' |\varepsilon|^2\, \vec{\nabla}N \tag{27}$$

but the photon number density is

$$N_\ell = \frac{\varepsilon^2}{2\pi h\nu} \tag{28}$$

So (27) and (28) imply that the average force per photon in a σ_+ beam can be written as the negative gradient of the potential

$$U_{\sigma+} = -\, 2\pi h\nu\, \alpha' N{\downarrow} \tag{29}$$

where $N{\downarrow}$ is the density of spin-down ground state atoms. An order-of-magnitude calculation of the splitting half angle θ in self circular birefringence can now be made. Let us regard the parent linearly polarized beam as two co-propagating, parallel but very slightly displaced beams of σ_+ and σ_- polarization. When the two beams are close to each other, a strong repulsive force exists, and as the beams deviate from each other and separate more and more, the repulsion tends to zero. θ can be expressed as an integral of the repulsive force. Alternatively, θ can be obtained by the potential in (29). Initially a σ_+ photon is very near the σ_- beam, so

$$U_{\sigma+}\ (\text{initial}) \approx -2\pi h\nu\, \alpha'\left(\frac{3N}{4}\right) \tag{30}$$

where we have set $N{\downarrow}$ as $3N/4$ near the surface of the $\sigma-$ beam, because atoms from within the beam are mainly spin down polarized, while atoms from outside the beam are mainly unpolarized. The σ_+ photon is finally very far from the σ_- beam, so

$$U_{\sigma+}\ (\text{final}) = -2\pi h\nu\, \alpha'\frac{N}{2} \tag{31}$$

Hence the potential difference is

$$U_{\sigma_+}\ (\text{initial}) - U_{\sigma_+}\ (\text{final}) = -\pi h\nu\ \alpha'\ \frac{N}{2} \tag{32}$$

The potential difference (32) is converted into kinetic energy of transverse motion of the σ_+ photon, i.e.,

$$\frac{p_\perp^{\ 2}}{(2m)} = \left(\frac{h}{\lambda}\theta\right)^2 \Big/ \left(2\frac{h}{\lambda c}\right) \tag{33}$$

where $m = h/(\lambda c)$ denotes the effective mass of a photon.

On equating (32) and (33), we obtain

$$\theta = \sqrt{-\pi\alpha'\ N} \tag{34}$$

A comparison of the simple formula (34) with the measured splitting angles is shown in Fig.7, for laser detuning being +1.5 GHz, when $|\alpha'|$ is taken as $10^{-18}cm^3$ [6] in formula (34).

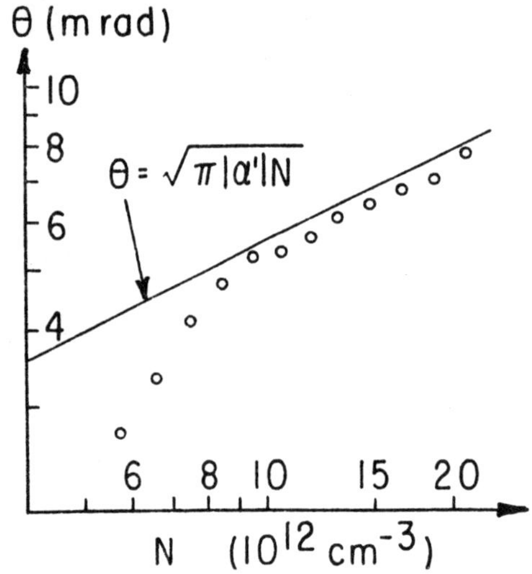

Fig. 7. Comparison of the observed deviation angles θ with the predicted values for various sodium densities N. Note that the asymptotic splitting angle between the daughter beams is 2θ.

We note that the agreement is remarkably good except for low atomic number densities where we suspect that the beams may not have reached asymptotic splitting before emerging from the sodium vapor.

We will conclude with an interesting suggestion by OKA and RADFORD [7]. For some time now the existence of circularly polarized maser emission from OH molecules has been known [8]. There are certain similarities between the structure of the Λ-doubling transition of the OH molecule and the sodium D_1 line. Perhaps the curious circular polarization of OH radiation has some connection to the self circular birefringence which we observe in sodium vapor.

References

1. See e.g. S. S. Schweber, "An Introduction to Relativistic Quantum Field Theory", Harper & Row, N.Y. 1961, p. 593.

2. A. C. Tam and W. Happer, Phys. Rev. Lett., 38, 278 (1977).

3. J. E. Bjorkholm and A. Ashkin, Phys. Rev. Lett., 32, 129 (1974).

4. D. Grischkowsky, Phys. Rev. Lett., 24, 866 (1970); A. J. Campillo, S. L. Shapin and B. R. Suydam, Appl. Phys. Lett., 23, 628 (1973); V. I. Bespalov and V. I. Talanov, JETP Lett., 3, 307 (1966).

5. W. Happer, Rev. Mod. Phys., 44, 169 (1972).

6. B. S. Mathur, H. Y. Tang, and W. Happer, Phys. Rev. A 2, 648 (1970).

7. T. Oka and H. E. Radford, private communication.

8. See e.g. D. teer Haar and M. A. Pelling, Rep. Prog. Phys., 37, 481 (1974).

PHASE MATCHING OF 2-PHOTON RESONANT 4-WAVE MIXING PROCESSES

G.C. Bjorklund, J.E. Bjorkholm, R.R. Freeman, and P.F. Liao

Bell Telephone Laboratories
Holmdel, NJ 07733, USA

Recently, 4-wave sum and difference frequency mixing processes in gases and vapors have received considerable attention as possible sources of tunable, coherent radiation in new regions of the electromagnetic spectrum [1]. The efficiencies of these processes have previously been enhanced either by phasematching by addition of a buffer gas of compensating dispersion or by two-photon resonant enhancement of the nonlinear susceptibility. The buffer gas technique works well for nonresonant mixing processes, but has the disadvantages of requiring a highly uniform gas mixture and of introducing additional absorption due to pressure broadening. This pressure broadening is more serious for two-photon resonant processes since broadening of the two-photon resonance reduces the nonlinear susceptibility [2].

We have recently developed a new technique for phase matching 4-wave mixing processes wherein collinear phasematching and two photon resonance may be simultaneously achieved in a pure nonlinear medium simply by properly choosing the frequencies of the fundamental beams [3]. The phasematching buffer gas is thus not required and the deleterious effects of pressure broadening are eliminated. In addition, for plane waves, this technique is insensitive to the spatial distribution of the nonlinear medium and thus sophisticated cell designs which produce very uniform gas mixtures, such as concentric heat pipe ovens [4], are not needed.

For plane waves, the generated power of a 4-wave frequency mixing process of the type

$$\omega_1 \pm \omega_2 \pm \omega_3 \rightarrow \omega_4 \tag{1}$$

is given by the relation

$$P_4 \propto \frac{N^2 L^2}{A^2} \chi^2 P_1 P_2 P_3 \left[\frac{\sin(\Delta kL/2)}{\Delta kL/2}\right]^2 \tag{2}$$

where P_i, λ_i and k_i are the power, wavelength, and wavevector of the beam at ω_i; N is the atomic number density of the nonlinear medium, L the cell length, A the beam area, χ the atomic third order nonlinear susceptibility, and Δk the phase mismatch. The nonlinear susceptibility, χ, is resonantly enhanced if

$$\omega_1 \pm \omega_2 = \Omega \qquad (3)$$

where Ω is the equivalent frequency of a two-photon transition. The phasematching factor in brackets is optimized for plane waves if

$$\Delta k = k_4 - k_1 - k_2 - k_3 = \frac{1}{c}\left(n_4\omega_4 - n_3\omega_3 - n_2\omega_2 - n_1\omega_1\right) = 0, \qquad (4)$$

where $n_i = n_i(\omega_i)$ is the index of refraction at frequency ω_i.

For a given desired generated frequency, ω_4, and a given two photon level with equivalent frequency Ω, Eqs. (1, 3, and 4) can be solved simultaneously for values of ω_1, ω_2, and ω_3. Real solutions for the input frequencies can be found for wide ranges of ω_4 provided that the frequencies can span a region of anomalous dispersion. In many instances, and particularly for the alkali metals, the index of refraction at all 4 frequencies can be accurately determined by making the approximation that all of the oscillator strength for transitions from ground is contained in the resonance transition. Thus only one term of the standard Sellmeier equation is kept and

$$n_i(\omega_i) \cong 1 + N 2\pi c^2 r_e \; f/(\omega_o^2 - \omega_i^2), \qquad (5)$$

where $r_e = 2.818 \times 10^{-13}$ cm, f is the oscillator strength, and ω_o is the frequency of the resonance transition. This approximation breaks down when any of the frequencies is very near to

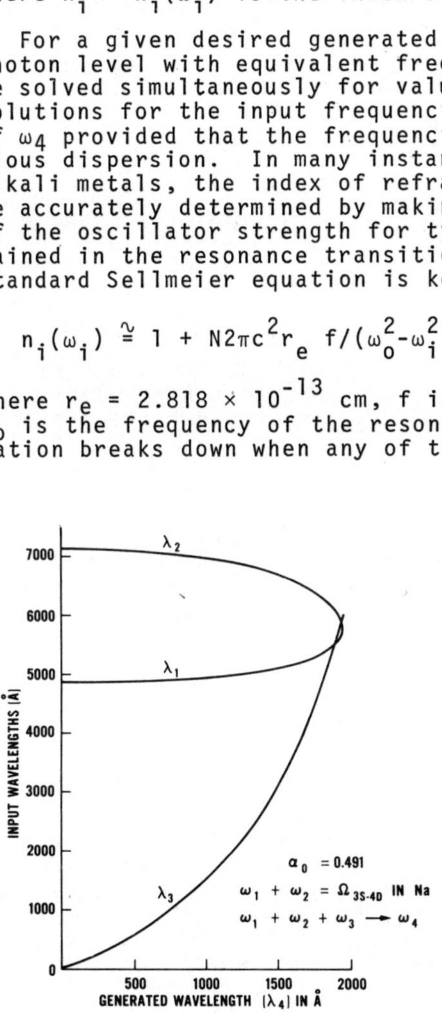

Fig. 1. Theoretical phasematching wavelengths for the process $\omega_1 + \omega_2 + \omega_3 \rightarrow \omega_4$ in Na with $\omega_1 + \omega_2$ resonant with the 4D channel.

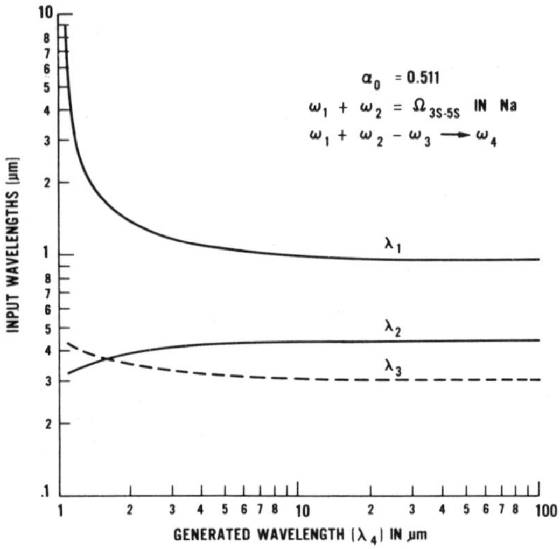

Fig. 2. Theoretical phasematching wavelengths for the process $\omega_1 + \omega_2 - \omega_3 \rightarrow \omega_4$ in Na with $\omega_1 + \omega_2$ resonant with the 5s channel. The ω_3 radiation (shown by the dashed line) experiences gain.

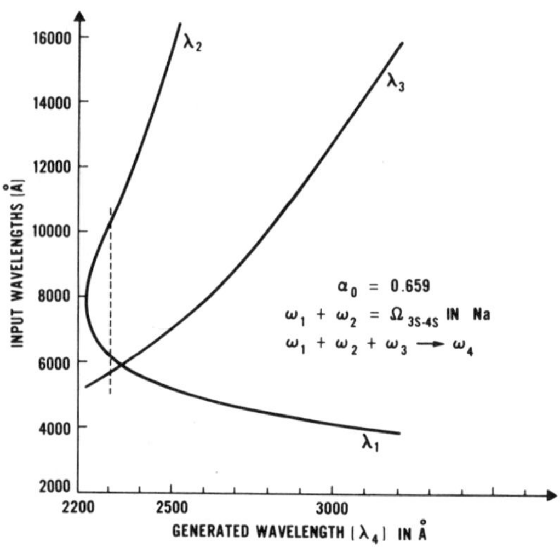

Fig. 3. Theoretical phasematching wavelengths for the process $\omega_1 + \omega_2 + \omega_3 \rightarrow \omega_4$ in Na with $\omega_1 + \omega_2$ resonant with the 4s channel. The intersection of the vertical dashed line with the solid curves indicates the experimental operating point.

one of the other (weak) transitions from the ground state or when any of the frequencies is within a few fine-structure splittings of a doublet main resonance line. Figs. 1, 2, and 3 show tuning curves calculated using this approximation for phasematching of several types of 4 wave mixing process in Na vapor. It is apparent from the figures that wavelengths from the VUV to the far IR can be produced by this method.

In order to verify the accuracy of the phasematching calculations and to demonstrate the insensitivity to the spatial distribution of the nonlinear medium, we conducted a series of experiments for the process $\omega_1 + \omega_2 + \omega_3 \rightarrow \omega_4$ in a cell which contained a highly nonuniform distribution of Na vapor. We constrained $\omega_1 + \omega_2$ to be resonant with the 3S-4S two photon transition. (See Fig. 3 for the theoretical phasematching wavelengths.) A Q-switched Nd:YAG laser (10 nsec pulse duration) and two dye lasers (pumped by the second harmonic of the YAG laser) provided polarized radiation at frequencies ω_1, ω_2, and ω_3. The beams were spatially filtered, collimated, combined with parallel polarizations, passed through the Na cell, and then the generated radiation at ω_4 was isolated by means of a quartz prism. Our experiment consisted of keeping λ_1 (612 nm) and λ_2 (1.064 μm) constant and two photon resonant with the 3S-4S transition, while λ_3 was varied in the vicinity of the theoretical phasematching value of 570 nm and P_4, the generated power at $\lambda_4 \cong 231$ nm, was monitored. Fig. 4 shows the energy level diagram for this process and Fig. 5 shows the experimental set up. Since λ_4 in this case lies in the smoothly varying Na continuum, χ would be expected to remain fairly constant as λ_3 was varied and any variation in P_4 as a function of λ_3 could be attributed to phasematching.

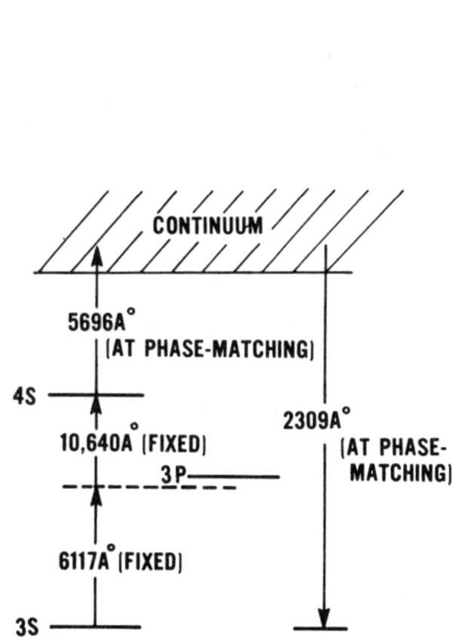

Fig. 4. Level diagram for 4-wave mixing experiment in Na.

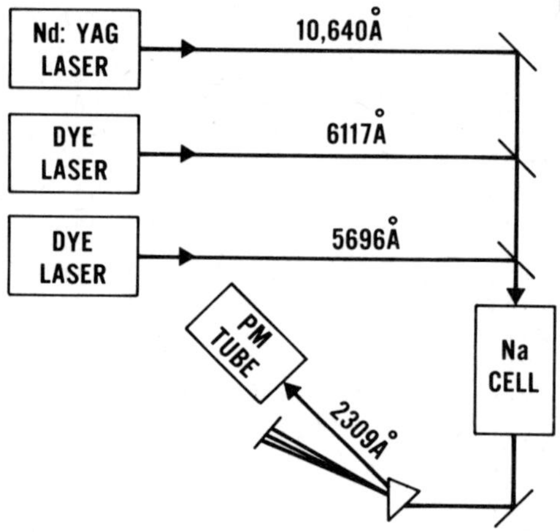

Fig. 5. Experimental set-up for Na experiment.

Typical experimental results are shown in Fig. 6. The location and shape of the observed peak is in excellent agreement with theory. Defining (NL), the integrated number density, as

$$NL = \int_0^L N(z)dz; \tag{6}$$

it can be shown theoretically that the amplitude and half width of the peak should vary as $(NL)^2$ and $(NL)^{-1}$ respectively. We observed this behavior experimentally for low and moderate values of NL. At higher values of NL ($NL>5\times10^{17}$ cm^{-2} with L=90 cm), Na_2 molecular absorption was sufficient to cause significant attenuation of the ω_1 and ω_3 beams and the peak amplitude began to vary more slowly than $(NL)^2$. The behavior we observed in this region was still in agreement with theory when the effects of loss were included.

We have also applied this phasematching technique to enhance the efficiency of the production of tunable VUV radiation by 4-wave mixing in Sr vapor. Sr has a much more complicated energy level structure than any of the alkalis and the oscillator strengths of many of the transitions from the ground state are not well known, although the oscillator strength of the main

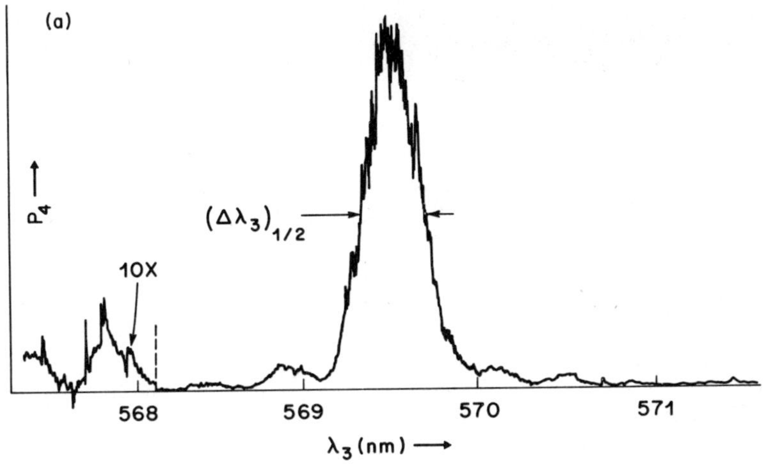

Fig. 6. Phase-matching data for Na experiment.

resonance line has been determined to be large. The most striking difference between Sr and the alkalis lies in the region above the ionization potential. The continuum of the alkalis is smoothly varying and featureless, while the continuum of Sr is dominated by autoionizing resonances. It was not initially clear to us that the single term approximation used in the calculation of the phasematching wavelengths for the alkalis would be accurate for Sr when the generated wavelength was near one of these autoionizing resonances. However it was important to find out, since it is just this condition of near resonance with A-I features which yields the largest nonlinear susceptibility [5].

We found that one particular channel, utilizing the 38444 cm^{-1} $5s^{2\,1}S_0$-$5s7s\,^1S_0$ two photon transition, was particularly efficient. We used this channel in all of our Sr experiments. Fig. 7 shows the energy level diagram for this experiment. Fig. 8 shows the theoretical phasematching wavelengths, calculated using the single term approximation, for the process $\omega_1 + \omega_2 + \omega_3 \rightarrow \omega_4$ with $\omega_1 + \omega_2$ two photon resonant with this level. In order to check the accuracy of the theoretical phasematching wavelengths, a series of experiments with plane waves and two or three Nd:YAG pumped dye lasers was conducted.

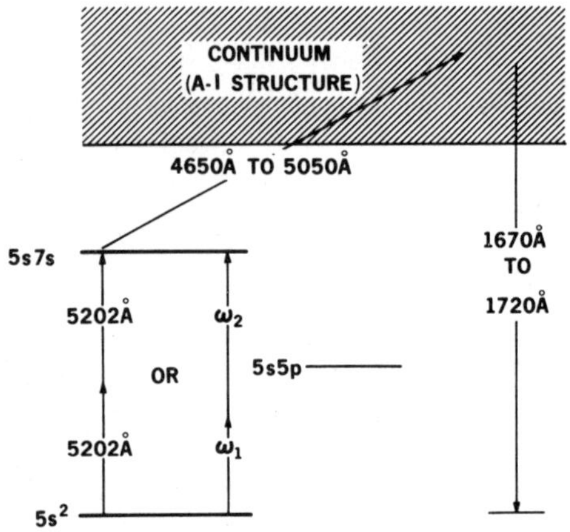

Fig. 7. Level diagram for 4 wave mixing experiment in Sr.

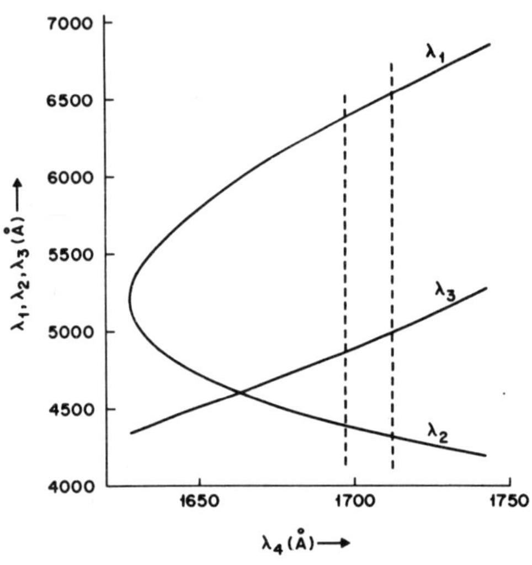

Fig. 8. Theoretical phasematching wavelengths for the process $\omega_1 + \omega_2 + \omega_3 \rightarrow \omega_4$ in Sr with $\omega_1 + \omega_2$ resonant with the 5s7s channel.

In the first experiment, the two photon transition was pumped with two equal energy photons ($\omega_1=\omega_2$) at 520.2 nm and the generated VUV power monitored as the ω_3 dye laser was swept between 470.0 nm and 504.0 nm. From Fig. 8 it can be seen that no phasematching could have occurred in this experiment and thus that any variations in the P_4 vs. λ_3 curve must arise from dispersion of χ. Fig. 9(b) shows the experimental results. The observed peaks correspond to resonant enhancements of χ due to autoionizing features in this region.

Two peaks, corresponding to λ_3 = 488.2 nm, λ_4 = 169.7 nm and λ_3 = 501.2 nm, λ_4 = 171.2 nm were chosen for detailed study. We next set up the configuration involving three separate dye lasers and pumped the 5s7s level using two unequal energy photons ($\omega_1+\omega_2=\Omega$, $\omega_1\neq\omega_2$). The generated power at λ_4 = 169.7 nm and λ_4 = 171.2 nm was monitored as ω_1 and ω_2 were varied subject to the constraint of two-photon resonance. The experimental values of λ_1 and λ_2 necessary to produce the maximum generated power at these wavelengths were found to be within 1.0 nm of the values in Fig. 5. Thus even though λ_4 is nearly resonant with autoionizing transitions, it appears valid to neglect the dispersion due to the continuum in cases where two of the fundamental wavelengths are near the resonance line.

The effect of phasematching on the P_4 vs. λ_3 spectrum was to impose a multiplicative enhancement envelope on the non-phasematched spectrum of Fig. 9(b). We observed that the width of this envelope decreased as NL decreased, while the enhancement varied as $(NL)^2$ between NL = 1.5×10^{16} cm^{-2} and NL = 6×10^{18} cm^{-2} with L = 30 cm. Above this number density significant linear absorption of one of the fundamental beams set in. Fig. 9(a) shows the P_4 vs. λ_3 spectrum we observed with λ_1 and λ_2 set at 440.0 nm and 636.2 nm to phasematch λ_3 = 488.2 nm, λ_4 = 169.7 nm and T = 1050°C (NL = 1.5×10^{19} cm^{-2}). In this case the phasematching enhancement is over 10^3 and the phasematching peak width is about 0.5 nm. All of the other autoionizing features are, by comparison, suppressed. Any of the other peaks in the non-phasematched spectrum could alternatively have been enhanced by using different values for λ_1 and λ_2.

We next attempted to obtain the maximum possible VUV power by using focused high power beams. The conversion efficiency was found to strongly saturate with respect to the powers of the two beams which pumped the two-photon resonance. Fig. 10 shows P_4 for λ_1 = 636.7 nm, λ_2 = 439.8 nm, λ_3 = 488.2 nm, λ_4 = 169.7 nm, NL = 4×10^{18} cm^{-2} as each of the fundamental beam powers was varied while the other powers were maintained at their maximum values. (These optimum values for λ_1 and λ_2 were biased from the plane wave values by focusing effects [6].) There are several possible two-photon saturation effects which could have been occurring; however, we observed one of these effects, induced absorption of one of the fundamental beams, in a striking manner. Under conditions of maximum P_1 and P_2, we observed at least 75% induced absorption of the weaker λ_2 beam. Our best values for generated power were P_4 = 1.6 W at λ_4 = 169.7 nm with P_1 = 97 kW at λ_1 = 636.7 nm, P_2 = 2.7 kW at

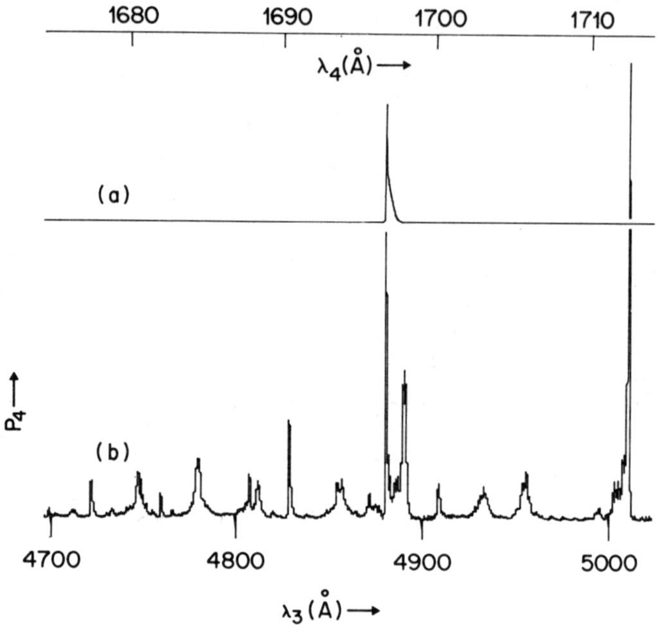

Fig. 9. Generated VUV power vs. λ_3. (a) Phasematched configuration; (b) nonphasematched configuration. These curves have not been normalized for the roll off in P_3 at either extreme value of λ_3.

λ_2 = 439.8 nm, P_3 = 4.3 kW at λ_3 = 488.2 nm and T = 900°C. We also obtained P_4 = 3.2 W at λ_4 = 171.2 nm with P_1 = 39 kW at λ_1 = 651.7 nm, P_2 = 800 W at λ_2 = 432.9 nm, P_3 = 3.6 kW at λ_2 = 501.2 nm and T = 900°C.

From Fig. 10 it can be seen that at lower laser powers the generated power is linear in each of the 3 input beam powers. Thus a figure of merit coefficient, C, can be defined for a given experimental configuration where

$$P_4 = CP_1P_2P_3 \, , \tag{7}$$

all powers are in watts, and C is in units of watts^{-2}. Using spatially filtered, low peak power beams focused to b = 30 cm, we obtained C = 2.4×10^{-10} W^{-2} for λ_4 = 169.7 nm with λ_3 = 488.2 nm, λ_1 = 638.1 nm, λ_2 = 439.1 nm, T = 770°C, and $NL = 5.7 \times 10^{17}$ cm^{-2} and we obtained C = 1.6×10^{-8} W^{-2} for

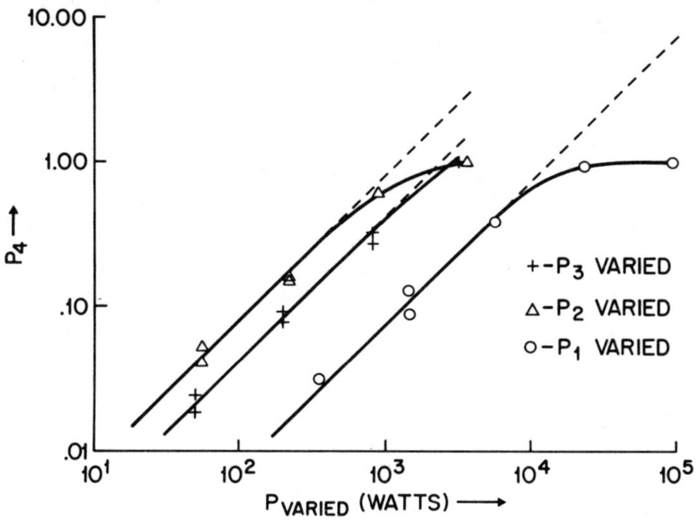

Fig. 10. Generated VUV power vs. input powers. The solid lines are drawn through the data points to aid the eye in following data. The dashed lines are drawn with unity slope and indicate linear extrapolations from unsaturated values.

λ_4 = 171.2 nm with λ_3 = 501.2 nm, λ_1 = 651.7 nm, λ_2 = 439.2 nm, T = 870°C and NL = 2.7×10^{18} cm^{-2}. These values for C represent an increase of at least 10^4 over the best previous values in Sr vapor.

It might even be possible to generate an appreciable amount of cw VUV radiation by this technique. Extrapolation using the C coefficients to cw power levels of 0.1 W in each of the fundamental beams yields cw VUV power levels of 2.4×10^{-13} W at 169.7 nm and 1.6×10^{-11} W at 171.2 nm.

References

1. R. B. Miles and S. E. Harris, IEEE J. Quantum Electron. QE-9, 470 (1973); J. F. Young, G. C. Bjorklund, A. H. Kung, R. B. Miles, and S. E. Harris, Phys. Rev. Lett. 27, 1551 (1971); R. T. Hodgson, P. P. Sorokin, and J. J. Wynne, Phys. Rev. Lett. 32, 343 (1974); S. C. Wallace and G. Zdasiuk, Appl. Phys. Lett. 28, 449 (1976); P. P. Sorokin, J. J. Wynne, and J. R. Lankard, Appl. Phys. Lett. 22, 342 (1973).

2. E. A. Stappaerts, G. W. Bekkers, J. F. Young, and S. E. Harris, IEEE J. Quantum Electron. QE-12, 330 (1976).

3. See G. C. Bjorklund, J. E. Bjorkholm, P. F. Liao, and R. H. Storz, Appl. Phys. Lett. 29, 729 (1976) and references therein.

4. C. R. Vidal and J. Cooper, J. Appl. Phys. 40, 3370 (1969).

5. J. A. Armstrong and J. J. Wynne, Phys. Rev. Lett. 33, 1183 (1974).

6. G. C. Bjorklund, IEEE J. Quantum Electron. QE-11, 287 (1975).

LASER SPECTROSCOPY OF BOUND NaNe AND RELATED ATOMIC PHYSICS [1]

D.E. Pritchard, R. Ahmad-Bitar,[2] and W.P. Lapatovich

Physics Department and Research Laboratory of Electronics
Massachusetts Institute of Technology
Cambridge, MA 02139, USA

I. Introduction

Tunable lasers are revolutionizing diatomic molecular spectroscopy. Their high resolution (anything <0.02 cm^{-1} = 600 MHz by conventional standards) can resolve the rotational spectra of the heaviest diatomics and reveals a constellation of fine and hyperfine magnetic and electric structure; their high intensity permits observation of two quantum absorption, saturation labeling of spectra (e.g. double resonance), optical pumping of molecules, and the observation of extremely low concentrations of molecules. Finally the frequency of laser light can be determined interferometrically or by heterodyne techniques--both being far more accurate than the conventional methods of grating spectroscopy.

The present study of transitions between the lowest two potential curves of NaNe is the first complete sub-Doppler study of a transition between two electronic states of a molecule in the sense that the laser was tuned over the entire frequency range of observable transitions between the two potential curves. It is basically an absorption experiment although the absorption is detected by its associated fluorescence rather than the undetectably small amount of absorption. In this type of experiment the fluorescence is not spectrally analyzed; the wavelength specificity is obtained by measuring the wavelength of the exciting light. In our experiment this light is from a single mode laser and its wavelength is determined interferometrically. The details of the experiment will be discussed in Sec. II which follows. The analysis of the spectrum, which yields precise information about the interatomic potentials, is discussed in Sec. III.

The weak binding of van der Waals molecules in general, and NaNe in particular (less than 10 degrees Kelvin in the ground state) has several implications. First, it makes the potentials extremely difficult to calculate from an ab initio basis--for example, the excellent ab initio calculation for NaAr by Saxon et al [1] required for the Π states alone a basis with 4235 orthonormal configuration state functions, each a linear combination of N-particle Slater determinants! This has given rise to a number of pseudopotential methods for calculating these potentials [2,3,4,5] which can be checked by this experiment. Secondly, the weakness of the binding means that Na and Ne do not form molecules when mixed under normal conditions.

---

[1] Supported by N.S.F. Grant No. PHYS-75-15421-A01

[2] University of Jordan Scholar, Present Adress: University of Jordan, Physics Dept., Amman, Jordan.

Thus a wide variety of processes involving unbound Na and Ne can be, and have been, studied experimentally. The theories which have been developed for these processes all relate interatomic potentials and the observations: our potentials make it possible to subject these theories to a clearcut test. The comparison of our results with other atomic physics experiments and with various theories will be taken up in detail in Sec. IV.

II. The Experiment

Campargue [6] has recently demonstrated that by using relatively high pressures (≳100 atmospheres) it is possible to produce supersonic molecular beams with Mach numbers greater than 50 in vacuum systems with moderate pumping speeds (∿100 ℓ/s). Such flows exhibit kinetic temperatures on the order of 1 K and offer the possibility of producing weakly bound molecules in sufficient quantities to be studied using laser fluorescence spectroscopy--the work of Smalley, Levy, and Wharton on HeI_2 [7] and NaAr [8] is an example.

The only sophisticated part of our molecular beam apparatus was the source, which is a refinement of earlier ovens developed in our lab to make KAr molecules [9]. It consists basically of a tube containing a small boat for the Na, several apertures, and a sintered stainless steel filter. It has a 30μ hole less than 40μ long (the nozzle) in the end and two separately heated regions. The gas mixture (70% neon, 30% helium) is introduced into the oven at ∿120 atmospheres pressure, and ∿0.1 torr of Na is mixed in at our operating temperature of 300 - 400°C. The gas throughput with a 30μ diameter nozzle is ∿10 torr liters/sec, and our 4" Stokes "Ring Jet" booster diffusion pump backed by two 15 cfm mechanical pumps connected in parallel maintains a background pressure ∿0.1 torr. At this pressure the mean free path is ∿1/2 mm; nevertheless the expanding gas is so dense that it pushes the background gas back ~1 cm from the nozzle under our operating conditions. Along the centerline of the expanding gas jet is a zone of silence with temperatures ≲1K: our laser beam and collection optics probed this zone. No skimmers were required for our experiment.

The experimental setup is shown in Fig. 1. The molecular beam was probed at right angles with a Spectra Physics Model 580 single mode cw dye laser whose etalon was locked to the cavity as described earlier [10] and whose prism was tilted in proportion to the voltage driving the etalon. To sweep the laser frequency a sawtooth voltage was applied to the cavity length driver. The amplitude of the resulting motion of the cavity was adjusted to be exactly $\lambda/2$. The time constant of the locking loop for the etalon was long enough so that it could not follow the cavity length flyback--thus the etalon frequency (and consequently the laser frequency) increased linearly with time. We conducted single mode scans of 5 cm^{-1} (the free spectral range of the etalon) with this setup, adjusted the prism tilt angle manually to the next etalon mode, and scanned the next 5 cm^{-1}. Typical scan rates were 0.5 and 1.0 GHz/sec. The fluorescence from such a scan is shown in Fig. 2.

Fluorescence from laser-induced transitions originating from bound ground state molecules in the beam (both Na_2 and NaNe) was collected with an f 1.0 Fresnel lens, focused through a baffle, and detected by a cooled photomultiplier. Typically 10mW of laser power was focused approximately 5mm from the oven. The laser light entered and left the apparatus through tubular arms containing annular baffles which reduced stray scattered light to ≲200 Hz. The principal background in the experiment arose from near-resonant scattering from Na in the beam.

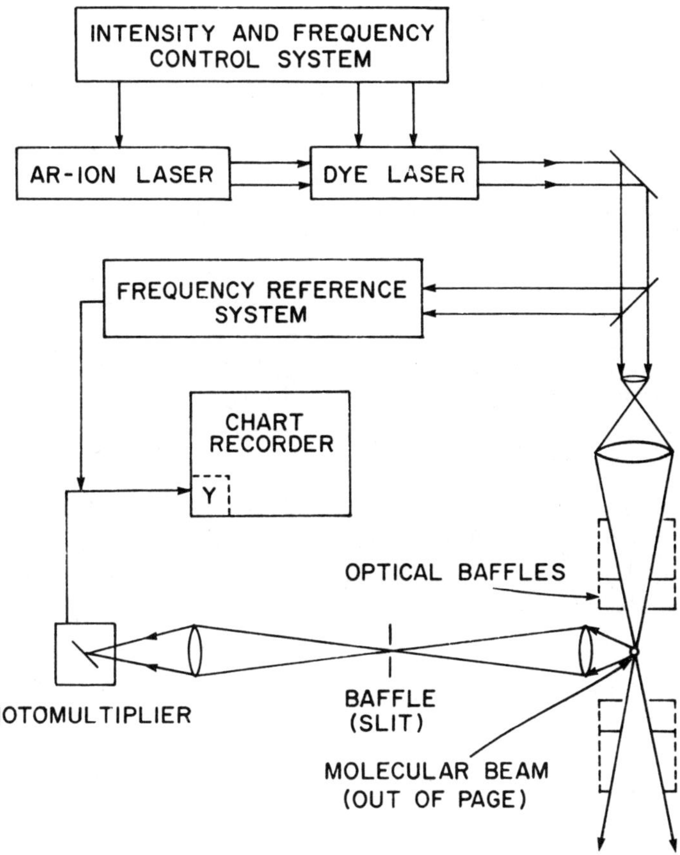

Fig. 1 Schematic view of apparatus

As the dye laser was swept in frequency, its frequency was established by passing a portion of the laser beam through two air-spaced Fabry Perot etalons with free spectral ranges of 0.5254 cm^{-1} and 0.0633 cm^{-1} (as determined from the D-lines of Na). The transmitted light was detected and electronically processed to record frequency reference marks simultaneously with the fluorescence spectrum. These marks permit the frequency of any observed spectral feature to be determined to ∿0.002 cm^{-1} relative to the atomic Na D lines. These marks were suppressed in Fig. 2.

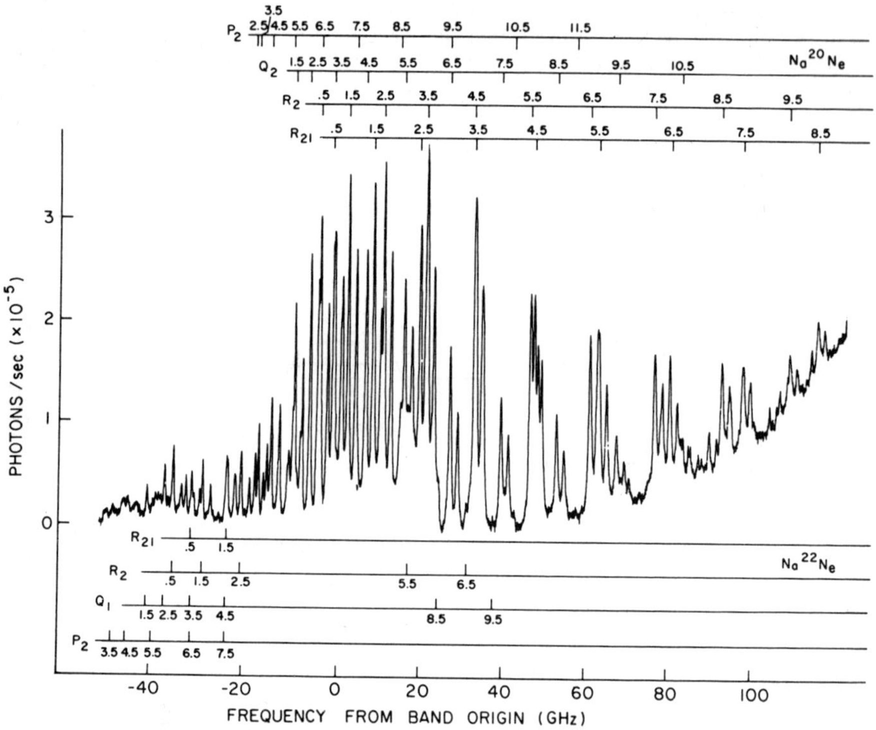

Fig. 2 The $A^2\Pi_{3/2}$ (v=5) ← $X^2\Sigma^+$ (v=0) band with frequencies measured from the band origin. Weak $^{23}Na^{22}Ne$ lines are at the left, the Lorentz tail of the atomic $P_{3/2}$ line is at the right.

III. Analysis

The molecular spectrum which was observed in the frequency domain from 16925 to 16975 cm^{-1} is a mixture of both $A^2\Pi \leftarrow X^2\Sigma$ of NaNe and $A^1\Sigma_u^+ \leftarrow X^1\Sigma_g^+$ of Na_2 in addition to the D-lines of sodium. The NaNe lines are discriminated from the rest by their doublet hyperfine structure [7] and because increases in neon pressure suppress nearly all of the Na_2 lines while dramatically enhancing the strength of NaNe lines. The entire NaNe spectrum consisted of about 380 hyperfine doublets with intensities from 4 x 10^5 Hz to 2 x 10^3 Hz. (Roughly 20% of the doublets had one component obscured by other lines.)

The spectrum is readily grouped into a number of vibronic bands each of which displays four branches with rotational quantum number changes "ΔN" = 0, ±1, ±2 (but always the total angular momentum change ΔJ = 0, ±1). We analyzed each band by forming the first and second combination differences [11] for the ground state levels. These agreed within the experimental measurement accuracy (0.003 cm^{-1}) and permitted us to assign rotational quantum numbers. Figure 2 shows such an assignment. The term eigenvalues for each $X^2\Sigma$ and $A^2\Pi$ vibrational level were deduced from the combination

Table I Long range fits to band origins and rotational constants of $A^2\Pi_{3/2}$ using Eqs. 9 and 33 in Ref. 12. The 5* is for $^{23}Na^{22}Ne$; appropriate fits were obtained using formulae in Ref. 13. The band origins are measured from $3^2S_{1/2} \rightarrow 3^2P_{3/2}$ of Na.

$A^2\Pi_{3/2}$

v	Band Origin from $P_{3/2}$	Fit	$B_v x10^5$	Fit	$D_v x10^8$	Splitting From $A^2\Pi_{1/2}$
3	-31.64	-31.68	13068	13067	5684	11.86
4	-16.31	-16.32	10480	10649	4605	11.81
5*	- 7.14	- 7.07	7939	7916	5711	not seen
5	- 6.06	- 6.06	8230	8231	6325	12.8(1)
6	0.31	.28	5826	5829	6652	15.02
v_D	---	5.23	---	∞	---	---
-1/2	---	-143.0	---	21530	---	---

$A^2\Pi_{1/2}$

v	Band Origin from $P_{1/2}$	$B_v x10^5$	$D_v x10^8$	Λ Doubling x 10^6
3	-26.30	12848	4509	7547
4	-10.92	10368	4759	11594
5	- 1.68	---	---	---
6	2.49	5733	6826	---

$X^2\Sigma^+$				$B^2\Sigma$		
v	Dissociation Energy D_{oX}	$B_v x10^5$	$D_v x10^5$	v	Band Origin from $P_{3/2}$	B_v
0	-5.2	4890	3	?	2.82	.0186

differences of the ground state and the position of the assigned lines. These eigenvalues were fit using the appropriate rotational eigenvalue equations [15,11]. Our linear least square fit yielded the molecular spectroscopic constants shown in Table I which reproduce the data to within ±0.003 cm^{-1} except near perturbations. For the $\Pi_{1/2}$ state Λ doubling was observed which means that the effective rotational constant for Π^+ and Π^- states differed somewhat, the values in Table I for B_v and D_v are the average values. These constants should be thought of as effective constants because of the interaction with the excited $B^2\Sigma$ state as well as the observed perturbation between $^2\Pi_{1/2}$ and $^2\Pi_{3/2}$ substates belonging to different vibrational levels. A more accurate non-linear least-square fit including self perturbation and interaction with $B^2\Sigma$ is in progress.

In all, 10 bands were analyzed in this manner. We observed transitions only from the v" = 0 level of the ground $X^2\Sigma$ state (which should have a bound v" = 1 level). Two weak bands slightly to the red of similar strong bands were identified as $^{23}Na^{22}Ne$ (neon contains 9% ^{22}Ne) by their position and rotational constants. Bands with v' = 6 for both $^2\Pi_{1/2}$ and $^2\Pi_{3/2}$ were found to have only two branches; the $^2\Pi_{1/2}$ had P and R branches, the $^2\Pi_{3/2}$ had R_{21} and R_2. This may result from interaction with the nearby $B^2\Sigma$ state and may indicate the use of a more atomic basis for these levels (e.g. Hunds case e). A band which we identify as $B^2\Sigma \leftarrow X^2\Sigma$ was observed whose origin was 2.82 cm^{-1} above the $3^2P_{3/2}$ transition in Na. It has a low rotational constant B = 0.018 cm^{-1} and prominent P and R branches. The large uncertainty in the $^2\Pi_{1/2}$ v' = 5 level reflects the presence of the intense $^2S_{1/2} \rightarrow {}^2P_{1/2}$ atomic transition close to the bandhead.

The key to extracting information about the interatomic potentials from the observed spectra was analysis based on the assumption that the $A\Pi$ potential varies as $-C_6R^{-6}$ at long range. Long range analysis methods have been developed extensively in recent years [12,13,14] and are generally more suitable for van der Waals molecules than the conventional Dunham type analysis which is based on an expansion about the minimum of the potential.

The physical basis of long range analysis is that molecules which are highly vibrationally excited move in very anharmonic potentials which are hard at the inner turning point and soft at the outer turning point (due to the long range R^{-n} tail of the potential). Consequently a molecule in a high vibrational level, v, spends most of its time at large R and the radial wave function is largest near the outer turning point R_v^+. It is therefore reasonable to assume that the rotational constant can be approximated as

$$B_v = \frac{h}{4\pi c\mu}(R_v^+)^{-2} \tag{1}$$

where μ is the reduced mass and c the speed of light (B_v is in cm^{-1}). Since the potential is assumed to have its long range asymptotic form

$$V(R) = -C_6R^{-6}\left[1 + \frac{C_8}{C_6}R^{-2} + \frac{C_{10}}{C_6}R^{-4}\right] \tag{2}$$

In this region, the corresponding vibrational energy G_v satisfies

$$D - G_v = V(R_v^+) \tag{3}$$

where D is the dissociation energy of the potential curve. Combining Eqs. 1-3 yields

$$D - G_v = -C_6\beta_v^3\left[1 + \frac{C_8}{C_6}\beta_v + \frac{C_{10}}{C_6}\beta_v^2\right] \tag{4}$$

where $\beta_v = 4\pi\mu\ B_v/h$. A more exact treatment by LeRoy [12] shows that a better value of β_v is

$$\beta_v = 6.54\pi\mu\ B_v/h \tag{5}$$

In addition he shows that if the C_8 and C_{10} terms in Eq. 4 are negligible then the rotational constant B should decrease linearly with v', reaching 0 at v'_D, and the vibrational eigenvalues G should be represented by [15]

$$[D - G(v')]^{1/3} = 19.94\mu^{-1/2}|C_6|^{-1/6}(v_D-v') \tag{6}$$

where μ is the reduced mass, and v_D the (non-integral) dissociation vibrational number. Due to the spin orbit interaction, long range analysis applies well only to the $^2\Pi_{3/2}$ state, which is analyzed in Table I. Eq. 6 was used to compare the data from different isotopes [12] yielding the absolute numbering for v shown in Table I.

The value of D above differs from the $3^2S_{1/2}\rightarrow 3^2P_{3/2}$ transition in Na by the binding energy of the v"=0 level of the $X^2\Sigma$ state which is thereby found to be $D_{0X}=5.2\pm.8\text{cm}^{-1}$. (The error was estimated by analyzing, using the above equations, eigenvalues computed from several plausible potential forms.) In addition, it is possible to extrapolate to v'=-1/2 to estimate the well depth of the $A^2\Pi 3/2$ potential. This yields D_{eA}=140±3cm^{-1} and r_{eA}=5.1±.1a_o. The isotopic analysis also gives the difference between D_{oX} for the two isotopes of the $X^2\Sigma$ state to be 0.064cm^{-1} +.016cm^{-1} - a value consistent with the estimate ω_{eX} = 6.511cm^{-1} based on a Morse potential which fits the observed rotational eigenvalues. Thus $D_{eX} \cong D_o + 1/2\ \omega_{eX} - 1/4\ \omega_e x_e = 8.1 \pm 0.9$cm^{-1}. We obtain r_{eX} = 10.0 ± .2a_o from the Morse fit.

IV. Relationship with Other Atomic Physics

Molecular spectroscopy is the definitive method for determining interatomic potentials; there is no question that this theory works well for our NaNe spectrum. This permits a clearcut comparison of our results with the four extant pseudopotential calculations for NaNe. Furthermore our results bring into sharp question some aspect of the experiments or of their interpretation for all previous experiments which have given information about the excited state potential of NaNe. These experiments have involved diverse phenomena: differential scattering of excited Na from Ne [16], fine structure changing collisions involving Na (3P) and Ne [17,10,18], far wing emission of excited Na in a Ne buffer [19], and shifts and widths of the Na D-lines in a Ne buffer gas [20].

Table 2 shows a compendium of theoretical and experimental values for the well positions and depths for the NaNe XΣ, AΠ, and BΣ states. Our results show that the recent pseudopotential calculations of Malvern and Peach [5] are much better able to predict the excited state interaction potentials for NaNe than earlier pseudopotentials of Baylis [2], Pascale [3], and Bottcher [4].

Table II Depth and location of minima for $A^2\Pi_{3/2}$ and $X^2\Sigma^+$ potential curves of NaNe

Method	$X^2\Sigma^+$		$A^2\Pi_{3/2}$	
	D_{eX}(cm^{-1})	$r_{eX}(a_o)$	D_{eA}(cm^{-1})	$r_{eA}(a_o)$
Ref. 2	1.8	12.9	13.7	8.5
Ref. 3	1.8	13.13(·12)	6.2	9.00
Ref. 4	< 40	> 9.45		
Ref. 5	17.8	10	89.5	6.0
This work	8.1±1	10.0±.2	140±3	5.1±.1
Scattering	11^{+3}_{-4}	9.1±.4	120±15	8.0±.3

The X state parameters found from earlier scattering data from our lab [16] are in reasonable accord with this more definitive work: D_{eX} agrees within error, and although r_{eX} deduced from the scattering data is too low we feel this disagreement reflects the departure of the true potential from the MSV shape assumed earlier. (The period of the rapid oscillations used to determine r_{eX} depends more sensitively on the location of the repulsive wall of the potential than on the location of the minimum.) Our value for $D_{eX}\, r_{eX} = 81 \pm 10\text{cm}^{-1}$ confirms earlier results obtained from measurements of the total cross section [21]. The excited state well depth determined from differential scattering agrees with the result reported here but r_{eA} definitely does not. We feel that this discrepancy most probably arises from failure of the elastic approximation used in the scattering analysis: Although Saxon, Olson, and Liu [1] have shown that this approximation works well for the AΠ state of NaAr (with $D_{eA} \approx 500\text{cm}^{-1}$); Reid [22] has shown that it fails miserably for NaHe (using $D_{eA} \approx 40\text{cm}^{-1}$). Apparently NaNe is an intermediate case; close coupling calculations using the potential parameters proposed here would be valuable.

The present experiment calls into question the ability of spectral line shape data to yield useful information on interatomic potentials for weakly bound systems. The conclusion of Lwin, McCartan, and Lewis [23] based on line shifts and broadening that the NaNe potentials are more repulsive than those of Baylis [2] or Pascale and Vandeplanque [3] is completely wrong, as is the finding of McCartan and Farr [20] that $D_{eB} = 0.5\text{cm}^{-1}$. The failure of York, Scheps, and Gallagher [19] to observe a pressure dependence of the far red wing fluorescence is puzzling in view of the success of their analysis for NaAr. It seems most probable to us that they did not take emission profiles at sufficiently low perturber pressure for NaNe to see anything except the high pressure (thermalized) emission. This suggests that the failure of Gallagher and coworkers to observe pressure dependence on LiHe, LiNe[24] and NaHe [19] may stem from the same problem, and it raises the possibility that these systems may also have excited state well depths of order kT. The failure of the preceeding line shape experiments to indicate that the excited state of NaNe is many times more attractive than predicted by pre-1977 calculations does not invalidate those experiments--rather it raises the challenge of finding what went wrong in their interpretation together with the hope of finding out more about line broadening and/or parts of the NaNe potential curves which are not determined in this experiment [e.g. the $B^2\Sigma$ state and repulsive regions of the $X^2\Sigma$ state].

We gratefully acknowledge many helpful discussions with I. Renhorn and R. Field, and help analyzing the data from R. McGrath. The laser system was purchased by the M.I.T. Sloan fund for basic research; laser modifications and development of the frequency reference system were supported by the Joint Services Electronics Program (Contract No. DAAB07-76-C-1400).

References

1. R. P. Saxon, R. E. Olson and B. Liu, J. Chem. Phys. to be published.
2. W. E. Baylis, J. Chem. Phys. 51, 2665 (1969).
3. J. Pascale and J. Vandeplanque, J. Chem. Phys. 60, 2278 (1974).
4. C. Bottcher, A. Dalgarno, and E. L. Wright, Phys. Rev. A. 7, 1606 (1973).
5. R. Malvern and G. Peach, private communication 1977.
6. R. Campargue, J. Chem. Phys. 52, 1795 (1970).
7. R. E. Smalley, et al., J. Chem. Phys. 64, 3266 (1976).

8. R. E. Smalley, et al., J. Chem. Phys. 66, 3778 (1977).
9. R. Freeman, et al., J. Chem. Phys. 64, 1194 (1976).
10. J. Apt and D. E. Pritchard, Phys. Rev. Lett. 37, 91 (1976).
11. G. Herzberg, Spectra of Diatomic Molecules (Van Nostrand, N.Y., 1950)
12. R. J. LeRoy, Molecular Spectroscopy, (The Chemical Society, Burlington House, London, W1V 0BN), Vol. 1.
13. W. C. Stwalley, J. Chem. Phys. 63, 3062 (1975).
14. R. J. LeRoy, and R. B. Bernstein, J. Chem. Phys. 52, 3869 (1970).
15. R. Mulliken, Rev. Mod. Phys. 3, 89 (1931).
16. G. M. Carter, et al., Phys. Rev. Lett. 35, 1144 (1975).
17. J. Pitre and L. Krause, Can. J. Phys. 45, 2671 (1967).
18. W. D. Phillips, et al., Phys. Rev. Lett. 38, 1018 (1977).
19. G. York, et al., J. Chem. Phys. 63, 1052 (1975).
20. D. G. McCartan, et al., J. Phys. B. 9, 985 (1976).
21. R. Duren, G. P. Raabe, Ch. Schlier, Z Phys. 214, 410 (1968).
22. R. H. G. Reid, J. Phys. B. 8, L493 (1975).
23. N. Lwin, et al., J. Phys. B. 9, L161 (1976).
24. R. Scheps, et al., J. Chem. Phys. 63, 2581 (1975).

VIII. Laser Sources

OPTICALLY PUMPED CONTINUOUS ALKALI DIMER LASERS

H. Welling and B. Wellegehausen

Institut für Angewandte Physik, Technische Universität Hannover
3000 Hannover, FRG

Introduction

Gases and vapors of simple molecules are promising candidates for efficient, scalable and tunable lasers in the ultraviolet, visible and near infrared. This paper reports on laser investigations of optically pumped alkali dimers, which are of special interest, due to their simple molecular structure, stability, ease of production and favorable absorption and emission cross sections. Pulsed laser oscillation in molecular sodium was first observed by Henesian et al. [1] and Itoh et al. [2]. In a recently published paper [3] we reported first continuous laser operation of Na_2 molecules. In the meantime this system was studied in more detail and laser operation with other alkali dimers was achieved. Furtheron successful pulse laser operation of sulphur dimers [4] and cw oscillation of J_2 [5] was reported.

Laser system

The laser system described in the earlier publication [3] consists of a heat pipe oven and a three element resonator with an internal tuning element. The pump radiation is focussed collinearly into the vapor zone.

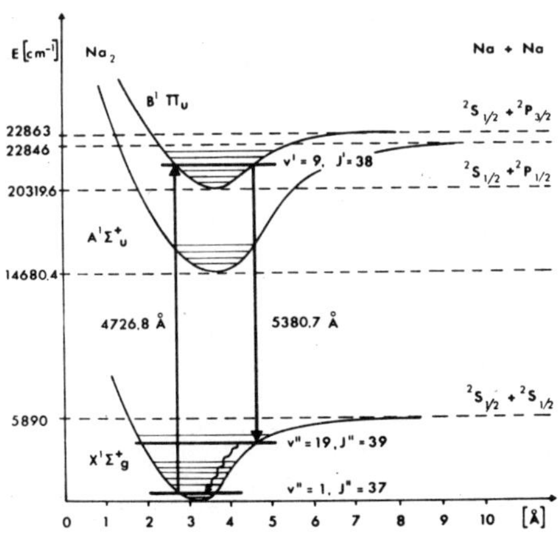

Fig.1 Energy level diagram of Na_2 and optical pump and laser cycle

A pump- and laser cycle within the energy level diagram of Na_2 is shown in Fig.1. By optical pumping of a certain rotational-vibrational level of the electronic ground state with a suitable pump laser line, population inversion between a rotational-vibrational level of an excited electronic state and several higher lying levels of the electronic ground state can be achieved. Depopulation of the lower laser levels occurs by collisional processes. The dimer laser emission spectrum typically consists of several lines belonging to a certain fluorescence series [6]. The number of lines and the intensity ratios are dependent on the Franck-Condon coefficients and the operation conditions of the laser system. Due to the great density of rotational-vibrational levels and the spectral width of 5-1o GHz of the argon or krypton pump lasers, in general several molecular transitions can be pumped simultaneously, which further increases the number of laser lines.

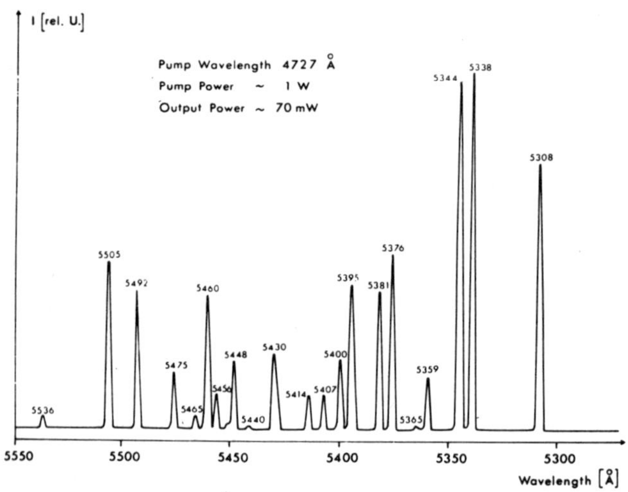

Fig.2 Multiline laser spectrum of Na_2 with $B^1\Pi_u \rightarrow X^1\Sigma_g^+$ transitions

Fig.2 shows a typical multiline laser spectrum of Na_2. These laser lines belong to $B^1\Pi_u \rightarrow X^1\Sigma_g^+$ transitions. About four molecular transitions are pumped with the 4727 Å line of the argon laser [6].

Operating Conditions

The output power of the laser system is strongly dependent on the temperature and total pressure within the heat pipe oven.

The temperature behaviour is characterized by two effects. The concentration of dimer molecules within the atomic vapor [7] and with it the output power increases strongly with increasing temperature. However, at higher temperatures the population density of the lower laser levels increases too, resulting in self absorption and higher thresholds. Therefore an optimum temperature and dimer concentration exists for laser operation. Fig.3 shows a typical dependence of the output power on the temperature for Na_2 $B^1\Pi_u \rightarrow X^1\Sigma_g^+$ transitions, with an optimum temperature around 55o °C. The optimum temperature is however slightly dependent on the pump- and laser transitions and decreases with increasing length of the vapor zone.

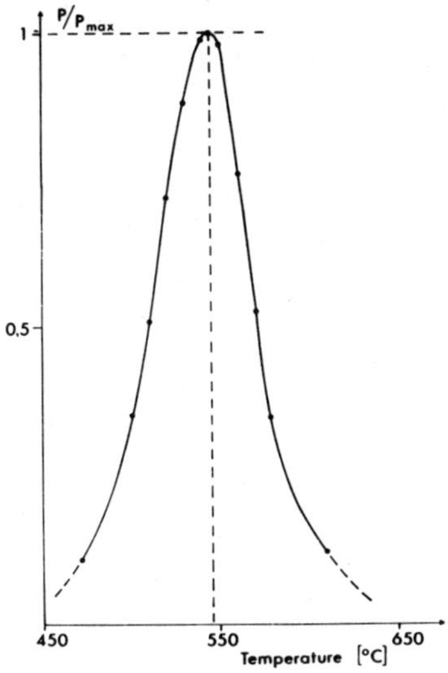

Fig.3 Temperature dependence of the normalized output power. Pump wavelength 488o Å. Laser wavelength 5251 Å. Total pressure 1o Torr. Length of vapor zone 3o mm.

The pressure dependence is in principle more complex as the partial pressure of the dimer molecules and the atomic vapor pressure can not be chosen independently. As radiative processes between rotational-vibrational levels of the same electronic state are not allowed, the lower laser level can only be depopulated by collisional relaxation. Therefore a necessary condition for cw laser oscillation is $K_2 > A_{32}$, where K_2 is the collisional relaxation rate of the lower laser level and A_{32} the spontaneous emission rate of the laser transition. This condition requires a minimum pressure of about 1 Torr, corresponding to a vapor temperature of 4oo °C for the Na-system. With increasing pressure the relaxation conditions of the lower laser levels are improved, whereas the population of the upper laser level is reduced due to various relaxation and energy transfer processes [8]. Therefore the output power decreases with increasing buffer gas pressure for buffer gas pressures above the operating vapor pressure, corresponding to the optimum dimer concentration. Laser oscillation could be observed up to buffer gas pressures of about 1oo Torr. So far a specific influence of the type of buffer gas could not be observed.

The discussed pressure and temperature behaviour as well as saturation effects which occur at higher pump powers can be described with a rate equation model including collisional processes. A more detailed analysis and further measurements will be published [9].

Data

So far continuous laser oscillation was obtained for Na_2 and Li_2 both on $B^1\Pi_u \rightarrow X^1\Sigma_g^+$ and $A^1\Sigma_u^+ \rightarrow X^1\Sigma_g^+$ transitions. Table I summarizes data and operation conditions. The output power and threshold values are not optimized for all the pump laser lines. The number of laser lines and the output power values are strongly dependent on the spectral overlap between the pump laser and different absorption transitions. Especially the output power

Table I Data of Li_2 and Na_2 Dimer Lasers

Molecule	Excitation Line [Å]	Laserlines [Å]	Output Power (all lines) [mW]	Pump Power [W]	Typical Thresholds [mW]	Operation Conditions
Na_2		$B^1\Pi_u \longrightarrow X^1\Sigma^+_g$ Transitions				
	4579	5564 5587 5597 5619	3	1	< 15	Temp. 500 °C
	4658	5454 5459 5486 5492 5602 5623 5643	8	1	< 30	
	4727	5308 5329 5338 5344 5365 5376 5381 5395 5400 5407 5413 5414 5440 5448 5456 5460 5465 5475 5492 5505 5536	70	1	< 20	Length of Vapor Zone 120 mm
	4765	5317 5339 5376 5421	25	2,5	< 70	
	4880	5251 5290 5326 5340 5348 5357 5365 5375 5411	22	2,5	< 140	
		$A^1\Sigma^+_u \longrightarrow X^1\Sigma^+_g$ Transitions				
	6471	7930 7993 8004 8018 8027 8030 8065 8069 8076 8082 8101 8136 8148	10	1	< 170	500 °C 120 mm
Li_2		$B^1\Pi_u \longrightarrow X^1\Sigma^+_g$ Transitions				
	4579	5502 5584	4	1	< 50	1000 °C 50 mm
	4727	5500 5522 5582 5604	3	1	< 100	
	4765	5336 5423 5510	15	2	< 80	
	4880	5319 5598	1	2	< 200	
		$A^1\Sigma^+_u \longrightarrow X^1\Sigma^+_g$ Transitions				
	6471	7913 8264 8454 8846 8865 8963 9037 9047 9068 9122	30	1	< 200	1000 °C 50 mm

Wavelength accuracy to within $\pm$ 1 Å

for individual lines can be improved if narrowband pump laser radiation is used, for example single frequency lasers tuned to the exact absorption transition. Most of the laser lines can be identified as $B^1\Pi_u \rightarrow X^1\Sigma_g^+$ and $A^1\Sigma_u^+ \rightarrow X^1\Sigma_g^+$ transitions using the molecular constants given in Ref. [6,1o]. The threshold values given in Table I are typical values for systems with an output coupling of about 16 % and a focal length of the focussing lens of up to 5oo mm. In special systems minimum thresholds below 1 mW, corresponding to pump intensities below 1oo Wcm^{-2} for a focal length of 1oo mm, were observed. The gain was typically of the order of o,1 cm^{-1}, with a gain bandwidth of about 1 GHz. With an internal tuning element oscillation on single lines is achieved with output powers up to 2o % of the multiline values. It should be possible however to further improve this conversion rate by suitable optimization of the laser parameters. In single line operation the laser radiation normally consists of several longitudinal modes. With a short resonator, including only the tuning element, single frequency operation with output powers up to 1 mW can be easily obtained.

Future possibilities

In a similar way other alkali dimers should be operated, extending the range of laser lines throughout the visible and near infrared. Especially the Cs_2-system presently under investigation is of interest due to a greater density of lines and possible near infrared emission with a Nd-YAG pump laser source. Calculations further indicate that conversion efficiencies of more than 1o % should be feasible, so that output powers of several 1oo mW are expected using available pump sources.

The observed low threshold values offer now possibilities of flashlamp or discharge excitation of these systems which may yield a simple inexpensive multiline laser. Preliminary experimental investigations of low pressure sodium discharges indicate that laser oscillation around 8oo nm should be possible. The alkali dimer laser systems discussed here are useful for various spectroscopic investigations on the alkali molecules itself, for example as a diagnostic tool in molecular beam experiments and for the study of collisional processes. As these lasers can be easily operated in single frequency stabilized to the molecular transition, these systems are of interest for precision measurements and as a reference system.

References

1. M.A.Henesian, R.L.Herbst, R.L.Byer : J.Appl.Phys. 47,1515(1976)
2. H. Itoh, H.Uchiki, M.Matsuoka: Opt.Comm. 18,271(1976)
3. B. Wellegehausen, S.Shahdin, D.Friede, H.Welling: Appl.Phys. 13,97(1977)
4. S.R. Leone, K.G. Kosnik: Appl.Phys. Lett. 3o,346(1977)
5. R.W. Field: private communication,to be published
6. W. Demtröder, M.Stock: J.Mol. Spectrosc. 55,476(1975)
7. M. Lapp, L.P.Harris: J.Quant. Spectrosc. Radiat.Transfer 6,169(1966)
8. K. Bergmann, W.Demtröder, M.Stock, G.Vogt: J.Phys.B:Atom.Mol. Phys.7,2o36(1974)
9. B.Wellegehausen, K.H.Stephan,D.Friede: to be published
1o.D.C.Jain, R.C.Sahni: Trans.Faraday Soc. 65,897(1969)

TUNABLE INFRARED LASERS USING COLOR CENTERS

H. Welling, G. Litfin, and R. Beigang

Institut für Angewandte Physik, TU Hannover
3000 Hannover, FRG

Introduction

Certain color centers in alkali halide crystals are well suited as active materials for tunable lasers in the near infrared region. With $F_A(II)$ centers in KCl:Li and RbCl:Li and $F_B(II)$ centers in KCl:Na and RbCl:Na the spectral range from 2.25 μm to 3.05 μm can be covered so far [1,2]. Considering threshold values, output powers, tuning ranges and efficiencies color center lasers (CCLs) are essentially comparable with dye lasers. Comparing the frequency behavior of the above mentioned lasers with other tunable sources we find the advantage of CCLs in the ease of single mode operation and in the emission linewidth which is remarkably small [3]. On the other hand the mechanical system of a CCL is as simple as that of a dye laser [4]. Therefore CCLs seem to be very attractive for applications in laser spectroscopy and chemistry in the near infrared.

Emission and Absorption Bands of Color Centers

The emission and absorption bands of several types of color centers in different alkali halide crystals are shown in Fig.1. The range from 1 to 3 μm can be covered using F_2^+, $F_A(II)$ and $F_B(II)$ centers. With increasing lattice size the emission of these centers shifts to longer wavelengths. This behavior is indicated for $F_A(II)$ centers by the small figure inserted, where the peak wavelength increases going from KF:Li with the smallest lattice to KCl:Li and to RbCl:Li with the largest lattice. The largest lattice in which $F_A(II)$ centers are known to exist is RbCl. Considering the structure of $F_B(II)$ centers one can expect to obtain these centers in many other crystals [1]. Only the $F_B(II)$ centers in hosts like RbBr and RbJ give the chance to extend the tuning range of CCLs into the 3-4 μm region.

Experimental Results

We obtained for the first time $F_B(II)$ center laser oscillation with heavily sodium doped KCl and RbCl crystals [2]. The optical system we used for the CCL is very similar to the system of a dye laser where just the dye cell is replaced by a colored crystal [4]. It consists of a conventional astigmatically compensated three mirror arrangement with a collinear pump geometry.

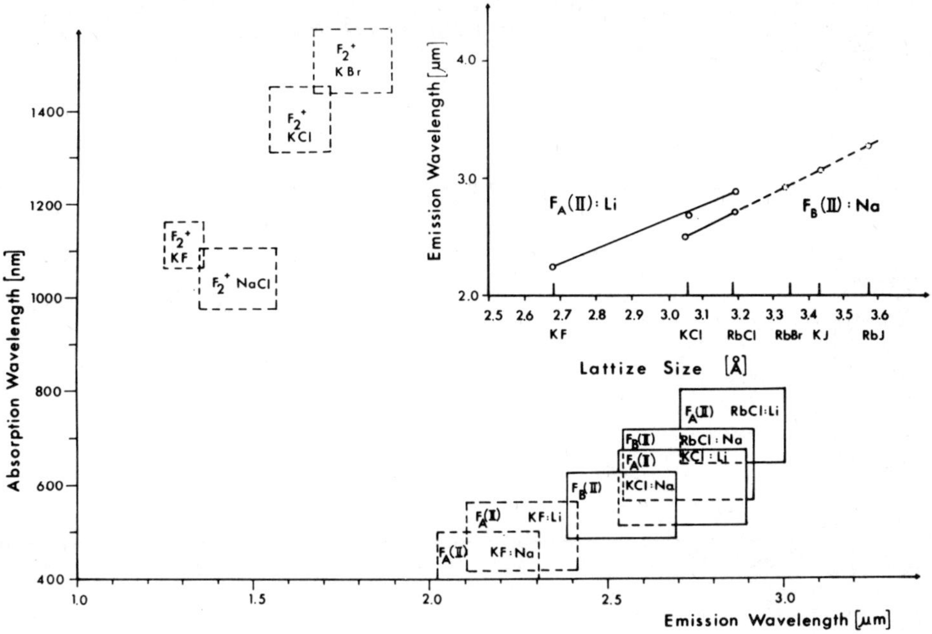

Fig.1 Emission and Absorption Bands of Various Color Center Crystals

The crystal is mounted on a cold finger which is cooled with LN_2. The cell containing the crystal is evacuated to provide thermal insulation. Several tuning elements have been tested to control the output wavelength of the CCL including prisms, a single birefringent plate and a complete birefringent tuner. A sapphire Brewster prism was preferred as gross tuning element causing only a slight power reduction due to its low insertion loss. It should be emphasized that in contrast to dye lasers the mode spectrum of the CCL is not affected by the special type of gross tuning element. Therefore the low dispersion of $3 \cdot 10^{-5}$ nm^{-1} of the sapphire prism is no disadvantage.

In order to get high output power with a CCL there are four points which have to be considered:

1. It has to be guaranteed that the crystal remains at the low temperature even at high pump power. This requires an efficient cooling arrangement.
2. The optical density of the active material has to be adjusted carefully so that for the pump wavelength an optimum absorption is achieved at the operating temperature.
3. Pump polarization and crystal orientation have to be chosen in such a way that the highest possible number of centers participate in the laser process.

4. The optical cavity has to be optimized with regard to the output coupling and to the mode matching between the pump and laser mode.

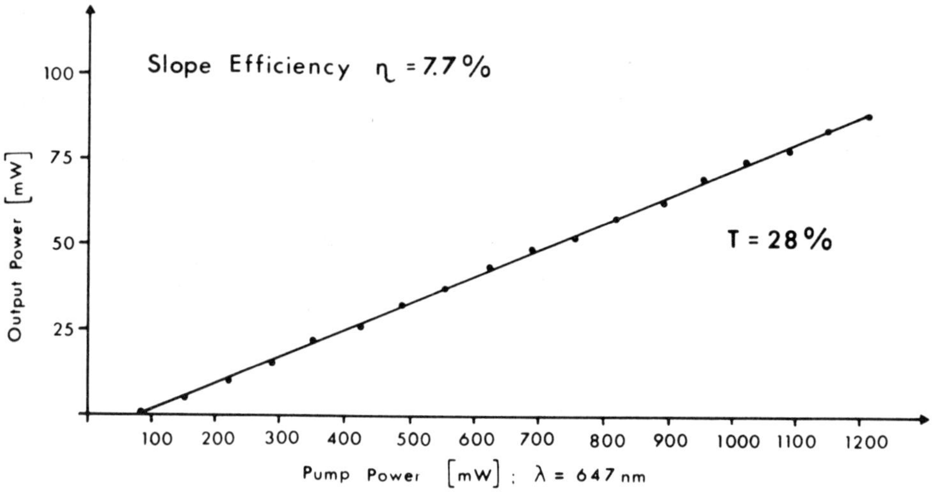

Fig.2 F_A(II) center laser with KCl:Li, λ = 2.7 μm

Fig.2 shows the power output versus the pump power for an optimized system where the active material is KCl containing F_A(II) centers. An output power of 85 mW has been obtained with a pump power of nearly 1.2 W, the maximum output power available from our krypton ion laser. The slope efficiency was measured with 7.7 % which is already close to the theoretically expected value of 9.2 % for the used laser system. No saturation or bleaching effects have been observed when the crystals were properly cooled [5].

A summary of the main characteristics we obtained for the four F_A(II) and F_B(II) center lasers is given below. Red and yellow krypton ion laser lines can be used to pump all lasers. The two KCl crystals can be pumped with argon ion lasers too. Pump powers at threshold of 2o mW and output powers of several 1o mW can be obtained. So far the total tuning range extends from 2.25 to 3.o5 μm. F_B(II) center lasers still have lower output powers and slope efficiencies than F_A(II) center lasers. This may be caused by a stress induced birefringence observed in most of our crystal samples.

Frequency Behavior and Linewidth

For some applications the frequency behavior of tunable lasers is quite important. The emission line of color centers in alkali halide crystals at liquid nitrogen temperature is homogeneously

Table I Main characteristics of CCLs

Crystal Type of Color Center	KCl : Li $F_A(II)$	RbCl : Li $F_A(II)$	KCl : Na $F_B(II)$	RbCl : Na $F_B(II)$
Pump Laser Pump Wavelength	Krypton;Argon 530,647;514nm	Krypton 647,676 nm	Argon;Krypton 470-530;568nm	Krypton 647,676nm
Pump Power at threshold	13 mW	100 mW	20 mW	26 mW
Output Power	85 mW	6 mW	20 mW	6 mW
Slope Efficiency	7.7 %	2 %	2.3 %	2.1 %
Tuning Range	2.5 - 2.9 µm	2.75-3.05µm	2.25 -2.60 µm	2.5-2.9µm

broadened. For that reason laser oscillation is observed in one primary and one or two hole burning modes [3]. Fig.3 shows the mode spectrum of a CCL without using frequency selecting elements.

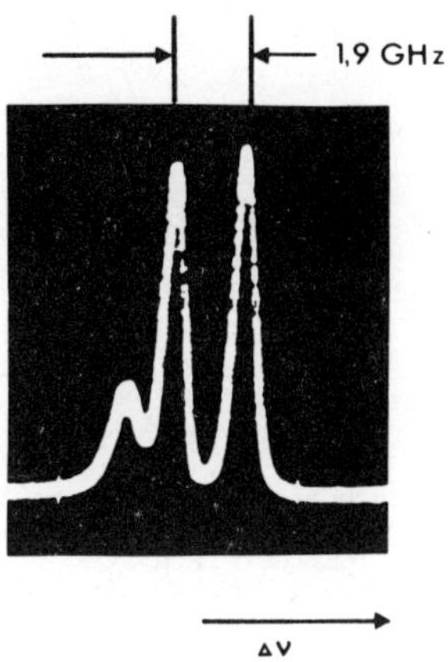

Fig.3 Mode Spectrum of the Free Running CCL

The emission spectrum is remarkably clean even at high pump power. The observed mode separation agrees with the theoretically expected value of 1.97 Ghz. Due to this clean frequency spectrum stable single mode operation can be achieved with one Fabry Pèrot Etalon only. According to the small loss of this F.P.E. a single mode output power of 75 % of the output power in multi-mode operation is obtained.

The emission linewidth of the cw CCL is determined by fluctuations of the optical path within the laser resonator [3]. These fluctuations are caused by mechanical instabilities of the laser system and by temperature fluctuations in the active zone of the CCL. Temperature variations arise from fluctuations of the pump power and from direct temperature variations of the cooling system. The linewidth of the CCL is given by

$$\Delta\nu = \sqrt{\Delta\nu_m^2+\Delta\nu_p^2+\Delta\nu_T^2}, \tag{1}$$

where $\Delta\nu_m$, $\Delta\nu_p$ and $\Delta\nu_T$ are the linewidth due to mechanical instabilities, pump power variations and direct crystal temperature fluctuations respectively. It is assumed that instability effects are statistically independent. The linewidth of the free running single mode laser was measured to be smaller than 26o kHz which is the frequency resolution of our measuring system.

For an estimation of the actual linewidth the frequency deviation of the CCL emission versus temperature fluctuations and pump power variations was measured (Fig.4). Assuming direct temperature fluctuations of o.5 mK the temperature linewidth can be calculated to 15 kHz.

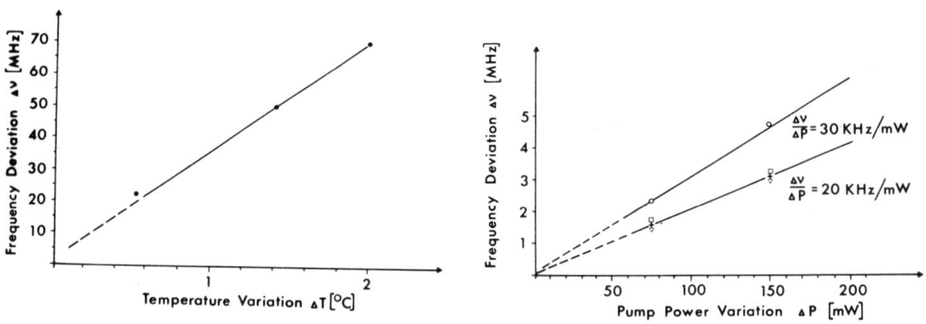

Fig.4 Frequency Deviation of the CCL Emission versus Temperature Variations and Pump Power Variations

Pump power fluctuations of 1 mW yield a linewidth of 2o kHz. With a mechanical linewidth of 1o kHz an estimated overall linewidth of 25 kHz results. The intrinsic small linewidth is a advantage if the CCl will be used in frequency stabilized systems.

Conclusion

F_A(II) and F_B(II) centers which are mainly discussed in this paper are only a small group of color centers which should be appropriate for laser action. In figure 5 the estimated pump powers at threshold and the tuning ranges of different color center crystals are summarized. The calculations are based on spectroscopic data such as absorption wavelength, emission wave-

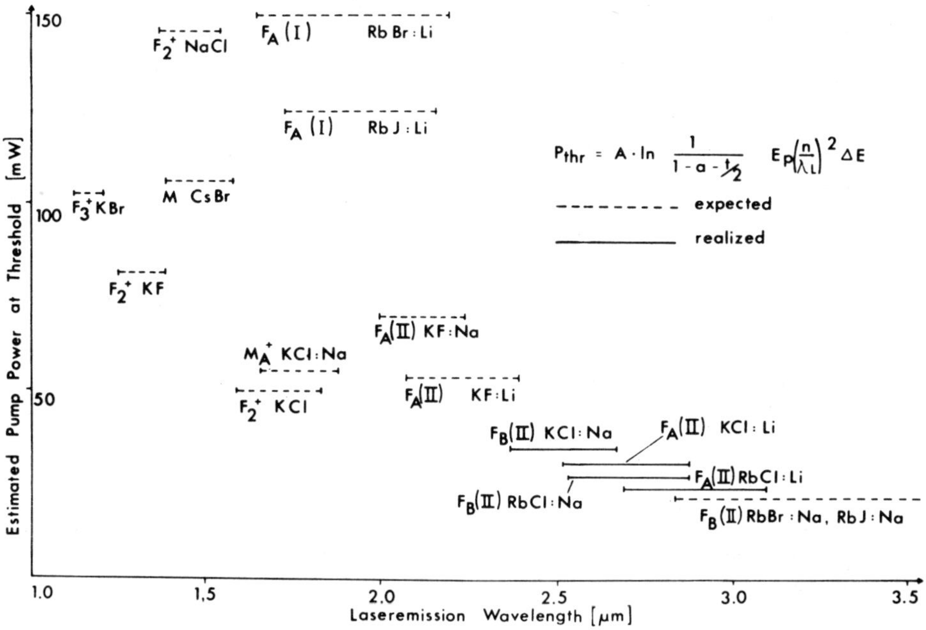

Fig.5 Estimated Pump power at Threshold for Various Color Center Crystals

length and halfwidth of the emission band. A single pass loss of 1o % is assumed. Saturation and self absorption effects are not considered. Figure 5 shows that it should be possible to cover the entire wavelength range from 1 μm to 4 μm.

In summary we have shown that CCLs are essentially comparable to dye lasers in nearly every aspect including the ease of operation. Considering the stability and the intrinsic small linewidth the CCL seems to be really attractive for high resolution molecular spectroscopy. Furthermore there is a fair chance to extend the tuning range of CCLs into the fingerprint region of molecules between 3 and 4 μm by use of F_B(II) centers in crystals such as RbBr and RbJ.

REFERENCES

1. L.F. Mollenauer and D.H. Olson: J. Appl. Phys. 46, 31o9 (1975)
2. G. Litfin, R. Beigang and H. Welling: to be published in Appl. Phys. Letters
3. R. Beigang, G. Litfin and H. Welling: to be published in Optics Comm.
4. H. Welling, 1977 IEEE/OSA Conference on Laser Engineering and Applications, Book of Digests, paper 7.7, p. 31.
5. H. Welling, 1977 IEEE/OSA Conference on Laser Engineering and Applications, Book of Digests, paper 13.10, p. 65.

IR SPECTROSCOPY VIA DIFFERENCE-FREQUENCY GENERATION*

A.S. Pine

Lincoln Laboratory, Massachusetts Institute of Technology
Lexington, MA 02173, USA

Introduction

In this paper we discuss recent refinements to a visible-to-infrared difference-frequency converter for application to high-resolution molecular spectroscopy. These refinements include improved stabilization and calibration techniques, increased electronic tuning range and computerized data-processing. The system is capable of precision Doppler-limited spectroscopy of complete rotation-vibration bands of many hydrogenic and other interesting light molecules.

The difference-frequency spectrometer operates by mixing two collinear cw beams from a stable single-mode argon laser and a tunable single-mode dye laser in the nonlinear optical crystal, $LiNbO_3$. The infrared power generated is proportional to the product of the incident laser powers, the infrared frequency squared, and the crystal length squared for 90° phasematching. The conversion efficiency from dye laser to infrared power at 3 μm for a 1 W argon pump laser optimally focussed into the 5 cm long $LiNbO_3$ crystal is calculated to be $\sim 2 \times 10^{-3}$. So for typical operating powers of $\sim$ 100 mW argon and $\sim$ 10 mW dye, we obtain a few microwatts of infrared which is 10^4 to 10^5 higher than the noise-equivalent-power of standard IR detectors.

The spectral range of the difference-frequency spectrometer is from 2.2 to 4.2 μm which can be completely covered simply by tuning the dye laser and the phase-matching temperature of the crystal. The short wavelength limit is imposed by induced optical damage to the crystal at operating temperatures below about 200^oC. Annealing occurs at higher temperatures since the shallow traps responsible for the damage are activated. The long wavelength limit is determined by the opacity of the crystal due to multiphonon absorption. In principle different primary lasers and other nonlinear materials such as the CO or CO_2 and spin-flip lasers mixed in various chalcopyrite crystals could extend the range much further out in the infrared; however the intrinsic ω_{IR}^2 dependence and the shorter interaction lengths due to off-axis phasematching make such systems less desirable. Other tunable lasers useful for spectroscopy within the range of the $LiNbO_3$ mixer include semiconductor diode lasers, color center and vibronic lasers, and the parametric oscillator. Each of these lasers provides more cw infrared power than the mixer, but the latter has the advantages of broader and more complete spectral coverage, no laser cryogenics, and alignment, calibration and mode-control conveniently in the visible. In previous papers [1,2] we described the basic operation of the difference-frequency converter and demonstrated its unique spectroscopic capabilities.

*This work was sponsored by the National Science Foundation.

Optical System

In Fig. 1 we show schematically the optical portion of the difference-frequency spectrometer. The single-mode argon laser and dye laser beams are rendered collinear by the dichroic mirror, M_1; then they are passed through a water cell to eliminate spurious infrared, chopped by a tuning fork for a-c synchronous IR detection, and focussed into a temperature-controlled $LiNbO_3$ crystal. The infrared emerging from the $LiNbO_3$ is split into sample and reference channels for ratio detection to compensate for fluctuations of the argon and dye laser amplitudes and for detuning of the phasematching response. For scans greater than the phasematching band width ($\sim 1\ cm^{-1}$ for the 5 cm long crystal), the $LiNbO_3$ temperature is electronically scanned along with the dye laser at a rate of $-0.12^{o}C/cm^{-1}$. The IR beamsplitter is a thin Si coating sputtered on CaF_2 [3] to allow for some visible transmission to help with alignment; the visible radiation is then rejected by the germanium filters in front of the InSb detectors.

A small fraction of the argon and dye laser beams are directed to 10 cm confocal Fabry-Perot reference cavities in order to servo control the laser frequencies. When ratio-locked to the edge of an interference fringe of these reference cavities, the Coherent CR-3 argon laser linewidth is reduced from $\sim$ 8 MHz free-running to $\sim$ 100 kHz and the Spectra-Physics 370/580 dye laser width is reduced from $\sim$ 10 MHz to $\sim$ 3 MHz. The resulting infrared linewidth is simply the convolution of the two visible lasers. The residual width of the dye laser arises from a jitter induced by the dye recirculator at frequencies higher than the piezoelectric servo-response. Both reference cavities are constructed of invar; that for the argon laser is temperature-controlled while that for the dye laser is pressure-scannable over several wavenumbers. A second servo-system in the dye laser maintains the mode-selecting etalon at the peak of the dye laser power thus preventing mode-hopping due to thermal drift or differential tuning rates. The dye laser can be uniformly and linearly scanned up to $7.5\ cm^{-1}$ with this etalon servo-control. The mode-selecting etalon in the argon laser is temperature controlled to prevent mode-hopping over hours of operation.

Drift Compensation and Absolute Stabilization

A multiplexing scheme has been devised for calibration, monitoring, drift compensation and absolute stabilization of the difference-frequency spectrometer. These functions are all performed by the broadband scan-calibration-interferometer (SCI), where a portion of the argon and dye laser beams are combined with a Lamb-dip stabilized He/Ne laser at 6328 Å. The SCI is repetitively swept and its transmission peaks are displayed on an oscilloscope as illustrated in Fig. 2. Gates G_{NE} and G_{AR} are generated to select only one each of the He/Ne and argon laser peaks. The gated argon laser peak triggers a boxcar gate, G_{DYE}, after a fixed delay, T_{DYE}. The boxcar then registers a marker pulse when the dye laser peaks scan through the gate, G_{DYE}.

This scan-delay-tracking scheme completely corrects for SCI drifts and partially compensates (by a factor of $\lambda_{AR}/\lambda_{DYE}$) for argon laser drifts as may be seen from the following arguments: 1) If the argon frequency is stable and the SCI drifts, the whole display pattern shifts uniformly. However the gate, G_{DYE}, which is fixed relative to the argon peak, always occurs at exactly the same interferometer spacing which maintains a constant calibration. 2) If the SCI is stable and the argon frequency drifts, the boxcar gate shifts the corresponding fraction of the argon laser interorder. However this displacement is a relatively smaller fraction of the dye laser free-spectral-range since the interorder spacing occurs for displacements of $\lambda/4$.

The drift compensation provided by this simple scan-delay-tracking scheme is

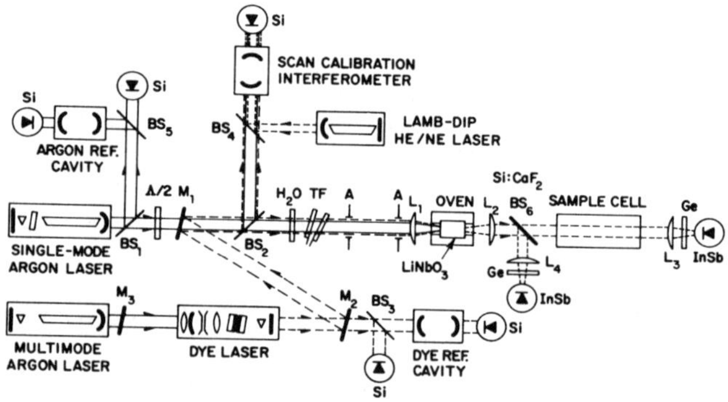

Fig. 1 Difference-frequency spectrometer optical system.

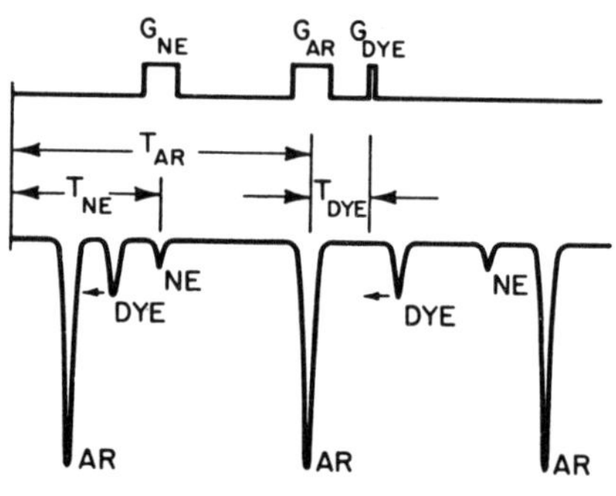

Fig. 2 Scan-delay-tracking scheme: scan-calibration-interferometer display and timing sequence for servo control.

adequate for most short-term experiments. However for longer periods of time where the argon laser may drift substantially or even jump modes, it is useful to have a method to reset its frequency in order to maintain a reproducible calibration. For this reason the argon laser is ultimately referenced to a Lamb-dip stabilized He/Ne laser by first adjusting the bias on the SCI to keep the delay, T_{NE}, constant Then the argon frequency is corrected via its reference cavity to maintain a constant delay, T_{AR}. With this technique we can reset the argon laser within 10 MHz, even after overnight shutdown, so that the compensated calibration is reproducible to within ~ 2 MHz. This absolute stabilization is extremely useful for recording the extended spectra of typical molecular rotation-vibration bands. Absolute calibration is accomplished by reference to infrared absorption line standards.

Computerized Data Processing

This high-resolution difference-frequency spectrometer is capable of generating an enormous amount of spectral data for even relatively simple molecules, and its precision is higher than can be conveniently read from chart recordings. Thus we have turned to computerized data acquisition and processing. Doppler-limited absorption traces are digitized and relayed to an IBM 370 computer for storage and data reduction. Programs have been developed to normalize the traces against baseline variations, logarithmically convert from absorption to spectral intensity, linearize and calibrate the frequency scale, read and tabulate the peak positions and intensities and plot the spectral data. The data are then in such a form as to be readily communicated or theoretically analyzed.

Molecular Spectra

We illustrate the stability, precision and tuning range of the difference-frequency mixer system by some sample Doppler-limited molecular spectra. In Fig. 3 we demonstrate the reproducibility of the calibration markers relative to the P(7) multiplet of the ν_3 band of methane at ~ 3.4 μm. These absorption traces, taken an hour apart, show no apparent marker drift. The extended electronic tuning range of 7.5 cm^{-1} is illustrated in Fig. 4 by a trace of the Q-branch of the ν_3 band of methane. The baseline variations arise from interference fringes from the windows of the sample cell and detectors and from the beamsplitter. Because these fringes shift with temperature, angle and time, the baseline is the most difficult part of the spectrum to reproduce. This channeling, most noticeable in these long traces, causes the major error in analyzing atmospherically-broadened spectra. In the case where the structure is sharp and well separated, the baseline can easily be interpolated by computer and the normalized spectrum can be plotted as shown in Fig. 5. This is the spectrum of the Q-branch of the $\nu_2 + \nu_3$ combination band of N_2O at ~ 3.6 μm. Here we see that the residual noise level and baseline variations are so small as to result in a nearly textbook example of the spectrum of a linear molecule. For this spectrum the dye laser was locked to the pressure-scanned reference cavity in order to demonstrate the high precision achievable with the difference-frequency spectrometer. The peak frequencies obtained by computer from this trace are listed in Table I and compared to the calculated values of AMIOT and GUELACHVILI [3] who studied this same band with a high-resolution Fourier-transform interferometer. The standard deviation between the observed and calculated frequencies is $\sim 1 \times 10^{-4}$ cm^{-1} or ~ 3 MHz which is within a factor of two of that achieved by AMIOT and GUELACHVILI. This precision is limited at this time by the dye laser jitter which is amenable to further improvement if necessary with a faster servo system[4,5].

Acknowledgments

The author is grateful to B. Palm and G. Dresselhaus for assistance with the computer programming and to P. Moulton and M. Coulombe for help with the circuit design.

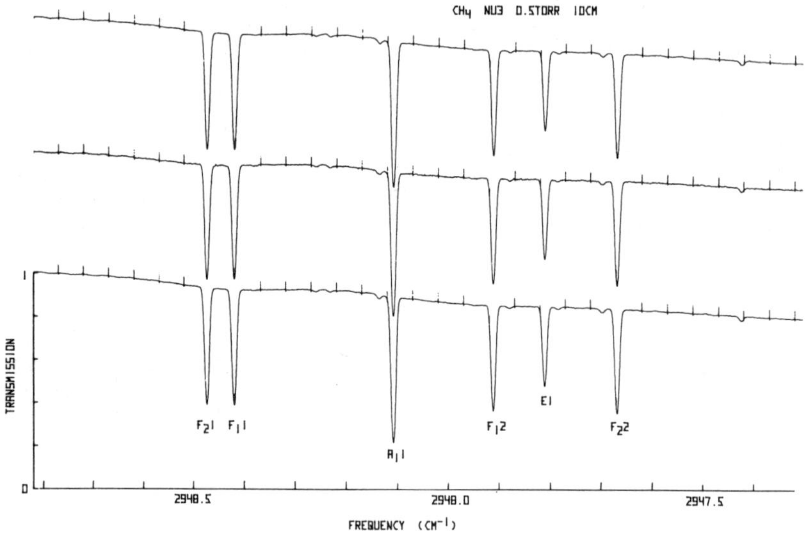

Fig. 3 Methane absorption spectrum, ν_3-band, P(7) multiplet, 0.5 Torr, 10 cm, marker spacing 0.050028 cm^{-1}.

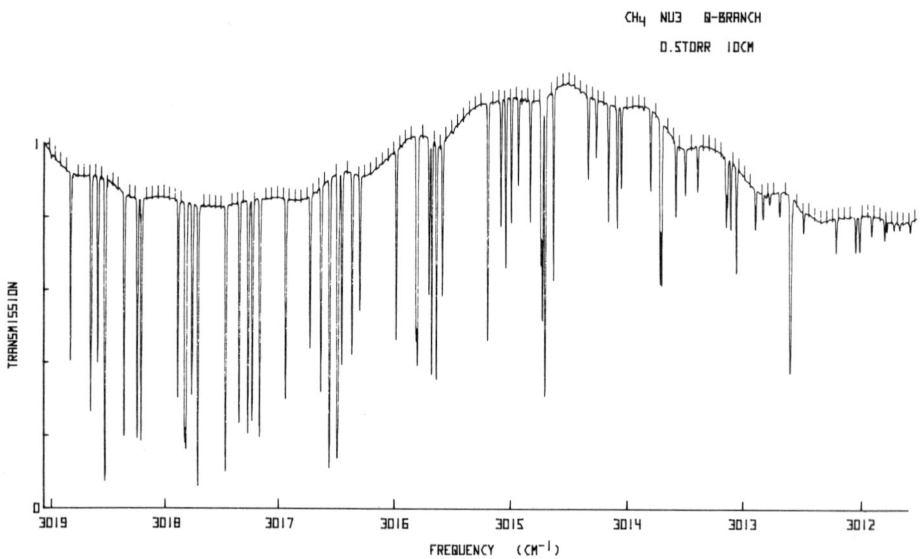

Fig. 4 Methane absorption spectrum, ν_3-band, Q-branch 0.5 Torr, 10 cm.

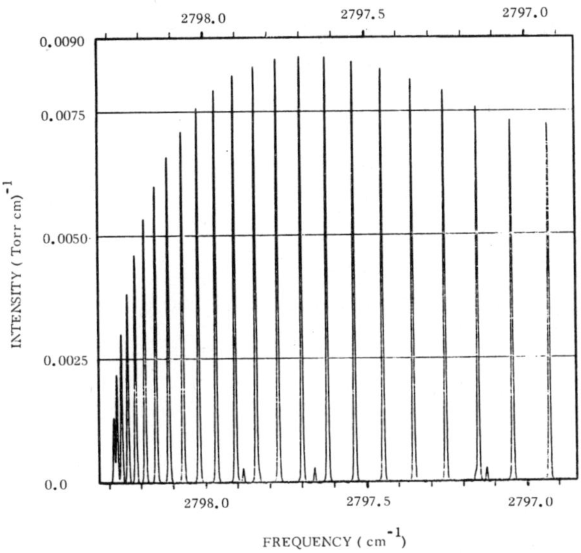

Fig. 5 Nitrous oxide spectrum, $\nu_2 + \nu_3$ combination band, Q-branch, 2 Torr, 100 cm.

Table I Observed and calculated frequencies for N_2O, $\nu_2 + \nu_3$ band, Q-branch.

J	Obs. Freq. cm^{-1}	Calc. Freq.[a)] cm^{-1}	O-C[b)] $10^{-4}cm^{-1}$	J	Obs. Freq. cm^{-1}	Calc. Freq.[a)] cm^{-1}	O-C[b)] $10^{-4}cm^{-1}$
1	2798.2881			13	2797.8447	2797.8447	0
2	.2782	2798.2783	-1	14	.7756	.7757	-1
3	.2635	.2635	0	15	.7018	.7018	0
4	.2438	.2438	0	16	.6228	.6229	-1
5	.2191	.2192	-1	17	.5388	.5391	-3
6	.1894	.1896	-2	18	.4504	.4504	0
7	.1550	.1551	-1	19	.3565	.3567	-2
8	.1157	.1157	0	20	.2580	.2581	-1
9	.0713	.0713	0	21	.1544	.1545	-1
10	.0220	.0221	-1	22	.0462	.0460	+2
11	2797.9679	2797.9679	0	23	2796.9325	2796.9326	-1
12	.9087	.9087	0				

a) Amiot and Guelachvili

b) $\langle O\text{-}C \rangle = -0.64 \times 10^{-4}\ cm^{-1}$; $\sigma = 1.0 \times 10^{-4}\ cm^{-1}$

References

1. A. S. Pine, J. Opt. Soc. Amer. 64, 1683 (1974).
2. A. S. Pine, J. Opt. Soc. Amer. 66, 97 (1976).
3. C. Amiot and G. Guelachvili, J. Mol. Spectrosc. 59, 171 (1976).
4. R. L. Barger, J. B. West and T. C. English, Appl. Phys. Lett. 27, 31 (1975).
5. F. Y. Wu and S. Ezekiel, Laser Focus 13, 78 (1977).

PULSED AND CW OPTICALLY PUMPED LASERS FOR NOVEL APPLICATIONS IN SPECTROSCOPY AND KINETICS

J. Brooke Koffend and R.W. Field [1]

Department of Chemistry, Massachusetts Institute of Technology
Cambridge, MA 02139, USA

and

D.R. Guyer and S.R. Leone [2]

Joint Institute for Laboratory Astrophysics, University of Colorado
and National Bureau of Standards and Department of Chemistry
University of Colorado, Boulder, CO 80309, USA

1. Introduction

There is a recent *explosion* of interest in optically pumped gas phase, *electronic transition,* diatomic lasers. Over the past year, the number of experiments performed on such lasers has increased tremendously. Table 1 is a brief historical outline of the development of diatomic molecular lasers. Both pulsed and cw optical pumped lasers have been demonstrated.

Optically pumped vibrational transition lasers [10] operating at wavelengths ranging from 10-1814 μm have been well known since 1970 and have found numerous spectroscopic and device applications. Photon conversion efficiencies in excess of 90% have been achieved for pulsed systems [11]. Several important differences exist between electronic and vibrational optically pumped lasers. Most significantly, the integrated cross section $\sigma^{\circ}\Delta\nu$ for vibrational transitions is seldom larger than 10^{-15} cm^2 whereas the maximum value for electronic transitions is 10^{-12} cm^2. This means that electronic transition lasers should be capable of higher gain per mode volume, lower pump thresholds, better photon conversion efficiency, and higher saturation output power. Because of the larger natural width, Doppler width, and more rapid broadening with pump power, shorter mode-locked pulses will be obtainable from electronic transition lasers, in perhaps some interesting spectral regions not reached by dyes. The relatively high energy of the electronically excited, optically pumped level makes it possible to observe, under identical pressure, pump frequency, and power conditions, transitions to lower levels belonging to several different electronic states or spanning a wide range of vibrational excitation. Chemically and spectroscopically interesting lower levels, particularly those near dissociation limits, may be selectively populated. At present, few of the potential advantages of electronic optically pumped lasers have been exploited.

---

[1] Alfred P. Sloan Fellow.

[2] Alfred P. Sloan Fellow and Staff Member, Division 274.00, National Bureau of Standards.

Table 1 Historical development of optically pumped molecular electronic transition lasers

Molecule-Transition	Pump Laser	Year	Remarks	Reference
I_2 $B^3\Pi_{0^+u} - X^1\Sigma_g^+$	doubled Nd	1972	pulsed	[1]
Na_2 $B^1\Pi_u - X^1\Sigma_g^+$	doubled Nd	1975	pulsed	[2]
$A^1\Sigma_u^+ - X^1\Sigma_g^+$	doubled Nd, dye	1975	pulsed	[2]
Br_2 $B^3\Pi_{0^+u} - X^1\Sigma_g^+$	doubled Nd	1976	pulsed	[3]
S_2 $B^3\Sigma_u^- - X^3\Sigma_g^-$	doubled dye	1976	pulsed	[4]
I_2 $B^3\Pi_{0^+u} - X^1\Sigma_g^+$	argon ion	1976	cw	[5]
Na_2 $B^1\Pi_u - X^1\Sigma_g^+$	argon ion	1976	cw	[6]
Te_2 $A0_u^+ - X0_g^+$	dye	1977	pulsed	[7]
Li_2 $B^1\Pi_u - X^1\Sigma_g^+$	argon ion	1977	cw	[8]
HgBr $B^2\Sigma^+ - X^2\Sigma^+$	ArF	1977	pulsed dissociation of $HgBr_2$	[9]

The subject of optically pumped diatomic lasers, at one time a laser-builders' laboratory curiosity, is rapidly developing into an important new field of research. There are three very general, significant areas of exploration: spectroscopy, kinetics, and new laser devices. Many transitions which are observable as spontaneous laser-induced fluorescence may also be observed as self-stimulated laser transitions with a tremendous solid angle detection advantage. Unobservably weak spontaneous transitions have been made to lase [5] and the spectrum of the laser output can be freed of interfering stronger transitions. The time duration, pump power, and pressure dependence of the optically pumped laser output is a direct measure of the rates of collisional transfer of populations out of upper and lower laser levels. Moreover, in the laser gain medium itself, various highly excited (and possibly metastable) rovibronic lower laser levels may be populated significantly in a cw or pulsed manner. The absolute population flux into such levels is determined by the laser output power. Practical applications of optically pumped lasers include frequency standards, multi-octave line tunable devices, efficient frequency up and down conversion, pulse shortening, improvement of spatial mode quality, gain diagnostics for new lasers, and development of lasers excited by more utilitarian techniques such as electric discharges, chemical reactions, flashlamps, or solar radiation.

To date, intermediate size molecules, triatomics and other polyatomic molecules which are not considered to be dyes, have not been shown to lase on electronic transitions. If optically pumped lasing experiments are successful for small polyatomic molecules, they could provide a tremendously powerful new tool to study the complex and dense spectra of these systems.

2. Experimental

A typical apparatus for optically pumped laser studies is shown in Fig.1. The pump laser may be pulsed or cw. Each has distinctly different advantages. The pulsed laser offers much greater peak intensity and produces population inversions in short 5 nsec to 10 μsec time scales. The high peak intensity is useful for exciting extremely weak optical transitions. Pulsed excitation is most effective for generating stimulated emission rates which compete successfully with collisional relaxation, quenching, and high spontaneous radiative or predissociative rates.

One important advantage of high peak power pumping is that the upper laser level may be power broadened so that the homogeneous width of the laser transition is as large as its Doppler width; consequently all molecules, regardless of velocity, will contribute to laser gain. Continuous lasers, in spite of their usually higher average power than pulsed lasers, would be useless as optical pumps if it were not possible to insert *all* of their output power into one spatial and longitudinal mode of the optically pumped laser. A single mode pump produces a sub-Doppler excitation profile. The stimulated emission cross section for such a profile is larger than that of a Doppler profile by the ratio of the Doppler to homogeneous line widths, typically 10^2 to 10^4. Unfortunately, the fraction of molecules that can contribute to laser gain is reduced by the same factor. Furthermore, there will be no gain at all if the frequency of a resonator mode does not match that of the sub-Doppler excitation profile. Interestingly, although more

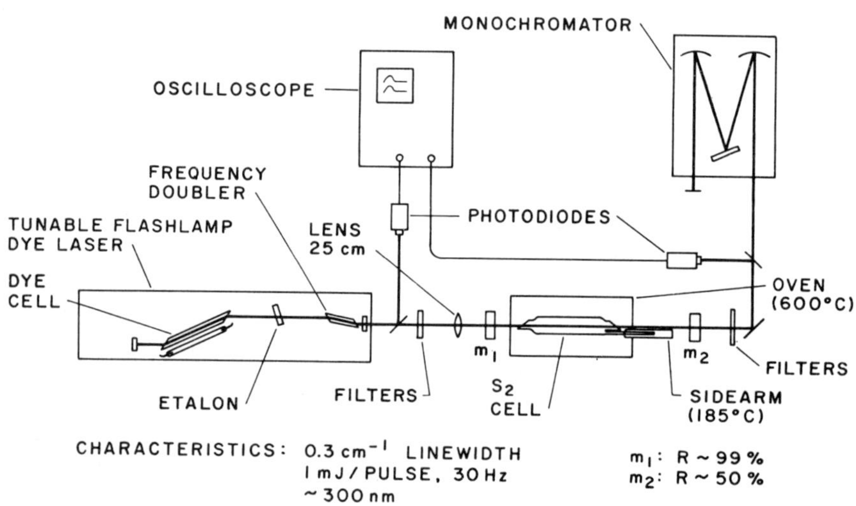

Fig.1 Apparatus for optically pumped laser studies on the sulfur dimer.

pump power is required to saturate an entire Doppler profile than a single velocity group, the magnitude of the stimulated cross section times inversion density (gain/cm), $\sigma\Delta N$, is approximately the same in both limits. Under all conditions, single-pass gain saturates at much lower pump power than does output power or conversion efficiency.

Referring to Fig.1, the appropriate pump laser is focused into a cell containing the diatomic molecule of interest. The most effective arrangement uses mode-matched longitudinal pumping. Pump power which is not inserted within the mode volume of the pumped resonator is wasted. By absorbing the pump light in a long path length of relatively low pressure vapor, collisional relaxation effects are minimized compared to the magnitudes of obtainable stimulated emission rates. Laser characteristics are most strongly affected by the pump energy actually inserted into the mode volume via a single excitation transition. Mode shapes of the pump and pumped lasers appear to be less important, but an optimum balance between pressure and absorbing length (mode volume) must be achieved. The optically pumped laser in some instances may operate in two regimes. The first is when laser stimulated emission only occurs from the single level, and even the single velocity group, which is excited by the pump laser. The second is when collisional relaxation occurs and the excitation of the single initial level is spread over a manifold of levels and gain becomes observable on collisional satellite transitions. The first is obviously a far easier condition to achieve and is observed most frequently. The latter case has been observed [12] for the cw pumped Na_2 dimer, and would be expected to be most important when collisional quenching rates are slow and the excitation rate is large enough to produce gain over many transitions at once.

In the example of the optically pumped S_2 laser in Fig.1, a pulsed, frequency-doubled dye laser was used. The pump characteristics are 0.3 cm^{-1} line width with 1 mJ energy in 2 μsec pulses near 300 nm. This excitation is sufficiently narrow to pump single lines of the S_2 spectrum, but is still many times broader than the Doppler line width of the S_2 molecules. The S_2 molecules are generated by heating sulfur in an all-fused quartz Brewster angle cell. The vapor pressure of S_2 is adjusted with an independently heated sidearm. The input mirror for the S_2 laser resonator transmits 90% of the 300 nm pump light and is highly reflective in the near ultraviolet (360-400 nm) or visible (400-600 nm) regions. The output coupler is varied from 50-95% depending on the experiment. A grating is substituted for the output coupler to achieve single line selection of the S_2 laser output. Photodiodes are used to enable display of the pump and lasing pulses simultaneously on a dual beam oscilloscope. The pump and lasing wavelengths are measured using a 0.5 meter monochromator.

Variations to this basic configuration include the use of room temperature cells [1,3,5] or heat pipes [6] to confine the species of interest. Gratings [4], prisms [3,5], and birefringent [5] tuning elements have been used to select single laser lines. Simple connection of the laser cell to a vacuum line allows the addition of various gases for quenching and relaxation experiments. Both fixed frequency and tunable lasers have been used as pumps. As would be expected, a frequency narrowed pump laser tuned to a single transition, as opposed to broadband excitation, is found to be the most efficient for pumping these diatomic lasers. Pump thresholds smaller than 1 mW [6] have been reported.

Typical lasing transitions and pump lines are shown on the S_2 potential energy curves of Fig.2. For excitation of a single vibrational and rotational

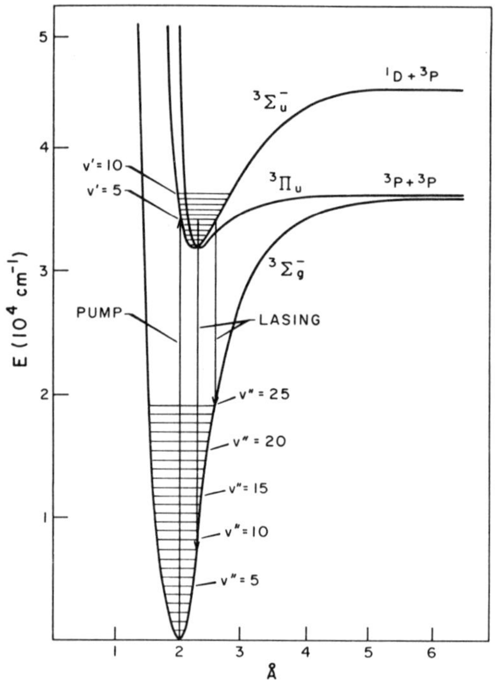

Fig.2 Potential energy surfaces for S_2, showing typical pump and laser transitions.

level in the $B^3\Sigma_u^-$ state, as many as 12 R- and P-branch doublets are observed lasing simultaneously to different vibrational levels in the ground state. The lower laser levels are sufficiently high in energy that they are not populated significantly even at the 600°C temperature required to generate the sulfur dimer. Several other prominent features that are important for successful optically pumped lasing may be noted. The shift of the excited electronic state to greater internuclear separation insures that there will be large Franck-Condon factors for transitions to high vibrational levels in the ground state. The $B^3\Sigma_u^- - X^3\Sigma_g^-$ electronic transition moment is large. The spontaneous radiative lifetime is 45 nsec [13], resulting in a large stimulated emission cross section of 5×10^{-14} cm^2 for Doppler broadened lines. Thus the optically pumped S_2 laser has very high gain and requires a low inversion density. At the opposite extreme, the Br_2 laser has the weakest transition strength which has been successfully pumped [3], a Doppler broadened cross section of 2×10^{-17} cm^2 which is, however, still comparable to those for many organic dyes. A distinct advantage of some dimers, such as S_2, is that the upper lasing levels are fully bound. Thus the molecule does not readily predissociate by collisions as do I_2 and Br_2. In other dimers the upper laser level is well above the first dissociation limit, but there are no predissociating curve crossings which could degrade laser performance either by providing a collision independent loss mechanism competitive with stimulated emission or by broadening the laser transition, thereby decreasing the peak stimulated emission cross section.

3. Spectroscopic Applications

The experimental arrangement and operation of optically pumped lasers is straightforward enough that they can be used to measure a variety of molecular spectral parameters with ease. Although analysis of dimer spectra is often difficult because small rotational and vibrational constants result in a tremendous number of closely spaced transitions and extensively overlapped bands, the spectrum of the output of an optically pumped laser can be extremely simple. Fig.3 shows the laser output of the I_2 molecule in the (43,56) band, on excitation to the J' = 16 and J' = 12 levels of the v' = 43 state [5]. This is exemplary of the typical kind of laser spectrum observed. The line frequencies are easily measured and assigned with the aid of a good spectrometer. The lasing spectrum has characteristic doublet progressions in v".

There are several differences between the spectra of laser output and of laser induced spontaneous fluorescence. Because of gain competition, the intensities of laser lines only approximately follow Hönl-London rotational intensity factors and Franck-Condon vibrational intensity factors. Lines with small intensity factors will not be observable in the spectrum of a laser with no intracavity tuning element, but a tuned laser can be made to oscillate exclusively on lines that would be unobservable in spontaneous fluorescence. The untuned laser spectrum is similar to a fluorescence spectrum, but is capable of greater simplification. Careful adjustment of single-mode pump laser frequency and power, and pumped laser cavity length could result in oscillation on transitions connected to only one of several simultaneously excited upper laser levels. For example, suppose pump transitions belonging to the (v',0) and (v'+1,1) bands coincide to within a Doppler width. Because of the smaller population in v" = 1, laser transitions out of v' will reach threshold at lower pump power.

The property of optically pumped lasers which is perhaps the most important to spectroscopists is the possibility of forcing lasing on very weak transitions. A dispersive element in the dimer laser resonator permits

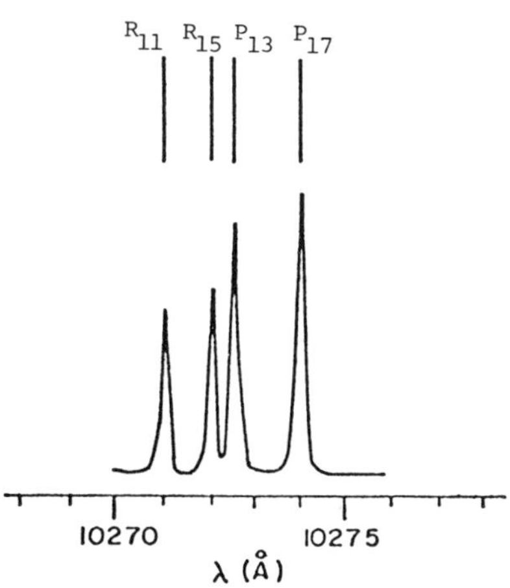

Fig.3 Lasing spectrum of I_2 in the B-X (43,56) band.

harnessing the entire inversion density to drive a single weak transition. For example the I_2 laser reaches threshold on the B-X (43,56) transition at an inserted cw pump power of 110 mW. This transition has the rather small Franck-Condon factor q = 0.005. As more pump power is inserted, transitions with smaller q-values will reach threshold. It appears that at least 100 times more power could be inserted before $\sigma\Delta N$ would saturate. Thus it should be possible to produce laser action on transitions with 1% the strength of (43,56). This is important for observing a number of weak transitions. Lasing can be achieved into the last few levels below a dissociation limit and on transitions into new electronic states. I_2 has five as yet unobserved gerade states which dissociate to ground state atoms in addition to the well-known $X^1\Sigma_g^+$ state. It should be possible to observe transitions on nominally forbidden branches such as $^3\Sigma$-$^3\Sigma$, $\Delta\Omega = 1$, crossover transitions, which are essential for determining spin-splittings [14], on weak extra lines at perturbations between two electronic states [15], and possibly on bound-free transitions.

The optically pumped laser also offers a unique way to determine some special molecular parameters such as the dipole transition strength, μ^2, as a function of internuclear separation. Consider the potential curves for the I_2 molecule shown in Fig.4. On excitation of the v' = 43 level with an

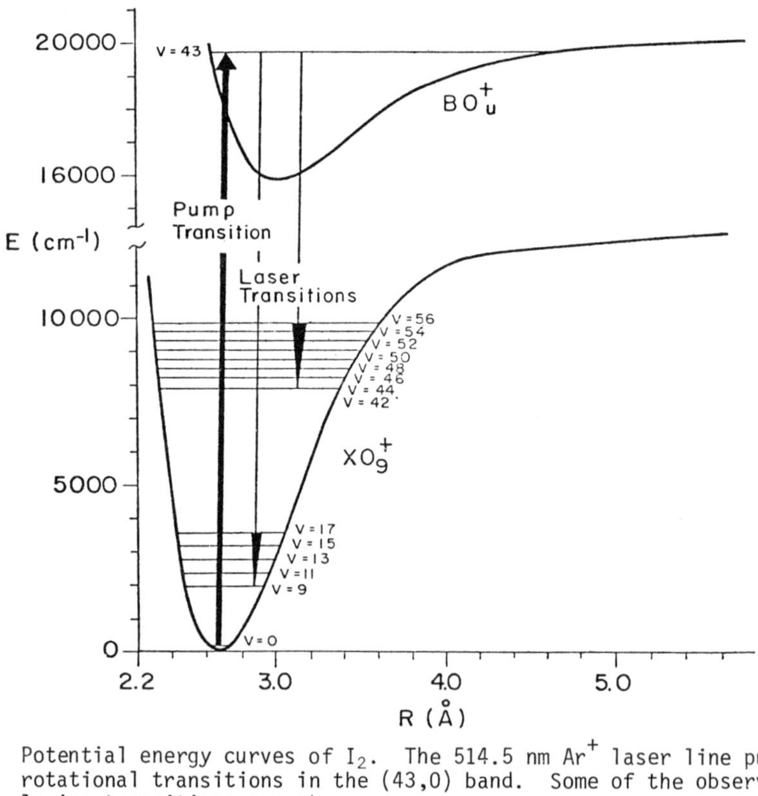

Fig.4 Potential energy curves of I_2. The 514.5 nm Ar^+ laser line pumps rotational transitions in the (43,0) band. Some of the observed lasing transitions are shown.

argon ion laser, lasing transitions have been observed to levels ranging from v" = 9 to v" = 83. The lasing lines sample a range of internuclear separation (r-centroid) from 2.8 to 4.3 Å. If the dipole transition strength remains constant over this range, the laser gain (and threshold) on different lines would be expected to vary with the strengths of the Franck-Condon factors. Laser gain was measured on the (43,56) and (43,83) bands corresponding to R-centroids R_{I-I} = 3.3 and 4.3 Å respectively. The gain was found to be 3% and 13% for the (43,56) and (43,83) transitions [5], while the calculated Franck-Condon factors are 0.0046 and 0.45 respectively [16]. The lower than expected gain on the (43,83) band is accounted for by a dramatic reduction in the transition dipole strength, μ^2, at larger R_{I-I}. As the atoms are separated, they come apart as neutrals. Thus at $R_{I-I} = \infty$, $\mu^2 = 0$. At R_{I-I} = 3.3 Å, μ^2 = 1 Debye [17]. Provided that intracavity losses are identical at the wavelengths of the (43,56) and (48,83) transitions, 1.0 and 1.3 μm respectively, the optically pumped laser gain experiments indicate that μ^2 must have decreased by 20-fold as the internuclear separation is increased by 1 Å to 4.3 Å.

4. Kinetic Applications

Optically pumped lasers are of interest to kineticists in at least two ways: modeling of a laser medium, particularly depopulation of lower laser levels; selective population, within the gain medium itself, of putative reservoir levels or of a chemically interesting series of levels possessing varying degrees of internal excitation.

Optically pumped lasers always have an optimum pressure range for operation with a given input pump power. At a lower pressure the inserted pump power decreases, causing a decrease in output power although not necessarily any decrease in single pass gain. At higher than the optimum pressure, the collisional quenching rate becomes larger than the stimulated emission rate, resulting in a decrease in both output power and gain. As pump power increases, the optimum pressure increases until the onset of gain saturation, at which point output coupling must be increased in order to obtain additional output power. Pulse pumped lasers typically operate in the saturating pump regime; in other words, the upper level is power broadened so that the homogeneous width of the laser transition is at least as large as its Doppler width. The pressure-path length is then adjusted to maximize single pass gain. Finally the output coupling is varied until maximum output power is obtained.

It is interesting to compare some lasing characteristics of the cw-pumped I_2 and pulse-pumped S_2 laser. Fig.5 is a plot of output vs. input pump power for a cw I_2 laser oscillating on low gain transitions near 1.0 μm. The laser has not been optimized with respect to pressure-path length or output cooling [5]. The rather low and nearly constant 0.1% conversion efficiency indicates the presence of intracavity collisional and absorptive loss processes. A similar plot of gain vs. inserted power confirms that gain saturation is not achievable with the available 4 watts of pump power. However, when oscillating on high gain transitions near 1.3 μm, side fluorescence intensity arising from the upper I_2 laser level is observed to increase by at least 25% when the rear laser mirror is blocked; thus the stimulated emission rate out of the upper laser level can be quite comparable to the combined collisional, predissociative, and spontaneous fluorescence loss rates. For the pulse-pumped S_2 laser, a 2% photon conversion efficiency was measured in

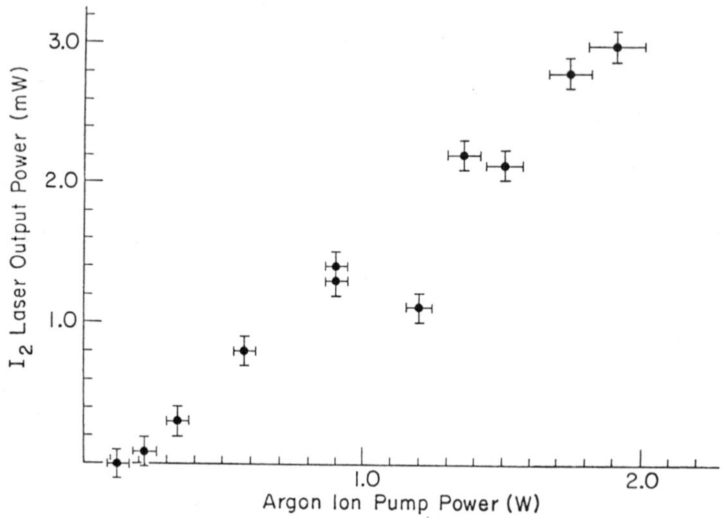

Fig.5 Broadband power output vs. input for the molecular I_2 laser.

spite of the fact that the pump laser line width exceeded the S_2 line width, the lasers were not mode-matched, and pressure-path length and output coupling were not fully optimized [4]. The higher conversion efficiency for the S_2 laser arises from two factors: a larger stimulated emission cross section and the high (>1 kW) peak pump power. With optimization, conversion efficiency of S_2 lasers should approach the theoretical limit.

Attempts to model optically pumped lasers have generally been successful in predicting gain parameters [1-3,5], but observed output powers are generally lower than expected, especially for cw lasers. The loss processes most likely responsible are thermal lensing, absorption from the upper laser level, population buildup in the lower laser level, and absorption from highly excited levels populated by collisional relaxation of the lower laser level.

The characteristics of optically pumped lasers can be similar to those of electrically or chemically pumped lasers oscillating on the same transition. Collisional relaxation of the lower laser levels is a most important problem for practical devices. Population bottlenecks in the lower level can cause lasing to terminate or to switch from line to line. The very fact that several optically pumped systems involving transitions to bound and nonradiating lower states have been made to lase cw indicates that collisional relaxation is unexpectedly effective. In fact the cw I_2 laser exhibits no significant gain bottleneck. Average output power with cw pumping is only 10% less than peak output power when the cw pump laser is chopped with a 0.1 μsec risetime [5]. Very high gain cw lasers (e.g. Na_2) and pulsed lasers pumped by high peak power may bottleneck. In Fig.6 are shown the saturating pump pulse and output lasing pulse of the S_2 molecule. The S_2 laser was restricted to a single transition with an intracavity grating. The 45 nsec radiative lifetime of S_2 is considerably shorter than the 2 μsec

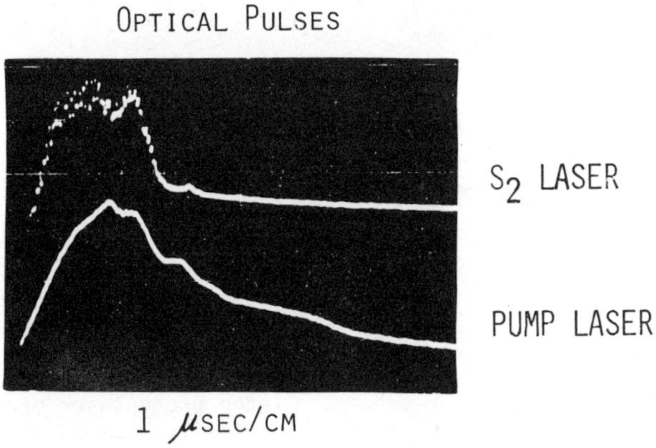

Fig.6 Output pulses of the pump and molecular S_2 lasers as function of time.

lasing pulse, and at the operating pressure the S_2 molecules in the volume of the beam undergo only 10 gas kinetic collisions per microsecond. Yet the lasing output follows the complete time duration of the pumping pulse with very nearly the same intensity profile. Evidently the lower laser level is efficiently relaxed during the saturating pump pulse.

Rather than simply observing the laser characteristics of optically pumped lasers, various pump-laser-probe experiments are capable of providing detailed information about V, R, T relaxation processes within a single electronic state, energy flow between nonradiating electronic states, vibrational dependence of relaxation rates especially for levels near dissociation limits (long range molecules), and effects of large and specific changes of internal excitation on chemical reactions. In the lasing process, the optically pumped laser delivers a known population to a single lower level determined by an intracavity tuning element.

A much wider range of excited levels may be selectively populated by this means than by other methods. For example, direct optical pumping is restricted to vibrational fundamentals and low overtones of heteronuclear diatomics; IR laser absorption plus collisional up-pumping is not selective; multiphoton IR pumping requires extremely high powers and is often not selective; cascade pumping by spontaneous fluorescence from higher electronically excited levels occurs to a complex series of states simultaneously. Stimulated emission pumping can be temporally controlled, state selective, and can produce a high population density within a well-defined volume.

A pump-laser-probe experiment in progress on I_2 is illustrated in Fig.7. I_2 is forced to lase on a transition into high vibrational levels of the $X^1\Sigma_g^+$ state. A probe laser, at a wavelength longer than those which are absorbed by unexcited molecules, interrogates I_2 molecules in or near the intracavity mode of the pumped laser. The resultant short wavelength fluorescence from within the interrogated intracavity mode volume of the probe laser is imaged onto the slit of a monochromator set to detect light at the wavelength of a specific (v',v") transition which is not present in the side fluorescence excited by the pump laser alone. As the wavelength of the

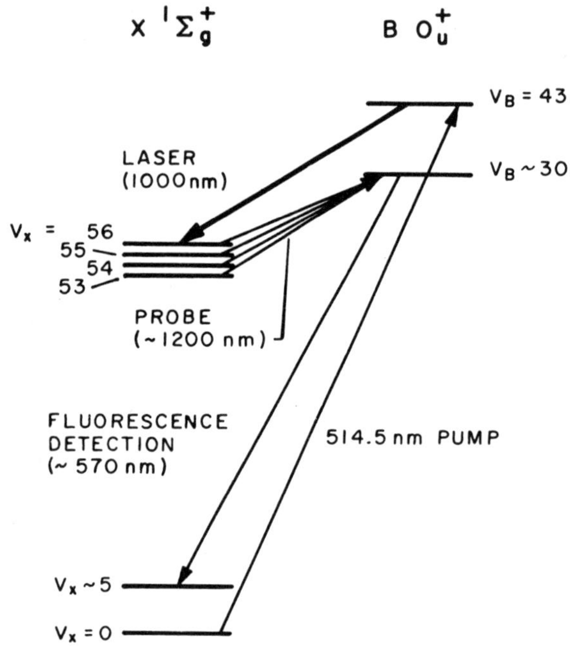

Fig.7 I_2 pump and probe experiment. The (43,0) transition is pumped, the (43,56) lasing transition populates $v_X''^{\dagger} = 56$, a tunable parametric oscillator probe laser excites partially relaxed molecules into the detection level of $v_B' = 30$, and a monochromator views fluorescence in the (30,5) detection band.

probe laser is scanned, the populations of various $v''^{\dagger}$ levels are sampled by excitation of $(v',v''^{\dagger})$ transitions. If the pumped laser is Q-switched, the population flux into the $v''^{\dagger}$ level will be pulsed. It is possible to measure the rate of appearance of population in any (v",J") level, as well as the vibrational and rotational dependence of diffusion coefficients.

5. Device Applications

An obvious motivation for optically pumped laser studies is to efficiently identify new, potentially important laser candidates. The ultimate goal would be then to find practical methods to pump these laser candidates by electric discharges, chemical reactions, etc. The optically pumped laser can provide a useful gain diagnostic tool for these more practical investigations. The insight gained from initial optical pumping experiments provides a mechanism for sifting over the many possible laser candidates for the few which should be pursued further.

As an example, the molecular S_2 laser has numerous features which make it potentially important to study in greater depth [4]. It has a broad tuning range throughout the visible and ultraviolet. Sealed-off operation is possible and the medium is non-degradable, in some sense self-healing. The optically pumped medium is quiescent and free of magnetic and electric fields. Consequently, good mode quality and narrow stable lines are obtainable regardless of the quality of the pump laser mode and line width.

High conversion efficiency may be obtained, since direct excitation of states 4-5 eV in energy will return photons of 2-3 eV. Single and narrow line operation is easily achieved. With the numerous isotopes of sulfur available, quasicontinuous tuning should be possible. Such laser devices may have applications, for optical frequency standards and measurements, photochemical isotope separation processes [18], frequency up-and-down conversion, narrow line, multiple frequency, or mode locked oscillators in an oscillator-amplifier setup, and many more.

6. Acknowledgments

We gratefully acknowledge the Office of Naval Research, the Air Force Office of Scientific Research, and the National Science Foundation for generous research support. We would like to thank our many fellow scientists for valuable discussions, especially J. L. Hall, J. Dallarosa, R. E. Drullinger, A. V. Phelps, J. Tellinghuissen, and J. I. Steinfeld.

References

1. R. L. Byer, R. L. Herbst, H. Kildal and M. D. Levenson, Appl. Phys. Lett. 20, 463 (1972).
2. M. A. Henesian, R. L. Herbst and R. L. Byer, J. Appl. Phys. 47, 1515 (1976) and H. Itoh, H. Uchiki and M. Matsuoko, Opt. Comm. 18, 271 (1976).
3. F. J. Wodarczyk and H. R. Schlossberg, "An Optically Pumped Molecular Bromine Laser," submitted to J. Chem. Phys.
4. S. R. Leone and K. G. Kosnik, Appl. Phys. Lett. 30, 346 (1977).
5. J. B. Koffend and R. W. Field, "cw Optically Pumped Molecular Iodine Laser," J. Appl. Phys. 00, 0000 (1977).
6. R. Wellegehausen, S. Shahdin, D. Friede and H. Welling, Appl. Phys. 13, 97 (1977).
7. D. R. Guyer and S. R. Leone, 5th Conference on Chemical and Molecular Lasers, St. Louis, MO (1977).
8. J. Dallarosa, R. E. Drullinger, J. L. Hall; H. Welling, private communication.
9. E. J. Schmitschek, J. E. Celto and J. A. Trias, Conference on Laser Engineering and Applications, Washington, DC (1977).
10. T. Y. Chang and T. J. Bridges, Opt. Comm. 1, 423 (1970); Appl. Phys. Lett. 17, 249 and 357 (1970).
11. V. I. Balykin, A. L. Golger, Yu. R. Kolomiiskii, V. S. Lethokhov and O. A. Tumanov, JETP Lett. 19, 256 (1974).
12. B. Wellegehausen, private communication.
13. T. H. McGee and R. E. Weston, Chem. Phys. Lett. 47, 352 (1977).
14. K. K. Yee and R. F. Barrow, J. Chem. Soc. Farad. Trans. 68, 1397 (1972); and F. Ahmed, R. F. Barrow and K. K. Yee, J. Phys. B 8, 649 (1975).
15. T. Ikeda, N. B. Wong, D. O. Harris and R. W. Field, "Argon Ion and Dye Laser Induced MgO Photoluminescence Spectra : Analysis of $a^3\Pi \sim X^1\Sigma$ Perturbations," J. Mol. Spectrosc. 00, 000 (1977).
16. J. Tellinghuisen, private communication.
17. J. Tellinghuisen, J. Chem. Phys. 58, 2821 (1973).
18. J. C. Vanderleeden, Laser Focus 13, No. 2, 51 (1977).

GENERATION AND APPLICATIONS OF 16 GHz TUNABLE SIDEBANDS FROM A CO_2 LASER

P.K. Cheo

United Technologies Research Center
East Hartford, CT 06108, USA

Abstract

A microwave-tuned infrared laser source has been developed to provide tunable sideband power from a CO_2 laser up to 200 mW. For each sideband, continuous tuning of 8 GHz has been obtained. Preliminary results on SF_6 transmission spectra at a resolution of 2.5 MHz are presented.

Tunable infrared laser sources for high resolution spectroscopy are inadequate at present. The high pressure CO_2 waveguide laser [1] is a useful source but has a very limited frequency tuning range [2]. The broadest tunable infrared source is the lead-salt diode laser [3]. However, diode laser technology has its drawbacks: (a) it is basically a low power device, ($\lesssim 1$ mW), (b) it requires cryogenic cooling, (c) it has limited lifetime due to thermal cycling, (d) it requires other lasers to provide frequency references, and (e) it needs spectrometer and/or etalon to separate various laser modes. This paper reviews a relative new technique [4] to obtain high power tunable laser radiation in the 9-11 μm region. This technique can be utilized to extend into other spectral regions by using appropriate laser source. This technique involves the generation of sideband power from an infrared laser source at microwave frequencies by using a very efficient infrared waveguide modulator [5]. This paper also presents some preliminary spectroscopic data on SF_6 molecules obtained by using a microwave-tuned [4] CO_2 laser source, MTLS. The resolution of 2.5 MHz is obtained by using a standard microwave frequency sweeper without the complication of super-heterodyne frequency synthesis [3]. Considerably higher resolution to within kHz range of CO_2 laser linewidth [6] can be achieved with this technique by using a finer microwave frequency synthesizer.

Efficient sideband generation is accomplished by mixing an infrared laser with a traveling microwave in an integrated optic modulator. The electrooptic interaction produces a phase retardation, $\Delta\phi$, of the infrared laser beam at microwave frequencies. The output power spectra of the infrared laser as a result of this interaction contains two sidebands at microwave frequencies f_μ, in addition to the laser carrier frequency f_o. The sideband power, P_s, can be expressed for the small angle approximation as:

$$P_s = \frac{(\Delta\phi)^2}{2} P_o \qquad (1)$$

where P_o is the laser power and

$$\Delta\phi = \frac{\pi}{\lambda t} n^3 r_{41} (P_\mu Z_o L)^{1/2} \left(\frac{1-e^{-4\alpha L}}{2\alpha} \right)^{1/2} \quad (2)$$

where λ and P_μ are the laser wavelength, and the microwave power respectively. n, r_{41}, α, t, L, and Z_o are the refractive index, electrooptic coefficient, absorption coefficient, thickness, length and characteristic impedance of the waveguide modulator, respectively. Equations (1) and (2) indicate that the sideband power increases linearly with increasing both the optical and microwave power. At 60 W microwave power input to a waveguide modulator, the optical power conversion from the carrier to the sideband of 2.2% has been obtained. The highest optical transmission through a GaAs waveguide is about 10W.

Figure 1 depicts the MTLS instrumentation for high resolution spectroscopy. The heart of this instrument is an infrared waveguide modulator [5]

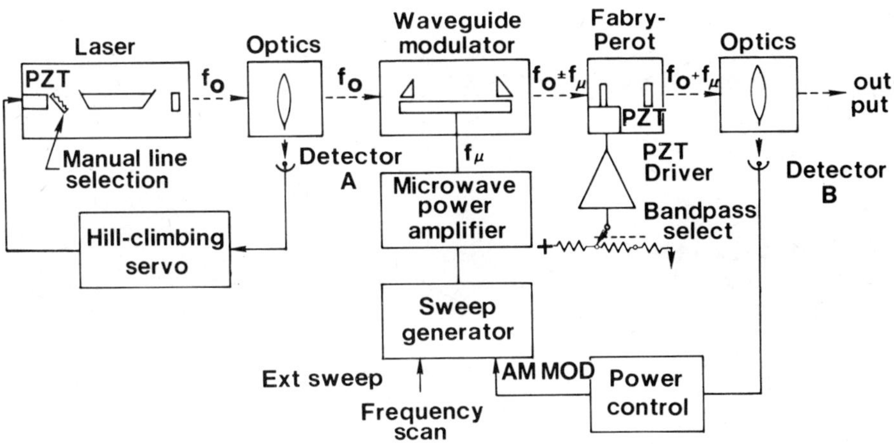

1. A Broadband Tunable CO_2 Laser-Block Diagram

integrated with a broadband microwave transformer network. [4] A stabilized CO_2 laser with a grating mirror is used to provide a high power source signal, f_o. The output of this laser beam is focused to a spot size of 1 mm at the front AR-coated surface of a germanium prism coupler. This coupler compresses the Gaussian beam to a guided-wave mode of the infrared waveguide, as shown in Fig. 1. Near the input prism coupler, a microwave power source at frequencies in X and Ku bands is fed into waveguide through a broadband transformer network, which has been built into the waveguide. This transformer consists of three discrete microstrip electrodes to provide a broadband impedance matching for the microwave from a 50-ohm source to a 2.75-ohm waveguide at these frequencies. The microwave power interacts with the optical-guided wave within a narrow channel of 1 mm in width. The length of the interaction channel is 2.78 cm. At the output end of the waveguide another prism coupler and a transformer network similar to those used at the input terminal are also installed in the waveguide.

The infrared waveguide is made in the form of a GaAs thin-slab with the typical dimensions: 1 cm wide, 5 cm long and 25 μm thick. This thin-slab is bonded onto an optically polished copper block, which provides both the mechanical strength and power dissipation. Two types of waveguide structures, as shown in Fig. 2, have been investigated. [5] The first kind

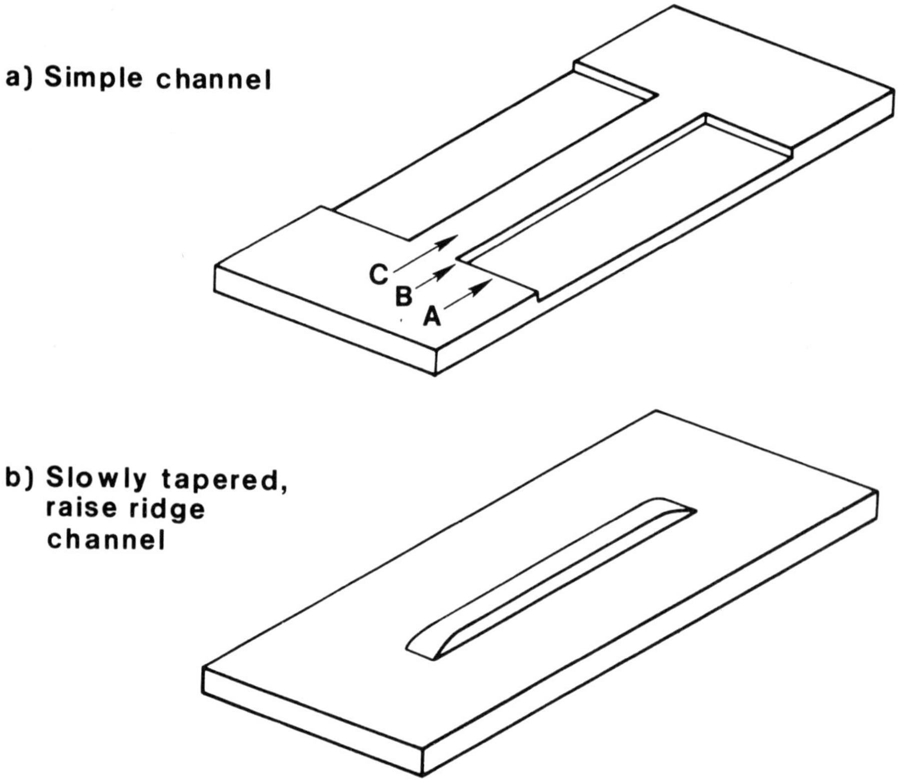

2. GaAs Thin-Slab Channel Waveguides

is configured in the form of a simple channel and the other has a more complex raised ridge structure. The typical optical transmission through a simple planar channeled waveguide modulator is 30% and that through a tapered ridge modulator structure is 60%. The modulated output signals consist of a carrier signal at frequency f_o, and two sidebands at frequencies $f_o \pm f_\mu$. This output is collimated through a double-pass, Fabry-Perot (DPFP) filter, which separates the sidebands from the carrier with a typical rejection ratio greater than 10^5 at designed frequencies. The optical transmission through the DPFP filter is about 30% and can be increased to about 55% by improving the collimating optics.

Figure 3 shows a GaAs thin-slab waveguide integrated with a broadband traveling microwave transmission line. The width of the interaction channel

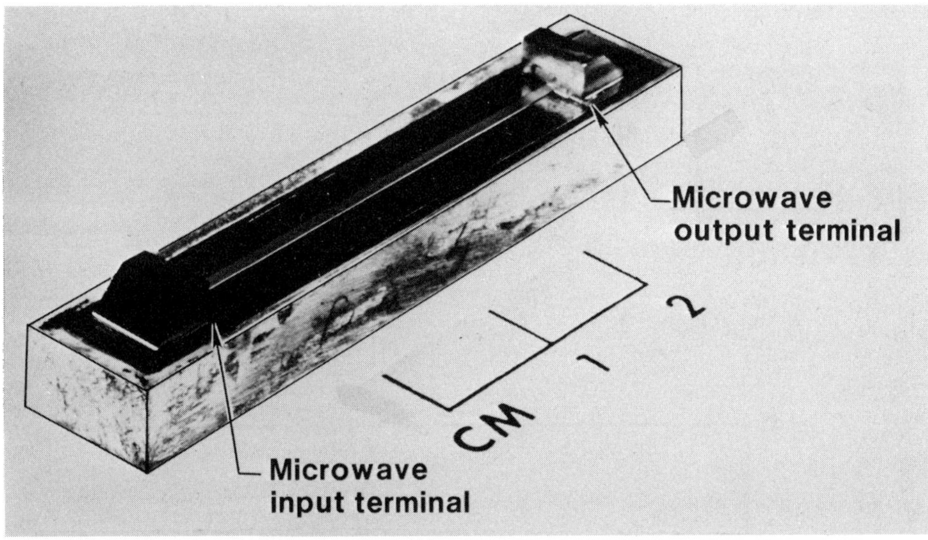

3. An Integrated Infrared Waveguide with a Traveling-Wave Microstrip Transmission Line

is 1 mm. The impedance of this channel at microwave frequencies is 2.75Ω. To couple a 50Ω microwave source into this channel, an impedance transformer is required. The modulator, as shown in Fig. 3, has a built-in three step transformer network which is made by adjusting both the electrode width and the GaAs slab thickness. Figure 4 shows the tunable infrared laser output at several microwave modulation frequencies. In each oscilloscope picture, there are two traces: the upper trace represents the voltage applied to the DPFP filter and the lower trace shows the output power spectrum. One full voltage scan produces a linear spatial variation to the filter, corresponding to a change in frequency from 0 to 50 GHz. At f_μ = 12.5 GHz the sideband-to-carrier ratio is about 8. As the offset frequency increases, the ratio increases rapidly. At 18 GHz, $P_S/P_O \geq 100$ by using a DPFP filter.

Figure 5 shows the measured sideband power as a function of microwave input power at 15.8 GHz. The results indicate that the GaAs waveguide modulator is capable of handling the microwave power at a level up to 60 W. As expected, the conversion efficiency increases linearly with increasing P_μ. At a typical microwave input level of 20 W, one expects a conversion

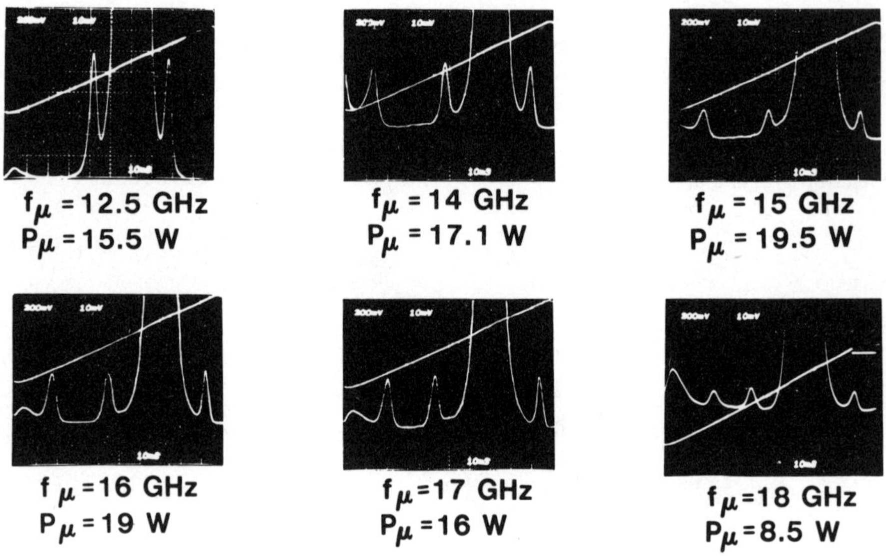

4. Sideband Power Spectra from 12.5 to 18 GHz

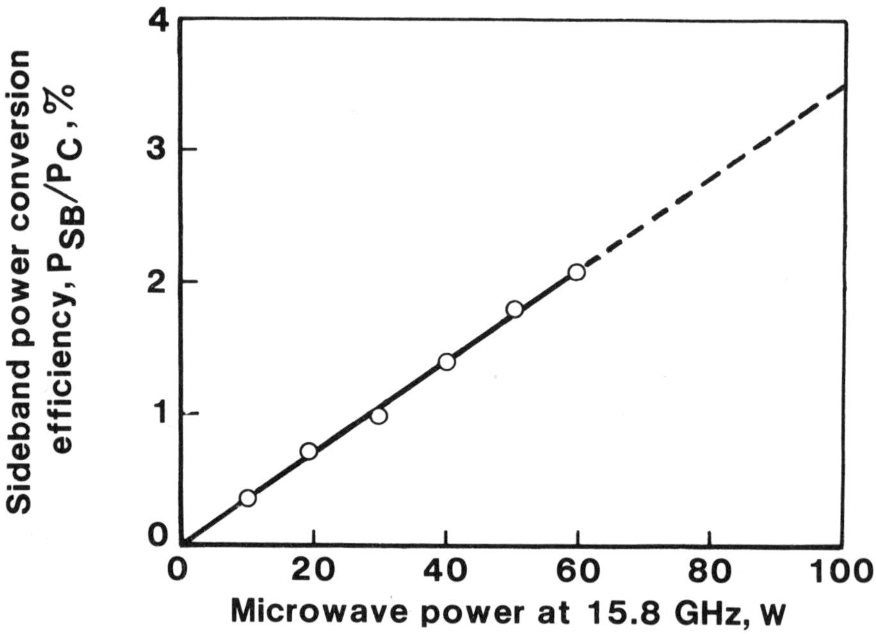

5. Power Dependence of Sideband Conversion Efficiency

of 0.7% from a modulator having an interaction length of 2.78 cm. A combined optical and microwave power up to 75 W has been applied to a waveguide modulator over a period of several hours. The measured increase in device temperature is only about 4° C. Figure 6 shows the measured sideband power as a function of microwave frequency over a range from 10 to 18 GHz for a specific waveguide modulator design. In the microwave transformer section,

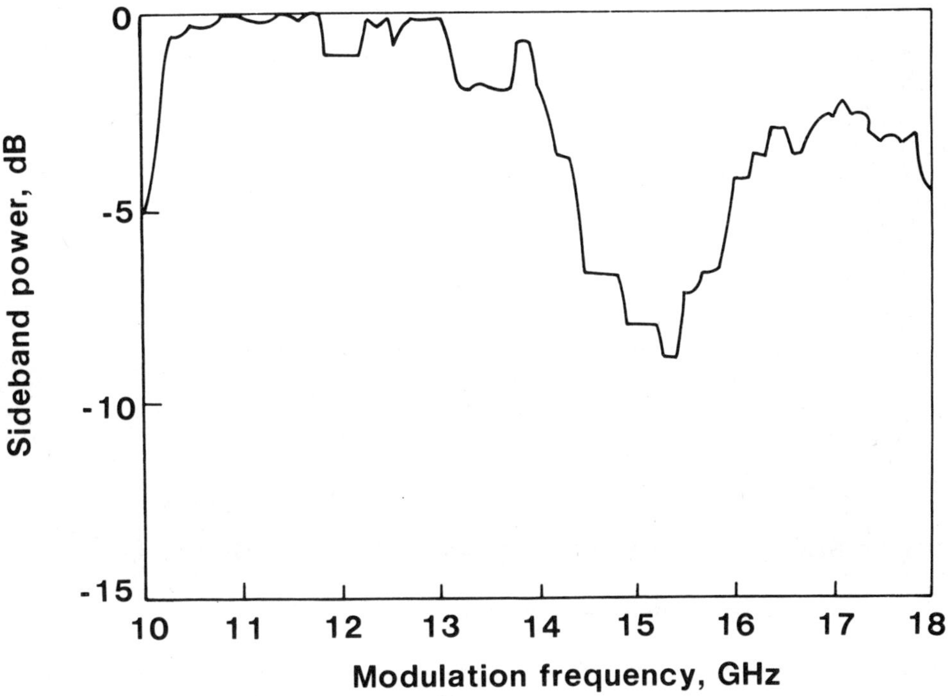

6. Frequency Tuning Characteristics of a CO_2 Laser Line

this modulator has a 3-mil thick GaAs strip bonded onto a 1-mil thick GaAs planar waveguide. The highest sideband power in Fig. 6 is measured to be 12 mW at the output of the modulator and about 4 mW at the output of the DPFP filter for an input CO_2 laser power of 4.5 W. It is possible to obtain significantly higher sideband power by using a tapered ridge modulator structure driven at a higher microwave power level. The dip in the curve in the proximity of 15 GHz is attributed to the microwave loss at the input transformer, which is a unique character of this particular modulator design, and it can be corrected with further experimentation. A slight decrease of sideband power at higher frequency is expected because the material loss increases with increasing frequency.

Transmission spectral measurements of SF_6 gas molecules have been made with this tunable IR source. The reason for choosing SF_6 is because the molecular structure of SF_6 is extremely complex and its spectra have extensively been analyzed, both experimentally and theoretically. [7]

These measurements should provide additional data for spectral characterization of SF_6 molecules. Figure 7 shows a typical transmission curve for the upper sideband of P(20) CO_2 laser line at 10.59 micrometers. Over

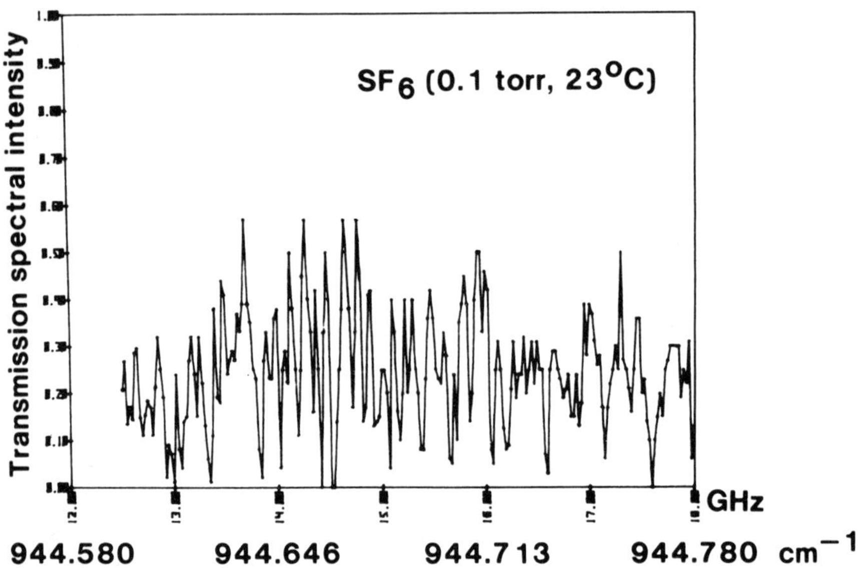

7. Transmission Measurements of SF_6 Molecules from 12.5 to 18 GHz

the frequency range from 12.5 to 18 GHz, there are a number of overlapping rotational spectra which is typical of the SF_6 molecular spectra at room temperature. These measurements are made at a frequency interval, Δf_μ of 25 MHz. To show the resolution capability of this device, a finer frequency sweep with Δf_μ equal to 2.5 MHz has been made in the neighborhood of 13 GHz. The measured linewidth is 30 MHz which represents a typical Doppler broadband line.

In conclusion, the capability of performing high resolution spectroscopy has been demonstrated with an efficient infrared modulator to generate sideband power from a CO_2 laser at microwave frequencies. This technique can provide tunable source in the spectral region from 9.2 to 10.8 μm region by using several isotopic CO_2 lasers.[8] This technique could also be extended into other spectral regions with help from other gas lasers. Other regions to be included are the near and far infrared by choosing appropriate laser sources. The attractive features of this

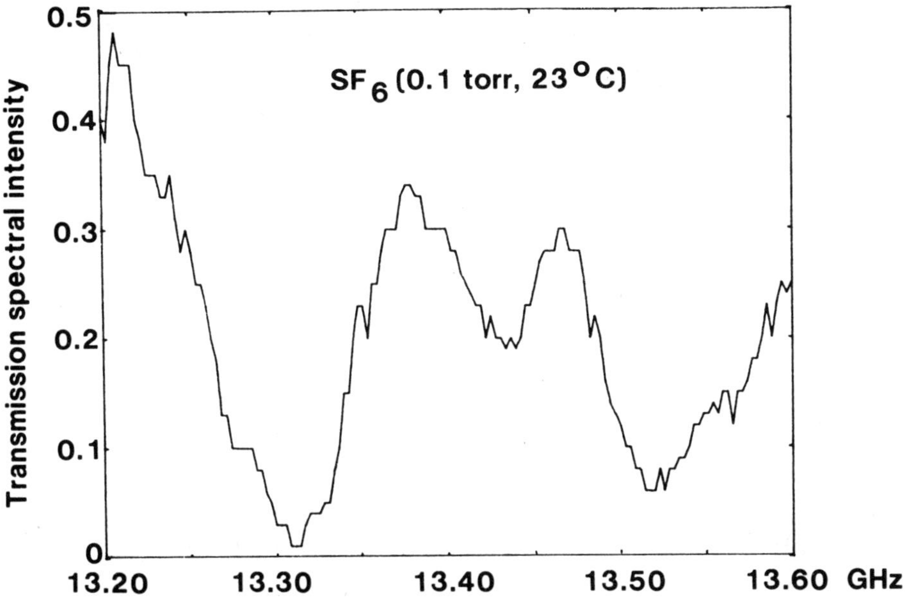

8. Transmission Measurements of SF_6 Molecules from 13.2 to 13.6 GHz

device are: (a) room temperature operation, (b) high power, as much as 200 mW, (c) high resolution, as high as 10^{-8} cm^{-1}, and (d) self-calibrated source with high accuracy.

References

1. Smith, P. W., Appl. Phys. Lett., 19, 132 (1971)
2. Abrams, R. L., Appl. Phys. Lett., 25, 304 (1974)
3. Hinkley, E. D., T. C. Harmans and C. Freed, Appl. Phys. Lett., 13, 49 (1968)
4. Cheo, P. K. and M. Gilden, Opt. Lett., 1, 38 (1977)
5. Cheo, P. K. and R. Wagner, IEEE J. Quant. Elect., QE-13, 159 (1977)
6. Freed, C. and A. Javan, Appl. Phys. Lett., 16, 53 (1970)
7. For example, see Herbert Flicker, Gordon Research Conference on Infrared and Raman Spectroscopy, August 23-27, 1976
8. Freed, C., R. G. O'Donnell and A. H. M. Ross, Conference on Precision Elect. Measurements, June 28--July 1, 1976, held in Boulder, Colorado

THE FREE ELECTRON LASER

D.A.G. Deacon, L.R. Elias, J.M.J. Madey, H.A. Schwettman, and T.I. Smith

Physics Department and High Energy Physics Laboratory, Stanford University
Stanford, CA 94305, USA

Introduction

In a free electron laser, optical radiation is made to interact with a beam of relativistic free electrons in the presence of a strong spatially periodic transverse magnetic field. This device is interesting because the operating wavelength can be set at will by variation of the electron energy and the device is potentially capable of operating at high average power.

There has been quite substantial progress in the development of this class of laser. A 10 μ free electron laser amplifier was operated at the High Energy Physics Laboratory in 1974. More recently, a 3 μ free electron laser oscillator has been run in a pulsed mode at the kilowatt level. Our purpose in this paper is to review the background of this development, the experimental progress to date and the direction of our present and future research.

The history of the concept can be traced back to 1951 when MOTZ analyzed the spontaneous radiation emitted by an electron in a periodic field. [1] MOTZ also carried out the first classical analysis of the amplification mechanism [2]. The first quantum analysis was published by MADEY in 1971 [3] and further contributions to the classical theory have recently been made by HOPF et al [4] and COLSON [5].

In the quantum theory, the electrons absorb virtual quanta from the periodic field and re-emit them as real photons. The transition rate for this scattering process is enhanced when radiation is present in the final state according to the customary rules for quantum electro-dynamics. In the classical theory, the radiation field modulates the velocity of the electron beam in the first part of the interaction region making it possible for energy to be extracted from the bunched beam in second half of region.

The equations for the magnitude and the dependence of gain on wavelength in the quantum and classical theories are now known to be identical to order $\left(\frac{h\nu}{\gamma mc^2}\right)$ and the differences in functional form are not believed to be significant at optical wavelengths. There is, however, a difference in the form of the results for the velocity distribution of the electrons [6].

The classical approximation is applicable in the limit of high power and long wavelengths. In the classical approximation, the commutators of the operators for the field and for the particle position and momentum are assumed to be zero so that the amplitude of the field and the coordinates of the electrons can be followed with arbitrary precision during the interaction. Careful analysis of the commutation relations for the field indicate that this approximation is valid only when the optical power density in the electron rest frame is large in comparison to $(h\nu/\lambda^3) \cdot c$.

In the classical approximation, the velocity of an electron in the interaction region is determined by the initial conditions, that is, by its momentum and the amplitude and phase of the field at the start of the interaction. If a classical ensemble of free electron lasers were assembled in which each system was prepared with the same initial conditions, the trajectories of the electrons in each system would be identical. Since, in fact, the amplitude of the field can not be specified with precision, in a real ensemble it would only be possible to specify the probability that an electron will follow a given trajectory and the possibility would have to be allowed that electrons entering the interaction region under identical circumstances could emerge with noticeably different momenta. In addition to the periodic variation of electron momentum with phase in the classical approximation, there will be a random spread in momentum due to the uncertainty in the field, a purely quantum effect. The uncertainty in the field is small at microwave wavelengths but becomes quite substantial in the ultraviolet [6].

Experimental Results

Our experimental work has had as its objective the verification of the small signal gain formula and the exploration of the saturation characteristics of the mechanism at high power. In the first experiments, which were run at 10 μ, a 24 MeV electron beam was used to amplify the radiation for a pulsed CO_2 TEA laser [7]. The electron beam and the radiation were run through a periodic field generated by a 5.2 meter bifilar superconducting helix (Fig. 1). The peak gain reached 7% at an instantaneous peak current of 70 mA and a magnetic field of 2.4 Kgauss (Fig. 2). No saturation was observed at a power density of 140 KW/cm^2 which was the highest power attainable in the experiment. At this power the mean electron energy after the interaction was reduced by 0.2%.

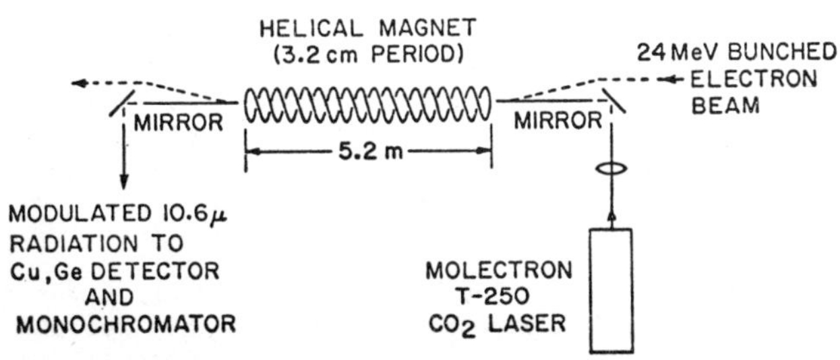

Fig. 1 The Free Electron Laser Amplifier. A 24 MeV electron beam was used to amplify the 10.6 μ radiation from a CO_2 TEA laser. The periodic magnetic field was generated by a 5.2 meter bifilar superconducting helical magnet. The helix period was 3.2 cm and the field strength 2.4 Kgauss.

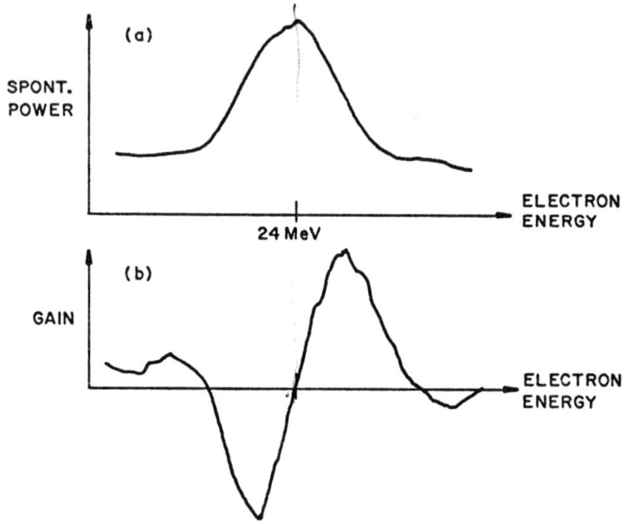

Fig. 2 Relationship of Gain and Spontaneous Emission. Figures 2(a) and 2(b) indicate the dependence on electron energy of the spontaneous power gain at 10.6 μ. The energy was swept over a 1% range in both panels. In Fig. 2(a), the spontaneous power radiated into a 5×10^{-6} stearadian cone on the optical axis and was run through a monachromator set at 10.6 μ. Figure 2(b) shows the measured gain on the same energy scale. Note the gain is positive on the high energy side of the spontaneous emission. The eletantaneous peak gain in this experiment reached 7% at an electron current of 70 mA.

More recently, the apparatus was modified for operation as a laser oscillator (Fig. 3). The modifications included the addition of a pair of spherical mirrors at the ends of the interaction region to provide feedback and the modification of the accelerator to raise the instantaneous peak current. The instantaneous peak current was raised by an order of magnitude by lowering the duty cycle. This was accomplished by pulsing the accelerator's electron gun at 11.8 MHz, the 110th subharmonic of the accelerator operating frequency. The resonator was designed to operate at 3 microns to take advantage of the reduced mode cross section at short wavelengths. At 10 μ specific account must be taken of the modifications to the propagation constants of radiation moving through the 10 mm bore of the helical magnet and coupling losses result in a significant reduction of cavity Q. At 3 μ, radiation can propagate through the bore as a Gaussian beam and cavity losses are determined primarily by mirror reflectance.

The oscillator was operated above threshold at 3.417 μ for the first time in January, 1977 [8]. A peak power estimated at 7 kilowatts was obtained from the oscillator. The average power was 360 milliwatts and the linewidth was 200 GHz. Spectra of the radiation emitted by the laser above and below threshold are shown in Fig. 4. Note that the line center above threshold is shifted to the long wavelength side of the lineshape for spontaneous emission consistent with the results in Fig. 2.

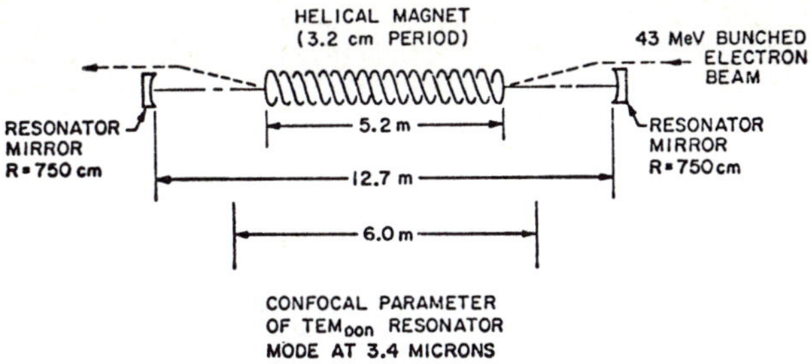

Fig. 3 The Free Electron Laser Oscillator. The apparatus used in the amplifier experiment was modified by addition of a pair of spherical mirrors at the ends of the interaction region. The mirror spacing was adjusted to match the 84.6 nsec spacing of the electron bunches.

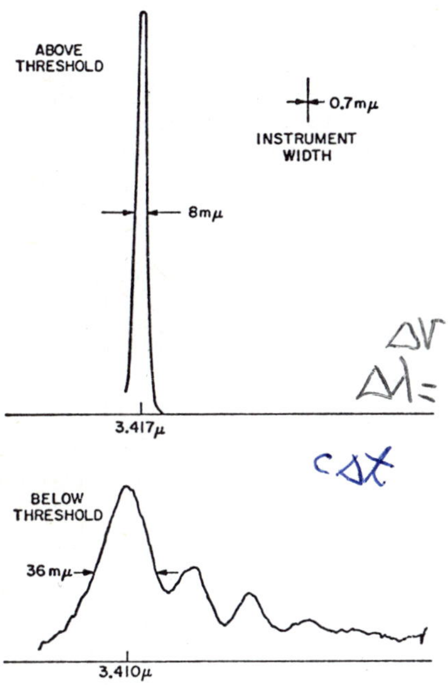

Fig. 4 (at left) FEL Oscillator Spectrum Above and Below Threshold. Note that the line center above threshold has moved to the long wavelength side of the lineshape for spontaneous emission consistent with Fig. 2. The linewidth below threshold is determined by the number of magnet periods $(\Delta\nu/\nu) \gtrsim 1/n$. By comparison the linewidth above threshold is limited by the length of the electron bunches, $\Delta\nu \gtrsim (1/\Delta t)$. The bunch $c\Delta t$ in the experiment is estimated to be 1.3 mm.

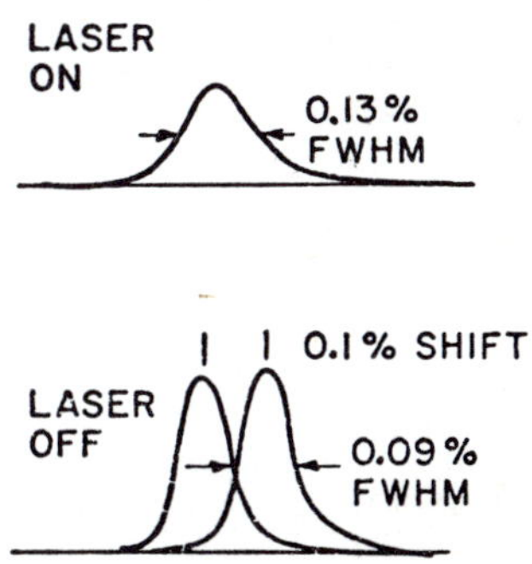

Fig. 5 Electron Momentum Spread Above and Below Threshold. The electron beam emerging from the interaction region was run through an analyzing magnet which converted the momentum spread into an angular dispersion. The momentum spread in the top trace, with the laser above threshold, was 0.13% FWHM. In the bottom traces, the laser cavity was blocked and the electron momentum was manually shifted by 0.05% between traces to ascertain the spectrometer dispersion and resolution. Laser operation broadened the momentum sprectrum by 0.09%.

The electron beam emerging from the interaction region was run through a magnetic spectrometer to analyze the spread in momentum, a quantity of some interest for applications involving the recirculation of the electron beam. The data from the momentum measurement is shown in Fig. 5 and indicates that the interaction in the laser has contributed .09% to the width of the momentum spectrum.

The power obtained from the oscillator is lower by an order of magnitude than the power expected from the earlier data from the gain measurements at 10 μ. In the oscillator experiment, approximately .01% of the electron energy was extracted as radiation in the laser whereas, at 10 μ, 0.2% of the electron energy was converted to radiation with no sign of saturation. It is possible that the problem was due to inadequate control of the resonator length. The resonator length has to be set to match the optical transit time to the space between electron bunches in the beam from the accelerator. In the experiment, it was observed that the oscillator output changed ~ 4x when the length was changed by 2.5 microns, the smallest increment possible with the gear drive and stepper motor used in the experiment. The drive has since been modified for higher resolution and we will have a chance to test this hypothesis in the next run. An alternative explanation would invoke optical pulse shortening due to the strong modulation of the amplifying medium. A short optical pulse would extract energy only from the electrons with which it overlapped and the net power radiated by the electron beam would be reduced. To evaluate this alternative we have planned an experiment in which an intra-cavity etalon will be used to prevent the collapse of the optical pulse.

Electron Beam Requirements

The free electron laser requires a high current, high quality electron beam. The gain varies as $i \times \lambda^{3/2} \times F_f$ where i is the current and F_f is a wavelength dependent filling factor inversely proportional to the cross section of the optical mode in the interaction region. For the helical magnet in our present experiment a current of 0.1 - 1 Amps is required for operation in the infrared while 1 - 10 Amps would be required for operation in the visible and ultraviolet. There are also restraints on the angular divergence and radius of the electron beam. For the present magnet, the emittance of the electron beam has to be maintained below 0.1 mm-mrad and at

Table 1 Summarizes the wavelength range and laser power output estimated to be obtainable using some existing electron accelerators.

Accelerator	Energy	W/L Range	Power (Peak)	Power (Avg)
Van de Graaf	0-4 MeV	to 100 cm^{-1}	10^3 watts	0.1 watt (100 cm^{-1})
Microtron	3-30 MeV	to 1000 cm^{-1}	10^3 watts	0.1 watt (10^3 cm^{-1})
Superconducting Linac	0-100 MeV	to 10^4 cm^{-1}	10^5 watts	10 watts (10^4 cm^{-1})
Storage Ring	?-5 GeV	to 10^5 cm^{-1}	10^6 watts	10^5 watts (10^5 cm^{-1})

Power Output ~ 0.1% EI

Linewidth: $\Delta\omega\Delta t \sim 1$

$\Delta t \sim$ 1 nsec - 10 μsec (Van de Graaf)
1 nsec (Storage ring)
10 psec (Linac and Microtron)

10^{-8} m $\quad \omega = k = \frac{2\pi}{\lambda}$

10^{-4} m

10 μ and would have to be kept below .01 mm-mrad at 1000 Å [9]. These are quite stringent requirements and there are only a limited number of accelerators capable of supplying a suitable electron beam. The characteristics of some suitable accelerators are listed in Table 1.

A Van de Graaf generator with a low duty cycle electron gun would be a natural accelerator to use for energies up to 4 MeV and it would appear that such a system would be capable of delivering useful power at wavelengths down to 100 μ. At higher energy, the superconducting accelerator and the microtron are the most promising candidates.

Our present experiments use a superconducting accelerator and this machine offers the advantages of unexcelled electron beam quality and stability. However, it is not a machine which could easily be duplicated.

The microtron is capable of delivering an electron beam of high peak current and low energy spread but the orbit plane emittance is poor. It is likely that the emittance of the beam could be improved, albeit at the expense of the current, by collimation, and there is a good chance that the microtron could be adapted to drive an infrared FEL. Microtrons are relatively compact and economical and would be an attractive alternate if the emittance problems could be resolved.

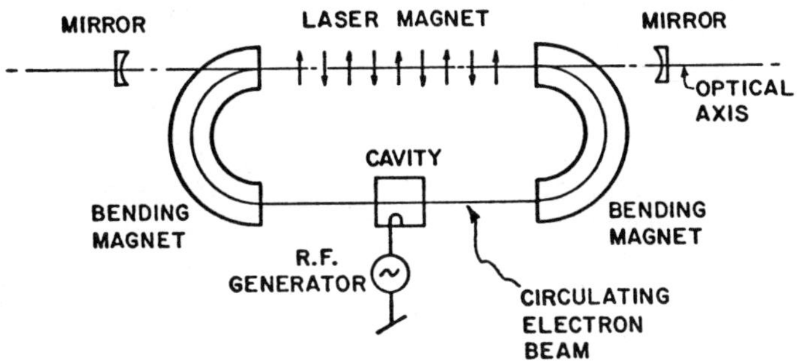

Fig. 6 A Storage Ring Free Electron Laser. The idea is to install the periodic magnet within a storage ring and to re-accelerate the beam after passage through the interaction region to replace the energy lost in radiation.

Storage Ring Lasers

In many ways, the electron storage ring is the ideal source for the free electron laser. The idea would be to install the periodic magnet within the ring and to re-accelerate the electron beam after passage through the interaction region (Fig. 6). Storage rings have the proven capability to store an electron beam with a peak current ~ 10A at the energy required for laser operating in the visible and ultraviolet and there is a potential for very high average power output and good efficiency.

We are presently analyzing the effect of laser operation on the circulating electron beam in a high energy storage ring. Laser operation will increase the magnitude of both the fluctuations in the radiated energy and the rate at which the fluctuations are damped and the problem is to identify the effect of laser magnet geometry and storage ring design on the characteristics of the circulating electron beam to determine the circumstances under which laser operation can be attained.

In our recent theoretical work, we have determined the equilibrium energy spread in a storage ring in which the laser is operated at a fixed and predetermined optical power density while the electron beam is repeatedly re-accelerated and cycled through the interaction region. Comparable results have been obtained with a linearized analytic approximation [6] and a monte-carlo simulation (Fig. 7). The result shown in Fig. 7 is for a constant period helix similar to our present magnet. For operation at 10 μ at a power density of 10^5 W/cm^2, the monte carlo simulation yields an equilibrium spread of 0.16% as compared to 0.11% from the linearized approximation. An electron energy spread of 0.3% would be adequate for laser operation [9]. These results are quite encouraging and we are proceeding with a more general analysis of storage ring operation.

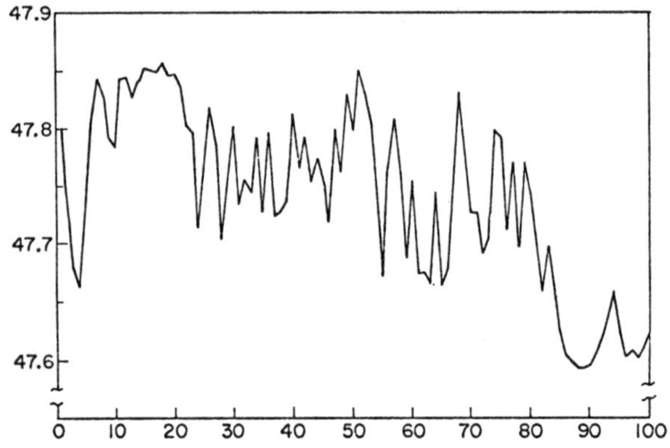

Fig. 7 Electron Energy in a Storage Ring with a Constant Period Helix. The figure traces the evolution of the energy of an electron which is made to pass repeatedly through the interaction region. The energy radiated by the electron in the periodic field was calculated by numerically integrating the classical equations of motion assuming constant optical power density and an initial phase which varied randomly from one pass to the next. The results are for a 160 period helical magnet operated at 2.4 Kgauss, and 10.6 μ radiation at a power density of $10^5 \omega/cm^2$. The electron was accelerated by 2.5 KeV after each pass through the magnet. The figure shows $\gamma = (E/mc^2)$ as a function of the orbit number. The Rms spread in γ for the 100 orbits in the figure is 0.16%.

Work supported in part by the Air Force Office of Scientific Research under Contract #F49620-76-C0018, ERDA Contract #EY-76-S-03-0326.

References

[1] H. Motz, J. Appl. Phys. 22, 527 (1951); H. Motz and M. Nakamura, Ann. Phys. (N.Y.) 7, 84 (1959); H. Motz, W. Thon and R. N. Whitehurst, J. Appl. Phys. 24, 826 (1953).

[2] H. Motz and M. Nakamura, Proceedings of the Symposium on Millimeter Waves, Microwave Research Institute Symposia Series, Vol. IX, (New York: Interscience, 1960) p. 155.

[3] J. M. J. Madey, J. Appl. Phys. 42, 1906 (1971).

[4] F. A. Hopf et al., Optics Commun., 18, 413 (1976); F. A. Hopf, et al., Phys. Rev. Letters 37, 1342 (1976).

[5]W. B. Colson, submitted to Physics Letters.

[6]J. M. J. Madey and D. A. G. Deacon, "Free Electron Lasers" in Cooperative Effects in Matter and Radiation (Plenum Press, N.Y., 1977) p. 313.

[7]L. R. Elias, et al., Phys. Rev. Letters 36, 717 (1976).

[8]D. A. G. Deacon, et al., Phys. Rev. Letters 38, 892 (1977).

[9]L. R. Elias, et al., "A Discussion of the Potential of the Free Electron Laser as a High Power Tuneable Source of Infrared, Visible and Ultraviolet Radiation," published in the Proceedings of the Synchrotron Radiation Facilities Quebec Summer Workshop, (Université Laval, Quebec, Canada, 15 - 18 June, 1976).

IX. Laser Wavelength Measurements

LASER WAVELENGTH MEASUREMENTS, WHY? HOW? AND HOW ACCURATE?

P. Giacomo

Bureau International des Poids et Mesures, Sèvres, France

Since their advent, lasers have become marvellous tools for spectroscopists and for many other physicists. This involves a few "trivial" problems, such as frequency stabilization or frequency control over a large range ; one is still actual : wavelength (or frequency) measurement. Measurement and stabilization of the wavelength can be combined in one and the same device ; this combination has seldom been used, in spite of its obvious advantages.

With the extraordinary resolving power of laser spectroscopy, the usual wavelength standards become definitely inadequate regarding accuracy. Physicists have to rely upon very few absorption lines of CO_2, CH_4 and I_2, which have been carefully studied and measured in a few laboratories, with a checked accuracy of 10^9*. Transferring this accuracy - with more or less personal tolerance - to the wavelength of an adjustable laser becomes the own business of the user. For this purpose, some laboratories have developed refined and elegant devices. We will point out some features and possible pitfalls common to most of them.

Precision is governed by signal to noise ratio. We can expect various compromises between quickness of operation (reading and control) and precision of setting.

The above mentioned accuracy of 10^9 implies a precision $> 10^9$. It involves lengthy measurements, worth the effort when the target is an improvement of two or three orders of magnitude on the accuracy of the value of the speed of light.

For usual purposes, where a precision of say 10^6 is aimed at, a much quicker operation can be achieved. Time constants can be reduced to less than 1 s, and the same device can be used for measurement and for control.

Of course, the number of users tends to decrease as the reciprocal of the precision. We must admit that the time and money consumption is responsible for this correlation. To the reverse, the number of relevant applications in physics increases with the available precision and accuracy ; we can expect an increasing demand of precise and accurate measurement of wavelengths and devote some time to it.

Accuracy is smaller than precision, due to systematic errors.

---

* To characterize accuracy or precision, we use here the reciprocal of the corresponding uncertainty ; it is not usual, but it avoids some usual incoherences.

An estimate of these pertains to flair, helped by extended experience in physics.

Among the most common ingredients of inaccuracy regarding wavelength measurements we find alignment and diffraction.

The reference beam and the "unknown" beam ought to be perfectly superposed ; it is of course impossible if they have not the same wavelength. At least, they must be exactly aligned with respect to each other. Alignment defects introduce a bias by some $\cos\theta$ factor. To get an accuracy of 10^8, the tolerance on θ is of the order of 10^{-4} rd, a difficult goal with more or less gaussian light beams.

The diffraction correction can be theoretically expressed for a pure gaussian beam ; it introduces a factor $1-(\lambda/2\pi w_o)^2$ at the beam waist (radius w_o). How far this calculation does apply to real beams is open to question. However, we can use it as a landmark. For the accuracy of 10^8, $2w_o$ must be $> 10^4\ \lambda/\pi$, leading to $2w_o > 2$ mm or 30 mm for $\lambda = 0.63\ \mu m$ or $10\ \mu m$, respectively. Further, to avoid trouble from beam limitation, aperture diameters must be much larger (three times looks a minimum).

Increasing the beam (and aperture) diameters is likely to reduce the diffraction correction (and its uncertainty) to a negligible amount. But the alignment becomes the more difficult. A compromise must be found between the efforts devoted to increasing beam diameter and to improving alignment.

Optical defects (typically flatness defects) would usually cancel out if both beams were exactly superposed. Even perfectly aligned beams differ in terms of divergence, focussing and various imperfections of the wavefronts, enhanced by multiple reflections. Optical imperfections, combined with differences between the two beams, introduce a further bias ; motion or interchange of some optical parts interfere, without improvement in this respect.

Various combinations of cube-corners, cat's eyes and plane mirrors have been devised to eliminate part of the troubles arising mainly with movable-mirror interferometers : changing adjustment, alignment and feed-back to the laser. Each of these is a cause of random and systematic errors. To our knowledge, a combination that would suppress all of them at the same time is still to be found.

<u>Tests</u> of the accuracy are difficult, as usual. An elementary check can however be performed : compare the wavelengths of two beams derived from the same laser, or from two frequency-offset lasers. Note that this check is optimistic : all wavelength dependent effects will tend to cancel out. The idea can be extended : frequency doubling or "up-conversion" (preferably using sum and difference frequency) provide largely different wavelengths with an accurately known relation. Measuring such wavelengths would provide a direct and objective check of the accuracy of any wavelength measuring device. Such checks could definitely ascertain an accuracy, the estimation of which is currently based on questionable hypotheses.

DIGITAL WAVEMETER FOR CW LASERS

F.V. Kowalski, W. Demtröder, and A.L. Schawlow

Department of Physics, Stanford University
Stanford, CA 94305, USA

The wavelength of light can be measured by counting the number of fringes as an interferometer mirror is moved through a known distance. The known distance is best obtained by simultaneously counting the number of fringes from a standard wavelength source for the same change of optical path length. The arrangement in Fig. 1 permits rapid and precise comparison of wavelengths by moving a mirror smoothly over long distances. The interferometer is a two-beam, division of amplitude type related to the Michelson and Sagnac interferometers. In it, the standard and unknown beams travel identical paths in opposite directions. They emerge at separate detectors, so that there is no need for a dichroic beam splitter to separate them, and measurements can be made close to the wavelength of the standard if desired. Moreover, the visible spots of light on the mirrors can be brought into coincidence, ensuring that the two pathlengths are nearly identical, and simplifying the alignment of the unknown beam. The effects of vibrations are also minimized by the use of identical paths.

The corner cube reflectors are moved until the detector counting fringes of the unknown laser reaches a present number, say 10^6. A second counter monitoring the known laser's fringes is then switched off, and the ratio of the number of fringes is displayed. This number, apart from dispersion corrections [1], is the ratio of wavenumbers in air.

For more accurate work the fringe pattern of the standard laser is multiplied by a phase locked loop. With a signal-to-noise ratio of better than 100 to 1, multiplication factors of up to 100 can be expected.

The standard laser is a He-Ne laser stabilized to an Iodine^{129} component. The output power is around 50 μW and the detector is an amplified pin photodiode. We used corner cube reflectors either floated on an air track or pulled up an incline on bearings. For setting a laser quickly to 500 MHz, the air track and a simpler version of the interferometer [1] seem most efficient. Using bearings to move the mirrors reduced density fluctuations in the air, making more accurate absolute wavelength measurements possible. Also an equal path length, at the middle of the track, seems necessary if fractions of a fringe are to be counted. See Fig. 1.

Iodine lines at 5145Å were reproducibly measured to within one part in 10^8 but the absolute wavelength has yet to be verified to that accuracy.

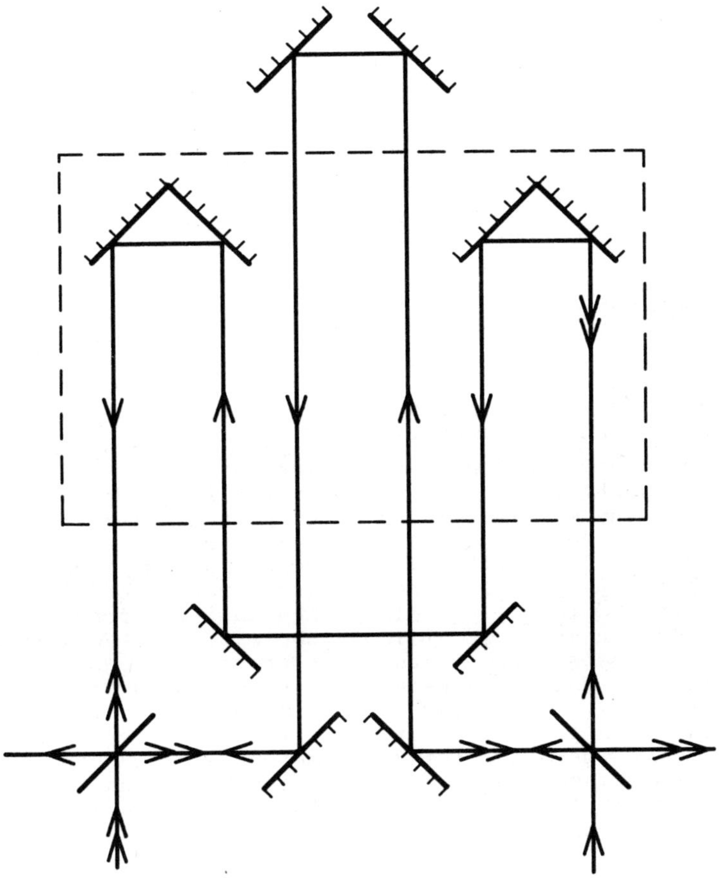

Figure 1

Reference

[1] F. V. Kowalski, R. T. Hawkins, and A. L. Schawlow, J. Opt. Soc. Am. 66, 965 (1976).

A WAVELENGTH METER

R.L. Byer, J. Paul, and M.D. Duncan

Applied Physics Department, Stanford University
Stanford, CA 94305, USA

Two years ago we set out to design and build a wavelength meter [1] that would complement our tunable source research.[2] The design goals were a device that measured wavelengths for both pulsed and cw lasers, displayed single pulse laser bandwidths, used inexpensive broadband optics and could be interfaced with a minicomputer.

The requirements led to a wavelength meter that uses a grating monochrometer followed by three Fabry-Perot etalons. Initially, the detector was a 256 element linear silicon diode array. However, recently a vidicon has replaced the diode array. The optical layout shown in Fig. 1 uses three Fabry-Perot etalons with free spectral ranges of 10 cm^{-1} 1 cm^{-1} and 0.1 cm^{-1}. Their fringe patterns are projected onto the diode array or vidicon tube with three observed fringe diameters. The measurement procedure consists of preselecting the desired wavelength with a grating monochromator to within $\pm$ 4 cm^{-1} . The monochromator transmitted wavelength is then projected onto the 10 cm^{-1} free spectra range etalon where the fringe diameters are measured and a fractional fringe order ε calculated by an analogue electronic circuit to $\pm$ .1%. Since the etalon spacing is known and the wavelength is known to within one order, m , the unknown wave number ν is measured to be $\nu(cm^{-1}) = [2(m + \varepsilon)/d(cm)]$. The procedure is then repeated for the higher resolution etalons since ν is now determined accurately enough to know m .

The Fabry-Perot etalons are fused silica spaced, optically contacted, vacuum enclosed and silver coated with a Finesse of 20 at 632.8 nm. The etalon resolutions using silicon diode or vidicon display is 20 GHz , 2 GHz and 200 MHz. With photographic readout, the fringe center can be determined to $\pm$ 15 MHz which is fully one order of magnitude better than with electronic readout. However, the vidicon display allows single shot fringe observation and a measurement rate of 10 times per second into the computer. In addition, the vidicon is sensitive enough to allow detection of incoherent light sources and with an infrared sensitive cathode allows the wavelength meter to be used out to 2 μm .

In conclusion we have constructed and measured the parameters of a wavelength meter that is useful over a .4 - 2 μm spectral range, gives absolute wavelength measurements to one part in 10^7 (± 15 MHz) and is useful for both pulsed and cw laser sources. The device does not have any moving parts and once properly calibrated with an iodine stabilized HeNe laser remains in calibration.

Finally, we have used the wavelength meter in a cw CARS experiment to measure the absolute Raman frequency of D_2 Q(2) to be $2987.237_1 \pm .001\ cm^{-1}$. The measurement demonstrate the application of the wavelength meter to precision spectroscopy.

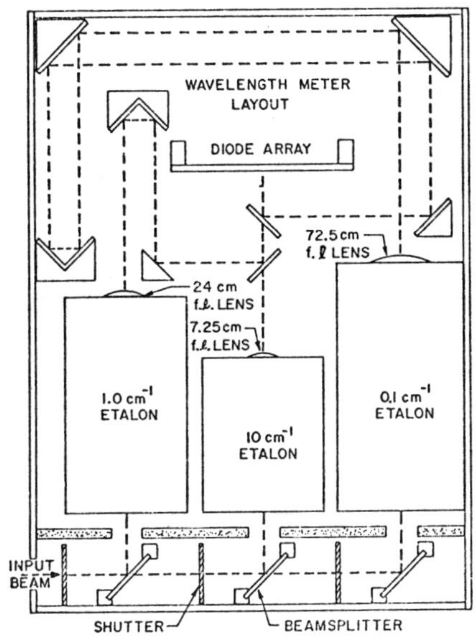

Fig. 1 Schematic of the wavelength meter optical layout showing the Fabry-Perot etalons and optical paths for imaging three fringes of nearby diameters onto the detector.

REFERENCES

1. J. Paul and R.L. Byer, "A Wavelength Meter", presented at the CLEOS Conference, May 1976, San Diego, California.

2. R.L. Byer, "Parametric Oscillators", published in Tunable Lasers and Applications, ed. by A. Mooridian, T. Jaeger, P. Stokseth, Springer-Verlag, Heidelberg, 1976.

3. M.A. Henesian, M.D. Duncan, R.L. Byer and A.D. May, "An Absolute Raman Frequency Measurement of the Q(2) Line in D_2 Using cw CARS", (to be published).

MOTIONLESS MICHELSON FOR HIGH PRECISION LASER FREQUENCY MEASUREMENTS: THE SIGMAMETER

P. Jacquinot, P. Juncar, and J. Pinard
Laboratoire Aimé Cotton, C.N.R.S. II, Bât. 505
91405 Orsay, France

This method permits to measure the frequency of a single mode tunable dye laser with an accuracy of a few MHz. The apparatus does not include any moving part, it gives a direct visualisation of the wavenumber and is able to work as well in a cw as in a pulsed laser regime. On the other hand, it can be used, as a reference, to servocontrol and pilot the frequency of the dye laser ; with an appropriate electronic system, it is able to scan the laser frequency step by step.

The basic element of the apparatus is a two channel Michelson interferometer shown in fig. 1 in a compact and achromatic version. As demons-

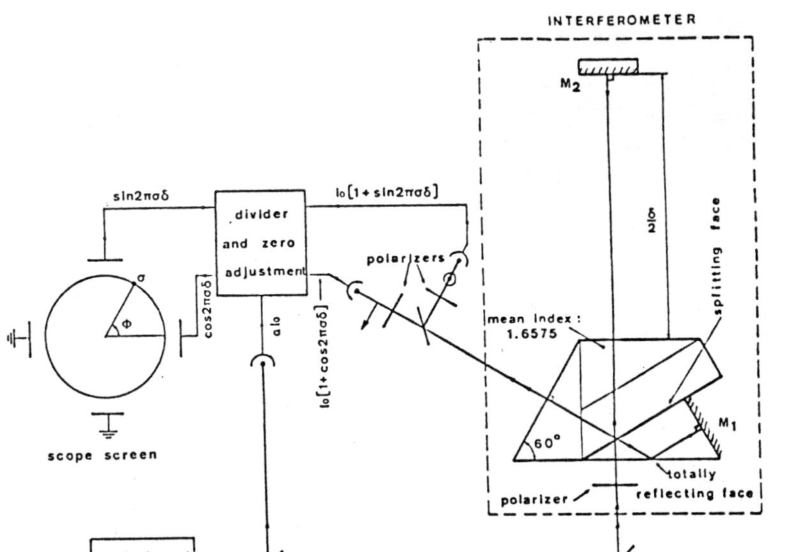

Figure 1

trated in a previous paper [1], we can extract from the output beam, using two polarizers, two signals expressed by :

$$\cos 2\pi\sigma\delta \qquad \text{and} \qquad \sin 2\pi\sigma\delta$$

where δ is the path difference of the interferometer and σ the wave-number of the laser light. Putting this two signals on the horizontal and vertical plates of a scope it is easy to determine σ modulo $\frac{1}{\delta}$. In order to overcome the indetermination $\frac{k}{\delta}$ one can use several interferometers of the same type having path differences in geometric ratios such as 100 cm, 10 cm, 1 cm, .1 cm. The wavenumber of the radiation can be deduced with an accuracy determined by the interferometer of the highest path difference that can be estimated to one hundredth of the circumference : $\frac{.01}{100}\ cm^{-1} \cong 3$ MHz with an indetermination given by the interferometer of the lowest path difference that is $\pm k \times 10\ cm^{-1}$ which can be overcome using a simple spectrometer.

All the path differences have to be determined with precision using at least 3 reference radiations, the ultimate precision being given by only one laser : a 6328 nm HeNe laser stabilized on iodine which is itself used to servocontrol the path difference of the interferometers.

Using an electronic system described in the reference we are able to servocontrol a dye laser and scan its frequency step by step , each step being a fractional integer of one circumference (.01cm^{-1}). The fig. 2

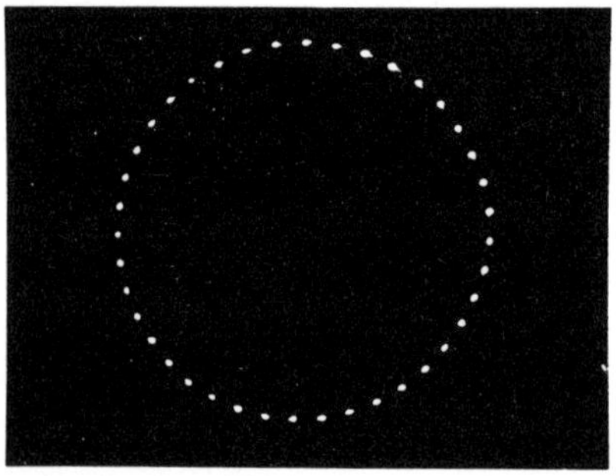

Figure 2

shows a photograph of the oscilloscope screen when the dye laser frequency is scanned step by step. Each step corresponds to $\frac{.10}{40}\ cm^{-1} \cong 7.5$ MHz .

[1] P. Juncar, J. Pinard, Opt. Commun. 14 (1975) 438.

FIZEAU WAVELENGTH METER

J.J. Snyder

Optical Physics Division, National Bureau of Standards
Washington, DC 20234, USA

Our Wavelength Meter is a self-contained instrument that measures the wavelength of radiation produced by lasers or other pulsed or cw sources of monochromatic light. The instrument is based on a Fizeau or "optical wedge" interferometer. The fringe pattern produced by the interferometer is digitized and stored in a small computer which converts the fringe pattern into the wavelength (in the desired units) of the interfering light.

The Fizeau interferometer divides the collimated input beam by reflection into two overlapping beams of equal intensity, and propagating nearly but not quite parallel to each other. The two beams generate the usual parallel, uniformly spaced Fizeau fringes. The fringe pattern is sampled periodically by an array of photodetectors, digitized, and loaded into the computer memory for processing. Because of the image storage capability of photodiode arrays, the laser source may be either pulsed or cw.

The computer program determines from the stored fringe pattern a) the spatial period of the pattern, and b) the location of a fringe intensity minimum. The known wedge angle (a calibration constant) allows the spatial period of the fringe pattern to be converted into an approximate wavelength. For proper operation, the uncertainty in this approximation to the unknown wavelength must be less than the free spectral range of the interferometer.

The next program step calculates the interferometer path difference at the fringe minimum determined in (b) above. This step utilizes both calibration constants: the wedge angle and the wedge spacing at a reference position.

Since the optical flats are uncoated, the path difference at a fringe minimum must contain an integer number of wavelengths of the interfering light. The computer calculates the quotient (path difference)/(approximate wavelength), yielding the interferometer order number plus a remainder. The necessary wavelength correction term is calculated from the ratio (remainder)/(order number). The resolution of the final wavelength value is limited only by the resolution of the measurement of the path difference at the fringe minimum. Moreover, since the order number of the interferometer is unambiguously determined from the fringe spatial period, the calculation of the wavelength has a unique solution. In effect, the free spectral range is infinite.

A working model of the wavelength meter has been constructed. The instrument consists of laser beam handling optics, a beam expanding collimator, a simple Fizeau interferometer, a self-scanned silicon photodiode array of 1024 elements, signal processing electronics, and a minicomputer to calculate and display the wavelength. This instrument has a wavelength range of $\sim$ 400 nm to ∿ 1.1 μm, and resolution of a few parts in 10^8. Figure 1 shows an optical schematic.

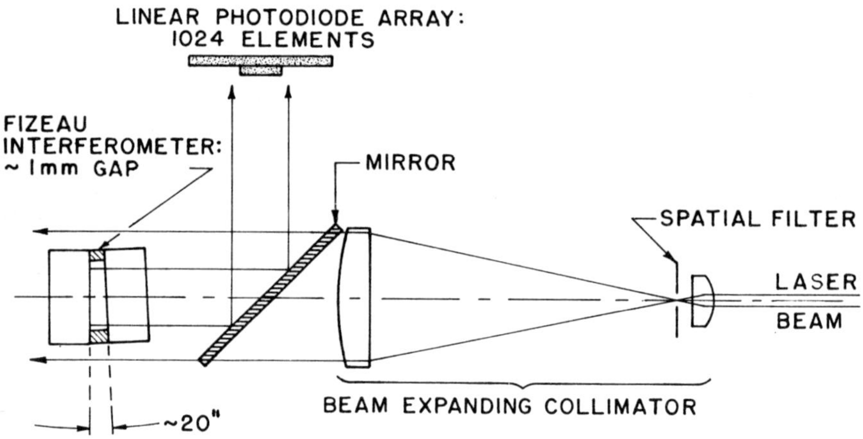

Fig. 1 Fizeau Wavelength Meter

The interferometer consists of a pair of uncoated optical flats separated by ∿ 1 mm and wedged ∿ 20". The separation is maintained by low expansion glass-ceramic spacers. The two interfering beams are produced by reflection from the inner surfaces. The interferometer is both compact and dimensionally very stable. In addition, there is no phase dispersion since the reflecting surfaces are uncoated.

Once the wavelength meter has been initially calibrated using a laser of known wavelength, its operation requires only that the input radiation be aligned to the optical system. The computer then calculates and displays the updated wavelength at ∿ 14 Hz.

Although the working model has been optimized for visible to near IR monochromatic laser sources, this restriction is not fundamental. The wavelength meter will work equally well for any source of radiation for which a similar interferometer is constructed, and for which there are means for digitizing the fringe pattern. For example, "image dissecting" photomultipliers or their equivalent will allow wavelength measurements of monochromatic non-laser sources. Photodetector arrays sensitive to other wavelength regions can be used to extend the wavelength meter's capabilities into either the UV or the far IR.

Among the advantages of this wavelength meter are 1) cw or pulsed source compatibility; 2) high speed; 3) completely static operation; 4) optical and mechanical simplicity; 5) high resolution; and 6) wide wavelength range.

A TRAVELING MICHELSON INTERFEROMETER WITH PHASE-LOCKED FRINGE INTERPOLATION

S.A. Lee and J.L. Hall*

Joint Institute for Laboratory Astrophysics
National Bureau of Standards and University of Colorado
Boulder, CO 80309, USA

The advent of single-mode cw dye lasers and their application to Doppler-free spectroscopy makes essential the capability of rapidly measuring the dye laser's wavelength with sub-Doppler absolute accuracy. Our wavelength measuring apparatus [1] is basically an automatic-scanning Michelson interferometer utilizing corner-cube retroreflectors, with phase multiplication for extending the resolution. The interferometer measures an unknown wavelength in terms of a reference laser wavelength. Motion of the carriage holding the corner-cubes lengthens one arm and shortens the other arm of this interferometer. For a given distance traveled by the carriage, a different number of fringes will be counted for the unknown laser and the reference laser (since their wavelengths are different). Naturally, a higher resolution level for the measured wavelength will require a longer travel, and consequently, a longer measuring time. Our resolution-extension concept, however, enables us to obtain wavelength information of a given resolution level in 100-fold less time than is required with direct fringe counting.

The fixed fringe rate generated by a uniform motion of the carriage allows one to use the technique of frequency metrology rather than distance metrology in the fringe interpolation algorithm. The resolution of each optical fringe into 100 distinct levels is obtained by phase-locking the $\nu/100$ digital output of an oscillator to the optical fringe rate. Digitally counting the phase-locked oscillator output, ν, provides the fringe interpolation. A basic optical fringe rate of ~250 kHz (corresponding to carriage velocity of 4 cm sec^{-1}), allows four independent wavelength measurements per second at 2×10^{-7} resolution level.

The velocity of the carriage needs to be reasonably uniform, so that the mechanically induced fringe-rate variations do not introduce significant phase errors between the optical fringe and its electrical counterpart. The velocity noise in our system is $\lesssim 6\%$ with its spectrum rapidly decreasing beyond ~100 Hz. This low-frequency noise introduces a peak phase error of ~1/100 fringe due to the fast settling time of the phase-lock (~100 μsec) and the appropriate phase servo-loop filter design.

In addition to resolution extension by counting a frequency that is 100 times the basic optical fringe rate, the phase-lock system also provides a useful improvement in signal-to-noise ratio over direct fringe counting: at

*Staff Member, Quantum Physics Division, National Bureau of Standards.

the 1/100 fringe resolution level, the short correlation time of the fluctuations in a cw jet dye laser would otherwise introduce a serious precision limitation. [Although amplitude stabilization of the dye laser can lead to good amplitude stability at low frequencies, it is increasingly difficult to provide control at higher frequencies (beyond ~50 kHz).] By contrast, the phase-locked system averages the jitter of a number of successive fringe zero crossings and provides a better fringe interpolation algorithm: effectively the noise bandwidth is reduced below the basic fringe rate.

We now employ a quadrilateral interferometer with an additional folding mirror added to one end of the triangular configuration originally described. This system has some interesting properties: the left/right inversion induced by the additional reflection makes the reference and test laser beams coaxial and antiparallel approaching the retroreflectors. Thus temporary removal of the carriage allows the reference beam to trace out the test beam's path, but direction-reversed. The externally available beam, thus created, greatly facilitates precise alignment of a test laser beam, especially if it is infrared or coming from a distant source. Other improvements over the published version include the use of metallic support rails and an attempt to place the center of mass of the carriage in the plane of support.

Three systematic effects influence the present apparatus: first, slight misalignment of the optical beams from parallelism with the direction of translation gives a cosine error which we estimate to be $\leq 2 \times 10^{-7}$ in $\delta\lambda/\lambda$. With the aid of a wedge-free plane mirror in the alignment procedure, we should eliminate this problem to the 10^{-8} level. The diffraction correction, $\delta\lambda/\lambda = 0.06(\lambda/w_0)^2$, is $<10^{-9}$ for our $w_0 = 3$ mm beam and $\lambda \sim 6000$ Å. To avoid the problem of large corrections due to the dispersion of air, preparations are underway to enclose the apparatus in a He-filled environment. The total dispersion correction should then be $\sim 1 \times 10^{-8}$ even at atmospheric pressure.

In summary, we have built a scanning corner-cube interferometer with phase-locked resolution extender that provides a real-time wavelength display. The current instrument has a demonstrated absolute accuracy of $\sim 2 \times 10^{-7}$, a resolution capability of $\sim 2 \times 10^{-9}$, and an accuracy capability that we judge to be $\sim 10^{-8}$.

This work was supported in part by National Science Foundation grant no. PHY76-04761 through the University of Colorado.

Reference

1. J. L. Hall and S. A. Lee, Appl. Phys. Lett. 29, 367 (1976).

A SELF-CALIBRATING GRATING

T.W. Hänsch

Department of Physics, Stanford University
Stanford, CA 94305, USA

The wavelength of a single-mode cw laser can be accurately measured with the help of a fringe-counting traveling Michelson interferometer, but unfortunately this method cannot easily be extended to pulsed lasers, such as the widely used nitrogen-pumped dye laser systems. Today, most laboratories seem to employ conventional grating spectrographs to measure the wavelength of pulsed lasers. The resolution of such instruments can actually be quite sufficient in view of the typical larger bandwidth of pulsed lasers. But because of mechanical imperfections of the grating drive, the available digital wavelength readout can only be used for rough estimates, and any absolute wavelength measurement with the help of spectral lamps and wavelength catalogs is at best tedious.

This paper proposes a novel scheme for the absolute calibration of a grating spectrograph. A reference laser of known wavelength, such as a He-Ne laser, can be used together with a diffraction grating with special multiple rulings to superimpose an accurate, easy to read, ruler-like absolute wavelength scale directly onto the unknown spectrum. Each calibration line is in effect a "ghost-line," generated by deliberate irregularities in the grating.

One possible implementation of this scheme is illustrated in Fig. 1. The grating in the otherwise conventional spectrograph is ruled in a special fashion: by controlling the lifting mechanism of the ruling diamond stylus, every tenth groove, say, of the main grating is extended somewhat into a polished margin on the substrate, thus producing an auxiliary grating with a ten times larger grating constant. Likewise, every 100th groove is extended still further to produce a grating with 100 times larger spacing, and additional extensions are ruled every 1000 and perhaps every 10 000 grooves. The appearance of the grooves is thus in effect similar to the calibration lines of a decimal ruler.

Light from a reference laser enters the spectrograph through the entrance slit, and is diffracted by the system of auxiliary gratings. It produces a regular pattern in the image plane, corresponding to the Fourier-transform of a decimal ruler, which has again the appearance of a decimal ruler: The auxiliary grating with the finest ruling produces the coarsest calibration lines and vice versa. This ruler-like pattern appears superimposed upon any regular spectrum, such as that of a laser line of unknown wavelength, and the spacing of the calibration lines is equidistant on a wavelength scale. The unknown wavelength can thus be read out very conveniently in a visual, photographic or scanning instrument.

Since the calibration scale is, in effect, an integral part of the diffraction grating itself, the accuracy of a wavelength measurement is not affected by such factors as backlash of the mechanical grating drive or thermal expansion. The scheme can be easily implemented with existing grating spectrographs, and the manufacturing of the special grating by replication should not be much more expensive than that of a conventional grating, once a master is available.

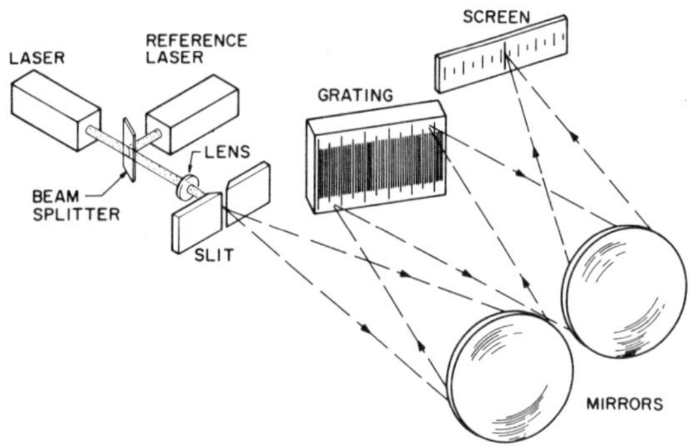

Fig. 1 Spectrograph with self-calibrating grating

Unfortunately, existing ruling engines are not easily modified for the proposed special ruling. But one can produce similar, though somewhat less accurate, marker lines with a mosaic of auxiliary gratings which are ruled separately on the same blank. By choosing the proper groove spacing for a given reference laser, one can then produce a wavelength scale which is directly calibrated in Angstroms or nanometers. Very encouraging results have been obtained with an experimental grating, ruled in this way by REINHARD ERDMAN of PTR OPTICS CORP., which produces calibration lines with 2 Angstrom intervals. By simple visual readout, using a magnifying glass with reticle, it has been possible to measure line positions easily to within 1/100 Angstroms. This is sufficient to set the wavelength of a tunable dye laser to within a typical visible Doppler line width.

The suggested scheme allows a number of interesting modifications. The auxiliary gratings could be directly superimposed on the main grating, for instance by skipping lines at predetermined intervals or by ruling them with different width or phase. The calibration lines would then be true ghost lines, and this approach would minimize the errors introduced by imperfect imaging and collimating optics. It should also be possible to manufacture such gratings holographically, by superimposing the hologram of a wavelength scale onto a regular holographic grating. This scheme would permit the addition of labels and numbers to the scale, thus facilitating visual or photographic readout.

The proposed self-calibrating grating clearly cannot improve the resolution of a grating spectrograph. But it can provide a simple and relatively foolproof way for absolute wavelength measurements. Unlike some other wavelength meters, it is not restricted to a single-line laser, but works equally well with multiple lines, incoherent sources, or absorption lines in a broadband spectrum.

A SIMPLE MOVING-CARRIAGE INTERFEROMETER FOR 1 IN 10^7 WAVELENGTH INTERCOMPARISON, AND A SERVOCONTROLLED FABRY-PEROT SYSTEM FOR 3 IN 10^{11} ACCURACY

W.R.C. Rowley, K.C. Shotton, and P.T. Woods

Division of Quantum Metrology, National Physical Laboratory
Teddington, Middlesex, U.K.

We have developed two forms of interferometer to intercompare the wavelengths of laser sources. The first is a simple, Michelson-type moving-carriage interferometer to enable the wavelengths of continuous-wave visible lasers to be measured rapidly against a standard laser wavelength, with an accuracy of about 1 part in 10^7. The second is a complex, Fabry-Perot interferometer which gives visible wavelength intercomparisons accurate to a few parts in

Several forms of Michelson interferometer, operating in fringe counting mode, have been developed to measure the wavelengths of dye lasers [1,2]. The device reported comprises a cube-corner interferometer assembled with a minimum of engineering, and standard laboratory instruments for the signal-processing electronics. The basic interferometer, comprising fixed and moving cube-corner reflectors and a beam splitter, is mounted on a 1-metre triangular optical bench with the moving cube attached to a standard saddle and slid along by hand.

The wavelength standard used for this instrument is a visible He-Ne laser, adapted from a commercial internal mirror device, and uses a stabilization loop based upon a heater wound around the discharge tube [3]. The beams from the two lasers are roughly collimated and directed into the interferometer by adjustable mirrors to form parallel beams a few centimetres apart. The beam splitter is adjusted by hand and then rigidly clamped so that the beams from both cube corners overlap. In practice, it was readily possible to attain a visibility of 90%. Also by visually observing the separation between the two beam centres over a 4 metre path, a parallelism of better than 3 mrad was obtained, resulting in a systematic error of wavelength measurement of less than 1 part in 10^7. A further potential source of systematic error arises from the finite size of the laser beams within the instrument. Calculations of the diffraction correction indicate that for the beam sizes used ($\omega_0 \geqslant 1$ mm), the error due to this is less than 1 part in 10^8.

The two output beams are focussed onto PIN photodiodes, coupled to preamplifiers (frequency response 100 Hz - 100 MHz), with outputs compatible with the main electronic counting system. A frequency multiplier was inserted into one of these channels, and served two functions. It enabled both the positive and negative going edges of the fringe signal to trigger the counter, giving twice the counting accuracy, and it made the counter display more digits than necessary so as to avoid rounding errors. The frequency response of this device limited the maximum scan speed of the cube corner to about 300 mm s^{-1}. The main electronic system comprises a programmable frequency counter (Hewlett-Packard HP5345A) with two inputs, and a desk calculator (Hewlett-Packard HP9815A) which controls the counter and performs the simple analyses.

The counter monitors the fringe frequency from one photodiode and when this exceeds a predetermined value, the system changes to a frequency ratio mode for a fixed number of counts (corresponding to 0.8m of travel). The ratio is transferred to the calculator and stored. Several scans may be made, and the system computes the mean and its standard error, and applies wavelength and refractive index corrections [4], within a time of 1 s. The observed standard deviation of a single scan corresponds to 1 part in 10^7 of wavelength uncertainty.

More accurate absolute, visible and infrared wavelength measurements have been made using a plane-parallel servocontrolled Fabry-Perot interferometer [5]. We have recently measured, with this device, the wavelength ratio of visible radiation at about 679 nm, produced by up-converting radiation from a stabilized CO_2 laser, to that of a visible iodine-stabilized laser, with an accuracy of 3 parts in 10^{11}. This measurement, coupled with the previously measured [6] frequency of the CO_2 laser yields an effective visible frequency of the iodine-stabilized laser accurate to 6 parts in 10^{10}. Equal contributions to this uncertainty arise from the ratio and frequency measurements, thus indicating that wavelength measurements of comparable accuracy to present absolute frequency measurements are practical for optical lasers.

References

1. F.V. Kowalski, R.T. Hawkins and A.L. Schawlow: J. Opt. Soc. Am. 66, 965 (1976).

2. J.L. Hall, S.A. Lee: Appl. Phys. Lett. 29, 367 (1976).

3. S.J. Bennett, R.E. Ward, D.C. Wilson: Appl. Opt. 12, 1406 (1973).

4. B. Edlen: Metrologia 2, 12 (1966).

5. P.T. Woods, K.C. Shotton, W.R.C. Rowley: Appl. Opt. to be published.

6. T.G. Blaney, C.C. Bradley, G.J. Edwards, D.J. Knight, P.T. Woods, B.W. Jolliffe: Nature 244, 504 (1973).

   T.G. Blaney, C.C. Bradley, G.J. Edwards, B.W. Jolliffe, D.J.E. Knight, W.R.C. Rowley, K.C. Shotton, P.T. Woods: Proc. Roy. Soc. London A355 61 (1977).

X. Postdeadline Papers

TUNABLE LASER INFRARED PHOTOCHEMISTRY AT CRYOGENIC TEMPERATURES

J.K. Burdett, P.G. Buckley, B. Davies, J.H. Carpenter, A. McNeish, M. Poliakoff, J.J. Turner, D.H. Whiffen, R.L. Allwood, and J.D. Muse
School of Chemistry, University of Newcastle, Newcastle-upon-Tyne, U.K.

S.D. Smith, H. MacKenzie, and T. Scragg
Department of Physics, Heriot-Watt University, Edinburgh, U.K.

We report the first use of the spin-flip laser (SFL) for infrared photochemistry. The tuning capability combined with high intensity (up to 1W c.w. in a single mode) has been used to promote chemical reactions of molecules held in matrix isolation; these laser induced processes display not only isotopic and stereochemical selectivity but also selectivity between molecules in different sites in the matrix.

Following extensive development of the SFL in a joint programme at Heriot-Watt and Newcastle Universities, the device, supplied by Edinburgh Instruments Ltd., is now in routine use for photochemistry and molecular spectroscopy in the School of Chemistry at Newcastle.

In Edinburgh, the emphasis has been on removal of coupling problems (more common in tunable laser work than is generally realised), and the linearisation and removal of mode jumps in the tuning characteristics of the SFL. The latter aims have been achieved by a combination of minimising the spontaneous gain linewidth (Γ_s) and maximising the cavity linewidth (Γ_c). Plane wave theory gives the mode jump in frequency as $\Delta\nu = \Delta\nu_c \Gamma_s/(\Gamma_s + \Gamma_c)$. Normal SFL operation often has mode jumps of around 500MHz, in agreement with theory. Using n-type InSb with $N = 3 \times 10^{14}$ cm^{-3} and anti-reflected cavities with $R < 0.02$, we have observed, and used, continuous tuning which is both linear and without mode jumps to better than 30MHz, i.e. within the Doppler width of a typical molecular line.

The important photochemical result that we report is the use of the SFL to induce isotope, structure and site-sensitive chemical reactions in iron carbonyl compounds trapped in inert solids at 20K (the so-called matrix isolation technique) [1]. Many of the IR absorption bands of isotopically enriched $Fe(CO)_4$ species are not coincident with molecular gas laser lines. Where necessary we have been able to promote reactions using the tunable output of the SFL to obtain coincidence.

The highly reactive species $Fe(CO)_4$ can be generated and maintained indefinitely by UV photolysis of the matrix-isolated stable molecule $Fe(CO)_5$. Irradiation with c.w. IR laser radiation into one (or more) of the very intense CO-stretching bands of the matrix isolated $Fe(CO)_4$, in the presence of appropriate substrates, promotes the reactions

$$Fe(CO)_4 + CO \xrightarrow{\text{laser}} Fe(CO)_5$$

$$Fe(CO)_4 + Xe \longrightarrow XeFe(CO)_4$$

$$Fe(CO)_4 + CH_4 \longrightarrow CH_4Fe(CO)_4$$

The last of these is <u>not</u> formed by heating, but in general we must prove that the laser is not just a more efficient heating agent. This is done by isotopic substitution of ^{13}CO. Selective excitation is then observed which proves that the process is laser-specific. This is shown in Fig.1 where irradiation in either bands a or b (of $Fe(CO)_4$) independently causes the corresponding product bands A and B (of $CH_4Fe(CO)_4$) to grow. After detailed analysis we see that isotopic selectivity is also accompanied by stereochemical selectivity.

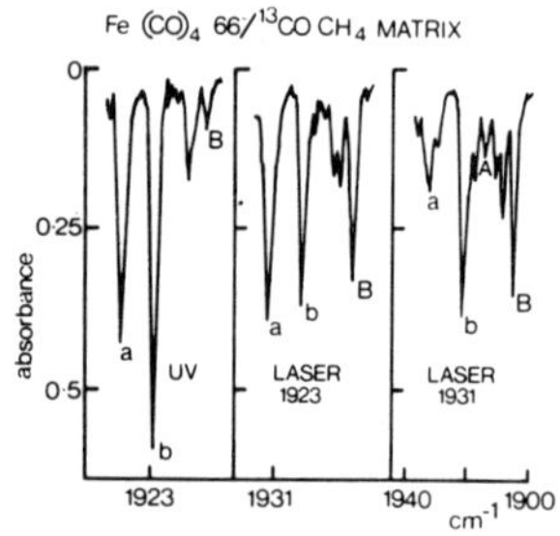

Fig.1 Selective excitation by laser

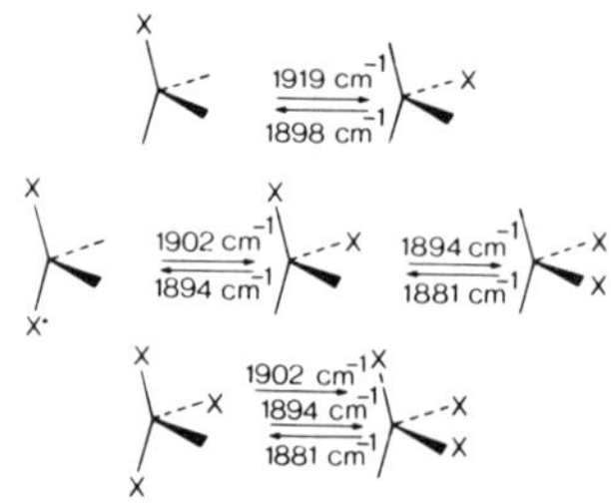

Fig.2 Laser induced isomerizations observed

The laser induced reaction of $Fe(CO)_4$ in argon shows intramolecular non-dissociative isomerization. The processes are depicted in Fig.2 and the stereochemistry can be followed. This is the first intramolecular ligand exchange to be characterized other than by dynamic N.M.R.

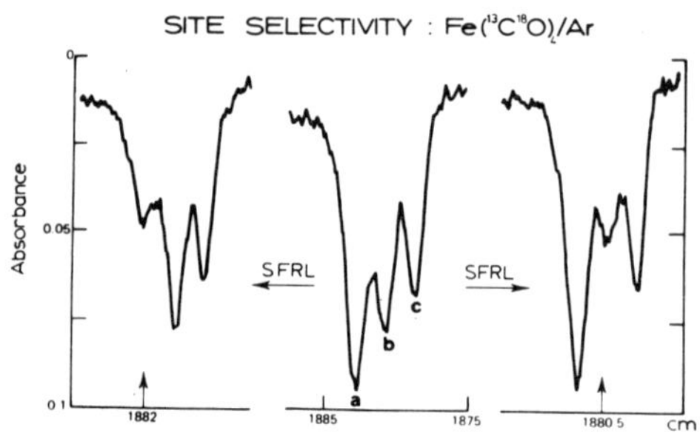

Fig.3 Site selectivity in matrix isolated $Fe(CO)_4$

Non-degenerate vibrational modes of $Fe(CO)_4$ give rise to several different bands in the matrix. These have been attributed to $Fe(CO)_4$ molecules in different substitutional sites as shown by the bands (centre) of Fig.3. Tuning the SFL (∿10mW power) to band a, reduces band a leaving b and c unchanged. Similarly tuning to band b leaves a and c unchanged. This indicates that site selective IR photochemistry is occurring.

Reaction yields are found to be proportional to total energy input and the reactions reported are therefore examples of 1-photon IR photochemistry at laser intensities dramatically smaller than those required for multiple-photon dissociations. The highly selective nature of this first SFL-induced photochemical process opens up many new possibilities.

Reference

1. M.Poliakoff, B.Davies, A.McNeish and J.J.Turner, Conf. on 'Lasers in Chemistry', Royal Institution, 1977.

MEASUREMENT OF A DIAMAGNETIC SHIFT IN ATOMIC HYPERFINE STRUCTURE

N.P. Economou, S.J. Lipson, and D.J. Larson

Lyman Laboratory of Physics, Harvard University
Cambridge, MA 02138, USA

We have measured a diamagnetic shift in the ground state dipole hyperfine structure of ^{85}Rb [1]. This is the first observation of such an effect in any atomic system. The shift appears as a departure, quadratic in the external field, of the level spacing from that predicted by the Breit-Rabi formula.

A high field optical pumping experiment was performed to measure the hyperfine splitting at fields of up to 75 kG. The pumping was done using a single mode cw dye laser tuned to the D_2 transition of rubidium at 780 nm. The tunability of the laser makes accessible the high field absorption lines which are displaced by as much as 0.3 nm from their zero field location. The narrow bandwidth allows frequency selective depopulation pumping of individual nuclear Zeeman levels which are spaced by 450-550 MHz.

Radiofrequency linewidths on the order of 20 Hz have been achieved for the $\Delta m_I = \pm 1$ transitions by confining the atoms in paraffin coated storage cells 1 cm in diameter. The centers of several of these transitions are determined at each magnetic field by monitoring the absorption of the light by the Rb vapor as a function of microwave frequency. The effects of laser amplitude noise are reduced by using nearly crossed linear polarizers before

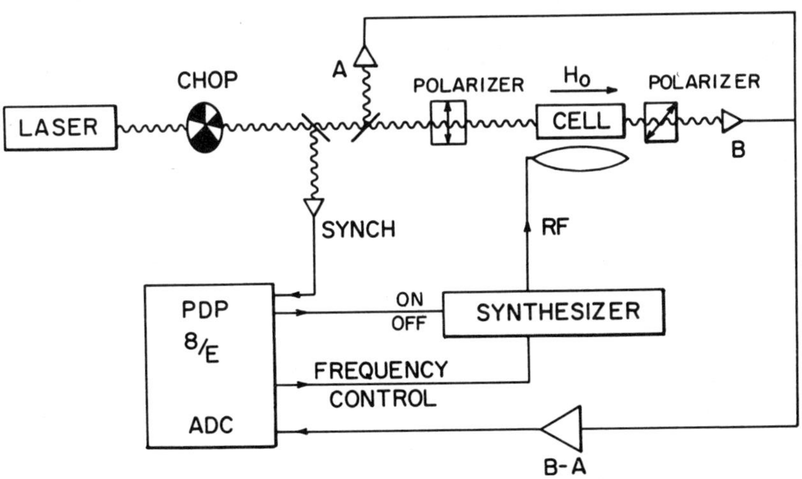

Fig. 1 Schematic diagram of the experimental apparatus

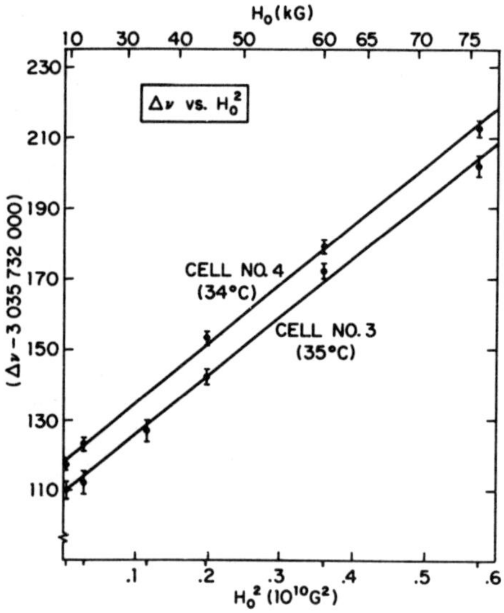

Fig. 2 Dipole hyperfine frequency measured in two sample cells at several magnetic fields. The diamagnetic shift is clearly exhibited.

and after the rubidium cell [1]. The light is chopped and the rf transition driven in the dark to eliminate light shifts (see Fig. 1). The rf is turned on every other time the light is off. The transmitted light level is recorded each time the pumping light is on. The signal is the difference between the levels when the rf has been on and when it has been off between light pulses. The experimental points are fit to a power series expansion about the line center since the chopped light and crossed polarizers make the real lineshape a complicated function of experimental parameters.

The hyperfine parameters are obtained at each field by fitting several of the nuclear Zeeman transition frequencies to a modified Breit-Rabi formula. The nuclear to electronic g-factor ratio, g_I/g_J, has been determined to be $-1.466\ 490\ 93(11) \times 10^{-4}$, independent of magnetic field. The data on the hyperfine frequency as a function of the square of the magnetic field from two sample cells are shown in Fig. 2. The data points have been fit to straight lines. The lines are displaced from one another due to a difference in the shift in the hyperfine frequency induced by collisions of the atoms with the coated walls. The average slope of the lines is $1.65(16) \times 10^{-8}$ Hz/G^2. This corresponds to a shift of 93 Hz at 75 kG out of a total hyperfine frequency of 3.035 GHz. We have increased the uncertainty in this result to allow for the possibility of the wall shift being dependent upon magnetic field. The final result is in agreement with a theoretical estimate of the diamagnetic shift [1].

Reference

1. N. P. Economou, S. J. Lipson and D. J. Larson, Phys. Rev. Lett. 38, 1394 (1977).

LIGHT-ASSISTED COLLISIONAL ENERGY TRANSFER

Ph. Cahuzac and P.E. Toschek

Institut für Angewandte Physik, Universität Heidelberg
6900 Heidelberg, FRG

We have observed collisional energy transfer from europium atoms in the $y\,^8P_{9/2,\,7/2,\,5/2}$ states to the $5p^2\ ^1D_2$ level of strontium, when light at λ_2 = 657 nm, 651 nm, and 644 nm provides for the energy difference (s.Fig.1). Processes of this type have been discussed by GUDZENKO and YAKOVLENKO |1|; recently HARRIS, LIDOW, and collaborators |2| have given evidence for light-induced inelastic collisions between Ca and Sr.

The reaction studied in the present experiment is

$$Eu^{*} + Sr + h\omega \rightarrow Eu + Sr^{**}.$$

A favorable feature of this system is the proximity of the intermediate level Sr 5p $^1P_1$ with the energy-storing europium levels, one of which is above and two of which are below the intermediate level.

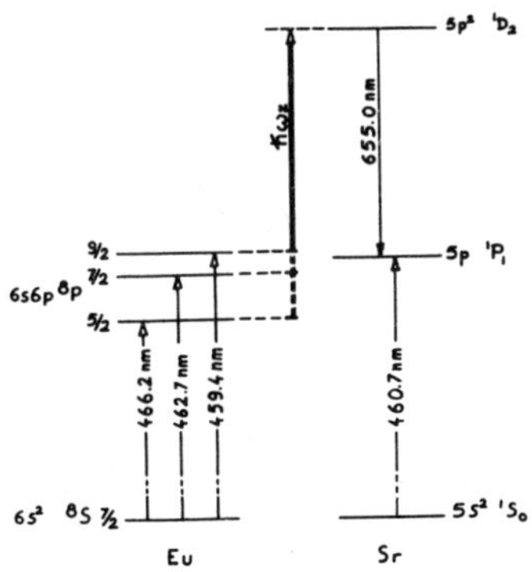

Fig. 1

The experimental setup consists of a heat pipe oven at a temperature of approximately 1200 K, which contains a mixture of Eu and Sr vapor. A nitrogen laser excites two dye lasers which include gratings and internal telescopes ($\tau_p \approx 2$ nsec): 1. a coumarine laser whose blue output selectively pumps one of the europium P levels, and 2. a continuously tunable cresyl violet/rhodamin 6G laser which is tuned across the spectral range of the energy difference of the excited Eu level and the Sr $^1D_2$ level. Transfer to the $^1D_2$ level is detected by its fluorescent decay at 655 nm.

The spectrum of the light-assisted collision process is separated from the dominating 2-photon excitation of Sr by either detuning the blue excitation light from its respective resonance, or by the application of a time delay between the two light pulses. With a delay of the order of the pulse length, both signals are shown in Fig. 2 vs. the wavelength of the inducing light, λ_2, for discrete values of the excitation wavelength, λ_1. With a slightly longer delay, the peak signals of the light-assisted collision process are shown vs. λ_1 in Fig. 3. The variation of the signal reflects the change in the Eu excitation efficiency. - With delays longer than the pulse lengths ($\Delta t \simeq$ 3.5 nsec), we observed tuning spectra (signal vs. λ_2, s. Figs. 4 and 5), where the 2-photon line is absent. This is demonstrated in Fig. 5 by detuning of λ_1, which would permit the two contributions to spectrally separate. The spectral asymmetry of the line shape is inverted for a Eu level below the intermediate Sr level as compared with excitation of a Eu level above it.

For the Eu $^8P_{9/2}$ - Sr $^1D_2$ energy transfer, the cross section derived from a measurement of fluorescence yields is ~ 2 Å^2 at c. 100 kW/cm^2 flux of the inducing light, λ_2.

References

|1| L.K. Gudzenko and S.I. Yakovlenko, Zh. Eksp. Teor. Fiz. 62, 1686 (1972) Sov. Phys. JETP 35, 877 (1972)

|2| S.E.Harris, R.W. Falcone, W.R. Green, D.B. Lidow, J.C. White, and J.F. Young, "Tunable Lasers and Applications", Ed. A. Mooradian, T. Jaeger, and P. Stokseth, Springer Series in Optical Sciences, Springer-Verlag Berlin, Heidelberg, New York, 1976

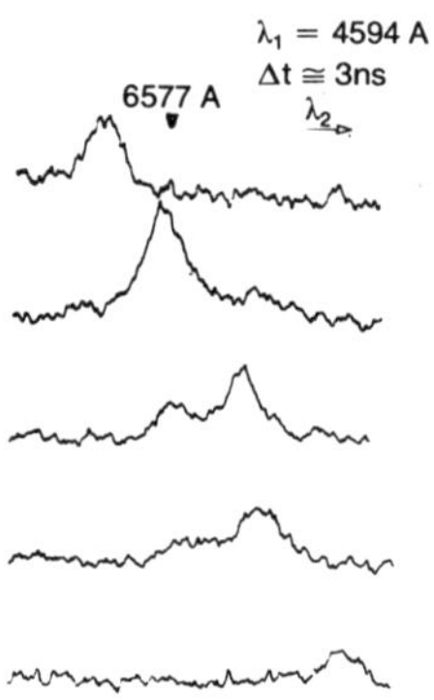

Fig.2

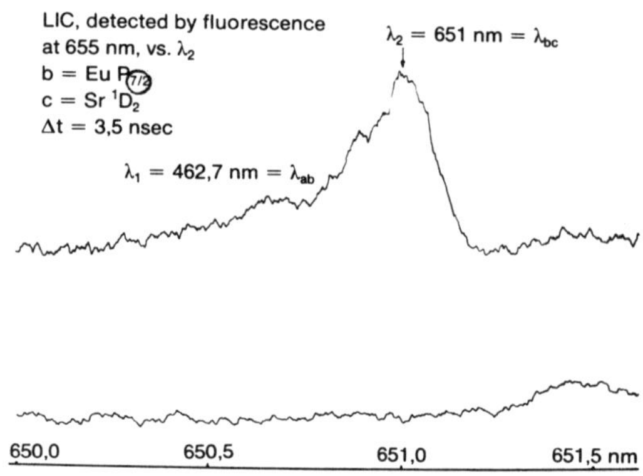

Fig. 4

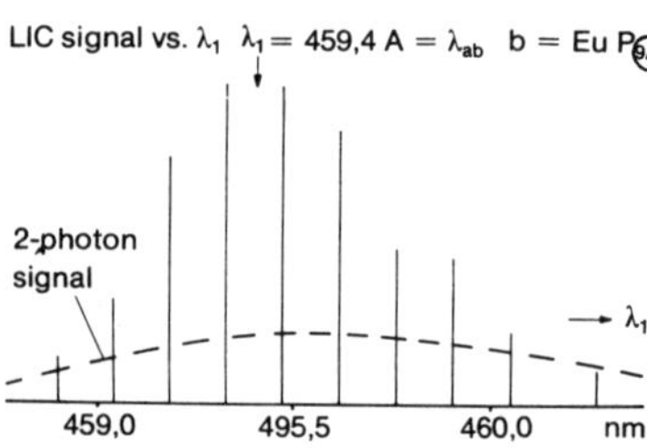

Fig. 3

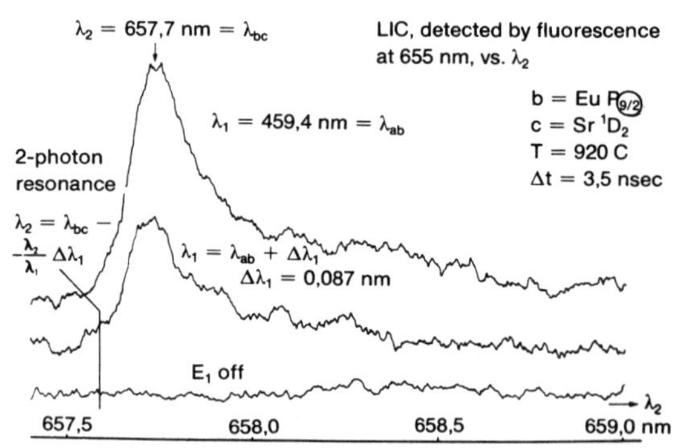

Fig. 5

TWO-PHOTON OPTICAL FREE-INDUCTION DECAY IN ATOMIC SODIUM VAPOR

P.F. Liao, J.E. Bjorkholm, and J.P. Gordon

Bell Telephone Laboratories
Holmdel, NJ 07733, USA

We describe the first use of the Stark-switching technique [1] to observe two-photon optical free-induction decay (FID). Our measurements were made with visible dye lasers and sodium vapor. The transients occurred on a nanosecond time scale and were obtained with both a resonant and a nonresonant intermediate state. In the latter case we observed simple FID signals which were in good agreement with theory. The case of a resonant intermediate state produced different behavior. This situation has not been previously studied experimentally and a complete theory has not yet been developed.

The outputs of two single mode cw dye lasers were focused and passed in opposite directions between Stark electrodes ($\sim$ 2 mm spacing) in a 2.5 cm long cell containing sodium vapor at a density of about 10^{12} cm^{-3}. One laser was tuned such that it was resonant or nearly resonant with the $3S_{1/2}$ ground state $3P_{1/2}$ intermediate state transition (590 nm) while the other (568 nm) was tuned such that the sum frequency was approximately centered on the $3S(F=2) \rightarrow 4D_{3/2}(m=\pm 3/2)$ two photon transition. A dc Stark field of 25 kV/cm was applied so that the transition could be selectively excited. The transmission of the 590 nm laser beam was monitored with a fast photodiode to observe the optical transients.

The FID signals were generated by applying an additional voltage in the form of a step voltage to the Stark plates. A step of 160 volts caused the $4D_{3/2}(m=\pm 3/2)$ levels to suddenly shift by 120 MHz. The shifts of the $3P_{1/2}$ and $3S_{1/2}$ states which were directly a result of the voltage were negligible. The FID signal appeared as a beat note whose frequency was given by the shift of the $4D_{3/2}(m=\pm 3/2)$ states. This beat was the result of interference between the 590 nm laser radiation and Stark shifted radiation from the two-photon coherently prepared atoms.

In the case of a nonresonant intermediate state the two-photon transition can be reduced to a two level problem [2] and hence these two-photon coherent transients will be exactly analogous to those observed in single photon transitions. We have obtained a complete and closed form expression which includes all Doppler effects for the free decay signal, and the observed signals are in good agreement with this theoretical expression.

The case of a resonant intermediate state is more complex since the phenomena can be complicated by single photon transients. At low intensities the FID signal is essentially identical to the nonresonant intermediate state situation. At higher intensities the phase of the beat note shifts and another signal component which has been tentatively ascribed to a single-photon optical nutation appears. This signal can be suppressed by increasing the intensity of the laser at 590 nm.

In summary, two-photon optical free-induction decay has been observed in atomic sodium vapor with the Stark switching technique. By using a resonant or nearly resonant intermediate state to greatly enhance the two-photon transition, large signals are easily obtained. However, more theoretical study is required to completely understand the phenomena observed in the case of a resonant intermediate state.

REFERENCES

1. R. G. Brewer and R. L. Shoemaker, Phys. Rev. Lett. 27, 631 (1971).

2. D. Grischkowsky, M. M. T. Loy and P. F. Liao, Phys. Rev. A 12, 2514 (1975).

DOPPLER-FREE COHERENT TWO-PHOTON TRANSIENTS BY STARK SWITCHING

M.M.T. Loy

IBM Thomas J. Watson Research Center
Yorktown Heights, NY 10598, USA

The study of transient coherent effects has long been most fruitful in yielding precise and detailed information on the relaxation times of the physical systems since the early work in magnetic resonance. The advent of laser sources made it possible to observe analogs of these effects in optical transitions. While precise relaxation time measurements were initially hampered by the difficulties in controlling pulsed laser outputs, the introduction of the Stark switching and subsequent frequency switching techniques substantially bypassed these difficulties [1]. Recently, there has been much interest in coherent effects in two-photon transitions, and the use of these effects to study two-photon relaxation times should be very exciting especially since the experiment can be performed under near Doppler-free condition. Here we report on the application of the Stark switching technique to the two-photon problem and with it, the first measurement of the collision-induced relaxation time of a two-photon transition in NH_3.

The experimental configuration consists of two counter-propagating beams interacting with a two-photon transition that is near resonant to the sum of the laser frequencies. Application of a Stark voltage brings the two-photon transition through the sum frequency of the lasers. A coherent two-photon polarization is generated, and, in terms of the two-photon vector model [2],[3], the polarization vector precesses about the effective field. At low laser intensities, this characteristic precession frequency is simply the time-dependent frequency offset between the Stark-shifted transition frequency and the laser sum frequency. The precessing two-photon polarization, depending on the relative phases, induces alternating absorption and emission of light. The decay of the precession signal is a direct measurement of the dephasing time of the coherent two-photon polarization. The experiment is closely related to that first suggested in the theoretical paper of BREWER and HAHN [2] and also the recent observation of LOY [4] where the transition was shifted via the optical Stark effect.

The NH_3 transition used was $(v_2,J,K,M)=(0^-,5,4,\pm5)\rightarrow(2^-,5,4,\pm5)$ with the uniquely defined intermediate state $(1^+,5,4,\pm5)$. This two-photon transition is 294 MHz away from the sum of the 10μm CO_2 P18 and P34 lines, and the Stark field required to shift on resonance is 5253 volts/cm. The 10 cm long Stark plates were separated by precision 2 mm spacers, to ensure Stark field homogeneity. Tne weaker (1 kW/cm^2) of the two input beams, at the P34 line, had a pulse duration of 100μs and can be considered cw. The strong laser beam, at the P18 line, was from a transverse-excitation atmospheric (TEA) CO_2 laser operating in single-longitudinal mode. For this experiment, the laser was adjusted to give an output pulse of 1μs duration and 100 kW/cm^2 intensity. Since the time scale of the experiment was about 100 nsec, this beam could also be considered cw and the time-dependent frequency shift was effectively controlled only by the external electrical pulse. At this intensity, the optical Stark

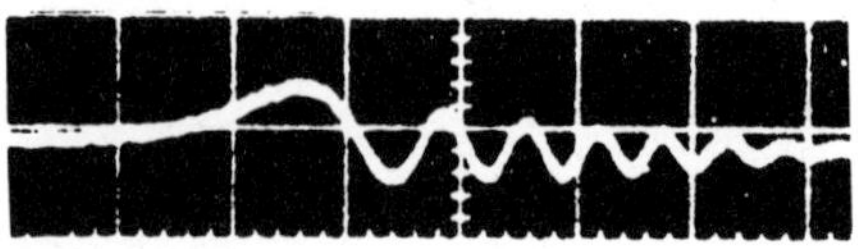

Fig. 1 Two-photon coherent precession signal, at the time scale of 10 ns/div. The NH_3 pressure was 200 mTorr. The Stark pulse used had an amplitude of 1200 volts with an exponential fall time of 47.5 nsec.

shift only contributed to a slowly-varying background less than 10% of the total frequency shift. Also, the inhomogeneous dephasing due to spatial intensity fluctuation discussed in an earlier experiment [4] became negligible. The two-photon precession signal was monitored by detecting the weak beam after it had passed through the cell using a fast photoconductor Ge:Cu (4.2 K) and a Tektronix 7904 oscilloscope, with an overall time constant of 2 nsec.

Figure 1 shows a typical Stark-switched two-photon precession signal. The signal was stable and repeatable from shot to shot. The precession decay time as a function only of the gas pressure, and the precession frequency was a function only of the Stark pulse shape. While changes in laser intensities affected the magnitude of the signal, neither the decay time nor the precession frequency were affected. The precession signal was near Doppler-free (the residue Doppler width is less than 2 MHz) and was found to be in excellent agreement with theoretical prediction.

Using this two-photon coherent precession effect, we obtained the collision-induced dephasing time T_2 of this two-photon transition as a function of NH_3 gas pressure.[6] We found that above 200 mTorr, the measured T_2^{-1} increased linearly with pressure. Below 200 mTorr, the decay was apparently limited by mechanisms other than collisions, most probably due to a combination of finite laser linewidths and the residue Doppler width. From the linear portion of the data, the relaxation time for this transition in NH_3 is determined to be $T_2p=10.5\pm2$ nsec-Torr, where p is the NH_3 pressure in Torr. Comparing this result with the earlier pressure-broadened linewidth measurement [5], we found that the simple relation between linewidth $\Delta\nu$ and relaxation time T_2, namely, $T_2 = 1/(\pi\Delta\nu)$ is still valid, even though the present physical system is significantly more complicated than the ideal two-level system for which this simple relation is derived.

We have also applied this technique to measure the effect of buffer gases on T_2 of this two-photon transition in NH_3. The buffer gases used, in the order of their effectiveness in dephasing this two-photon coherence, are Ar, He, N_2, CS_2, CO_2, SO_2 and $CHCl_3$.

In conclusion, the Stark-switching technique has been successfully applied to the observation of two-photon coherent transients. The experiment clearly suggests that this technique and the more recent frequency switching technique which have been shown to be of such importance in one-photon coherent optics, should occupy a similar place in the study of multi-photon coherent effects.

This work is partially supported by the U.S. Office of Naval Research.

References

1. R. G. Brewer, Physics Today 30, 50 (July, 1977).
2. R. G. Brewer and E. L. Hahn, Phys. Rev. A 11, 1614 (1975).
3. D. Grischkowsky, M. M. T. Loy and P. F. Liao, Phys. Rev. A 12, 2514 (1975).
4. M. M. T. Loy, Phys. Rev. Lett. 36, 1454 (1976).
5. W. K. Bischel, P. J. Kelly and C. K. Rhodes, Phys. Rev. A 13, 1829 (1976).
6. M. M. T. Loy, Phys. Rev. Lett. 39, 187 (1977).

SUB-DOPPLER SPECTROSCOPY OF NO_2

I.R. Bonilla, W. Demtröder, F. Paech, and R. Schmiedl

Fachbereich Physik, Universität Kaiserslautern,
6750 Kaiserslautern, FRG

The visible absorption spectrum of NO_2 is so dense that many lines overlap within their Dopplerwidth. Sub-Doppler resolution is therefore demanded to completely resolve the spectrum in order to gain more information about the excited electronic states and their mutual interaction. This paper describes such measurements with greatly reduced Doppler width in a molecular beam. Different argon laser lines are used for selective excitation of various hyperfine structure levels. The fluorescence from these levels is monitored to investigate the radiative and radiationless decay channels of the excited NO_2-molecules under collision free conditions.

The argon laser is locked to the transmission peak of a stabilized reference Fabry-Perot, which can be continuously tuned over the whole gain profiles of the argon laser transitions. The laser frequency stability is better than 500 KHz. The laser beam is crossed perpendicularly with the NO_2-beam and the laser excited fluorescence is monitored while the laser frequency is tuned across the absorption spectrum. Due to a collimation ratio of 1:100 for the molecular beam the Doppler width of the molecular absorption lines is reduced to about 10 MHz. This allows the resolution of most of the hyperfine-structure components. The upper trace of Fig. 1 shows as an example such an "excitation spectrum." The lower trace is obtained when not the total fluorescence, but only one vibrational band terminating on the (0,1,0)-vibrational level is monitored as a function of laser frequency.

In order to identify the various absorption lines the laser is stabilized on one of the lines and the resulting fluorescence spectrum is recorded through a monochromator. This procedure has been performed for most of the prominent absorption lines within the tuning range of the argon laser. The analysis of these fluorescence spectra yields the surprising results that in many cases the emitting state differs from the primarily excited state. This implies that radiationless transitions between different excited levels take place even without collisions [1].

With a Pockel-cell the laser beam could be pulse-modulated, which allowed monitoring of the time-resolved fluorescence. Using single photon counting techniques and delayed coincidence electronics the decay curves resulting from selective excitation of individual hfs-levels could be obtained. Many of these levels show nonexponential decays. This again can be explained by the assumption that the fluoresence is due to a superposition from several emitting levels while only a single level has been excited by the laser.

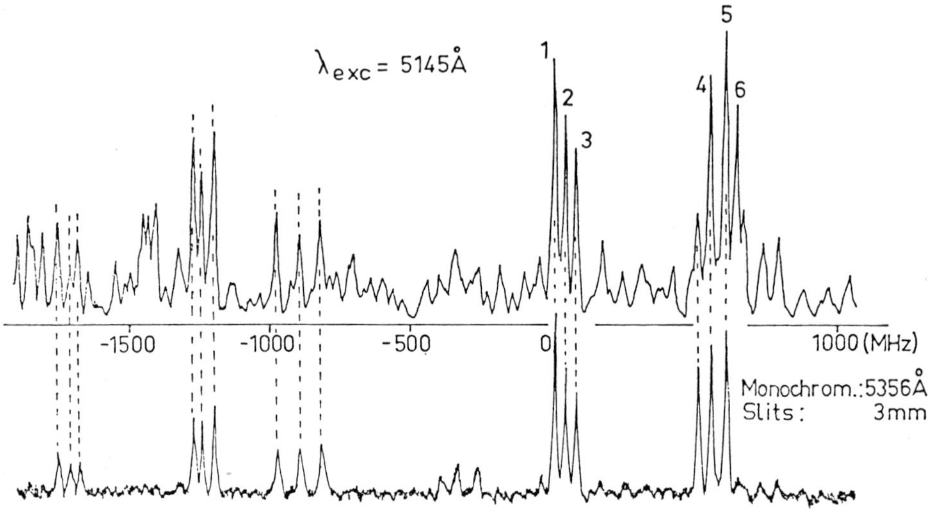

Placing the excitation region in the molecular beam into an external magnetic field, the Zeeman splitting of the different hfs-components could be observed. Since the Landé-factor g of the ground state can be calculated, the line-splittings yield the g-value of the upper state.

In the next experiment the polarization P of the fluorescence as a function of a weak magnetic field H (Hanle effect) was measured. From the half-width $\Delta H_{1/2}$ of the resulting Lorentzian P(H) the product $g \cdot \tau_c$ of Landé-factor g and coherence-lifetime τ_c was obtained. With the g-value known from the Zeeman-splitting the coherence lifetime could be extracted. The interesting result is that the coherence time is between one and two orders of magnitude shorter than the radiative lifetime. This again shows the predominant influence of perturbations which may cause a fast dephasing of the wave-function originally prepared by the selective excitation [2].

Polarization spectroscopy [3] and optical-rf-double-resonance experiments, which are currently prepared at Stanford, will further help to elucidate these problems.

References

[1] I. R. Schmiedl, F. Paech, I. R. Bonilla, and W. Demtröder, J. Mol. Spectrosc. in print (1977).

[2] I. R. Bonilla, and W. Demtröder, Chem. Phys. Letters, to be published (1977).

[3] W. Demtröder and R.Teets, to be published.

DISSOCIATION OF MATRIX-ISOLATED MOLECULES BY INFRARED LASER RADIATION

R.V. Ambartzumian, Yu.A. Gorokhov, G.N. Makarov,
A.A. Puretzky, and N.P. Furzikov

Institute of Spectroscopy, Academgorodok, Podolski r-n
Moscow, 142092, USSR

1. Introduction

Selective dissociation of polyatomic molecules in a gaseous phase by intense infrared laser pulses has now been well examined; the mechanism of the process and the application of the phenomenon are understood [1]. The next step toward understanding the process is the investigation of unimolecular dissociation of molecules by infrared pulses when the molecules are deposited in low temperature matrices, that is, the interaction of intense infrared radiation with condensed matter.

The vibrational relaxation time for molecules isolated in liquid matrices is in the range of several picoseconds, and therefore it is difficult to expect that in liquid matrices the molecule can absorb large amounts of vibrational energy from typical CO_2 laser pulses and reach the dissociation limit. Indeed we failed to complete the dissociation by CO_2 laser pulses of SF_6 molecules dissolved in liquid nitrogen.

In solid matrices the V-T relaxation times allow enough energy (in vibrational degrees of freedom) to accumulate for dissociation. Dissociation of this type was performed for SF_6 molecules isolated in solid Ar matrices [2]. It was shown that the dissociation was isotopically selective. The absorption of intense infrared radiation by SF_6 isolated in Ar matrices was studied as a function of pulse energy, pulse length, and frequency.

2. Experimental

The apparatus included a helium optical cryostat, a TEA CO_2 laser, and a control system. The SF_6:Ar or SF_6:CO (1:300 to 1:2000) mixture was deposited on a CsI substrate which was mounted on the cold finger. The temperature of the CsI plate was 8 to 10°K. Matrix deposition was controlled by the infrared spectrum of SF_6. The line width of the $^{32}SF_6$ absorption centered at 938.5 cm^{-1} was $\sim$2 cm^{-1} for the Ar matrix and $\sim$4 cm^{-1} for the CO matrix. The SF_6:Ar (or SF_6:CO) matrix was irradiated by CO_2 laser pulses. The pulse length was 90 ns with 1 μs tail containing not more than 30% of pulse energy. The laser pulse could be shortened by optical breakdown to 40 ns without any tail.

Changes in peak absorption were registered for $^{m}SF_6$ molecules that occurred after irradiation by a CO_2 laser. The energy absorbed in matrix was also measured.

3. Results and Discussion

Figure 1a (curves 1 and 3) shows the infrared absorption spectrum of SF_6 (of natural isotopic abundance) in the Ar matrix. The solid curve (2) shows the spectrum after irradiation by a series of CO_2 laser pulses at 940.5 cm^{-1}. The laser radiation energy flux was ~0.5 J cm^{-2} ; the number of pulses was 100. Comparison of curves 1, 2, and 3 shows the preferential burnout of $^{32}SF_6$. The enrichment coefficient in this case was ~3.

In the case of deep burnout of $^{32}SF_6$ molecules, the IR absorption peak of $^{32}SF_6$ exhibited a red shift. Such a red shift is approximately 2 cm^{-1}. This, perhaps, is connected with the burnout of absorbing molecules in an inhomogeneously broadened absorption contour at the laser frequency, that is, the radiation dissociates only the molecules with a definite matrix surrounding.

Further irradiation usually led to the destruction of the matrix. The attempt to detect the dissociated products by IR spectroscopy failed, supposedly, because of their low intensity for identification.

Thus, these experiments confirm that irradiation of matrix-isolated SF_6 molecules leads to their dissociation.

One of the major questions concerns the absorption of IR radiation by a matrix-isolated molecule. Although molecules can be considered to be completely isolated in a low pressure gas, molecules in a low temperature matrix still interact with the matrix and thus are not completely isolated. This may ease the process of energy absorption, but, on the other hand, causes

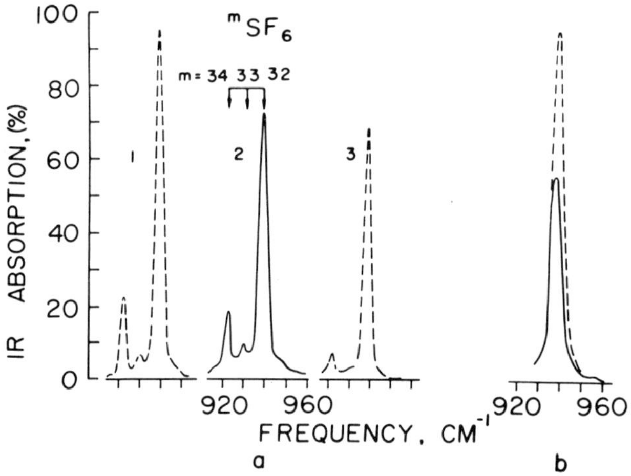

Fig.1 (a) Infrared absorption spectra of SF_6 molecules isolated in a solid matrix: curves 1 and 3 - before irradiation; curve 2 - after irradiation by 100 pulses of a CO_2 laser ($\nu = 940.5$ cm^{-1}, $E = 0.5$ J cm^{-2}). (b) The change of the absorption spectrum of $^{32}SF_6$ molecule under dissociation in the inhomogeneously broadened short wave spectrum wing.

some vibrational deactivation. The direct calorimetric measurement of absorbed energy shows that at energy flux ~0.2 J cm^{-2} with pulse length 100 ns, each molecule absorbs energy corresponding to n ~ 150 quanta of CO_2 laser radiation. Figure 2 shows the dependence of absorbed energy in the matrix versus the incident energy flux. The measurements were performed at 938.7 cm^{-1}. The upper curve of Fig.2 shows the results using a 100 ns pulse length and the lower curve was obtained using shortened, 40 ns pulse. The absorbed energy depends on the incident energy as $n \sim E^{0.5}_{incid}$. The fraction of energy absorbed for both types of pulses is equal approximately to the ratio of pulse lengths.

Figure 3 shows the dependence of the absorbed energy on the laser frequency. The measurements were performed with energy flux 0.2 J cm^{-2} and a 40 ns pulse. The maximum multiple photon absorption coincides with the maximum linear absorption, but the height of the peak in multiple photon absorption is much less compared with linear absorption.

The results can be explained qualitatively on the basis of a cluster model for dissociation of SF_6 isolated in a matrix. Here we use the term "cluster" for a single isolated SF_6 molecule with its closest surrounding Ar atoms. It is possible to evaluate the number of particles, N, forming such a cluster. The evaluation can be made within a statistical model that well describes a unimolecular SF_6 dissociation in the gas phase [3,4]. The constant of the rate of such a unimolecular reaction K is

$$K = \omega\, e^{-ms/n} \quad , \qquad (1)$$

where ω is the laser frequency, m is the number of quanta necessary for dissociation, s is the number of vibrational degrees of freedom, and n is the number of quanta absorbed by one SF_6 molecule. Comparing the number of quanta necessary for SF_6 dissociation with an equal rate in the gas phase ($n \sim 30$) and in the matrix ($n \gtrsim 150$), it is easy to obtain the number of vibrational degrees of freedom for such a cluster where the absorbed energy is dissipated. This kind of estimate gives $N \simeq 30$. If the intracluster vibrational energy exchange time is long compared with the laser pulse

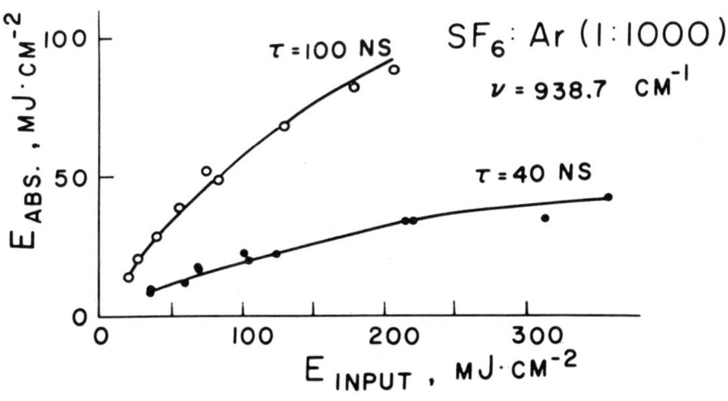

Fig.2 The dependence of the energy absorbed by matrix-isolated SF_6 molecules as a function of the incident energy flux.

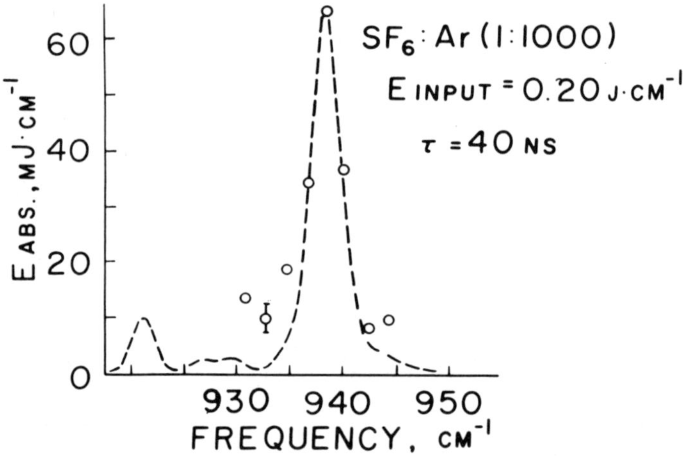

Fig.3 The energy absorbed by matrix-isolated SF_6 molecules as a function of laser radiation frequency. Dotted line shows a linear infrared spectrum of SF_6.

$T_{intr} \gtrsim \tau_{pulse}$, then the number of atoms forming such a cluster should depend on the length of the laser pulse. This will lead to a decrease of n by a factor which is roughly equal to the ratio of the pulse lengths.

The results presented here show that the dissociation of polyatomic molecules in a solid low temperature matrix is a new and interesting phenomena that needs further exploration both experimentally and theoretically.

Acknowledgments

The authors are indebted to Professor V. S. Letokhov for helpful discussions.

References

1. R. V. Ambartzumian and V. S. Letokhov, in Chemical and Biochemical Applications of Lasers, edited by C. B. Moore (Academic Press, New York, 1977), Vol. 3.
2. R. V. Ambartzumian, Yu. A. Gorokhov, G. N. Makarov, A. A. Puretzky and N. P. Furzikov, JETP Lett. 24, 287 (1976).
3. M. J. Coggiola, P. A. Schulz, Y. T. Lee and Y. R. Shen, A Molecular Beam Study of Multiphoton Dissociation of SF_6, (in press).
4. J. G. Black, E. Yablonovitch, N. Bloembergen and S. Mukamel, Phys. Rev. Lett. 38, 1131 (1977).

INFRARED SPECTROSCOPY OF MOLECULAR BEAMS

T.E. Gough, R.E. Miller, and G. Scoles

Guelph-Waterloo Centre for Graduate Work in Chemistry
University of Waterloo, Waterloo, Ontario, Canada

Supersonic Molecular beams are attractive samples for I.R. spectroscopic studies because the molecules in such a beam i) are in a collision-free environment ii) more in a common direction with a narrow spread in their velocities and iii) have very low vibrational and rotational temperatures. Features i) and ii) allow Sub-Doppler resolution to be achieved: feature iii) allows drastic simplification of the spectrum. The principal disadvantage of molecular beams is their very low optical density. The I.R. spectra may, however, be observed with excellent signal to noise ratio by monitoring bolometrically modifications in the internal energy of the beam as it is crossed by an IR laser tuned through vibrotational transitions of the beam molecules.

Using this technique spectra have been obtained of beams sampled from supersonic jets of carbon monoxide and carbon monoxide-helium mixtures.

An analysis of the intensities of the spectra leads to an estimate of the rotational temperature of the beams and to information on the rotational relaxation in these systems.

OBSERVATION OF INFRARED LAMB DIPS IN SEPARATED OPTICAL ISOMERS

E. Arimondo, P. Glorieux, and T. Oka

Herzberg Institute of Astrophysics, National Research Council of Canada
Ottawa, Ontario, Canada

The energy levels of optical isomers which are mirror images of each other have been assumed to be exactly the same. However there has not been accurate experimental check of this by high resolution spectroscopy of separated optical isomers. The technique of inverse Lamb dips using infrared lasers enables us to perform such an experiment with a high accuracy. This kind of experiment has a significance in relation to recent theories that the effect of parity non-conserving interaction may appear in atomic and molecular physics[1,2].

The absorption in the 9.5 μm region corresponding to the bending of the —C*—(CO)—C— group (C* denoting the optically active carbon atom) in d- and l-camphor was chosen because of its coincidence with 9.4 μm CO_2 laser lines (Fig.1). First, we observed several IR-RF double resonance signals which confirmed that vibration-rotation transitions of camphor have coincidences with laser lines and are efficiently saturated by the laser radiation.

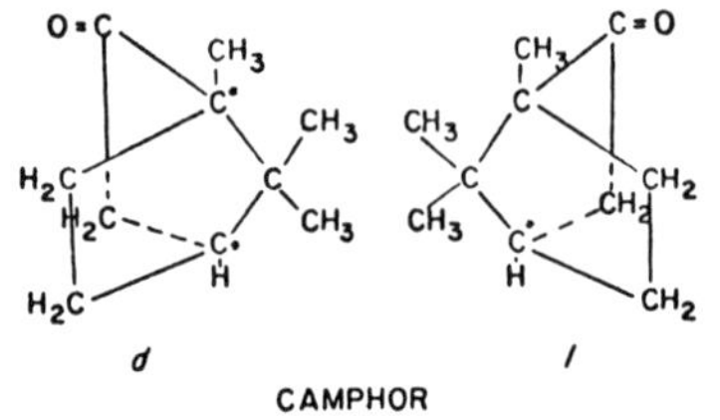

Fig.1

For the Lamb dip experiment a camphor sample at a pressure of ∿100 mTorr was placed inside the laser cavity and the laser output was observed while its frequency was scanned over the cavity profile. In order to increase the sensitivity of detection, Stark modulation and phase detection were used. Inverted Lamb dips were observed first for laser lines where double-resonance signals were observed and later also for neighbouring laser lines. In Fig.2 the inverted Lamb dip observed on a d-camphor sample using the R(24) laser line is shown. The high sample pressure was needed because the camphor absorption is weak due to large rotational and vibrational partition functions. The hwhm of the Lamb dips measured at a pressure of 55 mTorr was ∿400 KHz.

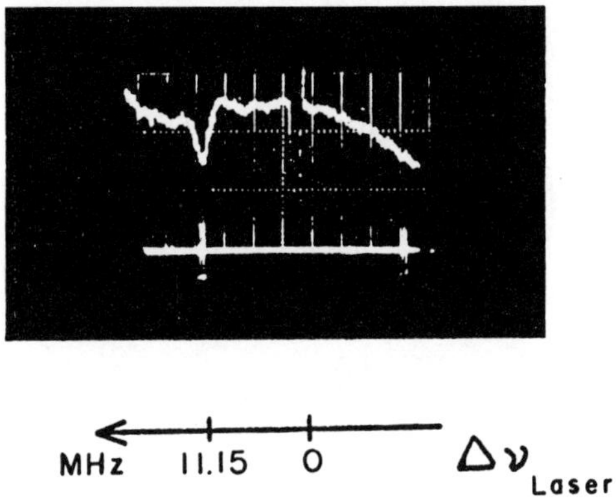

Fig.2 An example of oscilloscope traces of the d-camphor inverted Lamb dip and of the beat signal monitored through a tuned radio-receiver.

We observed that different laser oscillation conditions are required to obtain the best signal-to-noise ratio in each isomer. This behaviour has been qualitatively interpreted as due to the optical activity of our sample interacting with possible asymmetry of the optical system. The Lamb dip positions for d-camphor and l-camphor were compared by filling successively the sample cell with each optical isomer while keeping the same optical and electronic conditions.

The Lamb dip position was measured accurately by comparing the frequency of the CO_2 laser containing the camphor with that of another CO_2 laser stabilized to the CO_2 4.3 μm fluorescence Lamb dip. The beat note between the two lasers was measured at the camphor Lamb dip position. The accuracy in the comparison of transition frequencies in d- and l-camphor was limited by the laser stability and the linewidth of the observed signals when best conditions for both isomers are realized. For the measurements so far conducted, the transition frequencies of d- and l-camphor agree to within an uncertainty of $\sim$300 KHz which corresponds to an accuracy of 1.10^{-8}.

References

(1) M.A. Bouchiat and L. Pottier, "Laser Spectroscopy", J.L. Hall and J.L. Carlsten, eds., Springer-Verlag, 1977.

P.G.H. Sandars, "Laser Spectroscopy", J.L. Hall and J.L. Carlsten, eds., Springer-Verlag, 1977.

(2) V.S. Lethokov, Phys. Lett. 53A, 275 (1975).

O.N. Kompanets, A.R. Kukudzhanov, V.S. Lethokov and L.L. Gervits, Opt. Comm. 19, 414 (1976).

B.Y. Zel'dovich, D.B. Saakyan and I.I. Sobel'man, JETP Lett. 25, 106 (1977).

HIGH-RESOLUTION SPECTROSCOPY IN FAST ATOMIC BEAMS

R. Neugart, S.L. Kaufman[1], W. Klempt, G. Moruzzi[2]
E.-W. Otten, and B. Schinzler

Institut für Physik, Johannes Gutenberg-Universität
6500 Mainz, FRG

Narrow optical resonances have been observed in fast beams of Na and Cs atoms, obtained from ion beams by charge-transfer collisions with Na, K, or Cs. Corresponding to the narrowing of the velocity distribution, occurring by acceleration, the Doppler width along the beam direction is considerably reduced [1, 2].

The ion beam, emitted from a surface-ionization source, is superimposed with a single-mode cw dye laser beam. The beams, propagating together, cross an alkali-vapor cell in which the ions are neutralized by charge-transfer collisions. Optical excitation of the atoms is detected by counting fluorescence photons. Frequency-tuning of the absorbed light is accomplished by varying the acceleration potential and thus the large Doppler shift.

Residual Doppler widths of 20 MHz are achieved in the resonance lines of Na (3s $^2S_{1/2} \rightarrow$ 3p $^2P_{3/2}$), 5890 A and Cs (6s $^2S_{1/2} \rightarrow$ 7p $^2P_{3/2}$), 4555 A at beam energies between 3 and 10 keV. The observed spectrum is a superposition of the well-resolved hyperfine structure (hfs) and the energy-loss spectrum of charge transfer, corresponding to electron capture in different atomic states.

As an example, the spectrum of hfs components excited from the F = 4 ground state level in ^{133}Cs is shown in Fig.1. The structure of each component is due to charge transfer between Cs^+ and K at 7.6 keV. The explanation of this splitting is indicated in the figure: A change in the total internal energy of the collision partners appears as a change of kinetic energy. Essentially all of this kinetic energy is taken up by the fast partner which is practically undeflected in the laboratory frame. The fast atomic beam thus consists of groups having different velocities depending on the state in which the electron was captured. The observed peaks correspond directly to the distribution of Cs atoms over different final states. Systematic measurements of such distributions will be important in studies of the charge-transfer process.

---

[1] Present address: School of Physics & Astronomy, University of Minnesota, Minneapolis, Minnesota 55455

[2] Fellow of the Alexander von Humboldt Foundation, on leave from Instituto di Fisica, Università di Pisa, Pisa, Italy

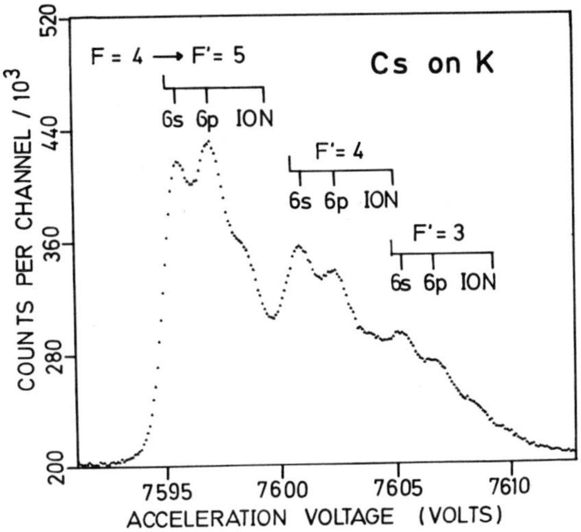

Fig.1 Doppler-tuned spectrum of Cs 4555 A containing the energy-loss spectrum of charge transfer between Cs^+ and K.

"Pure" hfs spectra are observed in beams neutralized by charge-transfer collisions between atoms and ions of the same species. In these collisions the resonant channel leading to the atomic ground state is predominant. Therefore, it may be preferably used for studies of isotopic shifts and hyperfine structure.

The use of a fast beam makes this method especially suitable for investigating isotopes far from stability available at on-line isotope separators. Direct use of their output beams avoids the loss of particles usually associated with stopping and re-evaporation of the atoms. Furthermore, the narrowing of the Doppler profile ensures excitation of most of the available atoms, combined with high resolution. When several isotopes are present in the beam, the observed lines are well separated due to the fact that the velocities of the accelerated ions depend on their masses.

We are preparing the application of this method to fission-produced isotopes of Cs available at the Mainz on-line isotope separator.

This work was sponsored by the Deutsche Forschungsgemeinschaft under its Laser Spectroscopy Program.

References

1 S.L. Kaufman, Opt. Commun. 17, 309 (1976)
2 W.H. Wing, G.A. Ruff, W.E. Lamb jr., J.J. Spezeski, Phys. Rev. Letters 36, 1488 (1976)

HIGH PERFORMANCE SPECTROMETER FOR DOPPLER-FREE SPECTROSCOPY: STUDY OF THE HYPERFINE PREDISSOCIATION IN IODINE MOLECULES

B. Couillaud and A. Ducasse

Laboratoire de Spectroscopie Moléculaire, Université de Bordeaux I
33405 Talence, France

A frequency- and intensity-locked cw dye laser is used in a saturated absorption spectrometer. The performances of this laser are illustrated by the following experiment : the laser was tuned in a saturated absorption experiment to the half maximum of an hyperfine component of I_2 and left free. The width of the hyperfine signal was 4.6 MHz. The recorded signal shows that the free-running laser frequency over time of 3 min. is held within a 800 kHz width frequency domain. When locked in frequency and intensity, the C.W. dye laser amplitude noise is less than 1% R.M.S., the jitter is less than 50 kHz but drifts of the order of 1 MHz/min. are present in spite of thermal stabilization of the reference cavity |1|.

This laser is used in a *Hänsch-Bordé* saturated absorption technique, in which the two counter propagating beams are perfectly colinear; a judicious use of polarizers suppressing the coupling between the laser and the spectrometer. A great care has been taken to minimize the causes of experimental broadening such as: pressure, transit-time, non colinearity of the two beams. The performances of the spectrometer are illustrated by the figure 1.

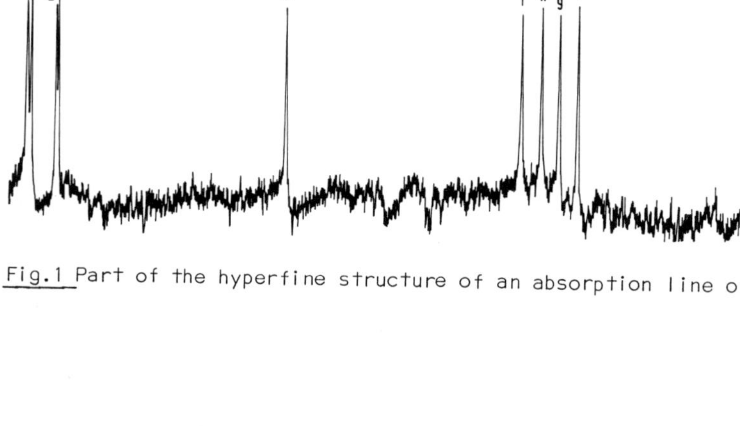

Fig.1 Part of the hyperfine structure of an absorption line of I_2 near 6000 Å.

The resolution is better than 1 MHz. The frequency marker used in the experiments limits the measure of the relative positions of the hyperfine lines to an accuracy of 1 MHz. Agreement between calculated and experimental line positions is then in the measured precision range of 1 MHz.

The hyperfine structure of I_2 has been investigated in the 9-3, 10-3, 11-3, 12-3 bands of the BX system for different values of J. The relative intensities of the hyperfine components of each line have been found to deviate from the degeneracy dependence. The observed deviation increases when v' decreases towards the value v' = 6 where the potential curves of the $^3\Pi_{0}^{+}$ and lu states are crossing . This perturbation has been attributed to the natural predissociation due to the rotational hamiltonian and the hyperfine hamiltonian |2|. The calculations of the hyperfine component intensities carried with this assumption in the case of a very small saturation parameter give a variation with J which is in good qualitative agreement with the experimental results. Particularly the theoretical ratio I(J-5) over I(J+5) can be easily shown to increase versus J with a similar dependence as the one imaged by the experimental results of fig. 2. This previous qualitative agreement leads us to think that a study of the saturation dependence of the intensity of each hyperfine line, allowing the extrapolation to null saturation, will lead to a quantitative exploitation of the experimental results.

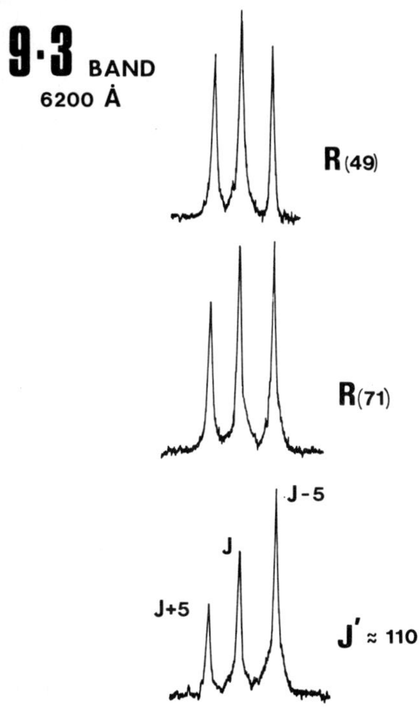

Fig.2 Evolution versus J of the intensities of three hyperfine components (F= J+5, J, J-5) in the 9-3 band.

References

1. B. COUILLAUD, A. DUCASSE, Opt. Comm. 21-2-199 (1977).
2. M. BROYER, J. VIGUIE, J.C. LEHMANN, J. Chem. Phys. 64, 11, 4793 (1976).

POLARIZATION SPECTROSCOPY OF CONDENSED PHASES

J.J. Song, J.H. Lee, and M.D. Levenson

Department of Physics, University of Southern California
Los Angeles, CA 90007, USA

The various techniques of saturation spectroscopy have been widely used to eliminate Doppler broadening and to determine the splittings and homogeneous widths of spectral lines of vapors. Less attention, however, has been paid to the problem of removing inhomogeneous broadening from absorption lines of *condensed phases*. In many cases, the physically relevant information about the system can be obtained by time-domain techniques, but these techniques fail for levels with short relaxation times [1].

We report here a technique in which the dichroism and birefringence induced in an absorbing sample by one laser field is detected by another. Our technique is similar to the polarization spectroscopy technique of Wieman and Hansch [2], and is related to the Rayleigh resonance work of Yajima [3] in the same way that our previous Raman-Induced Kerr Effect Spectroscopy (RIKES) is related to Coherent Anti-Stokes Raman Scattering (CARS). As in the case of RIKES, the present technique requires no phase matching and therefore is easier to execute experimentally and less sensitive to experimental artifacts. Initial demonstrations were performed on organic dye molecules in liquid solution and our results parallel those obtained by the time resolved absorption technique using picosecond lasers [4].

The experimental apparatus is similar to that employed in RIKES [5] or polarization spectroscopy of gases. A linearly or circularly polarized pump laser beam (ω_1) crosses a nearly counter progagating linearly polarized probe laser beam (ω_2) within the sample material. A nonlinear interaction produces a polarization component at ω_2 which is orthogonal to the initial linear polarization of the probe beam and passes through a crossed polarizer. We detect this orthogonal polarization component with a photomultiplier while ω_1 is varied over the region centered around ω_2. The laser frequencies, ω_1 and ω_2 are chosen to lie within the absorption peak of the sample solutions.

The experimental traces of the dye solutions we studied show a distinct peak centered at $\omega_1-\omega_2=0$, on top of an enhanced background signal, and the half-width of the central peak was found to vary with absorbing species. We also found that our signals exhibit linear and square law dependence on probe and pump laser intensity respectively. Although the host material in which the absorbing species is embedded also makes contributions to the detected signal, any interference or complications due to this solvent signals can be excluded from an analysis for the following reasons. First, these contributions have different characteristics and can be experimentally identified. Second, these signals are not resonantly enhanced through one-photon absorption and decrease as the concentration of the dye molecules is increased; therefore, polarization spectroscopy signals usually swamp these effects.

Information on relaxation times can be obtained from our experimental data as follows. When the equations of motion of the density matrix are solved, the detected intensity expressed as a function of difference in laser frequency Δ becomes

$$I_{\perp} \propto f(\omega)\left|\frac{T_1}{2T_2^{-1}-i\Delta}+\frac{1}{T_1^{-1}-i\Delta}\,\frac{1}{2T_2^{-1}-i\Delta}\right|^2 \qquad (1)$$

where T_1 and T_2 denote longitudinal and transverse relaxation times, respectively in a single Bloch equation representation of the two-level system. The factor $f(\omega)$ is the inhomogeneously broadened lineshape function. The first term in Eq. 1 results from saturation of the absorption found by the probe beam and its width is $2T_2^{-1}$(HWHM). The second term is responsible for the Rayleigh resonance of Yajima and caused by a population difference oscillating at frequency Δ. If $T_1 \gg T_2$, its width is approximately T_1^{-1}. Due to the limited resolution of our lasers, relaxation times must be shorter than 10 picoseconds to be measured in our experiments. In our preliminary results, T_1 values were deduced to be 1 psec for malachite green oxalate dissolved in water and 5 psec for a cyanine dye (1,3'-Diethyl - 2,2' quinolythia-carbocyanine iodide) in ethanol. Effective T_2 values were found to be equal to the inverse of the absorption bandwidth, which are in the region of 10^{-1} - 10^{-2} psec.

In spite of our simplified theoretical model, our values reasonably agree with the results obtained using picosecond lasers in malachite green solution [4]. To our knowledge, no relaxation rate measurement has been reported on the cyanine dye we mentioned above. Experiments are in progress to investigate lineshape changes with various input laser wavelength, polarization conditions and etc. Efforts are also being made to incorporate more realistic systems in our theory. Finally, we like to emphasize that our technique can be extended to crystals or glasses which exhibit inhomogeneously broadened spectral lines with fast relaxation rates.

References

1. A. Z. Genack, R. M. Malfarlane and R. G. Brewer, Phys. Rev. Lett. 37, 1078 (1976).
2. C. Wieman and T. W. Hansch, Phys. Rev. Lett. 36, 1170 (1976).
3. T. Yajima, Opt. Comm. 14, 378 (1975).
4. C. V. Shank and E. P. Ippen, Proceedings on the Second International Conference on Laser Spectroscopy, p 408, (1975).
5. M. D. Levenson and J. J. Song, J. Opt. Soc. Am. 66, 641 (1976).

QUANTITATIVE ANALYSIS OF RESONANT THIRD HARMONIC GENERATION IN Sr-Xe MIXTURES

H. Scheingraber, H. Puell, and C.R. Vidal

Max-Planck-Institut für Physik und Astrophysik
Institut für extraterrestrische Physik, 8046 Garching, FRG

Recently R.T. HODGSON et al. [1] reported resonant third harmonic generation and frequency mixing in a Sr-Xe mixture by tuning the output of an N_2-laser pumped Rhodamine 6G laser (λ = 5757 Å) to the two-photon transition $5s^2$-5s5d of Sr. From their experimental data they concluded that third harmonic generation in Sr should be very efficient due to resonances of the harmonic radiation with nearby autoionizing levels.

We have investigated the same nonlinear system with a flashlamp pumped Rhodamin 6G laser looking for the absolute conversion efficiency. The experiments were performed by focusing the fundamental beam with an f = 100 cm lens to a focal spot area of 0.01 cm^2 within a concentric heatpipe containing 8 torr Sr vapor and a variable amount of Xe. Optimum phasematching occurred at a pressure ratio of p(Xe)/p(Sr) = 53 as shown in Fig.1, which is in good agreement with the theoretical value of 53.1 calculated from the refractive indices of Sr and Xe. Increasing the fundamental power from 1 kW up to 40 kW we observed an increase of generated third harmonic power from $4\cdot10^{-5}$ W to 3 W (see Fig.2), consistent with the third power law expected from small signal theory. A maximum conversion of $4\cdot10^{-5}$ was achieved.

From these results we conclude that saturation effects are negligible within this intensity range. However, from the phasematching curve a significantly smaller effective length of the nonlinear medium than the experimental length of 35 cm is estimated. This indicates that due to absorption only the last few centimeters of the Sr vapor column contribute to the observed third harmonic. This was verified by measuring the absorption cross section at λ = 1919 Å which turned out to be $\sigma = (3.3 \pm 1.0) \times 10^{-18} cm^2$. It shows that in our experiment the nonlinear medium is optically thick for the third harmonic radiation. We estimate an optical thickness of $\tau = NL\sigma \simeq 9$ (absorption of the fundamental wave was found to be negligible). In this case the intensity conversion at optimum phasematching can be written as

$$\Phi_3/\Phi_1 = (24\,\pi^2 \omega \chi_T{}^{(3)} \Phi_1 / c\sigma)^2 \qquad (1)$$

From (1) it follows that the conversion becomes independent of particle density and length of the nonlinear medium, which was verified experimentally. The nonlinear susceptibility $\chi_T{}^{(3)}$ was calculated taking into account the lowest lying S, P, and D energy levels of Sr (a total of 42). The corresponding matrix elements were evaluated with the Coulomb approximation

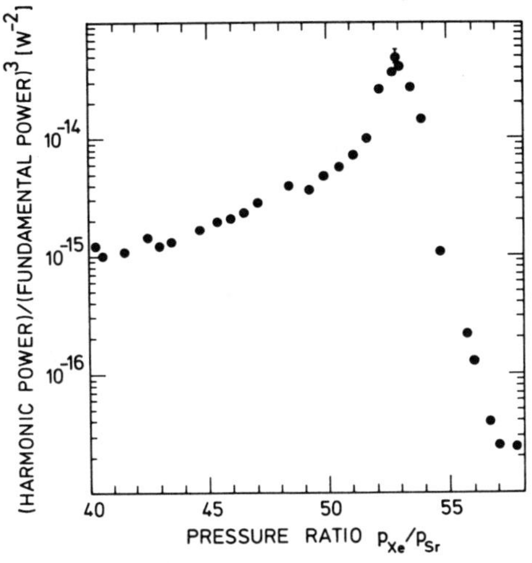

Fig.1 Phasematching curve for 8 torr of Sr at 40 kW input power

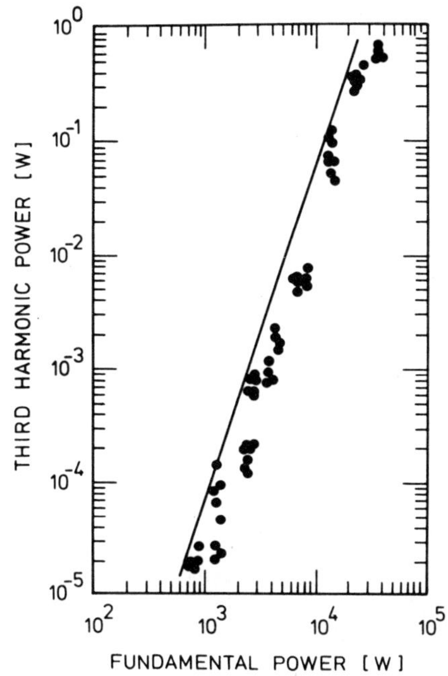

Fig.2 Third harmonic power vs. fundamental power at an Xe/Sr pressure ratio of 424/8. Theoretical curve (——) based on (1)

which gave excellent agreement even with the first resonance line. Inserting in (1) the calculated value of $\chi_T^{(3)} = 2.6 \cdot 10^{-31}/\Gamma \simeq 1.7 \cdot 10^{-30}$esu ($\Gamma \simeq$ 0.15 cm^{-1} is the measured linewidth originating from a convolution of the laser- and two-photon resonance linewidth) we get for our experimental conditions the solid line shown in Fig.2, which is in good agreement with the measured harmonic power (assuming a Gaussian radial intensity distribution).

From this good quantitative agreement we conclude that due to missing saturation phenomena one should easily be able to extrapolate the results of Fig.2 to significantly higher conversion efficiencies. The important result, however, is that in contrast to the suggestion of R.T. HODGSON et al. [1] the autoionizing levels may even deteriorate the conversion efficiency because due to absorption the effective length and density of the Sr vapor column is limited to an optical depth of the order of unity.

Reference

[1] R.T. Hodgson, P.P. Sorokin, and J.J. Wynne, Phys. Rev. Letters 32, 343 (1974)

NEW SPECTRAL INFORMATION BY LAMB-DIP SPECTROSCOPY

W. Radloff, V. Stert, and H.-H. Ritze

Central Institute of Optics and Spectroscopy, Academy of Sciences of the German Democratic Republic

Varying the physical parameters in the absorption cell (e.g., pressure, temperature, intensity, or polarization of the laser field, external electric or magnetic fields), one obtains changes of the nonlinear absorption signals, which give information about the spectral parameters of the investigated molecules. This method will be demonstrated by two examples: (1) Lamb-dip spectroscopy of ethane at 3.39 μm; and (2) polarization spectroscopy of molecules at 10.6 μm.

The linear absorption spectrum of ethane in the spectral region of a Zeeman-tuned He-Ne laser at 3.39 μm shows overlapping lines on a relatively strong background. Therefore, nonlinear spectroscopy is necessary to determine exactly the absorption of each single line as well as their variations due to changes of the physical parameters. Measuring the nonlinear absorption signals at weak and strong saturation, respectively, we have obtained relative values for the linear absorption and the saturation intensities of the thirteen strongest ethane lines. We show that saturation intensities obtained by intensity dependence of nonlinear absorption signals agree well with those values achieved from power broadening in line width measurements. Furthermore, from nonlinear absorption measurements at room temperature and at 210°K, we have obtained the ratio of corresponding linear absorption values. The comparison of these results with appropriate theoretical estimates for the symmetric ethane molecule has allowed a preliminary qualitative assignment of the measured lines.

The method of polarization spectroscopy published first by HÄNSCH [1] and demonstrated experimentally in the visible was used for spectroscopic investigations of molecules utilizing a line-tunable CO_2 laser with different isotopes. As is known, the light-induced birefringence is a function of the rotational quantum number J and it is small for the Q-branch and large for the R- or P-branch of the rotational vibrational band. In this way we are able to distinguish, for example, between various Q-branch and P- or R-branch lines of SF_6 at the P(16) line of the $^{12}CO_2$ laser at 10.55 μm. Because the Q-branch lines in this case belong to high J values ($J \approx 40$), no birefringence may be observed. On the contrary, the effect was clearly detected for a Q-branch line of $^{15}NH_3$ with $J = 5$, which is in good coincidence with the R(18) line of the $^{13}CO_2$ laser at 10.78 μm.

Reference

[1] C. Wiemann and T. W. Hänsch, Phys. Rev. Lett. 36, 1170 (1976).

DETECTION OF VERY WEAK ABSORPTION LINES WITH THE AID OF NEODYMIUM-GLASS AND DYE LASERS

V.M. Baev, T.P. Belikova, E.A. Svizidenkov, A.F. Sutchkov

Lebedev Institute, USSR

We describe a method for detecting weak absorption lines of gases whose absorption coefficients are as low as 10^{-8} - 10^{-9} cm^{-1}. The method is based on the multiple passage of light across an absorbing substance during broadband laser action. We used glass with Nd^{3+} and dye pulsed lasers. Then we used a cw dye laser of unusual construction.

With this laser the absorption spectra of the laboratory's air was investigated. We obtained an effective optical length of about 10^3 km. In the spectral interval of 585-600 nm, we obtained 700 lines. In the solar spectral atmosphere we find only 300 lines, 28 of which belong to nitrogen dioxide (NO_2). Fifteen strong lines are absent in the solar spectra because they coincide with iron lines in the solar atmosphere. Theoretically, the method has absolute sensitivity and temporal resolution for weak absorption or amplification of spectral lines in gases.

POLARIZATION DEPENDENCE AND FRANCK-CONDON FACTORS OF A VELOCITY-TUNED Na_2 TWO-PHOTON TRANSITION

J.P. Woerdman

Philips Research Laboratories, Eindhoven, The Netherlands

We report on (i) the polarization dependence and (ii) Franck-Condon factors of the Doppler-free Na_2 two-photon transition at $\nu_L/c = 16601.88\ cm^{-1}$ [1]. The novel aspect of this transition is its velocity tuning: from the occurrence of three-photon cross-resonances [2] it follows that there is a single dominant near-resonant ($\Delta \sim \Delta\nu_D$, Fig.1) intermediate level which is Doppler-shifted into resonance for two velocity packets of molecules when using two counterpropagating laser beams.

(i) Extending the work of BRAY and HOCHSTRASSER [3] we have calculated the ratio of two-photon transition strengths for linear and circular laser polarization for the case that only *one* intermediate rot-vib level contributes to the transition strength. In the classical limit ($J \to \infty$) this ratio is independent of the electronic symmetries of a, b and c; it only depends on J. Our experiments on the polarization dependence show that the two-photon transition belongs to the Q-branch, i.e. $J = 50$ for level a.

(ii) We observe that the Doppler-free two-photon line saturates very strongly. The saturation occurs mainly on bc and is governed by the Franck-Condon factor FC(b,c). We have determined FC(b,c) from the onset of the bc Lamb dip, using the Na-D_2 Lamb dip at 5890 Å as a reference, taking into account optical pumping of Na_2 and Na; this technique allows a direct measurement of FC(b,c). We find FC(b,c) $\approx 2.10^{-2}$; in addition we have determined FC($v' = 22, v''$) for $0 \leqslant v'' \leqslant 20$ by combining the relative line strengths in the fluorescence progression $\{v' = 22 \to v''\}$ with FC(b,c). These experimental FC values agree with those calculated by HESSEL and KUSCH within the experimental error [4].

References

1. J.P. Woerdman, Chem. Phys. Letters 43, 279 (1976).
2. J.P. Woerdman and M.F.H. Schuurmans, Opt. Commun. 21, 243 (1977)
3. R.G. Bray and R.M. Hochstrasser, Molec.Phys. 31, 1199 (1976).
4. M.M. Hessel, private communication.

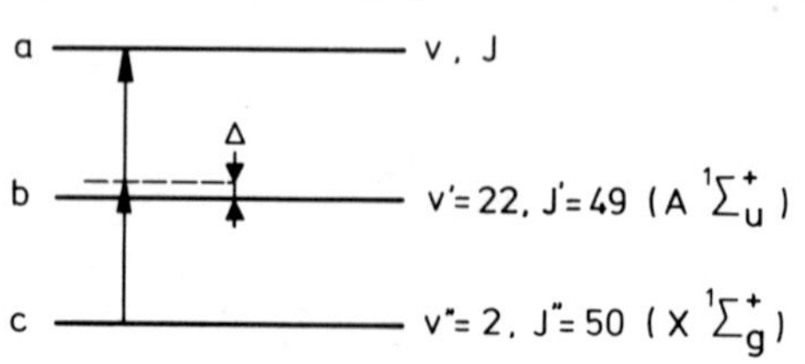

Fig. 1 Level scheme of the 16601.88 cm^{-1} Na_2 two-photon transition. $\Delta \equiv \frac{1}{2}\nu_{ac} - \nu_{bc}$ = 936 MHz, $\Delta\nu_D$ = 1.34 GHz

DOUBLE AND TRIPLE RESONANCE STUDIES OF ROTATIONAL RELAXATIONS IN NH_3-He AND NH_3-H_2 COLLISIONS

N. Morita, S. Kano, and T. Shimizu

Department of Physics, University of Tokyo
Bunkyo-ku, Tokyo 113, Japan

Collisional relaxation among vibration-rotation levels of NH_3 has been extensively studied in relation to an interpretation of the non-thermal stationary state distribution of NH_3 in interstellar space [1]. The laser-microwave double resonance method has attained a much higher sensitivity than that of microwave-microwave double resonance, and allows us to study the processes in detail. So far the NH_3-NH_3 collisions have been investigated in the presence of the pumping at the ν_2:Q(J=8, K=7) transition of $^{14}NH_3$ by the P(13) N_2O laser line [2,3,4]. In $^{14}NH_3$-$^{14}NH_3$ collision the double resonance signal arises from both processes of the collision induced transitions and the resonant energy transfer, while in the mixture of $^{14}NH_3$ and $^{15}NH_3$ only the excitation transfer gives rise to the double resonance signal. A well-defined preference rule for collision induced transitions is established and the existence of the efficient V-V excitation transfer process is clarified.

The changes in the intensities of inversion transitions due to the laser pumping were measured. An example of the observations in NH_3-He mixture is schematically shown in Fig.1. The results in the NH_3-He(H_2) collisions are markedly different from those in the NH_3-NH_3 collisions. In the present experiment, the ratio of the partial pressures was $P_{NH_3}:P_{He}(P_{H_2}) = 1:100$. Since NH_3 molecule mostly collides with He(H_2), only the effect of collision induced transitions appears in the double resonance signal. Furthermore in these cases NH_3 molecule is subjected to a strong collision in which the short range force works and the higher order terms in the interaction may become effective.

The preference rules in rotational quantum numbers, J and K, are not simple, but the principal features are (i) the strongest negative signals occur for the $\Delta K=0$, $\Delta J=\pm1$ transitions and the strong positive signal occurs for the $\Delta K=0$, $\Delta J=2$ transition, (ii) the strong signals occur for the $\Delta K=\pm3$ and any ΔJ (up to $|\Delta J|=5$) transitions with either negative or positive signs with even ΔJ or odd ΔJ, respectively, (iii) the weak negative signals for the $\Delta K=-6,-9$ transitions, and (iv) no signals for the $\Delta K=3n\pm1$ transitions.

The $\Delta K=3n$ rule, established by OKA in the microwave-microwave double resonance [5,6], is confirmed even for the transitions with very small probabilities. Ortho-para conversion is strictly forbidden even in the case of hard collision. The parity selection rule seems undefined, but with one exception all observed $\Delta K=0,\pm3$ transitions show the ($+\leftrightarrow+$, $-\leftrightarrow-$) changes for even ΔJ and the ($+\leftrightarrow-$) changes for odd ΔJ [5,7]. This suggests that a different type of interaction between the multipole and multipole-induced dipole may dominate in either even ΔJ or odd ΔJ collision-induced transitions. The weak

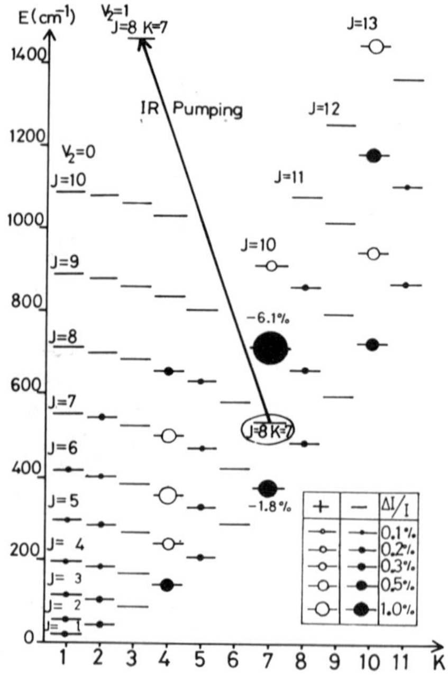

Fig. 1 Double resonance signals ($\Delta I/I$) appeared on the various inversion transitions of NH_3 in the presence of the N_2O laser pumping at the ν_2:Q(8,7) infrared transition. Black circles are for negative ΔI and blank circles are for positive ΔI. The partial pressures are P(NH_3) =20 mTorr and P(He)=2000 mTorr. The laser power is kept constant so that the value of $\Delta I/I$ on the J=8, K=7 inversion transition is 0.8.

negative signals may be due to an increase of effective rotational temperature after several collisions (thermalization processes). Drastic changes in the signals of types(i) and (ii) were observed when the J=8, K=7 inversion transition was saturated by the resonant microwave (the triple resonance). These signals were ascertained to be due to the direct transitions from the J=8, K=7 level pumped by the laser. Consistency has been obtained in the results of NH_3-He and NH_3-H_2 collisions [6].

References

1. A. C. Cheung, P. M. Rank, C. H. Townes, S. H. Knowles, and W. T. Sullivan III; Astrophys. J. 157 L13 (1969)
2. S. Kano, T. Amano and T. Shimizu, J. Chem. Phys. 64 4711 (1976)
3. N. Morita, S. Kano, Y. Ueda, and T. Shimizu, J. Chem. Phys. 66 2226 (1977)
4. K. Shimoda; "Microwave-Infrared Double resonance" presentation at TICOLS
5. T. Oka; J. Chem. Phys. 49 3135 (1968)
6. A. R. Fabris and T. Oka; J. Chem. Phys. 56 3168 (1972)
7. T. Oka; J. Chem. Phys. 47 13 (1967)

TIME RESOLVED SPECTROSCOPY OF THE 2p-1s TRANSITION IN NEON

L.A. Christian, C.G. Carrington, W.J. Sandle, and J.N. Dodd

Physics Department, University of Otago, Dunedin, New Zealand

Previous studies, both experimental [1,2,3] and theoretical [4], of the ten 2p levels (Paschen notation) of neon have yielded conflicting results with regard to lifetimes and coherence-destruction cross sections. In this study we aim to improve on and extend the previous measurements by using the technque of pulsed tunable dye laser excitation of neon atoms in a mild discharge.

A nitrogen laser pumped dye laser is used to excite atoms from one of the four 1s levels to a particular 2p level. The resulting fluorescence is monitored by a photomultiplier through a tuned monochromator and recorded with single photon counting equipment coupled to a multichannel analyzer. Analytic treatment [5] of the data is used to correct for the distortion caused by the inability of the equipment to record more than one photon per laser pulse. A fluorescence decay expression of the appropriate theoretical form is then computer fitted to the treated data.

A general expression for the measured fluorescence, for field B, is

$$a_0 e^{-\Gamma^{(0)} t} + a_2 e^{-\Gamma^{(2)} t} + b_1 e^{-\Gamma^{(1)} t} \cos(\omega t + \phi_2) + b_2 e^{-\Gamma^{(2)} t} \cos(2\omega t + \phi_2)$$

where the $\Gamma^{(k)}$ correspond in the usual way to relaxation of upper state population, orientation and alignment, and ω is the Larmor frequency. The production of modulation terms by applying a field is a most useful way for the experimenter to obtain a direct handle on $\Gamma^{(1)}$ and $\Gamma^{(2)}$, which can be difficult to accurately unravel from a sum of unmodulated exponentials (case $\omega = 0$).

The natural decay constant is most directly and accurately determined as follows. With the applied field zero, and with the linear polarizer in the excitation beam at the "magic" angle, only the a_0 term is non-zero. Fitting an exponential to the observations establishes a value for $\Gamma^{(0)}$ (Fig.1). To establish the natural radiative decay constant γ one must establish experimentally or theoretically the effects of radiation trapping and collisions. At the pressure (0.4 torr) and under the discharge condition (2.5 mA) used, the radiation trapping correction is about 2 or 3% of the decay constant (statistical errors are less than 1%); a theoretical estimate of the correction is sufficient to make the appropriate correction. The effect of $\Gamma^{(0)}_{coll}$ on the result is negligible. Table 1 lists measured values of γ.

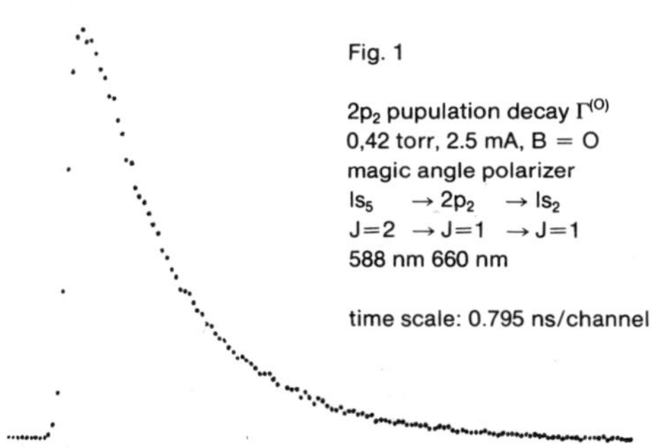

Fig. 1

In order to determine the velocity averaged cross sections σ_1 and σ_2, the decay constants $\Gamma^{(1)}$ and $\Gamma^{(2)}$ must be measured at neon pressures (up to 8 torr) such that $\Gamma^{(k)}_{coll}$ is large enough to make them significantly different from γ. The simplest way to extract $\Gamma^{(1)}$ and $\Gamma^{(2)}$ is to choose conditions such that a single modulation term in the above expression dominates. This can be done

Table 1 Summary of lifetime measurements (in nanoseconds)

Level	Present experiment	C.G.C. [3]	B & K [1]	Theory [4]
$2p_2$	18.67 ± 0.26	17.6 ± 0.6	18.8 ± 0.3	17.82
$2p_4$	19.26 ± 0.53	19.2 ± 1.0	19.1 ± 0.3	18.19
$2p_5$	19.97 ± 0.47	18.6 ± 1.3	19.9 ± 0.4	18.78
$2p_6$	19.40 ± 0.41	18.2 ± 0.7	19.7 ± 0.2	18.77
$2p_7$	19.65 ± 0.35	19.4 ± 0.9	19.9 ± 0.4	18.62
$2p_8$	19.6 (?)	19.6 ± 1.0	19.8 ± 0.2	19.05

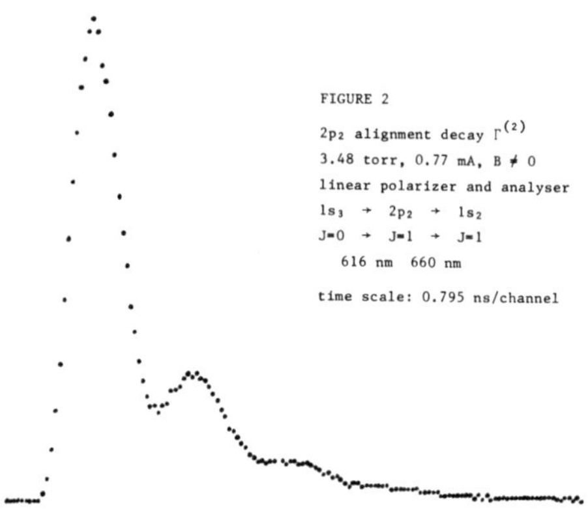

FIGURE 2
$2p_2$ alignment decay $\Gamma^{(2)}$
3.48 torr, 0.77 mA, B ≠ 0
linear polarizer and analyser
$1s_3 \rightarrow 2p_2 \rightarrow 1s_2$
$J=0 \rightarrow J=1 \rightarrow J=1$
616 nm 660 nm
time scale: 0.795 ns/channel

by appropriate choice of (i) polarizations for the excitation and fluorescence beams, (ii) the geometry of the optical system and the magnetic field, and (iii) the initial and final lower state J-values. Figure 2 shows a typical experimental result for $\Gamma^{(2)}$. Table 2 lists cross sections observed.

The advantages of using a tunable and pulsed dye laser are numerous. The tunability enables the selection of a particular level for study and this is especially important for a case like neon because of the great number of transitions involved. The technique directly displays the time-development of the ensemble density matrix as revealed through radiative emission. Any systematic error or trend, any unexpected behavior or distortion is directly displayed in a way that can be recognized or analyzed easily.

Table 2 Cross-section measurement

Level	Present experiment $\sigma_1/10^{-15}$ cm^2	C.G.C. [3] $\sigma_1/10^{-15}$ cm^2	Present experiment $\sigma_2/10^{-15}$ cm^2	C.G.C. [3] $\sigma_2/10^{-15}$ cm^2
$2p_2$	2.36 ± 0.46	2.73 ± 0.08	4.68 ± 0.88	(4.55)*
$2p_5$	6.03 ± 0.93	5.30 ± 0.15	8.90 ± 1.54	(8.85)*
$2p_7$	4.44 ± 0.73	5.20 ± 0.17		*theoretical
$2p_8$	9.00 ± 1.98	8.52 ± 0.28		prediction

References

1. W. R. Bennett and P. J. Kindlmann, Phys. Rev. 149, 38-51 (1966).
2. E. Fournier, M. Ducloy, B. Decomps and M. Dumont, C.R. Acad. Sci. Paris 268 B, 1495 (1969).
3. C. G. Carrington, J. Phys. B 5, 1572-1582 (1972).
4. S. Feneuille, M. Kapisch, E. Koenig and S. Liberman, Physica 48, 571-588 (1970).
5. P. B. Coates, J. Phys. E: Sci. Instrum. 5, 148-150 (1972).

List of Participants

R. L. Abrams
Hughes Aircraft Co.
3011 Malibu Canyon Road
Malibu, CA 90265

R. V. Ambartzumian
Institute of Spectroscopy
Moscow, U.S.S.R.

E. Arimondo
Herzberg Inst. of Astrophysics
National Research Council
Ottawa, Ontario K1A OR6 CANADA

R. L. Barger
National Bureau of Standards
Boulder, CO 80302

R. F. Barrow
Physical Chemistry Lab.
Oxford University
Oxford, ENGLAND

F. Bayer-Helms
Physikalisch-Technische Bundesanstalt
Bundesallee 100
3300 Braunschweig, FRG

L.-E. Berg
Institute of Physics
University of Stockholm
Vanadisv. 9
S-11346 Stockholm, SWEDEN

J. C. Bergquist
JILA, Univ. of Colorado
Boulder, CO 80309

A. F. Bernhardt
Lawrence Livermore Lab.
P.O. Box 808, L-258
Livermore, CA 94550

R. Bernheim
The Pennsylvania State Univ.
152 Davey Laboratory
University Park, PA 16801

M. Berry
Materials Research Center
Allied Chemical Corp.
P.O. Box 1021R
Morristown, NJ 07960

A. Betz
Dept. of Physics
University of California
Berkeley, CA 94720

S. Bialowski
Dept. of Chemistry
Univ. of Utah
Salt Lake City, UT 84112

J. H. Birely
Los Alamos Scientific Lab.
Group AP-3, MS-565
Los Alamos, NM 87545

J. E. Bjorkholm
Bell Telephone Labs
Room 4C-318
Holmdel, NJ 07733

G. Bjorklund
Bell Telephone Labs
Room 4C-316
Holmdel, NJ 07733

N. Bloembergen
Pierce Hall, 231
Harvard University
Cambridge, MA 02138

C. J. Bordé
Univ. de Paris-Nord
Ave. J. B. Clement
93430 Villetaneuse, FRANCE

M. A. Bouchiat
Lab. de Physique de 1/E.N.S.
24 rue Lhommond
75231 Paris, Cedex 05 FRANCE

R. G. Brewer
San Jose Research Lab. (IBM)
5600 Cottle Rd., Dept. K01-281
San Jose, CA 95193

J. Brossel
Lab. de Spectroscopie
24 rue Lhommond
75231 Paris, Cedex 05 FRANCE

J. E. Butler
U.S. Naval Research Lab., Code 6110
Washington, DC 20375

R. Byer
Hansen Physics Lab.
Stanford University
Stanford, CA 94305

B. Cagnac
Universite P. et M. Curie
Laboratoire de Spectroscopie
4 place Junnieu
75230 Paris, FRANCE

P. Cahuzac
Lab. Aime-Cotton
CNRS II bat. 505
91405 Orsay, FRANCE

C. D. Cantrell
Theoretical Molecular Physics Group
MS 531, Los Alamos Scientific Labs.
Los Alamos, NM 87545

J. L. Carlsten
JILA
Univ. of Colorado
Boulder, CO 80309

C. Chan
Spectra-Physics
1250 West Middlefield Rd.
Mountain View, CA 94042

P. K. Cheo
United Technologies Research Center
East Hartford, CT 06108

C. Cohen-Tannoudji
Laboratoire Physique ENS
24 rue Lhommond
75231 Paris, Cedex 05 FRANCE

R. Cone
Montana State University
Bozeman, MT 59715

B. Couillaud
Lab. de Spect. Moléc.
Université Bordeaux I
33405 Talence, FRANCE

G. Coutts
Lawrence Livermore Labs.
P.O. Box 808
Livermore, CA 94550

W. Demtröder
Fachbereich Physik
Univ. Kaiserslautern
FRG

R. D. Deslattes
A 141 Physics
National Bureau of Standards
Washington, DC 20234

R. DeVoe
IBM Research Laboratory
5600 Cottle Rd.
San Jose, CA 95193

L. S. Ditman, Jr.
NSWC, Code WR-43
Silver Spring, MD 20910

J. N. Dodd
Dept. of Physics
Univ. of Otago
P.O. Box 56
Dunedin, NEW ZEALAND

A. Donszelmann
Zeeman Laboratorium
Der Universiteit Van Amsterdam
Plantage Muidergracht 4
Amsterdam, HOLLAND

R. E. Drullinger
National Bureau of Standards
Time & Frequency Dept., 277.05
Boulder, CO 80302

M. Ducloy
Laser Physics Lab.
Université Paris-Nord
Avenue J. B. Clement
93430 Villetaneuse, FRANCE

M. Dufay
Universite Lyon 1
43 Boulevard de 11 November 1918
69621 Villeurbanne, FRANCE

P. Esherick
Sandia Laboratories
Albuquerque, NM 87115

K. M. Evenson
National Bureau of Standards
Boulder, CO 80302

S. Ezekiel
M.I.T., Room 33-214
Cambridge, MA 02139

M. S. Feld
Department of Physics
Mass. Inst. of Technology
Cambridge, MA 02139

M. L. Gaillard
Universite Lyon I
43 Boulevard 11 Novembre 1918
69620 Villeurbanne, FRANCE

H. Galbraith
Los Alamos Scientific Labs.
Los Alamos, NM 87545

F. L. Galeener
Xerox Palo Alto Research Center
3333 Coyote Hill Rd.
Palo Alto, CA 94304

J. Garrison
Lawrence Livermore Lab.
Univ. of California, Box 808
Livermore, CA 94550

A. Genack
IBM K01/281
5600 Cottle Road
San Jose, CA 95193

H. Gerhardt
FU Berlin, Inst. f. Atom-v.
Festhorperph.
Boltzmannstr. 20
D-1000 Berlin 33, FRG

P. Giacomo
Bureau International des Poids
et Mesures
Pavillon de Breteuil
F-92310 Sevres, FRANCE

H. M. Gibbs
1E, 224 Bell Labs.
Murray Hill, NJ 07974

P. Glorieux
Herzberg Inst. of Astrophysics
National Research Council of Canada
Ottawa, Ontario K1A OR6 CANADA

T. E. Gough
Dept. of Chemistry
Univ. of Waterloo
Waterloo, Ontario CANADA

J. Hall
JILA
Univ. of Colorado
Boulder, CO 80309

R. Hall
Exxon Research & Engineering
Box 45
Linden, NJ 07036

T. W. Hänsch
Dept. of Physics
Stanford Univ.
Stanford, CA 94305

W. Happer
Dept. of Physics
Columbia University
New York, NY 10027

D. O. Harris
Dept. of Chemistry
Univ. of California
Santa Barbara, CA 93106

W. Harter
JILA
Univ. of Colorado
Boulder, CO 80309

K. C. Harvey
Dept. of Physics
Univ. of Toronto
Toronto, Ontario M5S 1A7 CANADA

J. Helmcke
Physikalisch-Technische
Bundesaustalt
Bundesallee 100
3300 Braunschweig, FRG

G. W. Hills
Herzberg Inst. of Astrophysics
National Research Council
Ottawa, Ontario K1A OR6 CANADA

B. Hitz
Laser Focus
930 E. Evelyn
Sunnyvale, CA 94086

G. S. Hurst
Oak Ridge Natl. Laboratory
P.O. Box X
Oak Ridge, TN 37830

M. Inguscio
Istituto di Fisica
dell' Universita
Pisa, ITALY

G. R. Isaak
Dept. of Physics
University of Birmingham
Birmingham, ENGLAND B15 2TT

R. R. Jacobs
Lawrence Livermore Lab.
P.O. Bos 808, L-470
Livermore, CA 94550

A. Javan
Dept. of Physics
M.I.T., Bldg. 6-208
77 Massachusetts Ave.
Cambridge, MA 02139

D. A. Jennings
National Bureau of Standards
Boulder, CO 80302

D. Johnson
Spectra-Physics, Inc.
1250 W. Middlefield Rd., MS-2-00
Mountain View, CA 94042

T. F. Johnston, Jr.
Coherent, Inc.
3210 Porter Drive
Palo Alto, CA 94304

V. P. Kaftandjian
Universite de Provence
Centre Saint Jerome
13397 Marseille Cedex 04 FRANCE

J. C. Keller
Laboratoire Aimé Cotton, CNRS II
Batiment 505
Orsay 91405 FRANCE

P. L. Kelley
M.I.T. Lincoln Lab., C-128
P.O. Box 73
Lexington, MA 02173

S. H. Khan
Pakistan Inst. of Nuclear Science
and Technology
P.O. Nilore
Rawalpindi, PAKISTAN

B. Koffend
Mass. Inst. of Technology
Chemistry Dept., 2-077
Cambridge, MA 02139

K. L. Kompa
Projektgruppe fur Laser forschung
des Max-Planck-Gesellschaft
D-8046 Garching, FRG

B. B. Krynetzky
P. N. Lebedev Lab.
Moscow, U.S.S.R.

A. Kung
Lawrence Berkeley Lab.
Univ. of Calif., Bldg. 70, Room 140
Berkeley, CA 94720

N. A. Kurnit
Los Alamos Scientific Lab.
Group AP-2, MS 564, P.O. Box 1663
Los Alamos, NM 87545

W. E. Lamb, Jr.
Dept. of Physics
Univ. of Arizona
Tucson, AZ 85721

W. Lange
Technische Universitat Hannover
Institute fur Angewandt Physik
3000 Hannover
Welfengarten 1 FRG

D. Larsen
Mass. Inst. of Technology
Francis Bitter Natl. Magnet Lab.
170 Albany St.
Cambridge, MA 02139

D. J. Larson
Lyman Laboratory
Harvard University
Cambridge, MA 02138

H. P. Layer
National Bureau of Standards
Room B322, Bldg. 220
Washington, DC 20234

S. A. Lee
JILA
Univ. of Colorado
Boulder, CO 80309

S. R. Leone
JILA
Univ. of Colorado
Boulder, CO 80309

M. D. Levenson
SSC 422, Univ. of S. California
University Park
Los Angeles, CA 90007

L. A. Levin
Nuclear Research Centre-Negev
P.O. Box 9001
Beer Sheva, ISRAEL

J. Levine
Div. 277
National Bureau of Standards
Boulder, CO 80302

P. F. Liao
Bell Laboratories
Holmdel, NJ 07733

W. C. Lineberger
JILA
Univ. of Colorado
Boulder, CO 80309

Y. Liran
Laser Dept.
Nuclear Research Center
P.O. Box 9001
Beer-Sheva, ISRAEL

M. M. T. Loy
IBM Research Center
Yorktown Heights, NY 10598

T. B. Lucatorto
National Bureau of Standards
Room A251, Physics
Washington, DC 20234

R. M. MacFarlane
IBM Research Lab.
5600 Cottle Rd.
San Jose, CA 95193

J. M. J. Madey
High Energy Physics Lab.
Stanford University
Stanford, CA 94305

D. W. Magnuson
Union Carbide Nuclear Co.
ORGOP M.S. 322, P.O. Box P
Oak Ridge, TN 37830

A. Maki
National Bureau of Standards
Molecular Spectroscopy Section
Washington, DC 20234

A. D. May
Dept. of Physics
Univ. of Toronto
Toronto, Ontario M5S 1A7 CANADA

G. O. Meisel
Inst. F. Angew. Physik
Wegelerstr. 8
D 5300 Bonn, FRG

R. T. Menzies
Jet Propulsion Lab.
Calif. Inst. of Technology
4800 Oak Grove
Pasadena, CA 91103

A. Mooradian
MIT Lincoln Laboratory
P.O. Box 73
Lexington, MA 02137

G. L. McAllister
Exxon Research & Tech. Center
2955 George Washington Way
Richland, WA 99352

R. S. McDowell
Los Alamos Scientific Lab., MS 565
P.O. Box 1663
Los Alamos, NM 87545

A. R. W. McKellar
Herzberg Inst. of Astrophysics
National Research Council
Ottawa, Ontario K1A 0R6 CANADA

R. Neugart
Johannes-Gutenberg Universität
Institut für Physik
6500 Mainz, FRG

T. Oka
Herzberg Ins. of Astrophysics
National Research Council
Ottawa, Ontario K1A OR6 CANADA

J. A. Paisner
Lawrence Livermore Lab.
P.O. Box 808, L-259
Livermore, CA 94550

A. S. Pine
MIT Lincoln Laboratory
Lexington, MA 02173

S. P. S. Porto
Universidade Estadual de Campinas
Cidade Universitaria Barao Geraldo
Instituto de Fisica
13.100 Campinas, SP - BRAZIL

O. Poulsen
Institute of Physics
University of Aarhus
DK 8000 Aarhus, DENMARK

D. E. Pritchard
Mass. Inst. of Technology
Room 26-231
Cambridge, MA 02139

H. B. Puell
Max Planck Inst. für
Extraterrestrische Physik
8046 Garching bei Munchen, FRG

A. A. Puretzky
Institute of Spectroscopy
Moscow, U.S.S.R.

H. E. Radford
Smithsonian Institute
60 Garden St.
Cambridge, MA 02138

W. Radloff
Central Inst. of Optics &
Spectroscopy
Academy of Sciences of Alie GDR
1199 Berlin-Adlershof
Rudower Chaussee 6, DDR

N. F. Ramsey
Lyman Lab. of Physics
Harvard University
Cambridge, MA 02138

I. Rappaport
Cornell University
601 E. 20th St.
New York, NY 10010

E. A. Reinhart
Department of Physics & Astronomy
University of Wyoming
Laramie, WY 82071

I. Renhorn
Mass. Inst. of Technology
Dept. of Chemistry, 2-123
Cambridge, MA 02139

C. K. Rhodes
Stanford Research Inst.
333 Ravenswood Ave.
Menlo Park, CA 94025

S. Rockwood
Los Alamos Scientific Lab.
AP-2, Grp. Leader, MS-564
Los Alamos, NM 87545

M. Salour
Dept. of Physics
Harvard University
Cambridge, MA 02138

P. G. H. Sandars
Clarendon Lab.
Oxford University
Oxford, ENGLAND

B. Sansal
Centre de Sciences et de Technologie
3 Bld. Frantz Fanon
BL 1147 Alger. Gare, ALGERIA

A. Scalabrin
Universidad Estadual de Campinas
Ciadade Universitaria - Barao Geraldo
Campinas, SP-CEP 13.100 BRAZIL

Y. R. Shen
University of California
Physics Dept.
Berkeley, CA 94720

T. Shimizu
Dept. of Physics
University of Tokyo
Bunkyo-ku
Tokyo 113, JAPAN

K. Shimoda
Dept. of Physics
University of Tokyo
Bunkyo-ku
Tokyo 113, JAPAN

R. Shuker
JILA
Univ. of Colorado
Boulder, CO 80309

W. T. Silfvast
Bell Laboratories
Holmdel, NJ 07733

K. Siomos
Oak Ridge Natl. Laboratory
P.O. Box X
Building 45008, H-162
Oak Ridge, TN 37830

S. D. Smith
Physics Department
Heriot-Watt University
Riccarton
Edinburgh, EH14 4AS SCOTLAND

J. J. Snyder
National Bureau of Standards
Room A141, Physics
Washington, DC 20234

M. S. Sorem
Los Alamos Scientific Lab.
Los Alamos, NM 87544

B. P. Stoicheff
Dept. of Physics
Univ. of Toronto
Toronto, Ontario M5S 1A7

F. Strumia
University of Pisa
Pisa, ITALY

A. F. Suchtkov
P. N. Lebedev Lab.
Moscow, U.S.S.R.

S. Svanberg
Department of Physics
Chalmers University of Technology
S-40220 Götebarg, SWEDEN

A. Szabo
Electrical Engineering Div.
National Research Council
Ottawa, Ontario K1A OR8 CANADA

H. Takuma
Dept. of Engineering Physics
Univ. of Electrocommunications
1-5-1 Chofugaoka Chofushi
Tokyo 182, JAPAN

J. P. Taran
O.N.E.R.A.
29, Ave. de la Division Lecierc
92320 Châtillon, FRANCE

F. K. Tittel
Rice University
Electrical Engineering Dept.
P.O. Box 1892
Houston, TX 77001

P. E. Toschek
Institut for Angewandte Physik
Univ. Heidelberg
Albert-Uberle-Str. 3-5
D-69 Heidelberg, FRG

K. Uehara
Dept. of Physics
Univ. of Tokyo
Bunkyo-ku
Tokyo 113, JAPAN

P. Violino
Istitute di Fisica
Piazza Torricelli, 2
56100 Pisa, ITALY

H. Walther
Projektgruppe fur Laserforschung
der Max-Planck-Gesellschaft
D-8046 Garching, FRG

C. C. Wang
Ford Scientific Laboratory
Box 2053
Dearborn, MI 48121

H. Welling
Institut für Angewandte Physik
Technical University of Hannover
Hannover, FGR

K. H. Welge
Fakultat f. Physik
Universitaet Bielefeld
48 Bielefeld 1, FRG

T. J. Whitaker
Battelle Northwest
329 Bldg., 300 Area
Battelle Blvd.
Richland, WA 99352

C. E. Wieman
Physics Dept.
Stanford University
Stanford, CA 94305

D. J. Wineland
National Bureau of Standards
Div. 277.04; 325 Broadway
Boulder, CO 80302

W. H. Wing
Optical Sciences Center
University of Arizona
Tucson, AZ 85721

C. Wittig
Electrical Engineering Dept.
Univ. of Southern California
University Park
Los Angeles, CA 90007

K. Wodkiewicz
Inst. of Theoretical Physics
Warsaw Univ.
Warsaw, POLAND

J. P. Woerdman
Philip Research Laboratories
Eindhoven,
THE NETHERLANDS

P. T. Woods
Quantum Metrology Division
Natl. Physical Laboratory
Teddington, Middlesex,
ENGLAND

E. S. Yeung
Dept. of Chemistry
Iowa State University
Ames, IA 50011

W. Yu
Hama-Matsu
120 Wood Ave.
Middlesex, NJ 08846

A. H. Zewail
Dept. of Chemistry
Mail Code 127-72
Calif. Inst. of Technology
Pasadena, CA 91125

Titles of Related Interest

DYE LASERS
Topics in Applied Physics, Vol.1
2nd revised edition (1977)
ed. by *F.P. Schäfer*

F.P. Schäfer: Principles of Dye Laser Operation

B.B. Snavely: Continuous-Wave Dye Lasers

C.V. Shank, E.P. Ippen: Mode-Locking of Dye Lasers

K.H. Drexhage: Structure and Properties of Laser Dyes

T.W. Hänsch: Applications of Dye Lasers

F.P. Schäfer: Progress in Dye Lasers: September 1973 till March 1977

LASER SPECTROSCOPY of Atoms and Molecules
Topics in Applied Physics, Vol.2
ed. by *H. Walther*

H. Walther: Atomic and Molecular Spectroscopy with Lasers

E.D. Hinkley, K.W. Nill, F.A. Blum: Infrared Spectroscopy of Molecules by Means of Lasers

K. Shimoda: Double-Resonance Spectroscopy of Molecules by Means of Lasers

J.M. Cherlow, S.P.S. Porto: Laser Raman Spectroscopy of Gases

B. Decomps, M. Dumont: Linear and Nonlinear Phenomena in Laser Optical Pumping

K.M. Evenson, F.R. Petersen: Laser Frequency Measurements, the Speed of Light, and the Meter

HIGH-RESOLUTION LASER SPECTROSCOPY
Topics in Applied Physics, Vol.13
ed. by *K. Shimoda*

K. Shimoda: Introduction

K. Shimoda: Line Broadening and Narrowing Effects

P. Jacquinot: Atomic Beam Spectroscopy

V.S. Letokhov: Saturation Spectroscopy

J.L. Hall, J.A. Magyar: High Resolution Saturated Absorption Studies of Methane and Some Methyl-Halides

V.P. Chebotayev: Three-Level Laser Spectroscopy

S. Haroche: Quantum Beats and Time-Resolved Fluorescence Spectroscopy

N. Bloembergen, M.D. Levenson: Doppler-Free Two-Photon Absorption Spectroscopy

LASER MONITORING OF THE ATMOSPHERE
Topics in Applied Physics, Vol.14
ed. by *E.D. Hinkley*

E.D. Hinkley: Introduction

S.H. Melfi: Remote Sensing for Air Quality Management

V.E. Zuev: Laser-Light Transmission Through the Atmosphere

R.T.H. Collis, P.B. Russell: Lidar Measurement of Particles and Gases by Elastic Backscattering and Differential Absorption

H. Inaba: Detection of Atoms and Molecules by Raman Scattering and Resonance Fluorescence

E.D. Hinkley, R.T. Ku, P.L. Kelley: Techniques for Detection of Molecular Pollutants by Absorption of Laser Radiation

R.T. Menzies: Laser Heterodyne Detection Techniques

INFRARED GENERATION BY NONLINEARITIES
Topics in Applied Physics, Vol.16
ed. by *Y.R. Shen*

Y.R. Shen: Introduction

R.L. Aggarwal, B. Lax: Mixing of CO_2 Lasers in the Far-Infrared

R.L. Byer, R.L. Herbst: Parametric Oscillation and Mixing

V.T. Nguyen, T.J. Bridges: Difference Frequency Mixing via Spin Nonlinearities in the Far-Infrared

J.J. Wynne, P.P. Sorokin: Optical Mixing in Atomic Vapors

T.Y. Chang: Optical Pumping in Gases

ULTRASHORT LIGHT PULSES
Topics in Applied Physics, Vol.18
ed. by *S.L. Shapiro*

S.L. Shapiro: Introduction--A Historical Overview

D.J. Bradley: Methods of Generation

E.P. Ippen, C.V. Shank: Techniques for Measurement

D.H. Auston: Picosecond Nonlinear Optics

D. von der Linde: Picosecond Interactions in Liquids and Solids

K.B. Eisenthal: Picosecond Relaxation Processes in Chemistry

A.J. Campillo, S.L. Shapiro: Picosecond Relaxation Measurements in Biology

OPTICAL AND INFRARED DETECTORS
Topics in Applied Physics, Vol.19
ed. by *R.J. Keyes*

R.J. Keyes: Introduction

P.W. Kruse: The Photon Detection Process

E.H. Putley: Thermal Detectors

D. Long: Photovoltaic and Photoconductive Infrared Detectors

H.R. Zwicker: Photoemissive Detectors

A.F. Milton: Charge Transfer Devices for Infrared Imaging

M.C. Teich: Nonlinear Heterodyne Detection

BEAM-FOIL SPECTROSCOPY
Topics in Current Physics, Vol.1
ed. by *S. Bashkin*

S. Bashkin: Experimental Methods

I. Martinson: Studies of Atomic Spectra by the Beam-Foil Method

L.J. Curtis: Lifetime Measurements

O. Sinanoglu: Theoretical Oscillator Strengths of Neutral, Singly-Ionized, and Multiply-Ionized Atoms: The Theory, Comparisons with Experiment, and Critically-Evaluated Tables with New Results

W. Wiese: Regularities of Atomic Oscillator Strengths in Isoelectronic Sequences

W. Whaling: Applications to Astrophysics: Absorption Spectra

L.J. Heroux: Applications of Beam-Foil Spectroscopy to the Solar Ultraviolet Emission Spectrum

R. Marrus: Studies of Hydrogen-Like and Helium-Like Ions of High Z

J. Macek, D. Burns: Coherence, Alignment, and Orientation Phenomena in the Beam-Foil Light Source

I. Sellin: The Measurement of Autoionizing Ion Levels and Lifetimes By Fast Projectile Electron Spectroscopy

LASER SPECTROSCOPY, Proceedings of the 2nd International Conference, Mégève, France, June 23-27, 1975
Lecture Notes in Physics, Vol.43
ed. by *S. Haroche, J.C. Pebay-Peyroula, T.W. Hänsch, S.E. Harris*

Springer-Verlag
Berlin Heidelberg New York